REVIEWS IN
ECONOMIC GEOLOGY

(ISSN 0741-0123) Volume 4

# ORE DEPOSITION ASSOCIATED WITH MAGMAS

ISBN 0-9613074-3-9

**Volume Editors**

J. A. Whitney
Department of Geology
University of Georgia
Athens, GA 30602

A. J. Naldrett
Department of Geology
University of Toronto
Toronto, ON M5S 1A1
Canada

**Series Editor:** James M. Robertson
New Mexico Bureau of Mines & Mineral Resources
Campus Station
Socorro, NM 87801

SOCIETY OF ECONOMIC GEOLOGISTS

**The Authors:**

Philip A. Candela
Laboratory for Mineral Deposits Research
Department of Geology
University of Maryland
College Park, MD 20742

Alan H. Clark
Department of Geological Sciences
Queen's University
Kingston, ON
Canada, K7L 3N6

C. Jay Hodgson
Department of Geological Sciences
Queen's University
Kingston, ON
Canada, K7L 3N6

Jeffrey D. Keith
Department of Geology
University of Georgia
Athens, GA 30602

C. Mike Lesher
Department of Geology
University of Alabama
Tuscaloosa, AL 35487-0338

Ed A. Mathez
Department of Mineral Sciences
American Museum of Natural History
New York, NY 10024

Wim van Middelaar
Department of Geology
University of Georgia
Athens, GA 30602

Anthony J. Naldrett
Department of Geology
University of Toronto
Toronto, ON
Canada, M5S 1A1

Cheryl L. Peach
Lamont-Doherty Geological Observatory
Palisades, NY 10964

James A. Whitney
Department of Geology
University of Georgia
Athens, GA 30602

# CONTENTS

# Chapter 1

# Introduction: Magmatic Deposits Associated with Mafic Rocks

A. J. Naldrett

Magmatic sulfide ores are thought to form as the result of droplets of an immiscible sulfide-oxide liquid forming within silicate magma and then becoming concentrated in a particular location. Certain elements, notably the Group VIII transition metals Fe, Co, Ni, Pd, Pt, Rh, Ru, Ir and Os together with Cu and Au, partition strongly into the sulfide-oxide liquid, and thus become concentrated with it. A number of factors may influence the concentration of this liquid, but the dominant one is gravitational settling, since the liquid has a density of $>4$ in comparison with a value of $<3$ for its host silicate magma.

To help in the understanding of deposits of this type, in this book we first discuss the phase relations of simple sulfide-oxide liquids and activity-composition relations within them. We then discuss the solubility of sulfide in mafic and ultramafic melts, followed by the partitioning of elements between silicate magma and sulfide-oxide liquid. The oxidation state and volatile content of a silicate magma can have a major influence on the segregation of a sulfide-oxide liquid and the distribution of metals so that this forms the focus of a second chapter.

Magmatic sulfide deposits can be viewed in terms of their associated mafic or ultramafic bodies and the tectonic settings into which these were emplaced. The scheme shown as Table 1.1 is adapted from that of Naldrett, (1989). In it, bodies are divided into whether they were emplaced in a rifted continental environment (category II), a cratonic environment (category III) or an active orogenic belt (category IV). Archean greenstone belts still represent an enigma in terms of present-day tectonics. For example, were komatiites erupted through continental crust (Arndt, 1986a; Compston et al., 1986) or do they represent the floor of a primitive ocean (de Witt et al., 1987)? Thus a separate category (category I) has been created for the syn-volcanic activity in this environment.

Experience in Archean greenstone belts has shown that mafic and ultramafic bodies fall into two main classes, komatiites and tholeiites, and that the tholeiites constitute two distinct sub-classes, one with picritic average compositions and chilled margins and the other very rich in anorthositic gabbro. The komatiites are host to Ni sulfide ores in Western Australia, Zimbabwe and Canada; these ores and their origin are discussed by C.M. Lesher in this volume. Examples of mineralization associated with the picritic sub-class of tholeiites include those of the Pechenga region at the north-western end of the Kola peninsular. These deposits are associated with differentiated sills, emplaced in a $70 \times 35$ km tectonic trench (graben-synclinorium) containing 10,000 m of volcanic and sedimentary rocks. The deposits have been described by Gorbunov (1968), Gorbunov et al. (1985a,b) and Glaskovsky et al. (1977). The anorthositic gabbro sub-class of tholeiites is less commonly mineralized than the picritic sub-class, although the Montcalm deposit (Barrie and Naldrett, 1988) is a small deposit associated with a gabbro body of the same name.

Bodies associated with the rifted plate margin environment are divided into those floored by or closely associated with continental crust and those without this close association; the latter comprise ophiolites. The Circum–Superior rift zone (Baragar and Scoates, 1981) is an example of the first association. It contains the ultramafic bodies of the Thompson belt, the Fox River sill of the Fox River belt, the mafic and ultramafic bodies of the Cape Smith belt and the mafic bodies of the Labrador trough. Important deposits of Ni–Cu sulfides occur in both the Thomson (Zurbrigg, 1963; Peredery et al., 1982) and Cape Smith belts (Wilson and Kilburn, 1966; Barnes et al., 1982; Dillon–Leitch, 1986). The Kemi and Koillismaa schist belts are thought to have been originally one continuous belt (Gaal, 1985) that is interpreted as a rift bounding the northern edge of an Archean craton. Intrusions emplaced up deep-seated tensional fractures into this belt include the Penikat intrusion, in which 3 separate horizons of Platinum Group element (PGE) mineralization have been discovered (Alapieti and Lahtinen, 1986). Ophiolite complexes are not noted for magmatic sulfide mineralization, although the Cliffs PGE deposit in the Shetland (Unst) ophiolite (Pritchard et al., 1986) is a possible example. The Acoje Ni sulfide deposit in the Zambales ophiolite of the Phillipines (Rossman et al., 1989) is a definite example.

The third category, that of intrusions in cratonic areas, comprise those related to flood basalt magmatism and large stratiform complexes. Flood basalt-related intrusions include those hosting the very important Noril'sk–Talnakh Ni–Cu–PGE camp, and the Duluth Complex with its extensive but currently uneconomic zones of disseminated Cu–Ni sulfide, both of which are described in this volume by A. J. Naldrett. The stratiform complexes are divided into those that are sheet-like, such as the Bushveld and Stillwater Complexes, and dike-like, such as the Great Dyke of Zimbabwe. All are hosts to important PGE mineralization and are described

TABLE 1.1—Mafic and ultramafic bodies and related sulfide ores.

| Petro-tectonic setting | Ni–Cu ores | PGE ores |
|---|---|---|
| I. SYNVOLCANIC (largely Archean) | | |
| **Komatiites** | E. Goldfields, Australia; Zimbabwe; Abitibi, Ontario | |
| **Tholeiites** | | |
| *Picritic* | Pechenga, USSR | |
| *Anorthositic* | Montcalm, Ontario | |
| II. RIFTED PLATE MARGINS & OCEAN BASINS | | |
| **Associated with Continental Crust** | **Circum–Ungava Belt**<br>Thompson Ni Camp; Ungava Ni Ores | **Kemi–Koilismaa Belt**<br>Penikat Intrusion |
| **Ophiolites**<br>(No continental crust) | | **Unst Ophiolite**<br>Cliffs deposit |
| III. CRATONIC AREAS | | |
| **Flood Basalt-related** | **Siberian Traps**<br>Noril'sk–Talnakh<br>**Karoo**<br>Insizwa–Ingeli Intrusion<br>**Keweenawan**<br>Duluth Complex<br>Crystal Lake Gabbro | |
| **Large Stratiform Complexes** | **Sheet-like**<br>Sudbury | **Sheet-like**<br>Bushveld: Merensky Reef UG–2 Chromitite<br>Stillwater: J–M Reef<br>**Dyke-like**<br>Great Dyke: Main sulfide zone |
| IV. EMPLACED DURING OROGENESIS | | |
| **Synorogenic** | **Appalachia**<br>Moxie & Katahdin intrusions, Maine | |
| **Late Orogenic** | | **Alaskan Type**<br>Tulameen Complex |

here by A. J. Naldrett. The Sudbury Igneous Complex is also in the sheet-like category, and contains the greatest concentration of Ni–Cu sulfides in one place in the world.

Mafic intrusions of the syn-orogenic category can be subdivided into those emplaced during the active, compressive phase of orogenesis and those of the zoned "Alaskan" type (Irvine, 1974), emplaced during epi-orogenic re-adjustments following compressive orogenesis. The former are hosts to relatively minor Ni–Cu mineralization, including that of the Moxie pluton in Maine (Thompson and Naldrett, 1984). Alaskan-type intrusions have not proved important as sources of Ni–Cu ores.

In terms of their genesis, large rich concentrations of Ni–Cu sulfides (as distinct from weak sulfide concnetrations hosting PGE) do not seem to be the normal consequence of the emplacement in the crust of magma derived from the mantle, followed by its cooling and crystallization. Lesher emphasizes the importance that the thermal erosion of sulfide-bearing sediment has played in the case of komatiite-related ores, Naldrett points out that the introduction of crustal sulfur into a magma has been extremely important both at Noril'sk and Duluth. Citing the work of Naldrett et al. (1986b), Naldrett documents that the magma of the Sudbury Igneous Complex has undergone greater than 50 percent contamination by crustal rocks, and relates this to the separation of the sulfide ores, as was suggested originally by Irvine (1975). Thus all of the world's major nickel sulfide camps are the consequence of a variety of types of contamination of mafic or ultramafic magma.

PGE concentrations also appear to require special processes for their formation. There has been considerable discussion in the recent literature as to how important hydrothermal fluids are to these processes. While recognizing that such fluids were present and were responsible for modifying aspects of the ore horizons, Naldrett favours sulfide liquid immiscibility and the main ore-forming process. Mathez emphasizes deposit types where hydrothermal fluids have played a major role in ore genesis.

## REFERENCES


Arndt, N. T., 1986a, Komatiitites—a dirty window into the Archean mantle: Cognita 6, p. 59–66.

Barager, W. R. A., and Scoates, R. F. J., 1981, The Circum-Superior belt—a Proterozoc plate margin?; *in* Kroner, A. (ed.), Precambrian plate tectonics: Elsevier, Amsterdam, p. 293–294.

Barnes, S. J., Coats, G. J. A., and Naldrett, A.J., 1982, Petrogenesis of a Proterozoic nickel sulfide-komatiite association, the Katiniq Sill, Ungava, Quebec: Economic Geology, v. 77, p. 413–429.

Barrie, C. T., and Naldrett, A. J., 1988, The geology and tectonic setting of the Montcalm Gabbroic Complex and Ni-Cu deposit, Western Abitibi Subprovince, Ontario, Canada: Proc. of the 5th Magmatic Sulfides Conference, Harare, Zimbabwe, Spec. Pub. Inst. Min. and Metall.

Compston, W., and Amiens, P. A., 1968, The Precambrian geochronology of Australia: Can. J. Earth Sci., v. 5, p. 561–583.

de Witt, M. J., Hart, R. A., and Hart, R. J., 1987, The Jamestown Ophiolite Complex, Barberton mountain belt—a section through 3.5 Ga oceanic crust: J. of African Earth Sc., v. 6, p. 681–730.

Dillon-Leitch, H. C. H., Watkinson, D. H., and Coats, C. J. A., 1986, Distribution of platinum-group elements in the Donaldson West deposit, Cape Smith belt, Quebec: Economic Geology, v. 81, no. 5, p. 1147–1158.

Gaal, G., 1985, Nickel metallogeny related to tectonics: Bull. Comm. Geol. Finland, v. 333, p. 145–154.

Gorbunov, G. I., Zagorodny, V. G., and Robonen, W. I., 1985a, Main features of the geological history of the Baltic Shield and the epochs of ore formation: Geol. Surv. Finland, Bull., v. 333, p. 17–41.

Gorbunov, G. I., Yakolev, Yu. N., Goncharov, Yu. V., Gorelov, V. A., and Tel'nov, V. A., 1985b, The nickel areas of the Kola peninsula: Geol. Surv. Finland, Bull., v. 333, p. 42–122.

Gorbunov, G. I., 1968, Geology and origin of the copper-nickel sulfide ore deposits of Pechenga (Petsamo): Moscow, Nedra, 352 p. (in Russian).

Irvine, T. N., 1974, Petrology of the Duke Island ultramafic complex, southeastern Alaska: Geol. Soc. America, Mem. 138, 240 p.

McCallum, I. S., Raedeke, L. D., Mathez, E. A., 1980, Investigations in the Stillwater Complex—Part I. Stratigraphy and structure of the landed zone; *in* Irving, A. J., and Dungan, M. A. (eds.), The Jackson volume: American J. Sci., v. 280–A, p. 59–87

Naldrett, A. J., Rao, B. V., and Evensen, N. M., 1986b, Contamination at Sudbury and its role in ore formation; *in* Gallagher, M. J., Ixer, R. A., Neary, C. R., and Prichard, H. M. (eds.), Metallogeny of basic and ultrabasic rocks: Spec. Pub. of Inst. of Mining & Metallurgy, London, p. 75–92.

Peredery, W. V., and Geological Staff, 1982, Geology and nickel sulfide deposits of the Thompson belt, Manitoba; *in* Hutchinson, R. W. (eds.), Robinson Volume: Geol. Assoc. Can. Spec. Pap., no. 25, p. 165–209.

Prichard, H. M., Neary, C. R. and Potts, P. J., 1986, Platinum group minerals in the Shetland Ophiolite; *in* Gallagher, M. J., Ixer, R. A., Neary, C. R., and Pritchard, H. M. (eds.), Metallogeny of basic and ultrabasic rocks: Spec. Pub. of Inst. of Mining & Metallurgy, London, p. 455–470.

Rossman, D. L., Castañada, G. C., and Bacuta, G. C., 1989, Geology of the Zambales Ophiolite, Luzon, Philippines: Tectonophysics, v. 168, no. 1–3, pp. 1–22.

Thompson, J. F. H., and Naldrett, A. J., 1984, Sulfide-silicate reactions as a guide to Ni–Cu–Co mineralization in central Maine; *in* Buchanan, D. L., and Jones, M. J. (eds.), Sulfide deposits in mafic and ultramafic rocks: Inst. Min. Metall. Special Publ., p. 103–113.

Wilson, H. D. B., Kilburn, L. C., Graham, A. R., and Rambal, K., 1969, Geochemistry of some Canadian nickeliferous ultrabasic intrusions: Economic Geology, Mon. 4, p. 294–309.

Zurbrigg, H. F., 1963, Thompson mine geology: Can. Inst. Min. Met., Trans. 66, p. 227–236.

# Chapter 2

# Sulfide Melts—Crystallization Temperatures, Solubilities in Silicate Melts, and Fe, Ni, and Cu Partitioning between Basaltic Magmas and Olivine

A. J. Naldrett

## INTRODUCTION

Most magmatic sulfide ores contain 5 major constituents, Fe, S, O, Ni and Cu. It is possible to simplify the discussion of the relevant phase equilibria by considering melting relations in the three component Fe–S–O system, and then the effects of introducing small amounts of Ni and Cu.

For this reason, high-temperature phase relations in the system Fe–S–O are discussed first, followed by consideration of the effect of substitution of other components, and then equilibrium between Fe–S–O liquids and mafic silicate magmas.

## PHASE RELATIONS IN THE SYSTEM FE–S–O

For reasons given later in this section, natural magmas fall within a limited range of Fe/S and Fe/O ratios. Within this range, the important features of the system Fe–S are the congruent melting of pyrrhotite at 1190°C, and the two bounding phase regions, pyrrhotite + liquid + vapour on either side. The three-phase region terminates on the metal-rich side at the eutectic at 988°C and on the sulfur-rich side at 1083°C.

Of the Fe oxides, magnetite is the dominant mineral crystallizing from high-temperature magmas. More rarely wustite forms, but because of its lower stability limit of 560°C, it is rarely preserved in the natural environment. Above 900°C, magnetite shows considerable solid solution towards hematite.

Condensed phase relations at 900°C (15°C below the temperature of the first appearance of a liquid in the Fe-rich portion of the system Fe–S–O) are shown in Fig. 2.1, which outlines the compositional field occupied by almost all natural ore magmas. Phase relations on the (O + S)-rich side of the magnetite–pyrrhotite tie-lines drawn in the figure have not been determined experimentally because of the high vapour pressures associated with these compositions. It is likely, however, that these compositions involve a four-phase field of magnetite + hematite + pyrrhotite + vapour (in which pyrrhotite is somewhat richer in sulfur than that shown in equilibrium with magnetite in the figure) followed by the three-phase field of hematite + pyrrhotite + vapour.

Liquidus relations in the system Fe–S–O (in the presence of vapour) are shown in Figs. 2.2a and b. Fig. 2.2a is characterized by fields of iron, wustite, pyrrhotite and magnetite. The iron, wustite, and pyrrhotite fields meet at a ternary eutectic where these three phases plus liquid and vapour are all stable. The wustite, pyrrhotite and magnetite fields meet at a ternary reaction point, at which magnetite plus liquid react in the presence of vapour to form pyrrhotite (containing 62.8 ± 0.2 wt% Fe) and wustite. Fig. 2.2b is an enlargement of a portion of Fig. 2.2a. Tie-lines in Fig. 2.2b illustrate the composition of pyrrhotite in equilibrium with iron oxide and liquid at different temperatures along the wustite–pyrrhotite and magnetite–pyrrhotite cotectic lines.

The relationship between the solidus and liquidus is illustrated in Fig. 2.3. To the right of the eutectic composition, the section is drawn along the pyrrhotite–wustite and pyrrhotite–magnetite cotectic lines. Consequently, regions in which one solid phase coexists with a liquid and vapour do not appear in this section. In addition, many of the data points do not represent experiments with bulk compositions in the plane of the section, but have been projected onto this plane along tie-lines.

The ternary eutectic, iron–wustite–pyrrhotite, was reported by Naldrett (1969) to occur at 915 ± 2°C and the ternary eutectic melt has the composition 68.2 wt% Fe, 24.3% S and 7.5% O. The ternary reaction point occurs at 934 ± 2°C. As shown in Fig. 2.3, the solidus rises very steeply away from the reaction point in the direction of decreasing Fe content (a mixture of pyrrhotite containing 62.5 wt% Fe and magnetite would start to melt at 1010°C) and then flattens out as the pyrrhotite in equilibrium with the magnetite becomes less Fe-rich.

Wendlandt and Huebner (1979) investigated the effect of pressure on the position and temperature of the ternary eutectic in the Fe–S–O system. At 30 kb pressure, this has a temperature of 1000 ± 10°C, and the eutectic melt is composed of 72–73 wt% Fe and less than 0.75 wt% O. This is close to the binary Fe–FeS eutectic at this pressure.

Brett and Bell (1969), Ryzhenko and Kennedy (1973) and Usselman (1975) studied the effect of pressure on the Fe–FeS eutectic in the Fe–S system up to 60 kb. All workers found that pressure had either no or only a slight effect on the temperature of the eutectic, and causes it to move to more Fe-rich compositions (Ryzhenko and Kennedy report it to contain 22 wt% S at 60 kb. Sharp (1969) and Ryzhenko and Kennedy (op cit) investigated the effect of pressure on the congruent melting of pyrrhotite and found the melting point to increase at the rate of 13°C/kb up to 65kb).

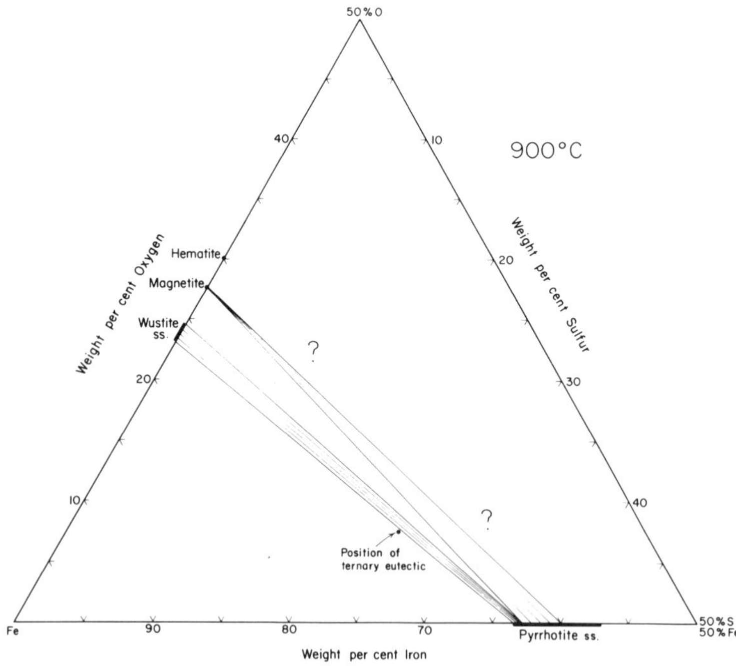


FIGURE 2.1—The Fe-rich portion of the Fe–S–O system at 900°C. From Naldrett (1969).

### Effect of other components on solidus temperatures

Naldrett (1969) pointed out that at temperatures immediately below the solidus, a pyrrhotite–pentlandite–chalcopyrite–magnetite ore in which the sulfides contain less than 15 wt% Ni and 4 wt% Cu would consist of two phases, a nickeliferous cupriferous pyrrhotite solid-solution and magnetite. Craig and Naldrett (1967) investigated the effect of the substitution of Ni and Cu for Fe in the pyrrhotite on the solidus temperatures of pyrrhotite–magnetite mixtures. They found that the substitution of up to 20 wt% Ni for Fe on an atom for atom basis did not lower melting temperatures measurably below those in the pure system Fe–S–O. The substitution of 2 wt% Cu on a similar basis lowers solidus temperatures 15 to 20°C.

Naldrett and Richardson (1967) investigated the extent to which water lowers the melting temperature of pyrrhotite–magnetite mixtures. Within the accuracy of their experiments (±10°C), they concluded that water does not act as a flux for sulfide ores (Fig. 2.4), or dissolve appreciably in a sulfide–oxide melt.

The above data indicate that, ignoring the pressure effect, melting temperatures in the Fe–S–O system are probably within 20°C of the melting temperatures of natural low-Cu (<2 wt%) sulfide–oxide magmas.

### Variation in $f_{O_2}$, $f_{S_2}$ and $a$FeO

The composition of an iron-sulfide–oxide liquid at any given temperature is a function of its oxygen and sulfur fugacities (Naldrett, 1969). This point is illustrated in Figs. 2.5a and b in which isobars of sulfur and oxygen fugacity are shown for the Fe–S–O system. The isobars are drawn on the basis of the data of Rosenqvist (1954), Bog and Rosenqvist (1958), and Nagamori and Kameda (1965).

Knowing activity–composition relationships for two of the components in a three-component system, the Gibbs–Duhem relation allows calculation of the chemical potential of the third component. This calculation has been done for the data given above on the system Fe–S–O (Shima and Naldrett, 1975) to obtain the activity of Fe. In turn, since the standard free energy for the reaction

$$\mathrm{Fe} + {}^1\!/_2\mathrm{O_2} = \mathrm{FeO}^{(\mathrm{liquid})} \qquad (1)$$

is known, it is possible to combine the $a$Fe and $f_{O_2}$ to calculate $a$FeO. These data then serve as the basis for the $a$FeO contours in Fig. 2.5a and b, for which the standard state of FeO is taken to be a supercooled liquid of stoichiometric FeO composition at 1200°C and a liquid of the same composition at 1450°C.

## Application of the system Fe–S–O to natural ore magmas

### Controls on the S and O content of ore magmas

Consider a basaltic magma at 1200°C, devoid of Ni, Cu, and Co but just saturated in sulfide, so that a small number of sulfide droplets have segregated from it and are in equilibrium with it. Since the droplets are few in number and widely dispersed throughout a large body of magma, their compositions are controlled by the composition of the host magma. This magma has a certain FeO content and a certain

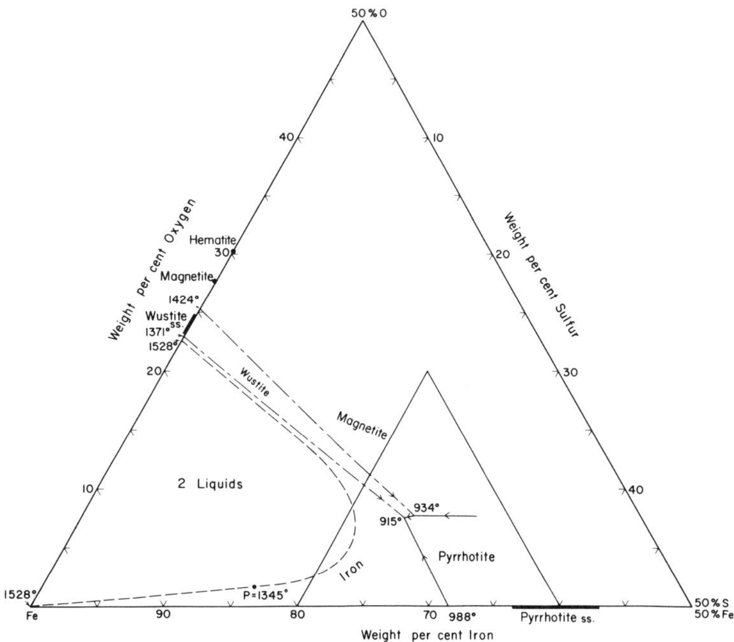


FIGURE 2.2—**A.** Liquidus diagram of the Fe-rich portion of the Fe–S–O system. **B.** Enlargement of the area outlined in A. Tie-line in the pyrrhotite field connect pyrrhotite compositions to those of co-existing liquids. Both diagrams from Naldrett (1969).

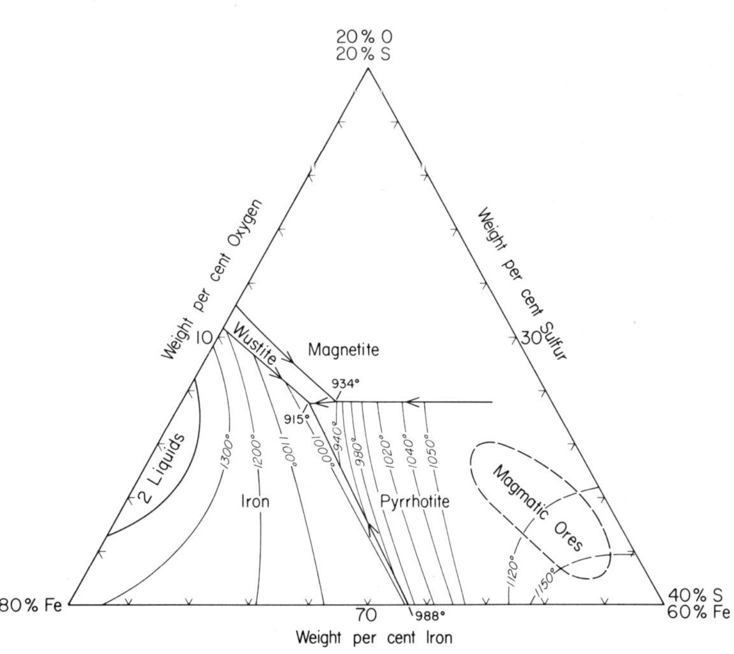

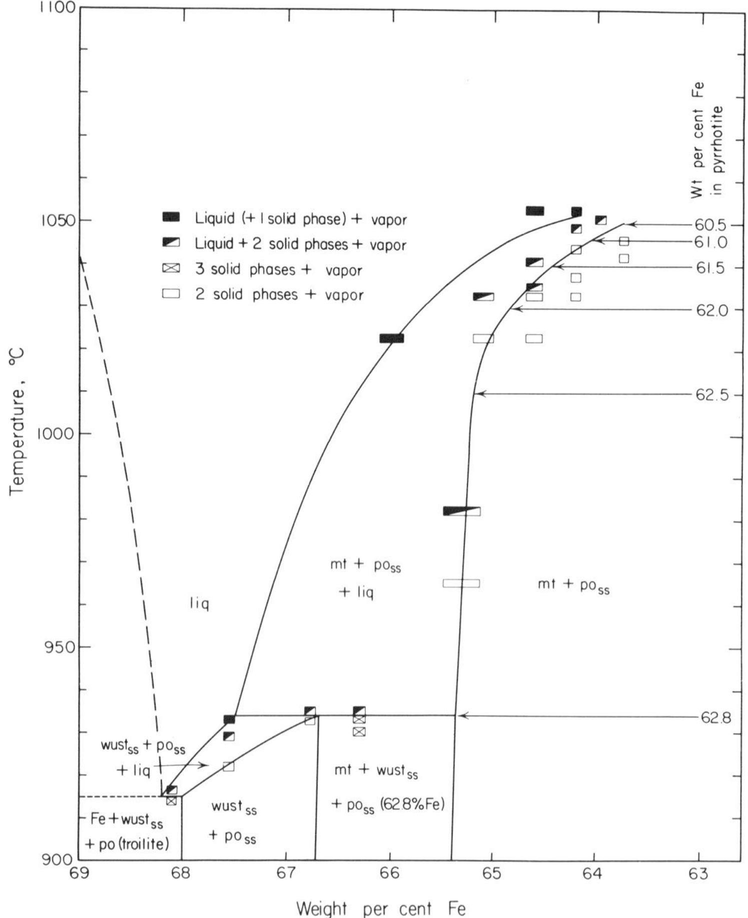


FIGURE 2.3—A T–X section showing the relationship between solidus and liquidus in the Fe–S–O system. The section is drawn along the pyrrhotite–wustite and pyrrhotite–magnetite cotectic lines. Many of the data points do not represent bulk compositions in the plane of the section, but have been projected to the plane along tie-lines. From Naldrett (1969).

$Fe^{3+}/Fe^{2+}$ ratio, which, together with other major elements, define the values of *a*FeO (obtainable from Roeder, 1974, or according to the relationship of Doyle and Naldrett, 1985) and $f_{O_2}$ in the magma. Since at equilibrium, the same values of *a*FeO and $f_{O_2}$ must apply to the sulfide droplets, the composition of the droplets must correspond to the values in Fig. 2.5a; i.e. in the hypothetical case under consideration, the composition of the basalt magma, in particular its FeO content and $Fe^{3+}/Fe^{2+}$ ratio, define the composition of the segregating sulfide ore and the $f_{S_2}$ prevailing at the time of segregation. Since most natural magmas have relatively restricted ranges of FeO content and oxidation state, the compositions of magmatic ores are similarly restricted their S and O contents.

In nature, Ni, Cu, and Co are also present in the basaltic magma and partition into the sulfide droplets. This tendancy complicates the simple relationship only to a limited extent. The composition of the sulfide liquid is still controlled by the *a*FeO and $f_{O_2}$ of the host magma.

Doyle and Naldrett (1987) pointed out that if, as Fudali (1965) has argued on the basis of experiments, the $f_{O_2}$ of the mantle sources to magmatic sulfide ore magmas are in the region of the FMQ buffer ($10^{-8}$ at 1200°C), the compositions of the ore magmas should fall on the oxygen-rich side of the magnetite–pyrrhotite cotectic in the Fe–S–O system, implying that they should crystallize magnetite before pyrrhotite. It is clear from polished section study that the majority crystallize pyrrhotite before magnetite.

Part of the answer to this problem may lie in the observation of Christie et al. (1986) on ocean ridge basalt glasses. They found these to have $Fe^{2+}/Fe^{3+}$ ratios consistent with oxygen fugacities two orders of magnitude less than the FMQ buffer. In addition, Cermer (1987) noted that, at a given oxygen fugacity, the oxygen content of an Fe–Ni–Cu–S matte depends strongly on the Fe/(Ni + Cu) ratio of the matte. He found that when the $f_{O_2}$ is $10^{-4}$ at 1250°C, the oxygen content decreases from 6 wt% in Fe-rich mattes to 0.1 wt% in Fe-free mattes. His data show that the substitution of 9 wt% Ni, 1 wt% Cu for Fe in a typical matte cuts the dissolved oxygen by nearly 1/3. Thus the presence of Cu and Ni in natural sulfide magmas may lower their oxygen contents, so that their bulk compositions fall closer to the pyrrhotite rather than the magnetite field than would be the case of pure Fe–S–O magmas.

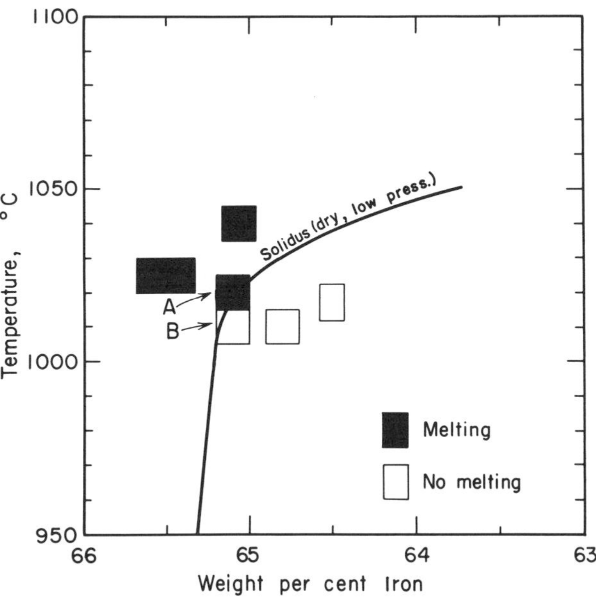


FIGURE 2.4—The effect of $H_2O$ on the solidus temperature of pyrrhotite of varying composition. Experimental data indicating those charges that have melted and those that have not melted after heating to the indicated temperature at 2 kb total pressure with excess water present in the system. From Naldrett and Richardson (1966).

### Crystallization of sulfide ores

The temperatures of the beginning of crystallization and final solidification of a sulfide ore are important when considering how far the sulfides can move away from their host intrusion as an ore magma, and whether they are likely to be mobilized as a liquid during high-grade metamorphism.

The region representing the compositions of most magmatic ores is superimposed on the Fe–S–O system in Fig. 2.2b. Although detailed information is not available on liquidus temperatures, most ores start to crystallize between 1160 and 1120°C. Since ores, like silicate magmas, can probably move as a mixture of crystals and liquid, the solidus of an ore provides a minimum temperature for its mobilization in a partly liquid state. Fig. 2.3 shows that the solidus temperature varies with the Fe content of the pyrrhotite forming the deposit. Almost all ores are made up of pyrrhotite containing between 62.5 and 60.5 wt% Fe (due to the controls on their composition discussed above) so that, on the basis of the system Fe–S–O, one would predict that their solidus temperatures would be between 1010 and 1050°C. As reported above, the substitution of up to 20 wt% Ni for Fe has no measurable effect on the solidus temperature. The substitution of 2 wt% Cu lowers the solidus temperature by 20°C. The substitution of more than 2 wt% Cu has a greater effect, so that temperatures are distinctly lower than those estimated above.

## THE SOLUBILITY OF SULFUR IN SILICATE MELTS

Knowledge of the solubility of sulfur in mafic and ultramafic silicate melts is important in understanding how magmatic sulfide deposits form and also in evaluating the potential of igneous bodies as hosts for ore deposits of this type. Early work on sulfur solubility was undertaken by metallurgists, notably Richardson and his co-workers (Fincham and Richardson, 1954; Abraham et al., 1960). Although the melt compositions studied by these workers were those of smelter slags rather than natural magmas, Fincham and Richardson's (1954) study provided valuable insight into the mechanism of sulfur solution and its important implications in nature. They showed that at low oxygen fugacities (less than $10^{-6}$ atm in melts studied at 1400 and 1500°C) sulfur dissolves primarily as sulfide and that a function termed the "sulfur capacity" of the melt, $C_S$, is constant for melts of the same composition and obeys the relationship:

$$C_S = (\text{wt\% S})\,[(p_{O_2}/(p_{S_2})]^{\frac{1}{2}}\ . \qquad [i]$$

$C_S$ increases with rise in temperature and changes with composition, generally increasing with increasing FeO, MgO, and CaO contents and decreasing with increasing $SiO_2$ and $Al_2O_3$.

It should be noted that the function "sulfur capacity" is somewhat misleading. It does not represent the overall capacity of a melt to dissolve sulfur, but is somewhat akin to the equilibrium constant for reaction [2] proposed below. It merely relates the amount of sulfur that will dissolve in a given melt in response to imposed sulfur and oxygen fugacities.

MacLean (1969), in his study of the system Fe–S–O–$SiO_2$, found that the sulfur content of a silicate melt in equilibrium with sulfide-rich liquid decreases with increasing oxygen content (this sulfur content is henceforth referred to as the "sulfur content at sulfide saturation" or "SCSS"). He attributed this observation to the fact that sulfur dissolves by displacing oxygen bonded to $Fe^{2+}$ and that increasing oxygen results in a increase in $Fe^{3+}$ at the expense of $Fe^{2+}$ in

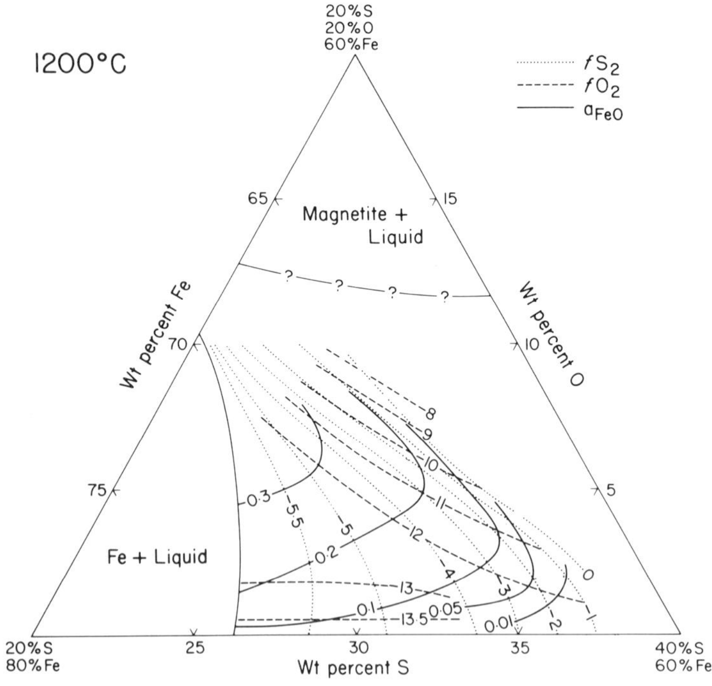


FIGURE 2.5—Contours of $f_{O_2}$, $f_{O_2}$ and FeO on the same portion of the Fe–S–O system as in Fig. 2.2B, **A.** 1200°C and **B.** at 1450°C. From Shima and Naldrett (1975).

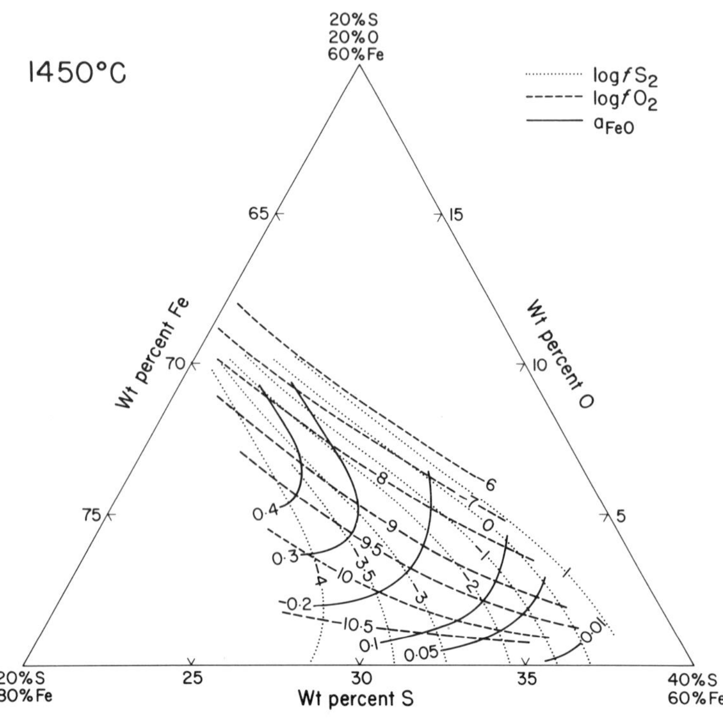

the melt. Connolly and Haughton (1972) confirmed this point.

Shima and Naldrett (1975) studied a komatiitic melt and obtained strong evidence supporting MacLean's contention as to the mechanism of sulfur solution.

Considering the reaction:

$$FeO^{(\text{in melt})} + 1/2S_2 = FeS^{(\text{in melt})} + 1/2O_2, \qquad (2)$$

the activities of the reactants are related by the expression:

$$a_{FeS} = \gamma_{FeS} N_{FeS} = K \times a_{FeO} \times [f_{S_2}/f_{O_2}]^{\frac{1}{2}}, \qquad (ii)$$

thus,

$$\log N_{FeS} = 1/2 \log f_{S_2} + [\log K + \log a_{FeO} - 1/2 \log f_{O_2} - \log \gamma_{FeS}], \qquad (iii)$$

where $\gamma_{FeS}$ is the rational activity coefficient for FeS in the melt, $N_{FeS}$ is the mole fraction of FeS, K is the equilibrium constant, and $f_{S_2}$ and $f_{O_2}$ are the fugacities of sulfur and oxygen respectively. If, for small changes in $N_{FeS}$, $\gamma_{FeS}$ is assumed to remain constant and also if the amount of FeO formed in a reaction such as [2] is sufficiently small that its formation has no appreciable effect on $a_{FeS}$, it follows from expression [ii] that a plot of log $N_{FeS}$ against log $f_{S_2}$ ($f_{O_2}$ and T kept constant) should produce a straight line relationship with a slope of 1/2, provided that sulfur is dissolving according to reaction [2]. This is a reasonable assumption, since the amount of Fe interacting with sulfur amounts to 0.2 to 0.6 in most melts containing 8 to 12 wt% FeO. Fig. 2.6, a plot of Shima and Naldrett's (1975) data, indicates that this relationship is obeyed. Furthermore, Fig. 2.6 shows that as log $f_{O_2}$ increases, log $N_{FeS}$ decreases. This also in accordance with expression [ii] above.

It should be appreciated that the preceeding discussion and data apply to the solution of S in a silicate melt from a superjacent gas phase in response to changes in the fugacity of sulfur in the gas. With the exception of Shima and Naldrett's experiment at log $f_{O_2}$ = −10.4 and log $f_{S_2}$ = −2, saturation in iron sulfide liquid was not achieved.

Haughton et al.. (1974) were the first to provide extensive data on the solubility of S in a silicate melt of naturally occurring composition in equilibrium with a sulfide liquid, (SCSS). They found a strong correlation of SCSS with FeO and, to a lesser extent, with $TiO_2$ contents, but no systematic change with the varying $f_{O_2}$ of their experiments. Buchanan and Nolan (1979) performed a similar series of experiments and were able to demonstrate the effect of varying $f_{O_2}$ (Fig. 2.7). The data at high FeO contents in this figure are not directly applicable to natural basalts, since FeO was increased in the experiments at the expense of all other elements, and as a result, at high FeO contents, $SiO_2$ fell well below the level to be expected in even the most Fe-rich natural basalts.

Mathez (1976) studied the S content of the glassy margins of submarine basalt pillows from the Atlantic and Pacific

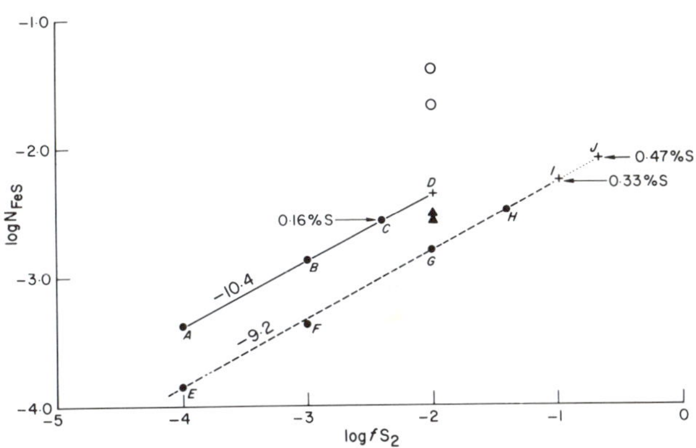


FIGURE 2.6—Relationship between $N_{FeS}$ and $f_{O_2}$ in a given silicate melt at two values of $f_{O_2}$. Solid dots are experimental values for unsaturated melts. Open circles are values for melt with immiscible sulfide droplets. Solid triangles are values for the sulfur content of melts containing immiscible droplets from which the droplets themselves have been removed. From Shima and Naldrett (1975).

that had been quenched at depths >1000 m. He concluded that these were saturated or nearly saturated when they were quenched. Their S content increases markedly and linearly with increasing FeO content from about 0.105 wt% S at 9.0 wt% FeO to 0.18 wt% at 12.9 wt%. These S concentrations are substantially higher than the experimental results illustrated in Fig. 2.7, an observation that Mathez attributed to the high pressure at which the submarine basalts were quenched, but which, in the light of Buchanan and Nolan's (1979) data, could be due to $f_{O_2}$.

### Effect of temperature

Buchanan et al. (1983) determined the solubility of sulfur as a function of $f_{S_2}$ in a basaltic melt containing 17 wt% FeO at 1200, 1300 and 1400°C. Data were obtained in the field of undersaturation which indicated that at constant $f_{O_2}$ and $f_{S_2}$, the dissolved sulfur content increases by a factor of 8.5 times per 100°C at 1000°C but that at 1400°C this factor is only 3 times per 100°C. Buchanan et al. also attempted to predict the saturation values of the sulfur content (i.e SCSS) but their inability to achieve saturation in most of their samples indicates that their predicted values should be treated with caution.

Relatively few data exist on the variation of SCSS with temperature. There seems to be little doubt that SCSS increases with temperature. Haughton et al. (1974) found that tholeiitic melts containing 10 wt% FeO dissolve about 0.04 wt% S when saturated with iron sulfide at 1200°C and $f_{O_2}$ = $10^{-10.5}$, whereas Shima and Naldrett (1975) found that a komatiitic melt containing the same FeO content dissolves between 0.16 and 0.27 wt% S at an $f_{O_2}$ of $10^{-10.4}$ at 1450°C,. This comparison is rendered inaccurate by differences in melt composition, but indicates an increase of between 0.05 and 0.09 wt% S per 100°C. Wendlandt (1982) found that the increase in temperature of a basaltic melt containing close to 8 wt% FeO from 1300 to 1460°C at 20 kb total pressure

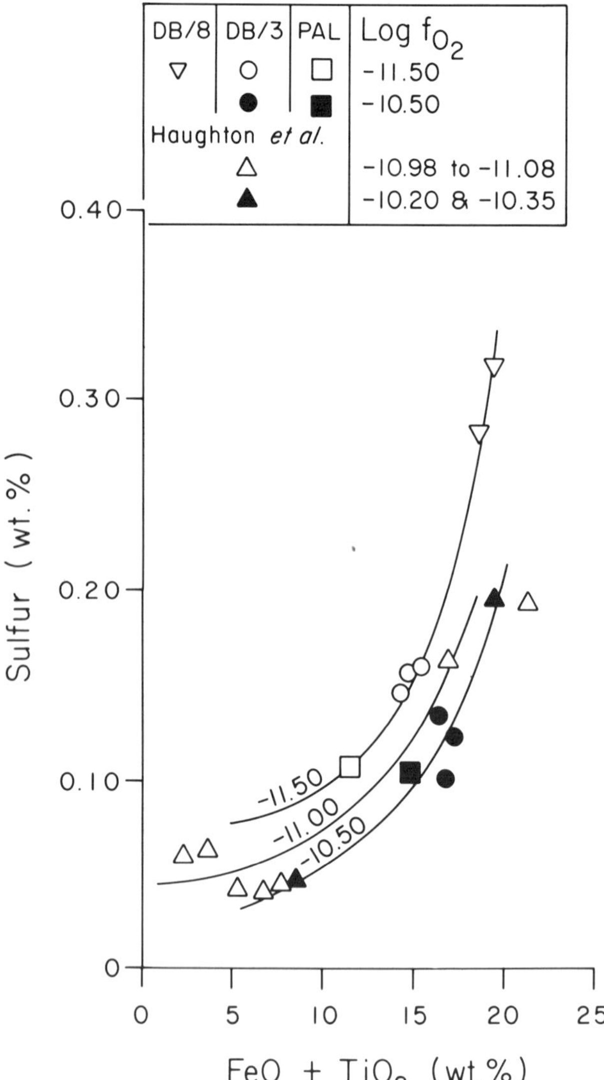


FIGURE 2.7—Sulfur content of silicate glass in equilibrium with an immiscible sulfide melt, illustrating the effect of varying $f_{O_2}$ and FeO content of the melt. Data shown as squares, circles and triangles with apices downwards from Buchanan and Nolan (1979), that shown as triangles with apices upwards from Haughton et al. (1973).

and an $f_{O_2}$ close to the C–$CO_2$–CO buffer (about one log unit above the quartz–fayalite–magnetite [QFM] buffer according to Wendlandt) caused an increase in SCSS of from 0.09 to 0.16 wt%, corresponding to an increase of 0.04 wt% per 100°C. Wendlandt's experiments on silicate melts of differing FeO contents produced similar results, given the errors inherent in his data. Possibly an increase in SCSS of 3 to 5 times from 1200 to 1450°C is of the order of that to be expected in nature.

### Effect of pressure

Huang and Williams (1980) investigated portions of the system Fe–Si–S–O at 32 kb and found that the miscibility gap between sulfide and silicate liquids expands with increasing pressure. Wendlandt (1982) studied the variation in SCSS in two basalts and an andesite at pressures between 12.5 and 30 kb. He found that SCSS increases with FeO and temperature and decreases with increasing pressure. Comparison with other studies was difficult because he relied on the C–$CO_2$–CO buffer which varies in $f_{O_2}$ with changing pressure, and yet the extent of these variations were uncalibrated between his experiments. However, his studies would seem to indicate that under natural conditions increasing pressure has a depressing effect on SCSS. This implies that as a magma rises to surface, its ability to dissolve sulfur increases and therefore it is not likely to approach saturation with sulfide. In Fig. 2.8, Wendlandt's data on the effects of pressure and FeO content have been combined (solid lines) to show the effect of pressure on SCSS in a magma containing constant FeO (Wendlandt found that due to the design of his experiments, the FeO content of the silicate portion of a particular starting composition did not remain constant); these relationships have been modified in Fig. 2.8 (dashed lines) to show the variation to be expected in a basalt rising from a depth equivalent to 30 kb, but which loses heat and stays at its liquidus temperature (assumed to vary from 1500°C at 30 kb to 1330°C at 12.5 kb). The reduction in SCSS due to cooling would largely counterbalance the effect of pressure. However, the uncertainties in the data are considerable.

### Experiments with melts in equilibrium with H–O–S and H–C–O–S fluids

Gorbachev and Kashirceva (1985) have conducted a series of experiments on basaltic to picritic melts in equilibrium with (H–O–S)- and (H–C–O–S)-bearing fluid phases. Their melts were in equilibrium with, and were enclosed in peridotite sheaths, that were themselves encased within doubled-walled Pt capsules. The oxygen fugacity was controlled by the assemblage olivine + orthopyroxene + chromite. Calculations indicate that the $f_{O_2}$ was between $10^{-8}$ and $10^{-9}$ atm. Their data indicate that at 15 kbars, and in equilibrium with both an H–O–S fluid and an iron sulfide–oxide liquid, the dissolved sulfur in a melt containing 12.5 wt% of both FeO and MgO increases from 0.45 wt % at 1250°C to 0.65 wt% at 1350°C. They found that in the presence of both an H–O–S and an H–O–C–S fluid, sulfur solubility in the melt increased slightly with increasing pressure up to 15 kb total pressure, but decreased progressively with increasing pressure above this.

Gorbachev and Kashirceva's (1985) results are compared with those of Wendlandt (1982) in Fig. 2.9. The latter's experiments were not conducted at pressures much below 15 kb, so that given the errors involved, they neither confirm nor refute Gorbachev and Kashirceva's finding of a change in the dependance of sulfur solubility on total pressure at 15 kb. The wide range of variables in Gorbachev's experiments preclude any exact analysis of the reasons for the variable dependance on pressure, but they open up an interesting area for future experimentation.

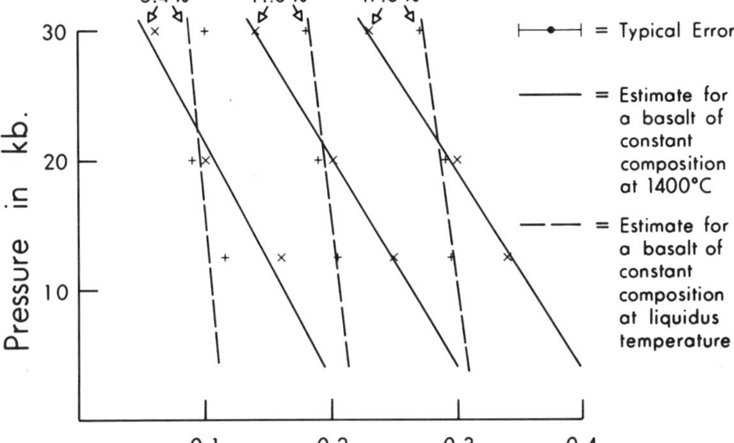


FIGURE 2.8—Solubility of sulfur in basaltic melts of different FeO content estimated as a function of pressure at 1400°C (solid line) and at the liquidus temperature of the basalt (dashed line). Based on the data of Wendlandt (1982).

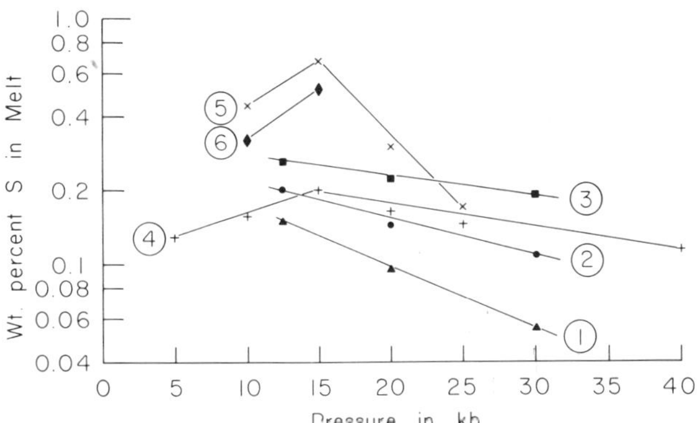


FIGURE 2.9—Comparison of the data of Gorbachev and Kashirceva and Wendlandt (1982) from Gorbachev and Kashirceva (1985). (1), (2) and (3) = experiments of Wendlandt (1982) (see cutline for Fig. 2.8). (4) = basalt + $H_2O$ + $CO_2$ at 1250°C. (5) = basalt + $H_2O$ at 1350°C. The composition of the melt changes with increasing pressure from about 18% MgO, 10% FeO at low pressure to 10% MgO, 14% FeO at high prssure, due to the lower degree of melting at the higher pressures. (6) = similar experiments to (5) at 1300°C.

## Variation of solubility of sulfide during fractional crystallization of a layered intrusion

The data on sulfide solubility discussed above, particularly that relating to dry systems, has been used (Naldrett and von Gruenewaldt, 1989) to construct a curve depicting approximately how sulfide solubilty will vary during the fractional crystallization of the magma of a layered intrusion. It has been shown above that, at constant pressure, three factors are extremely important to the solubility of sulfide in a silicate magma, temperature, the FeO content of the magma and oxygen fugacity. Mathez (this volume) argues that under closed conditions, and when not crystallizing large amounts of spinel, magmas follow buffer curves when cooling and crystallizing, which would tend to eliminate oxygen fugacity as a major control on sulfide solubility. Decreases in both temperature and FeO content of the magma decrease the solubility of sulfide.

During olivine crystallization, both temperature and the FeO content of the magma decrease, so that sulfide solubility decreases. Note that we are speaking of the FeO content of the magma here, not its FeO/MgO ratio. The latter will, of course, increase with olivine crystallisation. During bronzite crystallization, temperature continues to decrease sharply, but FeO content is relatively constant and then increases as the increase in the FeO/MgO beomes more pronounced. Once plagioclase joins either mafic mineral, the FeO content of thc magma will start to increase, and this will inevitably moderate the rate at which the solubilty of sulfide declines with continuing crystallization.

Mathez (1976) and Czamanske and Moore (1977) have discussed these general principles with respect to the differentiation of submarine basalts. Mathez pointed out that the crystallization of olivine alone drives submarine basalt magmas towards the sulfide saturation field, of plagioclase alone drives these magmas away from the field, while removal of plagioclase and olivine in the commonly observed proportions of 3 to 1 drives the magmas approximately parallel to the sulfide saturation surface.

Cawthorn and Davies (1983) have determined the crystallization sequence and temperature of appearance of liquidus phases of a marginal rock to the Bushveld Complex thought by them to be close to the composition of the Bushveld initial magma. Barnes and Naldrett (1986) have modelled the changes in composition of this magma with fractional crystallization. Five percent olivine crystallizes between 1325 and 1300°C, followed by 15% bronzite between 1300 and 1180°C. Barnes and Naldrett's calculations show that during this first 20% of crystallization, the FeO content decreases from 7.80 to 7.70 wt%. The decrease will be rapid during the crystallization of olivine, followed by a slight rise during bronzite crystallization, so that the reduction of 0.10 wt % is a net reduction. Plagioclase and augite join bronzite at 1180°C, and the resultant removal of these minerals in their cotectic proportions results in the magma being 50% crystallized at 1150°C, and the FeO content of the remaining magma increasing from 7.70 to 9.01 wt%.

These changes in temperature and FeO content have been coupled with the solubility data of Haughton et al. (1974), Buchanan and Nolan (1979) and Shima and Naldrett (1975) discussed above to construct the highly schematic sulfide solubility curve shown in Fig. 2.10. Experimental data important to this are Buchanan and Nolan's (1979) demonstration that basaltic magma containing 7.5 wt % FeO dissolves, at 1200°C and $f_{O_2} = 10^{-10.5}$, 0.035 wt% S (= 0.1 wt% sulfide), and that while it is dependant, this value is not critically dependant on the oxygen fugacity; this value therefore provides the approximate location of point C in Fig. 2.10. It has been argued above that between 1200 and 1450°C an increase in SCSS of between 0.04 and 0.09 wt perecent per 100°C increase in temperature is likely for basaltic magma of about this composition. This is equivalent to an increase of between 0.12 and 0.27 wt% sulfide. An intermediate value of 0.2 wt% sulfide per 100°C has therefore been used to extrapolate the solubility curve of Fig. 2.10 upwards from C towards B.

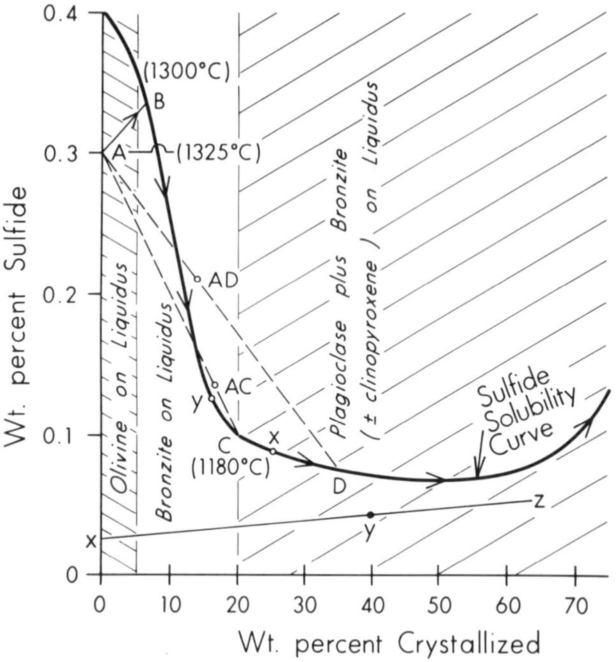


FIGURE 2.10—Schematic diagram illustrating the variation in the solubility of iron sulfide with the fractionation of Cawthorn and Davies's (1983) best representative (a sample of chilled marginal material) for the magma responsible for the Lower zone of the Bushveld Complex. Variations in solubility are based on the data of Haughton et al. (1974), Buchanan and Nolan (1979) and Buchanan et al. (1983) and the compositional calculations of Barnes and Naldrett (1986) as discussed in the text. After Naldrett and von Gruenewaldt (1989).

Because of the large temperature variations of the liquidus and the relatively small changes in FeO content of the Bushveld magma up to at least 50% crystallization, variations in temperature will have a greater influence on sulfide solubility over this crystallization interval than will variations in FeO content. The rate of change of temperature while olivine is crystallizing is 5°C per 1% magma crystallized; this rises to 8°C per 1% crystallized while bronzite is on the liquidus alone, and then falls to 1°C per 1 % crystallized between 20 and 50% crystallization. FeO content is increasing slightly during the bronzite crystallization, particularly during the final stages, so this will modify the effect of temperature, which is why the rate of descent of the solubility curve in Fig. 2.10 has been rounded off somewhat between points Y and C. Once plagioclase appears on the liquidus, FeO starts to increase markedly and this, coupled with the much less steep temperature gradient per unit crystallized is why the curve between C and D is drawn as flattening out. In the latter stages of crystallization FeO enrichment becomes marked and sulfide solubilities are likely to rise as is drawn above 60% crystallization.

It is stressed that the solubility curve of Fig. 2.10 is only approximate, although given the uncertainties inherent in attempting to estimate the composition and other intensive parameters such as temperature, oxygen fugacity and other volatile fugacities (all of which can affect sulfide solubility) in magma resposible for any layered intrusion, it is probably difficult to make any closer estimate of how saturation sulfide will vary with crystallization of the magma. When this curve is applied below to understanding the behaviour of sulfide during the fractional crystallization of the Bushveld Complex magma, the important point is the general shape of the curve, not whether point C corresponds to 0.1 wt% sulfide as drawn or to 0.2 or 0.05 wt%. Variations of this type will not affect the arguments that follow below. These depend solely on the shape of the curve, in particular on the difference in slope between crystallization before and after the appearance of plagioclase on the liquidus.

## PARTITIONING OF CHALCOPHILE ELEMENTS AMONG SULFIDES, SILICATE MELTS, AND SILICATE MINERALS

It is common to discuss the partitioning of a trace or minor element between two phases, A and B, in terms of a partition coefficient. Referring to equilibria between sulfide melts and silicate magmas, the Nernst partition coefficient $D_i$ for a metal $i$ is defined as:

$$D_i^{(\text{Sul. melt/Sil. magma})} = \frac{\text{wt\% metal } i \text{ in sulfide melt}}{\text{wt\% metal } i \text{ in silicate melt}}. \qquad \text{(iv)}$$

### Partitioning of nickel between sulfide and silicate liquids

In the case of metals such as Fe, Ni, Cu, and Co, it is believed (see Shimazaki and MacLean, 1976; Rajamani and

Naldrett, 1978) that they are bonded to oxygen in the silicate magma and to sulfur in the sulfide melts. Thus the reaction considered is (for example, in the case of Ni):

$$NiO^{(Sil.\ magma)} + 1/2 S_2 = NiS^{(Sul.\ melt)} + 1/2 O_2. \quad (3)$$

The equilibria involved in this reaction are related to the thermodynamic equilibrium constant $K_3$ by the expression:

$$K_3 = \frac{a_{NiS}}{a_{NiO}} \times \left[\frac{f_{O_2}}{f_{S_2}}\right]^{\frac{1}{2}} \quad (v)$$

$$= \frac{\gamma_{NiS}}{\gamma_{NiO}} \times \frac{N_{NiS}}{N_{NiO}} \times \left[\frac{f_{O_2}}{f_{S_2}}\right]^{\frac{1}{2}}, \quad (vi)$$

where $a$ refers to the respective activities, $\gamma$ to the activity coefficients, $N$ to the mole fractions and $f_{O_2}$ and $f_{S_2}$ are the fugacities of the gases. From comparison of expressions [iv] and [vi], one would expect $D_i^{Sul.\ melt/Sil.\ magma}$ for all of the metals Ni, Fe, Cu, and Co to be a function of $f_{O_2}$ and $f_{S_2}$ in addition to being a function of the temperature and pressure and the compositions of the two phases in question. However, reactions such as [2] and [3] may be combined to give exchange reactions of the type:

$$NiO^{(Sil.\ magma)} + FeS^{(Sul.\ melt)} = NiS^{(Sul.\ melt)} + FeO^{(Sil.\ magma)}. \quad (4)$$

The equilibrium constant for this reaction can be expressed by:

$$K_4 = \frac{a_{NiS}}{a_{NiO}} \times \frac{a_{FeO}}{a_{FeS}}$$

$$= \frac{\gamma_{NiS}}{\gamma_{NiO}} \times \frac{\gamma_{FeO}}{\gamma_{FeS}} \times \frac{N_{NiS}}{N_{NiO}} \times \frac{N_{FeO}}{N_{FeS}}. \quad (vii)$$

The equilibrium constant of this reaction is independent of the fugacities of sulfur and oxygen although these may affect the activities of components in the reaction. For example, $f_{S_2}$ will affect the activity coefficients of NiS and FeS. Citing the work of Scott et al., (1974) on the $Fe_{(1-x)}S$–$Ni_{(1-x)}S$ solid solution, Rajamani and Naldrett (1978) suggested that in liquids of the same composition as monosulfide solid solution, $\gamma_{NiS}$ and $\gamma_{FeS}$ have similar values despite the fact that both would decrease with increasing $f_{S_2}$. Thus, the ratio of these two functions would be close to 1 and show little change with variations in $f_{S_2}$. Variations in $f_{O_2}$ can affect $N_{FeO}$ by changing the $Fe^{3+}/Fe^{2+}$ ratio of the magma, but so long as the value of $f_{O_2}$ remains below about $10^{-8}$ atm (cf. Fudali, 1965), the effect on the $Fe^{3+}/Fe^{2+}$ ratio of basaltic magmas is small. Doyle and Naldrett (1986) emphasize that variations in $f_{O_2}$ also affect the oxygen content of sulfide–oxide liquids, and hence activity–composition relationships within them. Provided that the $f_{O_2}/f_{S_2}$ ratio (and hence the O/(O+S)) ratio in the sulfide/oxide liquid) remains approximately constant, this effect will not be noticeable with regard to element partitioning. An assumption of constant $f_{O_2}/f_{S_2}$ ratio is made in the following discussion. The effect of variable $f_{O_2}$ is considered below.

If a function, $K_D$ (the exchange partition coefficient), is defined (note the use of mole fractions in the definition) as:

$$K_D = \frac{N_{NiS}}{N_{NiO}} \times \frac{N_{FeO}}{N_{FeS}}, \quad (viii)$$

then from [vii] . . . . [viii] and noting that, as discussed above $\gamma_{NiS}/\gamma_{FeS} \approx 1$,

$$K_D = K_4 \times \frac{\gamma_{NiO}}{\gamma_{FeO}}. \quad (ix)$$

As $K_4$ is constant at given temperature and pressure, $K_D$ can be calculated from thermochemical data for any silicate magma and iron-rich sulfide melt, so long as the values of $\gamma_{NiO}$ and $\gamma_{FeO}$ are known for the silicate melt in question. Furthermore, if $K_D$ is determined for one particular silicate melt, knowledge of the temperature and compositional dependence of $\gamma_{NiO}$ and $\gamma_{FeO}$ and of the temperature dependence of $K_4$ permits new values of $K_D$ to be computed for melts of different composition and temperature.

Rajamani and Naldrett (1978) obtained values for $K_D$ for Ni, Cu, and Co using liquids with compositions of natural basalts, picrites and andesites (Table 2.1). Subsequently, Boctor and Yoder (1983) obtained similar values. The latter also demonstrated a dependence of $K_D$ on $f_{O_2}$.

Rajamani and Naldrett's (1978) data on sulfide–oxide liquid-silicate magma partitioning (with $K_D$ values for Ni of about 40) are compatible with the magmatic hypothesis, leading, as they showed, to reasonable predictions of the actual compositions of specific Ni-sulfide ores. The oxygen fugacity in their experiments was not buffered, but they predicted from the starting compositions of their materials that it was close to the quartz–fayalite–magnetite buffer, i.e., within the range of natural magmas. Boctor and Yoder's (1983) data confirm $K_D$ values of around 40 to 50 at temperatures of 1250 to 1300°C and $f_{O_2}$ values of $10^{-8}$ to $10^{-9}$ atm. Boctor and Yoder's data also demonstrate that $K_D$ decreases with increasing temperature of equlibration and increasing maficity of the silicate magma, as predicted by the calculations of Rajamani and Naldrett (1978) and Duke and Naldrett (1978), and the experimental studies of Campbell et al. (1979a).

Celmer (1987) studied the distribution of Ni between Ni–Cu–Fe matte and fayalite-rich slags. He found that the Nernst distribution coefficient for Ni was about 160 in $SiO_2$ saturated slags at 1250°C, $f_{O_2}$ of about $4 \times 10^{-9}$ and $f_{S_2}$ of about $5 \times 10^{-3}$. The coefficient decreased with decreasing Fe/Ni ratio in the matte and with increasing $f_{S_2}$ and $f_{O_2}$, and increased by a factor of 1.5 to 2 with the addition of up to 15 wt% $Al_2O_3$ and 10 wt% CaO to the slag. The slag compositions are far from those of basaltic melts, so that comparison of absolute values are not relevant.

TABLE 2.1—Experimental $K_D$ values for transition metal partitioning between sulfide melts and silicate magmas.

| Metal | $K_D$ | Temperature °C | Pressure | Log $f_{O2}$ | Melt composition |
|---|---|---|---|---|---|
| **Nickel** | | | | | |
| Rajamani & Naldrett (1978) | 42 | 1255 | very low | unknown* | basalt |
| | 38 | 1305 | very low | unknown* | basalt |
| | 34 | 1325 | very low | unknown* | olivine basalt |
| | 59 | 1255 | very low | unknown* | andesite |
| Boctor & Yoder (1983) | 49.4 ± 3 | 1300 | 1 atm | −9.0 | basalt |
| | 50.9 ± 2.6 | 1300 | 1 atm | −9.0 | basalt |
| | 43.1 ± 3.2 | 1400 | 1 atm | −8.0 | basalt |
| | 35.9 ± 2.5 | 1400 | 1 atm | −8.0 | basalt |
| | 22.4 ± | 1460 | 1 atm | −8.55 | Mafic 21% MgO |
| Gorbachev (personal communication 1988) | 35† | 1250 | 13 kb | Ni–NiO | Mafic 10.8 atom % MgO 7.1 atom % FeO |
| | 25† | 1250 | 15 kb | Ni–NiO | Mafic 21.9 atom % MgO 5.4 atom % FeO |
| | 27† | 1250 | 25 kb | Ni–NiO | Mafic 11.8 atom % MgO 7.7 atom % FeO |
| Gorbachev (personal communication 1988) | 32‡ | 1250 | 15 kb | Ni–NiO | Mafic 12.7 atom % MgO 8.0 atom % FeO |
| | 35‡ | 1300 | 10 kb | Ni–NiO | Mafic 14.6 atom % MgO 9.2 atom % FeO |
| | 42‡ | 1300 | 15 kb | Ni–NiO | Mafic 14.3 atom % MgO 10.0 atom % FeO |
| | 36‡ | 1350 | 25 kb | Ni–NiO | Mafic 16.4 atom % MgO 9.4 atom % FeO |
| **Copper** | | | | | |
| Rajamani & Naldrett (1978) | 35 | 1255 | very low | unknown* | basalt |
| | 24 | 1305 | very low | unknown* | basalt |
| | 48 | 1325 | very low | unknown* | olivine basalt |
| | 34 | 1255 | very low | unknown* | andesite |
| **Cobalt** | | | | | |
| Rajamani & Naldrett (1978) | 15 | 1255 | very low | unknown* | basalt |
| | 9 | 1305 | very low | unknown* | basalt |

*Rajamani and Naldrett (1978) state that from the initial compositions and proportions of their starting materials, they believe the oxygen fugacities in their experiments to have been close to the Quartz–Fayalite–Magnetite buffer.
†These data were obtained with excess $H_2O$ and $CO_2$.
‡These data were obtained with excess $H_2O$.

## Partitioning of nickel and iron between sulfide liquid and olivine

Nickel in a silicate magma also partitions preferentially into olivine according to the reaction:

$$NiO^{(Sil.\ magma)} + FeSi_{0.5}O_2^{(Olivine)} = FeO^{(Sil.\ magma)} + NiSi_{0.5}O_2^{(Olivine)}, \quad (5)$$

where the equilibrium constant $K_5$ is defined as:

$$K_5 = \frac{a_{FeO}}{a_{NiO}} \times \frac{a_{NiSi_{0.5}O_2}}{a_{FeSi_{0.5}O_2}} \quad (x)$$

and since $NiSi_{0.5}O_2$ and $FeSi_{0.5}O_2$ behave almost ideally in olivine (Campbell and Roeder, 1968; Williams, 1972),

$$K_5 = \frac{\gamma_{FeO}}{\gamma_{NiO}} \times \frac{N_{FeO}}{N_{NiO}} \times \frac{N_{NiSi_{0.5}O_2}}{N_{FeSi_{0.5}O_2}}. \quad (xi)$$

Reactions [4] . . . . [5] can be combined to express the partitioning between a sulfide melt and olivine:

$$FeS^{(Sul.\ melt)} + NiSi_{0.5}O_2^{(Olivine)} = NiS^{(Sul.\ melt)} + FeSi_{0.5}O_2^{(Olivine)}, \quad (6)$$

where the equilibrium constant $K_6$ is defined as:

$$K_6 = \frac{K_4}{K_5} = \frac{\gamma_{NiS}}{\gamma_{FeS}} \times \frac{N_{NiS}}{N_{FeS}} \times \frac{N_{FeSi_{0.5}O_2}}{N_{NiSi_{0.5}O_2}}. \quad \text{(xii)}$$

The exchange partition coefficient for the reaction is therefore equal to the equilibrium constant multiplied by the ratio of the activity coefficients of the two sulfide components in the sulfide melt. Values for the partition coefficient obtained in a variety of experiments involving Fe–Ni sulfides and olivine by Clark and Naldrett (1972), Fleet et al. (1977), Boctor (1981, 1982), and Fleet and McRae (1983, 1987) are given in Table 2.2, together with some data from actual deposits.

TABLE 2.2—$K_D$ value for Ni partitioning between sulfide melts and olivine.

| | $K_D$ | Temp. °C | Log $f_{O2}$ |
|---|---|---|---|
| **Experimental Data** | | | |
| Clark & Naldrett (1972) | 33.2 ± 3.4 | 930 | QFM |
| Fleet et al. (1977) | 33 | 1,160 | unknown |
| Boctor (1981) | 13.4 ± 1.4 | 1,400 | −9.0 |
| Boctor (1982) | 8.2 ± 1.4 | 1,400 | −8.0 |
| | 18.8 ± 1.4 | 1,300 | −9.0 |
| | 18.7 ± 0.8 | 1,300 | −9.0 |
| Fleet & MacRae (1983) | 27.2 ± 3.5 | 1,200 | unknown |
| Fleet & MacRae (1987) | 32–34 | 1,395 | −8.7 |
| | 30–38 | 1,385 | −8.87 |
| | 25–30 | 1,300 | unknown |
| | 25–27 | 1,200 | unknown |
| Gorbachev & Kashirceva (1986)† | 25 ± 3 | 1,250 | Ni–NiO |
| **Natural data from Thompson et al. (1984) and Fleet et al. (1983)** | | | |
| Katahdin | 3–9 | | |
| Burnt Nubble | 7–8 | | |
| Big Indian Pond | 8–16 | | |
| Black Narrows | 5–8 | | |
| Stillwater PGE Zone | 7–14 | | |
| Dumont Intrusion | 7–24 (av. 12.0) | | |
| Water Hen | 6 | | |
| Great Lakes Nickel | 8 | | |
| Perseverence W. Australia* | 23 ± 4 | | |
| Renzy Lake Quebec* | 21.6 ± 3.4 | | |

†Gorbachev's experiments were at pressures of 5 and 10 kbar in the presence of H-O-S fluid phase.
*Metamorphosed to upper amphibolite or higher grade and probably represents equilibrations at metamorphic temperature.

It is seen that there is a fundamental divergence between the experimental data. Fleet et al. (1977) and Fleet and MacRae (1983, 1987) obtained values between olivine and liquid sulfide ranging from 27 to 40, which are similar to that obtained by Clark and Naldrett (1973) between olivine and solid sulfide at a much lower temperature (915°C). (Fleet's most recent data, M. E. Fleet, personal communication 1987, is consistent with his earlier data). Their data have led Fleet and MacRae (1987) to suggest that $K_D$ does not vary appreciably with temperature.

On the other hand Thompson et al. (1984) have derived a relationship for $K_6$ (Fig. 2.11) from thermochemical data for solid sulfides and silicates which suggests that there should be a temperature dependance. If one assumes that $K_D$ approximates $K_6$ (i.e. that the ratio of the activity terms in equation [xi] is close to unity) and that the exchange reaction is not changed greatly by the involvement of a liquid rather than solid monosulfide solid solution, analogy with Fig. 2.11 indicates that $K_D$ should also decrease with increasing temperature.

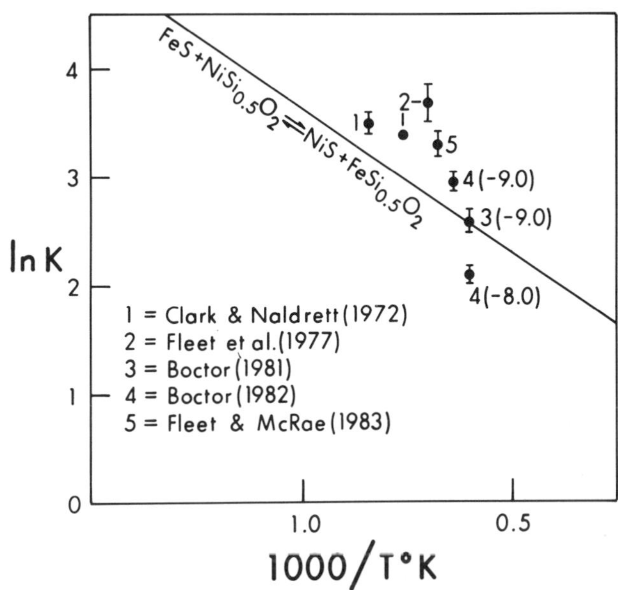


FIGURE 2.11—Experimental determinations of the distribution of Ni and Fe between olivine and sulfide. The sulfide phase was a solid solution in the experiment of Clark and Naldrett 1972; in all other cases it was molten. The error bars represent plus or minus two standard deviations of the mean of several determinations of the distribution coefficient at a given T and $f_{O_2}$. Where oxygen fugacities are known precisely, log $f_{O_2}$ is shown in parenthesis. The dashed line is the variation of the equlibrium constant for the exchange reaction with T, calculated with respect to the pure solid reactants as standard states. After Thompson et al. (1984).

As with Boctor and Yoder's (1983) experiments on co-existing sulfide and silicate liquids, Boctor's (1982) experiments on co-existing sulfide liquid and olivine indicate a dependance of $K_D$ on $f_{O_2}$, with $K_D$ increasing from 8.2 to 13.4 at 1400°C with a decrease in $f_{O_2}$ from $10^{-8}$ to $10^{-9}$ atm. Since the sulfide liquid is the common factor in both the sulfide-silicate liquid and sulfide-olivine experiments, it is logical to look to this for an explanation of the $f_{O_2}$ dependance. As was stated at the outset of this discussion, the fact that oxygen is present and the possible effect that this might have on the activities of the components such as NiS and CuS in the sulfide (-oxide) liquid have been ignored up to this point.

Doyle and Naldrett (1987) noted that the experimental data on Fe–O–S liquids indicate that as O increases at the expense of S, at a constant atomic proportion of Fe, the activity of Fe is not constant; i.e., the activity coefficient of Fe is a function of the O/(O+S) ratio, and thus of the $f_{O_2}/f_{S_2}$ ratio. This can be seen from fig. 1 of Shima and Naldrett (1975). On this, because of the near parallelism of the $f_{O_2}$ and $f_{S_2}$ isobars, a decrease in $f_{O_2}$ from $10^{-8}$ to $10^{-9}$ atm at a constant $f_{S_2}$ of $10^{-2}$ atm involves a change in composition corresponding to a decrease in *a*FeO in the coexisting sulfide and silicate melts of from 0.4 to 0.15. This decrease in *a*FeO can cause an increase in *a*FeS by a factor of 1.2. Analogous changes can be expected with respect to *a*NiS and the exchange partitioning will be reflected in the net sum of these changes.

Doyle and Naldrett observed that, when taken together, Fleet and MacRae's (1983) and Boctor's (1981, 1982) values of $K_D$ for Ni–Fe exchange between olivine and sulfide–oxide liquids increase progressively with decreasing O/(O+S) ratio of the sulfide–oxide liquids. This led them to suggest that, as noted above, the activity coefficient of Ni is also affected by the composition of the sulfide–oxide liquid, the coefficient for Ni decreasing more than that of Fe for a given decrease in the O/(O+S) ratio, so that $K_D$ for the exchange reaction between Ni and Fe decreases.

Fleet and MacRae's (in press) new data on the effect of $f_{O_2}$ and $K_D$ do not agree with Boctor's data, but indicate instead that $K_D$ decreases to 22 in a liquid containing 10 wt% $O_2$ instead of the value of about 10 predicted by Doyle and Naldrett (1987) from Boctor's (1982) data. Fleet and MacRae argue, contrary to the conclusions of Fudali (1965) but in accordance with the recent suggestions of Christie et al. (1986), that the oxygen fugacities in the igneous magmas giving rise to layered intrusions are two orders of magnitude lower than those of the FMQ buffer (see above). If correct, this would result in the oxygen content of natural sulfide liquids being much less than assumed by Doyle and Naldrett (1987), reducing even more the lowering of $K_D$. Observations on sulfides and olivines co-existing in nature (Table 2.2) reveal $K_D$ values from 3 to 24 with most falling in the range 5 to 15. Notable exceptions are the two deposits in Table 2.2 which have been subjected to the highest grades of metamorphism and which define values in excess of 20. Natural values in the range of 5 to 15 are consistent with the experimental results of Boctor (1981, 1982) but are different to those of Fleet and his co-workers. If the latter researchers are correct, and if $K_D$s are of the order of 30, problems arise for the magmatic model, which is what the latter have long maintained.

On the other hand, if Boctor's values are correct for natural conditions, and if Thompson et al.'s (1984) thermodynamic arguments as to the temperature dependance of $K_6$ and therefore $K_D$ are correct, a higher value is to be expected for sulfide-olivine pairs equilibrated at metamorphic as opposed to magmatic temperatures. This is consistent with the hypothesis that the Perserverance and Renzy Lake deposits have re-equilibrated during metamorphism as Binns and Groves (1976) claimed for Perserverance, but that this has not occurred in the other deposits.

**General remarks on the partitioning of nickel between sulfide liquids and silicate-rich phases**

At the time of writing, conflicting data exist for the partition coefficients that apply to Fe–Ni exchange between sulfide liquids and olivine. There is no direct conflict with respect to partitioning between sulfide and silicate liquids, although, as discussed below, the data for sulfide-olivine equilibria also carry implications with respect to sulfide-silicate liquid partitioning. In addition, variations in $f_{O_2}$ are likely to affect partitioning between sulfide and silicate melts. As related above, Cermer (1987) found that increasing $f_{O_2}$ lead to a decrease in the Nernst coefficient for matte-slag combinations, but this aspect has yet to be investigated for basaltic compositions. The uncertainty as to the magnitude of any possible effect should therefore be born in mind when the values tabulated in Tables 2.1 and 2.3 are put to use.

TABLE 2.3—Preferred partition coefficients.

| | Ni | Cu | Co | Pt | Pd |
|---|---|---|---|---|---|
| **Komatiitic Magma** | | | | | |
| 27% MgO | 100 | 250 | 40 | $10^5$ | $10^5$ |
| 19% MgO | 175 | 250 | 58 | $10^5$ | $10^5$ |
| **Basaltic Magma** | 275 | 250 | 80 | $10^5$ | $10^5$ |

Naldrett (1979) pointed out that there should be an interrelationship between the $K_D$ for nickel–iron exchange between olivine and silicate melts, olivine and sulfide melts, and sulfide and silicate melts. Experimental studies (Arndt, 1978) on the distribution of Ni between olivine and a range of sulfide-free silicate melts indicate that the $K_D$ varies from about 2 for a komatiitic to 8–14 for a basaltic melt. Given a $K_D$ between sulfide melt and basalt of about 50, the $K_D$ between sulfide and olivine should be about 5–7, which is far from the value of about 30 determined by Fleet and his co-workers. It is possible that (as Fleet et al., 1983 point out) the presence of sulfur in a silicate melt favours the retention of Ni in the melt more than Fe, thus reducing the olivine-silicate melt $K_D$ below the values determined for sulfide-free systems, but Cermer looked for this in his matte-slag experiments and was unable to document any such effect. The divergence between the $K_D$ of Fleet and his co-workers and natural observations thus poses a problem which extends beyond the area of co-existing olivine and sulfide melts and raises questions about the equlibria between sulfide and silicate melts.

A further complexity has been introduced by Cowden et al. (1987) with respect to the effect of oxidation on sulfide melt–silicate magma partitioning. They point out that, since the oxidation of a silicate magma is likely to change the number of octahedral sites (by reducing the $Fe^{3+}$ content) in the magma that are available for divalent cations, the tendancy for these cations to enter octahedral sites in sulfide melts will be increased, particularly those with high octahedral site preference energies. They predict, therefore, that oxidation should increase $K_D$(sulf. melt/sil. magma) between Ni and Fe. This is of course opposite to the effect of oxidation highlighted by Doyle and Naldrett (1987) for sulfide liquid-olivine partitioning. However Cowden et al. (1987) assume a constant sulfur/metal ratio for the sulfide melt, which Doyle and Naldrett showed not to be the case, and consider only the implications of oxidation on the silicate magma, not on the sulfide melt in addition.

Despite the fact that all of the experimental studies have been on the exchange reactions discussed above, because of simplicity in application it has been the custom in the literature involved with the modelling of sulfide liquid-silicate liquid partitioning to use the results of the experiments converted to simple Nernst-type coefficients. Naldrett (1981) warned of the errors involved in using Nernst coefficients at high concentrations of the elements involved, and where the partitioning also involves appreciable amounts of some third component such as copper. These warnings should also be born in mind when using the numbers tabulated in Table 2.3; the numbers here represent the preferred values of the Nernst coefficients, derived from the exchange partition coefficients of Table 2.1.

## REFERENCES


Abraham, K. P., Davies, M. W., and Richardson, F. D., 1960, Sulfide capacities of silicate melts, Pt. I: J. Iron and Steel Inst., 196, p. 309–312.

Barnes, S. J., and Naldrett, A. J., 1986, Geochemistry of the J–M (Howland) Reef of the Stillwater Complex, Minneapolis adit area. II. Silicate mineral chemistry and petrogenesis: Jour. Petrol., 27, p. 791–825.

Binns, R. A., and Groves, D. I., 1976, Iron–nickel partition in metamorphic olivine-sulfide assemblages from Perseverance, W. Australia: Amer. Mineral., 6l, p. 782–787.

Boctor, N. Z., and Yoder, H. S., Jr., 1983, Partitioning of nickel between silicate and iron sulfide melts: Carnegie Inst. of Washington, Yr. Book 82, p. 275–277.

Boctor, N. Z, 1982, The effect of $f_{O_2}$ and $f_{O_2}$, and temperature on Ni partitioning between olivine and iron sulfide melts: Carn. Inst. of Washington, Yr. Book 81, p. 366–369.

Boctor, N. Z., 1981, Partitioning of nickel between olivine and iron monosulphide melts: Carnegie Inst. of Washington, Yr. Book 80, p. 356–359.

Bog, S., and Rosenqvist, T., 1958, A thermodynamic study of the iron sulfide–oxide melts: Meddelebse Nr 12 fra Metallurgisk Komite, Trondheim, Norway.

Brett, R., and Bell, P. M., 1969, Melting relations in the Fe-rich portion of the system Fe–FeS at 30 kb pressure: Earth and Planet. Sci. Lett., 6, p. 479–482.

Buchanan, D. L., Nolan, J., Wilkinson, N., and de Villiers, J. R. R., 1981, An experimental investigation of sulfur solubility as a function of temperature in synthetic silicate melts: Geol. Soc. S. Africa, Spec. Publ. 7, p. 383–391.

Buchanan, D. L., and Nolan, J., 1979, Solubility of sulfur and sulfide immiscibilty in synthetic tholeiitic melts and their relevance to Bushveld-complex rocks: Canadian Mineral., 17, p. 483–494.

Campbell, F. E., and Roeder, P., 1968, The stability of olivine and pyroxene in the Ni–Mg–Si–O system: Amer. Mineral., 53, p. 257–268.

Campbell, I. H., Naldrett, A. J., and Roeder, P. L., 1979, Nickel activity in silicate liquids: some preliminary results: Canadian Mineral., 17, p. 495–506.

Cawthorn, R. G., Davies, G., 1983, Experimental data at 3 K bars pressure on parental magma to the Bushveld Complex: Contrib. Mineral. Petrol., 83, p. 128–135.

Celmer, R. S., 1987, The distribution of minor elements in nickel matte smelting: Unpub. Ph.D. Thesis, Univ. of Toronto, 341 p.

Christie, D. M., Carmichael, I. S. E., and Langmuire, C. H., 1986, Oxidation states of mid-ocean ridge basalt glasses: Earth Planet. Sci. Lett., 79, p. 397–411.

Clark, T., and Naldrett, A. J., 1972, The distribution of Fe and Ni between synthetic olivine and sulfide at 900°C: Econ. Geol., 67, p. 939–952.

Connolly, J. W. D., and Haughton, D. R., 1972, The volume of sulfur in glass of basaltic composition formed under conditions of low oxidation potential: Am. Mineral., 57, p. 1515–1517.

Cousins, C. A., 1964, The platinum deposits of the Merensky Reef; *in* Haughton, S. H. (ed.), The geology of some ore deposits in southern Africa, vol. II: Geol. Soc. S. Africa, Johannesburg, p. 225–238.

Cowden, A., and Woolrich, P., 1987, Geochemistry of the Kambalda iron–nickel sulfides: implications for models of sulfide-silicate partitioning: Canadian Mineral., 25, p. 21–36.

Craig, J. R., and Naldrett, A. J., 1967, Minimum melting of nickeliferrous pyrrhotite ores: Ann. Rep. Dir. of Geophys. Lab., Carn. Inst. of Washington, Yr. Book 66.

Czamanske, G. K., and Moore, J. G., 1977, Composition and phase chemistry of sulfide globules in basalt from the mid-Atlantic Ridge rife valley near 37°N. lat.: Geol. Soc. Amer., Bull., 88, p. 587–599.

Doyle, C. D., and Naldrett, A. J., 1985, Ideal mixing of divalent cations in mafic magma and its effect on the solution of ferrous oxide: Geochim. et Cosmochim. Acta, 50, p. 435–443.

Doyle, C. D., and Naldrett, A. J., 1986, Ideal mixing of divalent cations in mafic magma and its effect on the solution of ferrous oxide: Geochim. et Cosmochim. Acta, 50, p. 435–443.

Doyle, C. D., and Naldrett, A. J., 1987, The oxygen content of 'sulfide' magma and its effect on the partitioning of nickel between co-existing olivine and molten ores: Econ. Geol., 82, p. 208–211.

Duke, J. M., and Naldrett, A. J., 1978, A numerical model of the fractionation of olivine and molten sulfide from komatiite magma: Earth Plan. Sci. Letters, 39, p. 255–266.

Fincham, C. J. B., and Richardson, F. D., 1954, The behaviour of sulfur in silicate and aluminate melts: Proc. Roy. Soc., A223, p. 40–62.

Fleet, M. E., and MacRae, N. D., 1987, Partition of Ni between olivine and sulfide: the effect of temperature, $f_{O_2}$ and $f_{O_2}$: Contrib. Mineral. Petrol., 95, p. 336–342.

Fleet, M. E., and MacRae, N. D., 1983, Partition of Ni between olivine and sulfide and its application to Ni–Cu sulfide deposits: Contr. Mineral. Petrol., 83, p. 75–81.

Fleet, M. E., 1977, Origin of disseminated copper–nickel sulfide ore at Frood, Sudbury, Ontario: Econ. Geol. 72, p. 1449–1456.

Fudali, R. F., 1965, Oxygen fugacities of basaltic and andesitic magmas: Geochim et Cosmochim. Acta, 29, p. 1063–1075.

Gorbachev, N. S. and Kashirceva, G. A., 1985, Fluid-magmatic differentiation of basaltic magma and equilbrium with magmatic sulfides; *in*, Experiments in the study of important problems in geology: Nauka, Moscow, p. 98–1198.

Hilty, D. C., and Crafts, W., 1952, Liquidus surface of the Fe–S–O system: J. Metals, 4, p. 1307–1312.

Haughton, D. R., Roeder, P. L., and Skinner, B. J., 1974: Solubility of sulfur in mafic magmas: Econ. Geol., 69, p. 451–467.

Huang, W.–L., and Williams, R. J., 1980, Melting relations of portions of the system Fe–S–Si–O to 32 kb with implication to the nature of the mantle core boundary (abs.): Lunar and Planet. Sc. XI, Lunar and Planetary Inst., Houston, p. 486–488.

MacLean, W. H., 1969, Liquidus phase relations in the FeS–FeO–$Fe_3O_4$–$SiO_2$ systems and their application in geology: Econ. Geol., 64, p. 865–884.

Mathez, E. A., 1976, Sulfur solubility and magmatic sulfides in submarine basalt glass: J. Geophys. Res., 81, p. 4269–4276.

Nagamori, M. and Kameda, M., 1965, Equilibria between Fe–S–O system melts and $CO$–$CO_2$–$SO_2$ gas mixtures at 1200°C: Trans. Japan Inst. Met. 6, p. 21–30.

Naldrett, A. J., 1969, A portion of the Fe–S–O and its application to sulfide ore magmas: J. Petrol., 10, p. 171–201.

Naldrett, A. J., 1979, Partitioning of Fe, Co, Ni and Cu between sulfide liquid and basaltic melts and the composition of Ni–Cu sulfide deposits—a reply and further discussion: Econ. Geol., 74, p. 1520–1528.

Naldrett, A. J., 1981, Nickel sulfide deposits: Classification, composition and genesis: Econ. Geol., 75th Anniv. Vol., p. 628–685.

Naldrett, A. J., and Richardson, S. W., 1967, Effect of water on the melting of pyrrhotite–magnetite assemblages: Ann. Rep. Dir. Geophys. Lab., Carn. Inst. of Washington, Yr. Book, 66, p. 429–431.

Naldrett, A. J., and von Gruenewaldt, G., 1989, The association of PGE with chromitite in layered intrusions and ophiolite complexes: Econ. Geol. (in press).

Rajamani, V., and Naldrett, A. J., 1978, Partitioning of Fe, Co, Ni and Cu between sulfide liquid and basaltic melts and the composition of Ni–Cu sulfide deposits: Econ. Geol., 73, p. 82–93.

Roeder, P. L., 1974, Thermodynamics of element distribution in experimental mafic silicate liquid systems: Fortschr. Miner. Spec. Iss., IMA papers, 9th Meeting, Berlin–Regensburg.

Rosenqvist, T., 1954, A thermodynamic study of the iron, cobalt and nickel sulfides. Jour. Iron and Steel inst. 176, p. 37–57.

Ryzhenko, B., and Kennedy, G. C., 1973, The effect of pressure on the eutectic in the system Fe–FeS: American Jour. Sc., 273, p. 803–810.

Scott, S. D., Naldrett, A. J., and Gasparrini, 1974, Regular solution model for the $Fe_{1-x}S$–$Ni_{1-x}S$ (mss) solid solution: Int. Min. Assoc., Ninth General Meeting, West Berlin and Regensberg, collected abstracts, p. 172.

Sharp, W. E., 1969, Melting curves pf sphalerite, galena and pyrrhotite and the decomposition of pyrite between 30 and 65 kilobars: Jour. Geophys. Res., 74, p. 1645–1652.

Shima, H., and Naldrett, A. J., 1975, Solubility of sulfur in an ultramafic melt and the relevance of the system Fe–S–O: Econ. Geol., 70, p. 960–967.

Smith, F. G., 1963: Physical geochemistry: Addison–Wesley, Reading, U.S.A., p. 304–307.

Thompson, J. F. H., Barnes, S. J., and Duke, J. M., 1984, The distribution of nickel and iron between olivine and magmatic sulfides in some natural assemblages: Canadian Mineral., 22, p. 55–66.

Usselman, T. M., Hodge, D. S., Naldrett, A. J., and Campbell, I. H., 1979, Physical constraints on the characteristics of nickel sulfide ore in ultramafic lavas: Canadian Mineral., 17, No. 2, p. 361–372.

Usselman, T. M., 1975, Experimental approach to the state of the core: Part I. The liquidus relations of the Fe-rich portion of the Fe–Ni–S system from 30–100 kbars: Am. J. Sc., 275, p. 278–290.

Wendlandt, R. F., 1982, Sulfide saturation of basalt and andesite melts at high pressures and temperatures: Am. Mineral., 67, p. 877–885.

Wendlandt, R. F., and Huebner, J. S., 1979, Melting relations of portions of the system Fe–S–O at high pressure and applications to the compositions of the earth's core (abs.): Lunar Planetary Sc. X, Lunar and Planetary Inst., p. 1329–1331.

# Chapter 3

# Vapor Associated with Mafic Magma and Controls on Its Composition

E. A. Mathez

## INTRODUCTION

It seems a matter of faith that fluids are responsible for any number of nature's deeds that otherwise have no obvious explanation, a case in point being the petrogenesis of platinum group element (PGE) deposits in layered intrusions. One confusion that persists is that fluids associated with all magmas are similar. They are not, of course. Those associated with acid systems are aqueous, contain high concentrations of dissolved silica and salts and had in many instances a profound influence on metallogenesis. In contrast, the fluids associated with mafic magmas are initially $CO_2$-rich and evolve to more water-rich compositions, and their role in redistribution of metals in mafic layered intrusions is uncertain.

In and around many granitoid bodies, such as those that yielded porphyry coppers, both the geology and hydrothermal mineral assemblages lend themselves to theoretical deduction of the nature and petrogenetic role of fluids. However, in complexes such as the Stillwater and Bushveld, key relationships that would allow us to make similar deductions are lacking, and inferences must be based in part on knowledge gained by experiment and analogy to natural systems. Too, the high-temperature, fluid-rock interactions in mafic systems have not been so thoroughly scrutinized.

Sulfur must be included in a discussion of volatiles since it is a component of volcanic gas and sulfides are the major ore minerals. This is particularly so because the stability of sulfide-oxide liquids in equilibrium with silicate melts is sensitive to oxidation state of the system, and oxidation state also influences the way in which the fluid phase evolves.

There has been a tendency in the literature to treat oxygen fugacity ($f_{O_2}$) as a variable to be changed at will to accomplish a desired end; yet this particular Maxwell's demon is a well-behaved property. Since $f_{O_2}$ is basic to any discussion of magmatic fluids and sulfides, this review begins by exploring relationships involving $f_{O_2}$ and its significance in petrogenetic processes. Second, solubility and abundance constraints on the composition of vapor in equilibrium with mafic magma are examined. Third, a model is summarized for the evolution of fluid during cooling of a mafic intrusion emplaced in the shallow crust.

## IMPORTANCE OF OXIDATION STATE

### Historical background

The importance of $f_{O_2}$ as a variable in petrologic processes was first brought to the attention of petrologists by Kennedy (1948), who determined the relationship between $Fe^{3+}/Fe^{2+}$ and $f_{O_2}$ in silicate melts. The equilibrium (Fudali, 1965) is approximated as

$$FeO + {}^{1}/{}_{4}O_2 = FeO_{1.5}. \qquad [1]$$

Kennedy (1955) recognized that $f_{O_2}$ could affect the course of fractional crystallization because it influences the composition and temperature of crystallization of iron-bearing silicates and oxides. This was experimentally investigated by Osborn and colleagues (e.g., Osborn, 1959, Roeder and Osborn, 1966), who proposed that the Skaergaard and other intrusions which exhibit iron enrichment in the middle stages of fractionation (the Fenner trend on an FMA diagram) crystallized at low $f_{O_2}$. In these magmas the $f_{O_2}$ was maintained at low values by equilibrium among the iron-bearing condensed phases. In contrast, the calc-alkaline sequence basalt—andesite—dacite—rhyolite (the Bowen trend), which exhibits essentially no iron enrichment due to fractionation, was produced by fractional crystallization of a similar initial bulk composition but at relatively high $f_{O_2}$. The high $f_{O_2}$ of these systems were due to their high fluid contents. It is doubtful that the Earth's interior is anywhere so oxidized that calc-alkaline rocks could have originated in this manner, the effect of $f_{O_2}$ on magmatic evolution being much more subtle. However, recent data indicate that some andesitic and dacitic magmas are in fact considerably more oxidized than MORB magmas and that high $f_{O_2}$ does exert some control on the manner in which they fractionate (Rutherford et al., 1985; Johnson and Rutherford, 1987).

The perception of the importance of $f_{O_2}$ in petrogenesis changed as its enormous influence on the composition and relative stabilities of the volatile-element phases became apparent. First, it was recognized that the stabilities of graphite, diamond and carbonate and, as a result, the stability and composition of C–O–H fluids throughout the upper mantle and crust, are sensitive to $f_{O_2}$ (e.g., French, 1966). Second, $f_{O_2}$ strongly influences equilibrium between sulfide-oxide and silicate melts (Naldrett, 1969). The research emphasis changed from experimental studies to determination of oxidation states of natural systems. This was stimulated by the development by U.S.G.S. workers of an electrochemical device to measure $f_{O_2}$ directly in lavas (Sato and Wright, 1966) and "intrinsic $f_{O_2}$" of natural samples in the laboratory and by the development of the Fe–Ti oxide geobarometer (Buddington and Lindsley, 1964). Both types of measurements indicated that $f_{O_2}$ of mafic magmas are typically within an order of magnitude of the synthetic oxygen buffer QFM (fayalite + $O_2$ = quartz + magnetite). However, data accumulated over the last decade complicate the picture by indicating that some magmas and parts of the upper mantle are more reduced and that the Earth is heterogeneous in terms of oxidation state.

### Influence of temperature on redox equilibria

Nearly all oxidation equilibria in synthetic and natural mineral and melt assemblages are strongly dependent on temperature and weakly dependent on pressure. For example, the manner in which $f_{O_2}$ of melt changes with temperature at constant $Fe^{3+}/Fe^{2+}$ ratio according to reaction [1], which has been calibrated by Sack et al. (1980), is illustrated in Fig. 3.1 for a tholeiitic basalt composition. Also illustrated are two synthetic oxygen buffers. Note should be made of QFM. Because this assemblage is seldom observed in nature, there are few instances in which it actually controls $f_{O_2}$. However, the synthetic buffer is a suitable model for more complex reactions in nature, such as

$$\text{olivine} + O_2 = \text{orthopyroxene} + \text{spinel}, \qquad [2]$$

in the sense that if $f_{O_2}$ is controlled by this assemblage in a real rock, then the T–$f_{O_2}$ path followed by that system will be parallel to that of QFM. Furthermore, nearly all other equilibria involving $f_{O_2}$, including those for synthetic metal-oxide systems as well the homogeneous equilibrium for silicate melt (Fig. 3.1), are also subparallel to each other.

Most natural systems in fact exhibit this type of behavior. Two sets of observations (Fig. 3.1) illustrate the point. The first are the original measured $f_{O_2}$ of Sato and Wright (1966) through the crust of Makaopuhi lava lake, Kilauea Volcano. Second are a set T–$f_{O_2}$ data computed from ilmenite-titanomagnetite pairs in three flows recovered by drilling in the deep ocean (Gitlin, 1985). The latter reflect the manner in which individual flows up to 42 meters thick cooled in a closed system. Haggerty (1976) presented many other oxide data illustrating similar cooling behavior.

The point of this discussion is that closed natural systems should cool down T–$f_{O_2}$ paths approximately parallel to that of QFM. This knowledge is extremely important because it provides the basis for predicting how fluids evolve and interact with the rocks as they cool. It appears that some systems deviated from this behavior, however. Processes that might cause this are considered below.

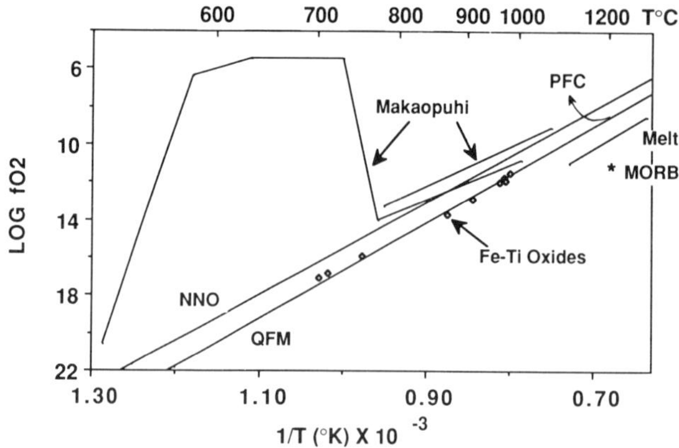


FIGURE 3.1—T–$f_{O_2}$ paths for some mafic systems. NNO and QFM are the synthetic buffers (see text). **Makaopuhi:** Intrinsic $f_{O_2}$ measurements in two drill cores through the crust of the lava lake (Sato and Wright, 1966). The oxidized zone encountered in one drill hole was interpreted to be due to the influx of meteoric water. Note that above about 800°C the Makaopuhi trends are subparallel to NNO. **Melt:** Tholeiitic basalt melt for which $Fe^{2+}/(Fe^{2+} + Fe^{3+})$ remains constant 0.95 computed using the calibration of Sack et al. (1980). **MORB:** The most reduced sample reported by Christie et al. (1986). **Fe–Ti Oxides:** Data from Gitlin (1985) for a thick MORB flow sampled by drilling during DSDP Leg 53. **PFC:** Oxidation path for perfect fractional crystallization (Carmichael and Ghiroso, 1986).

### Concept of relative oxygen fugacity

Obviously, the values of intensive variables such as $f_{O_2}$ depend on temperature and pressure. Since natural systems behave in a manner similar to the synthetic solid buffers, it has been found convenient to express the value of $f_{O_2}$ relative to one of them (Sato, 1978). Thus, NNO–2.5 refers to an $f_{O_2}$ of 2.5 orders of magnitude below that of the Ni–NiO buffer at the same temperature and, since most redox equilibria involving crystalline phases shift gradually and by approximately similar magnitudes to higher $f_{O_2}$ with increasing pressure (as computed from molar volumes of products and reactants of the buffer reactions), at the same pressure as well. Thus, $f_{O_2}$ of assemblages with identical compositions but at different temperatures and pressures are conveniently equated. The choice of NNO rather than QFM as a reference buffer (Carmichael and Ghiorso, 1986) is preferred because the former is more accurately determined.

### Relationship between oxidation state and $f_{O_2}$ of magma

Equilibrium [1] expresses the relationship between $f_{O_2}$ and oxidation state in the homogeneous melt. It has been experimentally calibrated by Sack et al. (1980) for a wide range of silicate melt compositions. With these data it is possible to compute magmatic $f_{O_2}$ from knowledge of the bulk composition and $Fe^{3+}/Fe^{2+}$ ratio, as determined by analysis of volcanic glass. Sack et al. showed that alkalis strongly effect the equilibrium, shifting equilibrium [1] to the right. Thus, alkali-rich magmas are more oxidized (i.e., their $Fe^{3+}/Fe^{2+}$ ratios are higher) than alkali-poor ones at identical $f_{O_2}$.

Equilibrium [1] is shifted to the left with increased pressure (Mo et al., 1983), but the effect of pressure on [1] is believed to be slightly greater than it is on NNO (Carmichael and Ghiorso, 1986). Therefore, a magma which decompresses but in which the $Fe^{3+}/Fe^{2+}$ ratio remains constant will become slightly more oxidized relative to the reference buffer.

Another feature of the equilibrium is illustrated in Fig. 3.2. It can be seen that the $Fe^{3+}/Fe^{2+}$ ratio increases progressively more rapidly as $f_{O_2}$ increases from IW (the iron-wütite buffer) to QFM and higher $f_{O_2}$. The result is that equilibrium [1] exerts a progressively stronger buffering influence on processes that tend to change $f_{O_2}$ as $f_{O_2}$ increases from reduced conditions near IW to oxidized conditions near NNO. Thus, reduced magmas are more easily oxidized than oxidized ones.

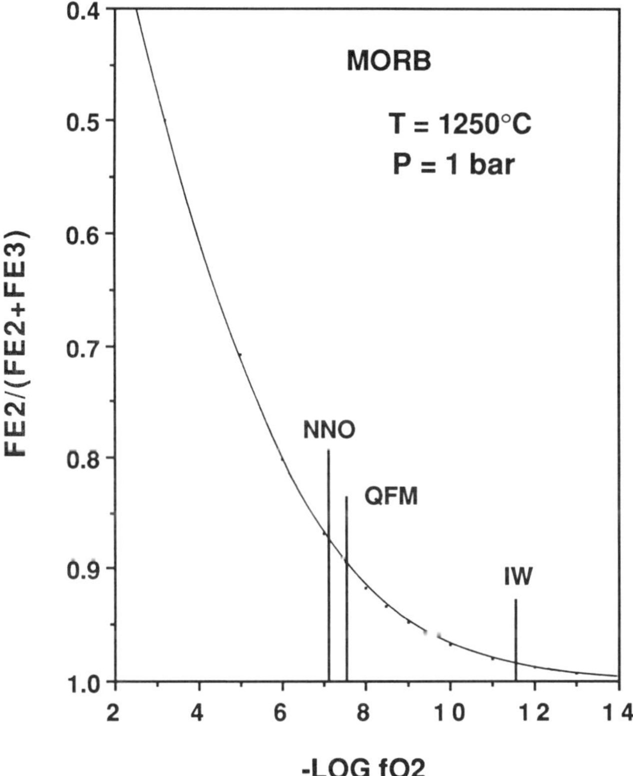


FIGURE 3.2—The relationship between $f_{O_2}$ and oxidation state of MORB melt computed with the calibration of Sack et al. (1980). Several synthetic buffers are shown for reference (after Mathez, 1984).

## Oxygen fugacities of natural systems

The oxidation states of magmas and their mantle source regions are not easily determined, and therefore the question continues to be debated. A current discussion of this issue, especially as it involves the veracity of intrinsic oxygen fugacity (IOF) measurements, is that of Ulmer et al. (1987). What is known about oxidation state is summarized as follows.

### Subaerial lavas

Subaerial tholeiitic basalts erupt at $f_{O_2}$ at or slightly above QFM. This is indicated by direct measurements with electrochemical probes (e.g., Sato and Wright, 1966) and by "reconstructed" compositions of volcanic gases (i.e., compositions corrected for atmospheric contamination, partial condensation, etc.) (e.g., Gerlach, 1980). However, degassing reactions may influence redox equilibria (discussed below), so oxidation states of volcanic gases and lavas need not reflect those of magma chambers or mantle source regions.

Some lava types are known to be generally more oxidized. Rutherford et al. (1985) found by experiment that an $f_{O_2}$ of $\approx$ NNO + 2 was required to reproduce the composition and phenocryst assemblage of Mount St. Helens dacite. Anhydrite is observed in some lavas, notably at El Chichon, and Carroll and Rutherford (1985) have shown that anhydrite crystallization proceeds at $f_{O_2}$ near HM. This is consistent with $Fe^{3+}/Fe^{2+}$ ratios of other sulfate-bearing, alkali-rich, mafic lavas, which yield computed $f_{O_2}$ in the range NNO + 1.5 to NNO + 3.5 (Carmichael and Ghiorso, 1986). Whether or not these oxidized values actually reflect those of the magma source regions or are due to processes resulting from decompression is not known.

### Submarine basalts

Christie et al. (1986) determined $Fe^{3+}/Fe^{2+}$ ratios of MORB glasses and found values that correspond to $f_{O_2}$ generally in the range NNO − 1.5 to NNO − 3.0. These glasses were rapidly quenched, relatively fresh and only moderately degassed, and therefore they may be regarded as representative of their source region.

### Xenoliths and megacrysts from alkali basalts and kimberlites

For kimberlites, new IOF measurements (Arculus et al., 1984; Virgo et al., 1988) on ilmenite megacrysts and computations based on olivine–orthopyroxene–ilmenite assemblages in xenoliths (Haggerty and Tompkins, 1983; Eggler, 1983) indicate $f_{O_2}$ of $\approx$ NNO–2 at pressures corresponding to mantle source regions. These values are consistent with $f_{O_2}$ being controlled by a mantle silicate–carbonate–graphite (or diamond) assemblage (Eggler, 1983). There is, however, no general agreement on oxidation states of spinel lherzolite and pyroxenite xenoliths from alkalic basalts. IOF measurements by Arculus and Delano (1981) and Arculus et al. (1984) indicate that some lherzolites are as reduced as IW. A similar conclusion was reached by Ulmer et al. (1987) based on additional experiments and detailed investigation of the applicability of the IOF technique to mantle materials. However, computations of $f_{O_2}$ from olivine–orthopyroxene–spinel assemblages in these rocks indicate $f_{O_2}$ more oxidized than NNO (Mattioli and Wood, 1986). The discrepency is not resolved.

Although the highly reduced IOF measurements have engendered considerable skepticism, mineral inclusions in diamonds also suggest that some parts of the upper mantle are reduced. These include inclusions of Cr-rich olivine (Ryabchikov et al., 1982), in which chromium is presumably dissolved as $Cr^{2+}$. Inclusions of moissanite (SiC) and FeO–MgO solid solutions are also known (Moore et al., 1986).

**Layered intrusions**

IOF data from the Skaergaard, Bushveld and Stillwater have been reported by Sato and Valenza (1980), Flynn et al. (1978), Elliot et al. (1982) and Buntin et al. (1985). Taken at face value, most of the measurements indicate that the initial undifferentiated magmas (as represented by rocks from lower stratigraphic positions) were intruded at $f_{O_2}$ between IW and MW and that samples from higher stratigraphic positions are more oxidized, between MW and QFM.

This writer doubts the veracity of magmatic oxidiation states deduced from IOF measurements on materials from slowly cooled rocks. An IOF is simply the equilibrium $f_{O_2}$ of a mineral or mineral assemblage at any specific temperature, and what is measured is the equilibrium $f_{O_2}$ path followed by a sample through heating and cooling cycles. No information is gleened from such a measurement about the temperature at which the final equilibrium $f_{O_2}$ was established. The problem in the layered intrusions is that most iron-bearing minerals re-equilibrate well below solidus temperatures. For example, olivine–phlogopite pairs from the Stillwater olivine-bearing zone I (nomenclature of McCallum et al., 1980) yield equilibration temperatures of $\approx$ 600°C (Boudreau et al., 1986), and chromite and olivine are known to re-equilibrate at subsolidus temperature (e.g., Irvine, 1967).

Observations from the rocks themselves do yield some information. For the Bushveld Complex, the presence of cumulate magnetite at the base of the Upper Zone requires that during crystallization the $f_{O_2}$ of those rocks was above that defined by MW (NNO–2.1 at 1100°C), and the presence of igneous graphite in the Critical Zone probably indicates a magmatic $f_{O_2}$ in the range of NNO–2.8 to NNO–2.2 (Mathez et al., 1988). Oxidation states of layered intrusions can only be approximated from direct observations of the rocks or by analogy with other mafic systems.

## Processes that perturb magmatic oxidation state

Sudden changes in $f_{O_2}$ have been invoked to explain the development of sulfide-rich horizons and chromitites in layered intrusions. Although these changes may obviously result from introduction of matter from country rocks or the addition of new magma to a pre-existing chamber, it is useful to consider what general processes will result in the deviation of a cooling magma from QFM-type behavior.

**Perfect fractional crystallization**

In modelling the evolution of tholeiitic basalt by fractional crystallization, Carmichael and Ghiorso (1986) observed that $Fe^{3+}/Fe^{2+}$ ratios of residual liquids increase because the iron-bearing crystallizing phases contain only ferous iron. Therefore, the T–$f_{O_2}$ paths of the residual liquids is such that $f_{O_2}$ increase instead of decrease with cooling (Fig. 3.1). In systems that do not evolve by perfect fractional crystallization, such as the basalt flow of Fig. 3.1 and to some extent the layered intrusions, this behavior is not to be expected. In these cases, residual melts may remain in partial contact and thus react with early-formed crystals—i.e., the oxidation state of residual melts may be buffered by the crystalline assemblage. However, Carmichael and Ghiorso pointed out that the basalt to rhyodacite sequence of Thingmuli Volcano, Iceland, is both precisely modeled by fractional crystallization and characterized by QFM-type cooling behavior. To explain this paradox, they suggested that $f_{O_2}$ is regulated by sulfur dissolved in the melt. (Sulfur is not considered in the modelling.) Sulfur dissolves primarily as $S^{2-}$ under $f_{O_2}$ more reduced than NNO and as $SO_4^{2-}$ under more oxidized conditions (Katsura and Nagashima, 1974; Carroll and Rutherford, 1985). Therefore, dissolved sulfur will have an enormous buffering capacity because on a mole for mole basis it may liberate or consume much more oxygen than any other species present in multiple oxidation states in the melt. Carmichael and Ghiorso emphasized that it is simply the homogeneous equilibrium between the oxidized and reduced forms of dissolved sulfur which should maintain melt composition near NNO during fractional crystallization.

**Degassing**

Degassing processes may strongly effect oxidation states of magmas. Thus, to explain the oxidized zone in the Makaopuhi lava lake (Fig. 3.1), Sato and Wright (1966) suggested that oxidation may result from continuous outward diffusion of $H_2$, causing dissociation of $H_2O$ and increase in $O_2$ content of the solidified crust according to the general reaction:

$$H_2O = H_2 + {}^1/_2O_2. \quad [3]$$

Oxidized zones similar to that at Makaopuhi are not observed in layered intrusions. However, Sato (1978) suggested that magmas may also become oxidized if $H_2$ escapes by diffusion through the melt. The potential sensitivity of $f_{O_2}$ to this process is illustrated by the fact that oxidation of MORB melt from IW to QFM can be accomplished by dissociation of only 0.07 wt.% $H_2O$ (Mathez, 1984). One can imagine how this process could operate in subsolidus environments where fluids are expected to be relatively $H_2O$-rich and $f_{H_2}$ high. However, in tholeiitic melt at 1250°C having an $f_{O_2}$ of NNO–2.5, $f_{H_2}$ is only about 1/10 that of $f_{H_2}$, so oxidation by this process may be limited by the miniscule amounts of $H_2$ that would be dissolved in such melts.

There are other possible degassing reactions. Mathez (1984) suggested that degassing of carbon species from reduced magma in the shallow crust will cause melt oxidation. Carbon dissolves in basaltic melt primarily as $CO_3^{2-}$, and at low pressure and reduced conditions coexisting fluid is a mixture of $CO_2$ and CO. Therefore, the decompression-driven degassing reaction of an initially reduced magma in the shallow crust is partly represented as

$$CO_3^{2-}(m) = CO(v) + O^{2-}(m) + {}^1/_2O_2(m) \quad [4]$$
$$(v = \text{vapor},\ m = \text{melt}),$$

which results in increase in the oxygen content of the residual melt. The process may operate for C–O–H fluids under mantle conditions as well. Thus, Holloway and Jakobsson (1986) showed experimentally that $H_2O/H_2$ and $CO_2/CO$ ratios of C–O–H fluids at 10–20 kb are lower than coexisting melt. As noted above and in Fig. 3.2, processes that tend to change magma oxidation state are less effective at $f_{O_2}$ near QFM or above, and therefore degassing should not cause extreme oxidation of magmas. Mathez (1984) argued that this behavior accounts for the fact that QFM conditions characterize most basalts erupted subaerially.

Submarine basalts may offer an example of the operation of an oxidation process. Although glass compositions indicate that submarine lavas are relatively reduced as they erupt, $f_{O_2}$ calculated from Fe–Ti oxides in crystalline flow interiors are always near QFM (Fig. 3.1). This must indicate that the lavas oxidize as they crystallize on the ocean floor.

#### Importance of $f_{O_2}$ and predictions for layered intrusions

In layered intrusions, the variable $f_{O_2}$ affects (a) the chemical equilibrium between silicate and sulfide-oxide melts, (b) the composition and temperature of appearance of the first oxide phase, (c) the temperature at which graphite becomes stable and (d) the manner in which the magmatic vapor evolves. From the above discussion, it should be clear that there are no reliable or precise means of estimating oxidation states of the initial magmas. IOF measurements and computations of $f_{O_2}$ from mineral equilibria suffer from the fact that most igneous minerals have reequilibrated at subsolidus temperatures; data from other systems suggest that the mantle is heterogeneous in terms of oxidation state; and the observed assemblages in layered intrusions provide only gross estimates. Knowledge of T–$f_{O_2}$ paths is also critical in deducing fluid evolution in layered intrusions. Although cooling paths should be approximately parallel to that of QFM, processes such as degassing may modify them, and some rocks, such as submarine basalts, do not follow the predicted behavior.

## NATURE OF THE MAGMATIC VAPOR

### Solubility and abundance constraints

The composition of the first vapor to exsolve from a magma will depend on relative solubilities and abundances of fluid species in the melt.

#### Carbon and hydrogen

Carbon dissolves mainly as $CO_3^{2-}$ (e.g., Fine and Stolper, 1985) and hydrogen as $OH^-$ and $H_2O$ (Burnham, 1979; Stolper, 1982) in mafic melts. Their solubilities at low pressures are compared in Fig. 3.3. It can be seen that the solubility of water in mafic silicate melt is nearly two orders of magnitude greater than that of $CO_2$ on a weight basis, and obviously it is more on a molar basis. Therefore, the pressure–temperature conditions at which vapor initially begins to exsolve from magma depend almost entirely on its carbon content, and the first vapor to exsolve will be relatively enriched in the carbon species (CO and $CO_2$).

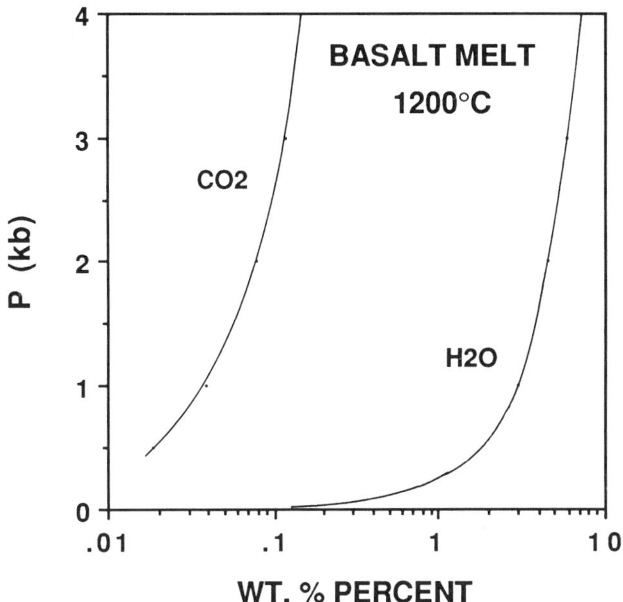


FIGURE 3.3—Comparison of the solubilities of $CO_2$ (Stolper and Holloway, 1988) and $H_2O$ (Hamilton et al., 1964) in basalt melt. The $CO_2$ curve is based on experiments up to 1500 bars, which, as explicitly noted by Stolper and Holloway, cannot be extrapolated to higher pressure with confidence.

No experimental data exist on C/H ratios of coexisting vapor and basaltic melt under the conditions of interest here, so it is not possible to specify precisely how water will distribute itself between $CO_2$–CO fluid and melt. However, the general character of magmatic fluid can be deduced from observations in nature. Pineau et al. (1976) and Moore et al. (1977) analyzed gases liberated by vacuum crushing at room temperature from vesicles in MORB glasses, including the Mid-Atlantic Ridge gas-rich "popping rocks." They concluded that the vesicle gases consist of >95 vol.% $CO_2$. [One other analysis presented by Pineau et al. contains less $CO_2$ but does not represent an equilibrium composition. Vesicle compositions reported by Pineau and Javoy (1983), in which water contents are estimated from oxygen isotopic compositions of $CO_2$, cover a wider range but are generally consistent with the above.] MORB glasses typically contain <300 ppm wt. $CO_2$ (e.g., Fine and Stolper, 1985) and <5000 ppm wt. $H_2O$ (Moore, 1970). Therefore, for coexisting basalt melt and vapor under several hundred bars total pressure,

$$(CO_2/H_2O)_v/(CO_2/H_2O)_m \approx 400\text{–}500 \qquad [5]$$

computed on a molar basis.

A second set of relevant observations comes from peridotite xenoliths in basalts. These rocks typically contain fluid inclusions consisting of nearly pure $CO_2$ or $CO_2$–CO trapped at pressures of >10 kb (e.g., Bergman and Dubessy, 1984).

Many of the rocks also contain interstitial glass (Irving and Mathez, 1982). The glasses are andesitic and yield analyses which usually total to 95–98%, implying that the difference consists of dissolved water. Assuming that the melt and vapor were in equilibrium, this suggests that $(CO_2/H_2O)_v >> (CO_2/H_2O)_m$ at upper mantle pressures.

### Sulfur

As noted above, sulfur dissolves in melt as $S^{2-}$ at low $f_{O_2}$ and $SO_4^{2-}$ at high $f_{O_2}$. The variation in the relative abundance of sulfur species in a C–O–H–S vapor as a function of $f_{O_2}$ is shown in Fig. 3.4. It can be seen that $SO_2$ and $H_2S$ are the major species for oxidized and reduced conditions, respectively, and that $S_2$ is important at intermediate $f_{O_2}$.

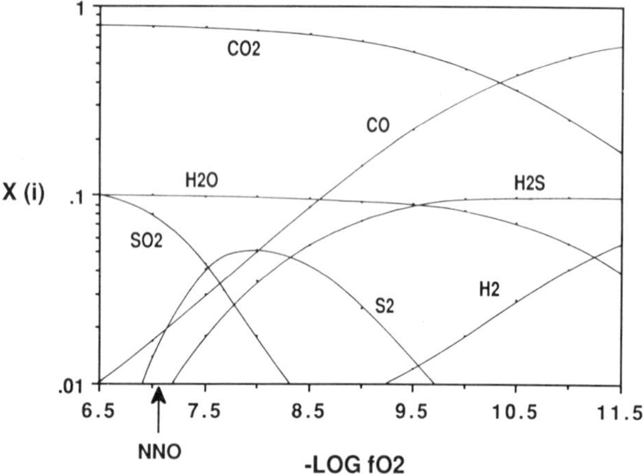


FIGURE 3.4—Mole fractions of major species [X(i)] of C–O–H–S vapor as a function of $f_{O_2}$ at T = 1250°C and P = 1 kb. Note the position of the NNO buffer. The composition is such that $X_{CO_2}+X_{CO}+X_{CH_4} = 0.8$, $X_{H_2S}+X_{S_2}+X_{SO_2} = 0.1$ and $X_{H_2O}+X_{H_2} = 0.1$ over the entire range of $f_{O_2}$, and therefore C/H and C/S ratios vary across the diagram. For this composition, $X_{CH_4}$ does not exceed 0.01 in the plotted range of $f_{O_2}$. Fugacities of pure species are computed by a modified Redlich–Kwong equation of state, mixing is ideal (after Mathez, 1984).

The solubility of sulfur in reduced melts increases dramatically as pressure increases from 1 to ≈ 200 bars (Moore and Schilling, 1973). At this and higher pressures, concentrations in natural melts reach 600–1800 ppm (Mathez, 1976), depending on the FeO content of the melt, and immiscible sulfide-oxide liquid is usually stable. Sulfur solubility is also strongly influenced by $TiO_2$ (Haughton et al., 1974; Danckwerth et al., 1979). The behavior of sulfur in melts is discussed in detail by Naldrett (this volume).

The amount of sulfur in vapor coexisting with silicate and sulfide-oxide melts at pressure is controlled primarily by the equilibrium between the two melts. The only relevant experiments are those of Rutherford and Fogel (pers. comm. 1988), who investigated equilibria in the Fe–C–O–S–silicate system. In their experiments the phases FeS melt, Fe melt, graphite and vapor were present. Because the assemblage is invariant (sulfide, metal, graphite and vapor are assumed to contain no silica or other components of the silicate melt except iron), the vapor composition can be derived from known compositions of the condensed phases. At 2000 bars pressure, the vapor was composed of nearly pure CO + $CO_2$, and the sum of the sulfur species was ≈ 1 bar. Except under unusual circumstances (e.g., Disko Island), coexisting sulfide and iron melts are not observed in natural basalts, and in consequence fugacities of sulfur and other sulfur-bearing species in magmatic vapors are generally higher than those in the experiments. Another observation also indicates that sulfur species do not make up substantial proportions of magmatic vapors. The walls of vesicles in MORB glasses are lined with sulfide droplets, which have been interpreted to originate by reaction of sulfur in the vapor with iron in the melt. Based on droplet abundances in individual vesicles, Moore et al. (1977) estimated that sulfur constituted 4–6 mole% of the original magmatic vapor. Thus, for a sulfide-saturated magma + vapor at shallow crustal pressures and at an $f_{O_2}$ below about NNO–2,

$$S_v/S_m \approx 40\text{–}80 \qquad [6]$$

on a molar basis. This contrasts with the generally higher sulfur abundances in volcanic gases associated with subaerial eruption (e.g., Gerlach and Nordlie, 1971). Again, it is emphasized that this dichotomy in fluid compositions in low- and high-pressure systems is mainly due to the strong dependency of solubility on pressure at pressures <200 bars.

### Chlorine

Chlorine is volatile and thus strongly partitions into the magmatic vapor (Iwasaki and Katsura, 1967; Kilinc and Burnham, 1972; Webster and Holloway, 1988). Although chlorine is generally not a major component of volcanic gas, Boudreau et al. (1986) reported chlorapatite and chlorine-rich phlogopite in the platinum ore zones and associated rocks of the Stillwater and Bushveld complexes and argued that the compositions required the presence of chlorine-rich fluid. These occurrences and the rationale for their argument are summarized by Mathez (this volume). The little that is known about chlorine abundances in magmas comes mainly from the study of glass (melt) inclusions in phenocrysts. Chlorine abundances of glass inclusions in MORBs are generally <200 ppm (Anderson, 1974; Unni and Schilling, 1978). However, those of inclusions in other mafic and intermediate lavas cover a wider range (Devine et al., 1984; Fallon and Green, 1986). For example, Devine et al. (1984) report inclusions with chlorine contents from 70 to over 4000 ppm. A systematic study aimed at deducing chlorine abundances characteristic of various mafic magma types is needed.

The solubility of chlorine in granitic melt–aqueous vapor systems has been experimentally investigated by Kilinc and Burnham (1972) and Webster and Holloway (1988) . These investigations showed that the chlorine content of the vapor is extremely sensitive to melt chlorine, alkali, and iron contents. The data are consistent with the formation of metal and alkali chloride complexes in both the melt and vapor. Therefore, it is likely that even small amounts of chlorine in magmatic systems will dramatically enhance the ability of vapor to dissolve and transport metals. For a melt of topaz rhyolite composition, Webster and Holloway (1988) reported chlorine melt–vapor distribution coefficients (wt. fraction melt/vapor) of 0.1 to 0.2. In mafic systems, the only experimental study of chlorine solubility is that of Iwasaki and Katsura (1967). Chlorine was introduced into their charges as NaCl, and thus the melts were probably peralkaline. Because of the strong dependency of chlorine solubility on alkalis and the fact that the melt compositions were not reported, it is difficult to use the data to extrapolate to alkali-poor compositions. The transport of metals and perhaps other magmaphile elements as chlorides in vapor is undoubtably important but will be not discussed further because no data are available for the compositions or conditions of interest here.

Mathez et al. (1988) computed the speciation in the C–O–H–Cl system for shallow crustal pressures. In the absence of alkalis, HCl is the only important vapor species under moderately oxidized conditions, but $CH_3Cl$ becomes dominant under reducing conditions. The effect of introducing chlorine to a C–O–H fluid is to reduce its $H_2O$ and $CH_4$ contents. This effect is probably partly mitigated by the presence of alkalis in the system; however, this cannot be evaluated because of lack of experimental data on mafic compositions.

### Other elements

The only other volatile elements that could conceivably exist in more than trace quantities in magmatic vapor are fluorine and nitrogen. The former is not particularly volatile, at least in acidic melts, the melt–vapor distribution coefficients being on the order of 10 (Hards, 1976). Nitrogen has only been studied in albite melts, in which its solubility is negligible (Kesson and Holloway, 1974). Since nitrogen is accepted in the structure of some micas, several phlogopite separates from the Bushveld and Stillwater Complexes have been analyzed (Mathez and Levinson, unpublished data). No significant concentrations were found, and therefore nitrogen was probably not important in fluids associated with these intrusions.

Silica may dissolve in $H_2O$-rich fluids in significant concentrations over a wide range of conditions (e.g., Kushiro et al., 1968). At high pressures the solubilities of silica and other solute species are strongly dependent on fluid $H_2O/CO_2$ ratio. For example, Eggler and Rosenhauer (1978) reported that $H_2O$ equilibrated with diopside melt at 20 kb contains $\approx$ 9.5 wt.% solute, whereas the solute content of $CO_2$ under the same conditions is only 0.2 wt.%. Although data at low pressures are lacking, this suggests that the initial $CO_2$-rich fluids in the layered intrusions were not likely to have contained significant amounts of silica or other solute components except for those combined as chlorides.

## CHEMICAL EVOLUTION OF FLUID FROM STILLWATER AND BUSHVELD MAGMAS

The above considerations suggest that first fluids to have exsolved from Stillwater and Bushveld magmas were essentially $CO_2$–CO mixtures containing 4–6% sulfur species and unknown but probably large amounts of chlorine species (Mathez et al., 1988). These fluids probably began exsolving from intercumulus melts trapped in the partially molten cumulate piles rather than at the tops of the magma chambers. How fluids evolve chemically during cooling in this type of an environment will depend in part on the temperature at which graphite becomes stable, which is in turn dependent on the $f_{O_2}$ of the system.

If $f_{O_2}$ are sufficiently high such that graphite does not become stable until the solidus temperature is reached, then with cooling fluid evolves in response to changing melt volatile solubilities, decreasing mass of melt and crystallization of volatile-bearing minerals. The precise way in which the fluid evolves will be governed by a complex set of reactions among fluid, melt, the hydrous and halogen-bearing major minerals (hornblende and phlogopite) and certain minor phases (e.g., apatite and sulfide). The evolution cannot be specified exactly due to the large number of compositional variables involved. In a general way, the compositional evolution is probably represented by the gases emitted from summit and rift eruptions of Kilauea volcano. The former are $CO_2$-rich and thought to represent the first vapor to escape from the magma chamber, whereas the rift gases are $H_2O$-rich and thought to represent vapor exsolved subsequently and in response to further decompression from the same magma (Gerlach and Graeber, 1985).

### Evolution of a model fluid in response to graphite precipitation

If graphite appears above the solidus, fluid evolution is more predictable. The graphite in the Critical Zone of the Bushveld Complex is thought to be magmatic (e.g., Ballhaus and Stumpfl, 1985). (This may be true of Stillwater graphite as well, although the metamorphism of the Complex precludes unique interpretation of the occurrence.) Mathez et al. (1988) examined this situation in detail by constructing a model of fluid evolution from the intercumulus environment of the Bushveld Critical Zone and Stillwater Lower Banded Series. This is summarized as follows.

In the model system, the first fluid to exsolve from intercumulus melt was defined as a $CO_2$–CO mixture containing 5% hydrogen and 10% chlorine. Sulfur was not considered in the analysis because its inclusion makes little difference to the outcome.

The initial $f_{O_2}$ of the system was arbitrarily chosen to be NNO–2.8, in which case the fluid becomes graphite saturated at 1050°C at a total pressure of 4 kb. It was assumed that the fluid evolves by cooling down the T–$f_{O_2}$ path NNO–2.8. It was noted above that although cooling along a path parallel to QFM (or NNO) appears not always to be followed, this behavior should be considered to be the general case. In any event, this assumption is extremely important for the sake of computing a fluid evolution path. The reason for this is illustrated in Fig. 3.5, which shows how the stability of graphite changes relative to oxygen isopleths in C–O–H fluids. With cooling, the field of graphite stability expands, the oxygen isopleths remain approximately stationary and, therefore, the graphite-fluid curve impinges on the isopleths. The result is that the composition of a fluid constrained to one of the isopleths is forced toward $H_2O$. The fluid also loses oxygen to its surroundings, which is

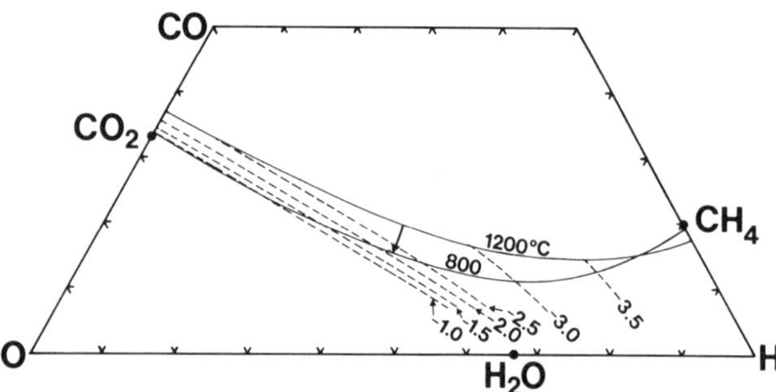


FIGURE 3.5—Graphite stability curve (solid line) and oxygen isopleths (dashed line) in the C–O–H system at 4 kb. The isopleths are labeled in log units relative to QFM (NNO – 0.8).

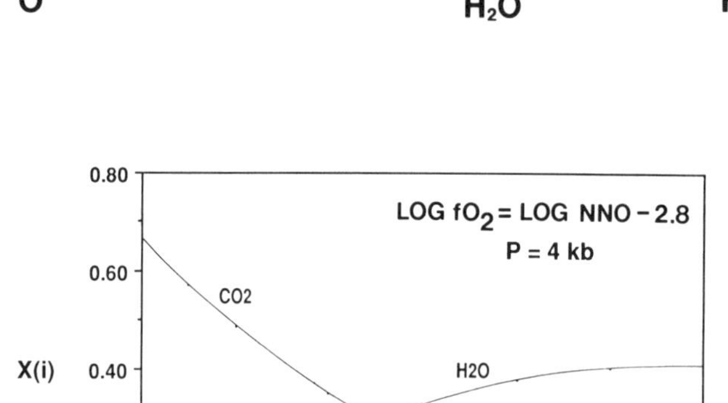


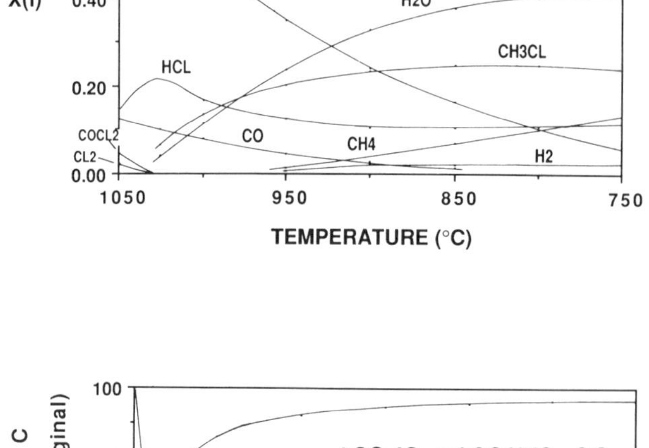


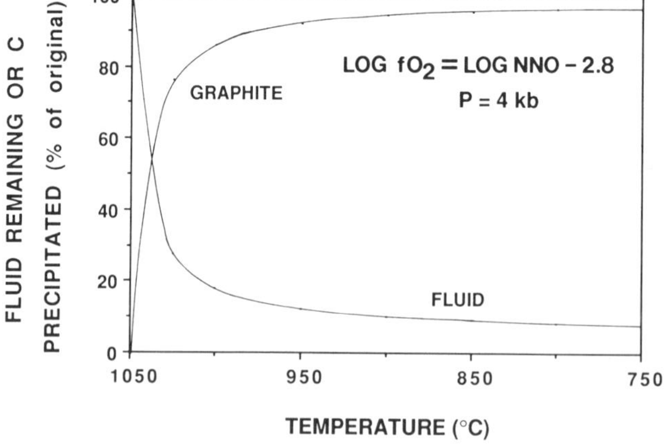


FIGURE 3.6—**A.** The composition projected from the Cl apex of l apex of a a C–O–H–10%Cl fluid at 4 kb as it evolves by graphite precipitation (see text). **B.** Composition of the fluid in terms of mole terms of mole fractions of the species present. **C.** The change in mass of fluid and s of fluid and graphite with cooling. The computations were performed by a free energy minimization procedure (Holloway and Reese, 1974), a modified Redlich–Kwong equation of state for the pure species and ideal mixing (after Mathez et al., 1988).

emphasized by the fact that its composition when constrained to an isopleth trends away from the C–O join rather than directly away from the C apex. Again, this means only that the $f_{O_2}$ of the system is controlled by the silicate-oxide assemblage, not by the fluid.

The evolving composition of the model C–O–H–Cl fluid is projected onto to C–O–H face in Fig. 3.6a. Here the fluid path is defined by intersections of the graphite stability curve and the NNO–2.8 isopleth at successively lower temperatures. Several features of this evolution should be noted:

(1) Since the bulk chlorine content exceeds that of hydrogen, the fluid initially contains no $H_2O$, all hydrogen being combined as HCl (Fig. 3.6b). The $H_2O$ content gradually increases but exceeds neither those of HCl or $CH_3Cl$ until the fluid has cooled to <975°C nor that of $CO_2$ until the temperature reaches 930°C. This behavior may account for the extremely Cl-rich apatite in the Critical Zone rocks. The presence of $CH_3Cl$ in the fluid also has the effect of depressing its $CH_4$ content.

(2) The relative orientations of the graphite stability curve and the oxygen isopleths (Fig. 3.5) causes the fluid composition to change rapidly in a narrow temperature interval. This reflects the fact that graphite precipitation rate is initially extremely rapid but progressively decreases with cooling (Fig. 3.6c). In the model composition, cooling from 1050 to 1025°C results in precipitation of 76% of the carbon originally in the fluid and a decrease in the mass of the fluid to 28% of its original.

A consequence of this behavior is that the residual species in the fluid will be driven toward much higher concentrations, which in turn will cause these components to flow from the fluid back into the coexisting melt. Extremely chlorine-rich melts could be generated in this way, particularly in the cumulate environment where the mass of melt may not be large compared to that of fluid. Limited quantities of sulfide should also precipitate from melt in response to graphite precipitation from vapor.

(3) It should be evident that fluid cannot evolve by graphite precipitation alone. With cooling, the mass of melt decreases due to crystallization and its $H_2O$ content must increase due to crystallization of anhydrous phases. Therefore, the evolution described above is hypothetical in the

sense that these factors are not taken into account. The crystallization effects should be overwhelmed when graphite initially begins precipitating, however.

(4) Recalling the data of Eggler and Rosenhauer (1978) cited above, a potentially important consequence of the rapid decrease in $CO_2/H_2O$ ratio is that the fluid solubilities of silica, alkalis, metals and other solute components may increase dramatically over a narrow temperature interval.

## SUMMARY

The point has been stressed that the composition and evolution of fluid in a crystallizing mafic magma must be examined in the context of its initial $f_{O_2}$ and how $f_{O_2}$ changes during cooling. Although neither intrinsic $f_{O_2}$ measurements nor computations from silicate-oxide mineral equilibria yield precise estimates of magmatic oxidation states, the occurrence of igneous graphite in the Critical Zone of the Bushveld Complex does provide an upper limit. If graphite became stable at $T > 1000°C$, the maximum $f_{O_2}$ was $\approx$ NNO–2.5. Such redox conditions are not unusual compared to redox conditions estimated for MORB magmas and their mantle source regions.

The available evidence indicates that magmas crystallize and continue to cool, at least through the high-temperature subsolidus regime, along T–$f_{O_2}$ paths parallel to synthetic mineral oxygen buffers such as QFM and NNO. This is to be expected if the cooling paths are dictated by equilibria among the condensed phases. At lower temperatures (<600°C), hydrothermal processes or metamorphic reactions (e.g., serpentinization) may control redox conditions.

Regardless of oxidation state, the first magmatic vapor to exsolve from a basaltic melt will be a $CO_2$- or $(CO_2+CO)$-rich mixture. Most of the water present will remain dissolved in the melt. It is estimated that the molar $(CO_2/H_2O)_v/(CO_2/H_2O)_m$ is in the range 400–500. Water contents of the fluid will be further reduced if there is sufficient chlorine to combine as HCl. The magmatic vapor will contain <10% sulfur species.

If graphite becomes stable at supersolidus temperature, it is possible to compute the manner in which fluid evolves from $CO_2$- to $H_2O$-rich compositions by assuming that the cooling path is parallel to QFM. In such a case, fluid evolution is controlled by graphite precipitation. The evolution is characterized by an extremely large decrease in fluid mass in a narrow cooling interval immediately after graphite becomes stable, which is a direct result of graphite precipitation. In a system constructed to model vapor evolution in the Bushveld and Stillwater environments, cooling from 1050 to 1025°C results in decrease in fluid mass to 28% of its original. It is thus possible to enrich horizons of partially molten cumulates in which relatively large fluid/melt ratios developed in the residual species of the fluid. The dramatic change in fluid composition over a narrow temperature interval is probably accompanied by large increases in the solubilities of silica, alkalis and other melt components.

ACKNOWLEDGEMENTS—Reviews of this paper by A. J. Naldrett, A. E. Boudreau and C. L. Peach are gratefully acknowledged. This work was supported by NSF grant EAR8720982.

## REFERENCES

Anderson, A. T., 1974, Chlorine, sulfur and water in magmas and oceans: Bull. Geol. Soc. Am., v. 85, pp. 1485–1492.

Arculus, R. J., and Delano, J. W., 1981, Intrinsic oxygen fugacity measurements: Techniques and results for spinels from upper mantle peridotite and megacryst assemblages: Geochim. Cosmochim. Acta, v. 45, pp. 899–913.

Arculus, R. J., Dawson, J. B., Mitchell, R. H., Gust, D. A., and Holmes, R. D., 1984, Oxidation states of the upper mantle recorded by megacryst ilmenite in kimberlite and type A and B spinel lherzolites: Contrib. Mineral. Petrol., v. 85, pp. 85–94.

Ballhaus, C. G., and Stumpfl, E. F., 1985, Occurrence and petrological significance of graphite in the upper Critical Zone, western Bushveld Complex, South Africa: Earth Planet. Sci. Lett., v. 74, pp. 58–68.

Bergman, S. C., and Dubessy, J., 1984, $CO_2$–CO fluid inclusions in a composite peridotite xenolith: implications for upper mantle oxygen fugacity: Contrib. Mineral. Petrol., v. 85, pp. 1–13.

Boudreau, A. E., Mathez, E. A., and McCallum, I. S., 1986, Halogen geochemistry of the Stillwater and Bushveld Complexes: Evidence for transport of the platinum-group elements by Cl-rich fluids: J. Petrol., v. 27, pp. 967–986.

Buddington, A. F., and Lindsley, D. H., 1964, Iron-titanium oxide minerals and synthetic equivalents: J. Petrol., v. 5, pp. 310–357.

Buntin, T. J., Grandstaff, D. E., Ulmer, G. C., and Gold, D. P., 1985, A pilot study of geochemical and redox relationships between potholes and adjacent normal Merensky Reef of the Bushveld Complex: Econ. Geol., v. 80, pp. 975–987.

Burnham, C. W., 1979, The importance of volatile constituents; *in* Yoder, H. S., Jr. (ed.), The Evolution of Igneous Rocks: Princeton University Press, Princeton, N.J., pp. 439–482.

Carmichael, I. S. E., and Ghiorso, M. S., 1986, Oxidation-reduction relations in basic magma: A case for homogeneous equilibria: Earth Planet. Sci. Lett., v. 78, pp. 200–210.

Christie, D. M., Carmichael, I. S. E., and Langmuir, C. H., 1986, Oxidation states of mid-ocean ridge basalt glasses: Earth Planet. Sci. Lett., v. 79, pp. 397–411.

Carroll, M. R., and Rutherford, M. J., 1985, Sulfide and sulfate saturation in hydrous silicate melts: J. Geophys. Res., v. 90, pp. C601–C612.

Danckwerth, P. A., Hess, P. C., and Rutherford, M. J., 1979, The solubility of sulfur in high-$TiO_2$ mare basalts: Proc. Lunar Planet. Sci. Conf., 10th, pp. 517–530.

Devine, J. D., Sigurdsson, H., Davis, A. N., and Self, S., 1984, Estimates of sulfur and chlorine yield to the atmosphere from volcanic eruptions and potential climatic effects: J. Geophys. Res., v. 89, pp. 6309–6325.

Eggler, D. H., 1983, Upper mantle oxidation state: Evidence from olivine–orthopyroxene–ilmenite assemblages: Geophys. Res. Lett., v. 10, pp. 365–368.

Eggler, D. H., and Rosenhauer, M., 1978, Carbon dioxide in silicate melts: II Solubilities of $CO_2$ and $H_2O$ in $CaMgSi_2O_6$ (diopside) liquids and vapors at pressures to 40 kb: Am. J. Sci., v. 278, pp. 64–94.

Elliot, W. C., Grandstaff, D. E., Ulmer, G. C., Buntin, T. J., and Gold, D. P., 1982, An intrinsic oxygen fugacity study of platinum–carbon associations in layered intrusions: Econ. Geol., v. 77, pp. 1493–1510.

Fallon, T. J., and Green, D. H., 1986, Glass inclusions in magnesian olivine pheoncrysts from Tonga: Evidence for highly refractory parental magmas in the Tonga arc: Earth Planet. Sci. Lett., v. 81, pp. 95–103.

Fine, G. J., and Stolper, E., 1985, Dissolved carbon dioxide in basaltic glasses: Concentrations and speciation: Earth Planet. Sci. Lett., v. 76, pp. 263–278.

Flynn, R. T., Ulmer, G. C., and Sutphen, C. F., 1978, Petrogenesis of the eastern Bushveld Complex: Crystallization of the middle Critical Zone: J. Petrol. v. 19, pp. 136–152.

French, B. M., 1966, Some geological implications of equilibrium between graphite and a C–O–H gas phase at high temperatures and pressures: Rev. Geophysics, v. 4, pp. 223–253.

Fudali, R. F., 1965, Oxygen fugacities of basaltic and andesitic magmas. Geochim. Cosmochim. Acta, v. 29, pp. 1063–1075.

Gerlach, T. M., 1980, Evaluation of volcanic gas analyses from Kilauea Volcano: J. Volcan. Geotherm. Res., v. 7, pp. 295–317.

Gerlach, T. M., and Graeber, E. J., 1985, Volatile budget of Kilauea Volcano: Nature, v. 313, pp. 273–277.

Gerlach, T. M., and Nordlie, B. E., 1971, The C–O–H–S gaseous system, Part I: Composition limits and trends in basaltic gases: Am. J. Sci., v. 275, pp. 353–376.

Gitlin, E., 1985, Sulfide remobilization during low temperature alteration of seafloor basalt: Geochim. Cosmochim. Acta, v. 49, pp. 1567–1579.

Haggerty, S. E., 1976, Opaque mineral oxides in terrestrial igneous rocks; *in* Rumble, D., III (ed.), Oxide Minerals: Mineral. Soc. America Short Course Notes 3, pp. Hg101–Hg300.

Haggerty, S. E., and Tompkins, L. A., 1983, Redox state of Earth's upper mantle from kimberlitic ilmenites: Nature, v. 303, pp. 295–300.

Hamilton, D. L., Burnham, C. W., and Osborn, E. F., 1964, The solubility of water and effects of oxygen fugacity and water content on crystallization in mafic magmas: J. Petrol., v. 5, pp. 21–39.

Hards, N., 1976, Distribution of elements between the fluid phase and silicate melt phase of granites and nepheline syenites; *in* Biggar, G.M. (ed.), Progress in Experimental Petrology: National Environment Research Council, Publication Series D, No. 6, pp. 88–90.

Haughton, D. R., Roeder, P. L., and Skinner, B. J., 1974, Solubility of sulfur in mafic magmas: Econ. Geol., v. 69, pp. 451–467.

Holloway, J. R., and Jakobsson, S., 1986, Volatile solubilities in magmas: Transport of volatiles from mantles to planet surfaces: J. Geophys. Res., v. 91, pp. D505–D508.

Holloway, J. R., and Reese, P. L., 1974, The generation of $N_2$–$CO_2$–$H_2O$ fluids for use in hydrothermal experimentation I. Experimental method and equilibrium calculations in the C–O–H–N system: Am. Mineral., v. 59, pp. 587–597.

Irvine, T. N., 1967, Chromian spinel as a petrogenetic indicator. Part 2. Petrologic applications: Can. J. Earth Sci., v. 4, pp. 71–103.

Irving, A. J., and Mathez, E. A., 1982, The origin of glass in ultramafic xenoliths (abs.): Terra Cognita, v. 2, pp. 243.

Iwasaki, B., and Katsura, T., 1967, The solubility of hydrogen chloride in volcanic melts at a total pressure of one atmosphere and at temperatures of 1200°C and 1290°C under anhydrous conditions: Bull. Chem. Soc. Japan, v. 40, pp. 554–561.

Johnson, M. C., and Rutherford, M. J., 1987, P,T, and X($H_2O$) conditions in the Fish Canyon Tuff, Colorado magma chamber as determined experimentally (abs.): American Geophysical Union, EOS, Transactions, v. 68, pp. 1542.

Katsura T., and Nagashima, S., 1974, Solubility of sulfur in some magmas at 1 atmosphere: Geochim. Cosmochim. Acta, v. 38, pp. 517–531.

Kennedy, G. C., 1948, Equilibrium between volatiles and iron oxides in igneous rocks: Am. J. Sci., v. 246, pp. 529–549.

Kennedy, G. C., 1955, Some aspects of the roles of water in rock melts: Geol. Soc. Am. Sp. Paper 62, pp. 489–533.

Kesson, S. E., and Holloway, J. R., 1974, The generation of $N_2$–$CO_2$–$H_2O$ fluids for use in hydrothermal experimentation II. Melting of albite in a multispecies fluid: Am. Mineral., v. 59, pp. 598–603.

Kilinc, I. A., and Burnham, W. C., 1972, Partitioning of chloride between a silicate melt and coexisting aqueous phase from 2 to 8 kilobars: Econ. Geol., v. 67, pp. 231–235.

Kushiro, I., Yoder, H. S., Jr., and Nishkikawa, M., 1968, Effect of water on the melting of enstatite: Bull. Geol. Soc. Am., v. 79, pp. 1685–1692.

Mathez, E. A., 1976, Sulfur solubility and magmatic sulfides in submarine basalt glass: J. Geophys. Res., v. 81, pp. 4269–4276.

Mathez, E. A., 1984, Influence of degassing on oxidation states of basaltic magmas: Nature, v. 310, pp. 371–375.

Mathez, E. A., Dietrich, V. J., Holloway, J. R., and Boudreau, A. E., 1988, Carbon distribution in the Stillwater Complex and evolution of vapor during crystallization of Stillwater and Bushveld magmas: J. Petrol. (in press).

Mattioli, G. S., and Wood, B. J., 1986, Upper mantle oxygen fugacity recorded by spinel lherzolites: Nature, v. 322, pp. 626–628.

McCallum, I. S., Raedeke, L. D., and Mathez, E. A., 1980, Investigations of the Stillwater Complex. Part 1. Stratigraphy and structure of the Banded zone: Am. J. Sci., v. 280A, pp. 59–87.

Mo, X., Carmichael, I. S. E., Rivers, M., and Stebbins, J., 1982, The partial molar volume of $Fe_2O_3$ in multicomponent silicate liquids and the pressure dependence of oxygen fugacity in magmas: Mineral. Mag., v. 45, pp. 237–245.

Moore, J. G., 1970, Water content of basalt erupted on the ocean floor: Contrib. Mineral. Petrol., v. 28, pp. 272–279.

Moore, J. G., Batchelder, J. N., and Cunningham, C. G., 1977, $CO_2$-filled vesicles in mid-ocean basalt: J. Volcanol. and Geothermal Res., v. 2, pp. 309–327.

Moore, J. G., and Schilling, J.-G., 1973, Vesicles, water and sulfur in Reykjanes Ridge basalts: Contrib. Mineral. Petrol., v. 41, pp. 105–118.

Moore, R. O., Otter, M. L., Rickard, R. S., Harris, J. W., and Gurney, J. J., 1986, The occurrence of moissanite and ferro-periclase as inclusions in diamond (abs.): Geol. Soc. Australia Abs. Series, v. 16, pp. 409–411.

Naldrett, A. J., 1969, A portion of the system Fe–S–O between 900 and 1080°C and its application to sulfide ore magmas: J. Petrol., v. 10, pp. 171–201.

Osborn, E. F., 1959, Role of oxygen pressure in the crystallization and differentiation of basaltic magma: Am J. Sci., v. 257, pp. 609–647.

Pineau, F., Javoy, M., and Bottinga, Y., 1976, $^{13}C/^{12}C$ ratios of rocks and inclusions in popping rocks of the mid-Atlantic ridge and their bearing on the problem of isotopic composition of deep-seated carbon: Earth Planet. Sci. Lett., v. 29, pp. 413–421.

Pineau, F. and Javoy, M., 1983, Carbon isotopes and concentrations in mid-oceanic ridge basalts: Earth Planet. Sci. Lett., v. 62, pp. 239–257.

Roeder, P. L., and Osborn, E. F., 1966, Experimental data for the system $MgO$–$FeO$–$Fe_2O_3$–$CaAl_2Si_2O_8$–$SiO_2$ and their petrologic implications: Am. J. Sci., v. 264, pp. 428–480.

Ryabchikov, I. D., Green, D. H., Wall, W. J., and Brey, G. P., 1982, The oxidation state of carbon in the reduced-velocity zone: Geochem. Int., v. 18, no. 1, pp. 148–158.

Rutherford, M. J., Sigurdsson, H., Carey, S., and Davis, A., 1985, The May 18, 1980, eruption of Mount St. Helens l. Melt composition and experimental phase equilibria: J. Geophys. Res., v. 90, pp. 2929–2947.

Sack, R. O., Carmichael, I. S. E., Rivers, M., and Ghiorso, M. S., 1980, Ferric–ferrous equilibria in natural silicate liquids at 1 bar: Contrib. Mineral. Petrol., v. 75, pp. 369–376.

Sato, M., 1978, Oxygen fugacity of basaltic magmas and the role of gas-forming elements: Geophys. Res. Lett., v. 5, pp. 447–449.

Sato, M., and Valenza, M., 1980, Oxygen fugacities of the layered series of the Skaergaard Intrusion, East Greenland: Am. J. Sci., v. 280–A, pp. 134–158.

Sato, M., and Wright, T. L., 1966, Oxygen fugacities directly measured in magmatic gases: Science, v. 153, pp. 1103–1105.

Stolper, E., 1982, The speciation of water in silicate melts: Geochim. Cosmochim. Acta, v. 46, pp. 2609–2620.

Stolper, E. A., and Holloway, J. R., 1988, Experimental determination of the solubility of carbon dioxide in molten basalt at low pressure: Earth Planet. Sci. Lett., v. 87, pp. 397–408.

Unni, C. K., and Schilling, J.-G., 1978, Cl and Br degassing by volcanism along the Reykjanes Ridge and Iceland: Nature, v. 272, pp. 19–23.

Ulmer, G. C., Grandstaff, D. E., Weiss, D., Moats, M. A., Buntin, T. J., Gold, D. P., Hatton, C. J., Kadik, A., Koseluk, R. A., and Rosenhauer, M., 1987, The mantle redox state; An unfinished story? *in* Morris, E. M., and Pasteris, J. D. (eds.), Mantle Metasomatism and Alkaline Magmatism: Geol. Soc. Am. Spec. Paper 215, pp. 5–23.

Virgo, D., Luth, R. W., Moats, M. A., and Ulmer, G. C., 1988, Constraints on the oxidation state of the mantle: An electrochemical and $^{57}Fe$ Mössbauer study of mantle-derived ilmenites: Geochim. Cosmochim. Acta (in review).

Webster, J. D., and Holloway, J. R., 1988, Experimental constraints on the partitioning of Cl between topaz rhyolite melt and $H_2O$ and $H_2O$ + $CO_2$ fluids: New implications for granitic differentiation and ore deposition: Geochim. Cosmochim. Acta (in press).

## Chapter 4

# THE GEOCHEMISTRY OF THE PLATINUM-GROUP ELEMENTS IN MAFIC AND ULTRAMAFIC ROCKS

E. A. Mathez and C. L. Peach

### INTRODUCTION

The platinum-group elements (PGE) ruthenium, osmium, rhodium, iridium, palladium and platinum together with iron, cobalt and nickel form Group VIII of Mendeleev's periodic table. All are strongly siderophile and chalcophile, and their distribution in the crust and upper mantle is largely determined by the distribution of sulfides. In the case of PGE, there are few specific data on how they distribute themselves among sulfides, silicates and other phases or on their solubilities in high-temperature, volatile-rich fluids. In fact, our understanding of how these elements behave in response to igneous differentiation or metamorphism is based on empirical, primarily geochemical, observations. For these reasons, models devised to explain the development of PGE-rich horizons in the layered intrusions are not well constrained.

Their generally siderophile and chalcophile nature should not obscure the fact that in some environments the PGE are volatile. This is clearly demonstrated by the enormous enrichments in iridium (along with gold, silver and certain other trace metals) in Kilauea volcanic gases relative to lavas (Olmez et al., 1986). In the Bushveld Complex, the presence of hydrothermal platiniferous pegmatite pipes (e.g., the Driekop) and the complexities of PGE distribution within the UG–2 chromitite and Merensky Reef imply some control by fluids. Again, lack of data reduce us to little more than speculation on the specific roles of fluids in the genesis of ores in the layered intrusions.

The purpose of this chapter is to assemble all the PGE data relevant to processes occurring in layered complexes. The available data on sulfide—silicate and crystal—melt distribution coefficients ($K_D$) and on concentrations of PGE in mineral separates from various rocks are summarized. In addition, heats of formation, entropies and heat capacities for PGE-rich phases are tabulated. The data are presented in the context of the issues involving the distribution of PGE in mafic and ultramafic rocks. It will become apparent that the hypotheses of ore formation and of PGE behavior are not, in general, founded in such information.

### DISTRIBUTION OF PGE BETWEEN SULFIDE AND SILICATE MELTS

#### The issue in layered intrusions

The facts that the Merensky Reef of the Bushveld Complex and Howland Reef (Howland et al., 1936) of the Stillwater Complex are sulfide-rich and confined to specific stratigraphic positions over the entire strike lengths of their respective intrusions have led to the general belief that they formed by accumulation of magmatic sulfides (e.g., von Gruenewaldt, 1979; Irvine et al., 1983; Campbell and Barnes, 1984; Kruger and Marsh, 1985). Detailed geochemical arguments have been developed to support the view that sulfides acted as collectors for the PGE in these as well as other large mafic bodies (e.g., Naldrett and Duke, 1980; Keays and Campbell, 1981; Barnes and Naldrett, 1985; Naldrett et al., 1986). An elegantly simple explanation for why some sulfide-bearing horizons are PGE-rich whereas others are not has been proposed by Campbell et al. (1983) and Campbell and Barnes (1984). They argued that sulfide melt–silicate melt $K_D$ are extremely large and, therefore, that sulfides become PGE-rich only if they equilibrate with large reservoirs of magma (e.g., sulfides remain suspended because of turbulent mixing). In contrast, sulfides which have seen only small magma reservoirs (e.g., because of rapid settling) will have low PGE contents, the latter being limited by the reservoir size. The hypothesis is described in detail by Naldrett (this volume).

Although sulfide melt–silicate melt $K_D$ for the PGE must be extremely high, they have never been determined. Consequently, the approach of several workers has been to derive $K_D$ from observed PGE/sulfide ratios in the rocks and estimated initial magma compositions. Thus, Barnes and Naldrett (1985) obtained values of $K_D$(Pt) and $K_D$(Pd) of $>10^5$ and $K_D$(Ir) of 24,000 from Howland Reef sulfides. Similar values derived by Sharpe (1982) from the Merensky are 4000 (Pd), 3900 (Pt) and 6600 (Ir) and from the UG–2 are 56,000 (Pd), 21,000 (Pt) and 174,000 (Ir). Applying the Campbell et al. hypothesis, one would postulate that the highest of these values for each element most closely approximates its true $K_D$. The alternatives are that estimated initial magma compositions are incorrect, that PGE contents are controlled by chromite or other magmatic phases as well as by sulfides, or that the PGE and/or sulfides have been redistributed to varying degrees by post-cumulus magmatic or metamorphic processes. The latter may include PGE transport and redistribution by vapor (e.g., Boudreau et al., 1986) or loss of sulfur if sulfides break down by reequilibrating with the silicate-oxide assemblage during cooling (von Gruenewaldt et al., 1986; Naldrett and Lehmann, 1987). In any case, without explicit knowledge of the $K_D$, hypotheses to account for PGE enrichments obviously cannot be tested.

### Data from submarine basalts

Submarine basalt glasses are good analogues for crystallizing mafic magmas in the shallow crust. They were rapidly quenched and thus preserve near-liquidus relationships; most are sulfide saturated, as indicated by the fact that the glasses typically contain spherical sulfide globules representative of sulfide-oxide melts immiscible in silicate melts (Mathez, 1976); and they have not lost sulfur by degassing (Moore and Schilling, 1973), which is why the globules are preserved in these particular rocks.

Sulfides from several glasses from the FAMOUS area of the mid-Atlantic Ridge have been separated and analysed by INAA for iridium, gold and several other chalcophile elements (Peach and Mathez, 1987). The data are presented in Table 4.1a, in which compositions of the bulk sulfide fraction of several samples, as well as similar data for *individual* sulfides from glass of a single sample (526–1), are reported. Sample 526–1 is particularly interesting because the sulfide in it contains much higher concentrations of iridium and gold than sulfide of other glasses. The 526–1 globules are relatively large (typically >50 μm), suggesting that they had remained suspended in the magma longer than usual. They are also partially resorbed. Since sulfide solubility in magma increases with decreasing pressure (Wendlandt, 1982), resorption is expected if the sulfides initially precipitated in a shallow magma chamber before eruption. Based on these features, Peach and Mathez argued that the 526–1 sulfide effectively equilibrated with a large reservoir of magma and thus was nearly in equilibrium with the melt with respect to the trace metals.

In the absence of trace element data for the 526–1 glass, it is necessary to estimate iridium and gold contents of the glass fraction in order to deduce $K_D$. PGE abundances in typical MORB are extremely low. For example, citing published and unpublished data, Hamlyn et al. (1985) reported average abundances of <0.02 ppb for iridium and 1.3 ppb for gold. The only data on glass separates are those of Hertogen et al. (1980). Included in their investigation is one of the FAMOUS samples (529–4), which, if its PGE contents are taken to be identical to those of 526–1, yields $K_D$ for gold and iridium of $3 \times 10^4$ and $2 \times 10^4$, respectively (Table 4.1b). We judge these estimates to be within a factor of 2 to 3 of the true $K_D$. Similar data for selenium indicate a $K_D$ of ~1000. The distribution of PGE between sulfide melt and silicate melt is currently being investigated experimentally as well as by the study of other submarine basalts.

## PGE IN CHROMITE, OLIVINE AND OTHER SILICATES

### Effect of fractionation on Pd/Ir ratios

It has been suggested that iridium, and to some extent osmium, act as compatible elements during magmatic differentiation owing to their inclusion in olivine. The suggestion comes primarily from investigations of komatiitic flows, in which it has been observed that iridium is concentrated in basal olivine-rich cumulates and relatively depleted in overlying and more differentiated spinifex-tex-

TABLE 4.1a—Compositions of sulfide globules separated from submarine basalt glass recovered from the FAMOUS area of the mid-Atlantic Ridge.

| Sample | Fe (%) | Ni (%) | Co (ppm) | Cr (ppm) | Se (ppm) | Au (ppm) | Ir (ppb) | Ave. Glob Mass (μg) |
|---|---|---|---|---|---|---|---|---|
| **Bulk sulfide fraction** | | | | | | | | |
| 526-1 | 38.3 | 13.4 | 1763 | 332 | 207 | 12.9 | 373 | 70 |
| 528-4 | 41.3 | 14.8 | 2175 | 399 | 210 | 1.3 | | 33 |
| 330-2-1 | 42.8 | 5.4 | 1062 | 204 | 89 | 1.0 | 44 | 25 |
| 521-1 | 46.8 | 5.0 | 1312 | 181 | 110 | 2.2 | 195 | 22 |
| 526-6 | 43.1 | 6.0 | 1754 | 282 | 150 | 1.8 | 49 | 8 |
| 518-3-1 | 41.8 | 8.7 | 1682 | 292 | 147 | 3.7 | 203 | 9 |
| **Individual sulfide globules from 526-1** | | | | | | | | |
| | 38.7 | 11.6 | 1624 | 386 | | 12.0 | 383 | 48 |
| | 39.9 | 15.3 | 1776 | 408 | | 8.9 | 309 | 73 |
| | 40.2 | 13.4 | 1782 | 366 | | 12.9 | 666 | 91 |
| | 39.8 | 14.1 | 1757 | 372 | | 16.3 | 285 | 67 |
| | 35.3 | 13.8 | 1628 | 252 | | 10.7 | 346 | 84 |
| | 39.8 | 12.7 | 1648 | 393 | | 12.6 | 443 | 25 |
| | 37.4 | 11.4 | 1651 | 374 | | 10.8 | 423 | 241 |
| | 41.2 | 13.5 | 1860 | 404 | | 12.7 | 441 | 136 |

TABLE 4.1b—Estimated partition coefficients for iridium and gold.

| | Au (ppb) | Ir (ppb) | Se (ppb) |
|---|---|---|---|
| 529-4 glass (Hertogen et al., 1980) | 0.38 | 0.0157 | 214 |
| 526-1 sulfide (Peach and Mathez, 1987) | 2,900 | 273 | 207,000 |
| Estimated $K_D$ | $3 \times 10^4$ | $2 \times 10^4$ | $1 \times 10^3$ |

tured rocks (Keays and Davison, 1976; Ross and Keays, 1979; Keays et al., 1981; Crocket and MacRae, 1986). Palladium (and presumably platinum, although there are fewer data) exhibits the opposite distribution, behaving incompatibly (Brügmann et al., 1987). Similar general trends of increasing Pd/Ir ratios with extent of differentiation have been observed in the Jimberlana Intrusion (Keays and Campbell, 1981), in ophiolite sequences (e.g., Oshin and Crocket, 1982; Barnes and Naldrett, 1987) and in boninitic lavas (Hamlyn et al., 1985).

### Iridium in olivine and other silicates

The compatible behavior of iridium has led to speculation that it dissolves in olivine. It is possible that iridium preferentially enters olivine over coexisting silicate melt (see below). However, it appears that iridium has an even greater affinity for other minerals. This is clearly indicated by the data of Mitchell and Keays (1981) on PGE contents of mineral separates from mantle-derived spinel and garnet lherzolites (Table 4.2a). The data show that iridium is concentrated in spinel and to a lesser extent clinopyroxene but that olivine is the most iridium-poor phase in all samples. In addition, no relationship has been found between iridium and MgO contents in analyses of bulk rocks from the Bushveld Lower and Critical zones (Lee and Tredoux, 1985), which would be expected if olivine exerted the main control on iridium distribution. In fact the highest iridium abundances were recorded in pyroxenites. If it is true that iridium is compatible in spinel and pyroxene as well as olivine, then the iridium enrichments in the basal cumulates of komatiites are understandable because olivine was the only phase accumulating; in other bodies in which olivine was crystallizing along with pyroxene or chromite, the distribution of iridium should be dependent on the detailed modal variations of the cumulate minerals.

A second explanation of the iridium distribution is that Ir–Os alloy precipitated along with olivine early in the fractionation history, with the metal particles then being incorporated as inclusions in the olivine (Keays et al., 1981; Crocket and MacRae, 1986; Barnes and Naldrett, 1987). In support of this idea, Keays et al. (1981) pointed out that sulfur fugacities of komatiitic magmas were probably low enough that the alloy rather than Ir–Os sulfide was stable. Incorporation of sulfide in early-formed silicates would not, in any case, account for the decreasing Ir/Pd ratio with differentiation since palladium is probably more chalcophile the iridium. In addition, they argued that the mantle contains Ir–Os alloy and therefore that melts derived from it should have been saturated in that phase.

This idea has not been seriously evaluated. It depends on whether or not an element present at the ppb concentration level can undergo homogeneous nucleation in the melt or inhomogeneous nucleation on a growing olivine. Although inclusions of Ir–Os alloy have been observed in chromite (e.g., Prichard et al., 1985), none have ever been found in olivine from any rocks. The argument that the alloys crystallized directly from magma rests primarily on the observations from chromitites and on the bulk geochemical data noted above.

### Iridium in chromite

The association of iridium with spinel appears to be universal. The first study to demonstrate the relationship was that of Greenland (1971), who reported a systematic and concomitant decrease in iridium whole-rock abundances with the chromite mode from the mafic base to the granophyric top of the Great Lake dolerite sill, Tasmania. Other studies, in addition to that of Mitchell and Keays (1981) on mantle xenoliths (noted above), serve to demonstrate that the association is universal. For example, Morgan et al. (1976) documented a well-defined correlation of chromium and iridium abundances in anorthosites from the Fiskenaesset Complex, which is thought to be a layered igneous body metamorphosed to granulite facies; Lee and Fesq (1986) found iridium contents to be related to modal chromite in Bushveld chromitites; from analyses of mineral separates, Gijbels et al. (1974) concluded that iridium, osmium and ruthenium are concentrated in chromite in Bushveld Lower and Critical zone rocks (Table 4.2b); and in the Thetford Mines ophiolite (Quebec), which has been subjected to greenschist to amphibolite facies metamorphism, Oshin and Crocket (1982) found iridium to be enriched in chromite from early-formed cumulates. The latter study further demonstrated that the PGE are localized on silicate-chromite grain boundaries (Table 4.2c).

Morgan et al. (1976) pointed out that ionic radius of $Ir^{4+}$ (0.71Å) is similar to that of $Cr^{3+}$ (0.70Å), so it is possible that iridium dissolves in significant amounts in spinel. It should be remembered, however, that this does not explain the relative enrichments of all the PGE in the chromitites of layered intrusions.

### Experimental data

Experimental data relevant to the above questions are extremely limited. First are those of Malvin et al. (1986), who determined crystal-melt $K_D$ for certain siderophile trace elements among phases in the pseudoternary forsterite–

TABLE 4.2—Radiochemical neutron activation data for gold and several PGE in oxide and silicate mineral separates from various rocks. Data in ppb.

**A. Spinel and garnet lherzolite mantle xenoliths (Mitchell and Keays, 1981).**

| | | Ir | Pd | Au |
|---|---|---|---|---|
| **Spinel lherzolites, Mt. Porndon, Victoria** | | | | |
| VSL-1 | mica | 1.1 | 1.1 | 0.6 |
| | spinel | 2.1 | 9.0 | 0.7 |
| | cpx | 0.4 | 6.7 | 0.2 |
| | opx | 1.8 | 2.4 | 0.3 |
| | oliv | 0.2 | 0.2 | 0.2 |
| VSL-3 | spinel | 6.3 | 8.1 | 0.6 |
| | cpx | 3.4 | 2.3 | 0.2 |
| | opx | 0.7 | 2.1 | 0.04 |
| | oliv | 0.5 | 0.5 | 0.04 |
| VSL-4 | spinel | 2.5 | 8.6 | 0.8 |
| | cpx | 1.9 | 1.7 | 0.5 |
| | opx | 0.8 | 2.8 | 0.8 |
| | oliv | 0.07 | 1.0 | 0.03 |
| **Spinel lherzolites, Kilbourne Hole, New Mexico** | | | | |
| KH1 | spinel | 2.7 | 1.2 | 0.4 |
| | cpx | 0.6 | 0.6 | 0.3 |
| | opx | 0.7 | 0.4 | 0.3 |
| | oliv | 0.2 | 0.1 | 0.2 |
| KH2 | spinel | 2.1 | 0.5 | 0.4 |
| | cpx | 0.7 | 0.6 | 0.3 |
| | opx | 0.4 | 0.1 | 0.06 |
| | oliv | 0.07 | 0.02 | 0.1 |
| KH3 | spinel | 3.1 | 1.3 | 1.2 |
| | cpx | 0.8 | 1.1 | 0.3 |
| | opx | 0.4 | 0.4 | 0.4 |
| | oliv | 0.3 | 0.3 | 0.07 |
| **Garnet lherzolites, Matsoku, Lesotho** | | | | |
| M7 | garnet | 0.03 | $<0.05$ | 0.1 |
| | cpx | 6.9 | 3.5 | 0.2 |
| | opx | 1.0 | 3.1 | 0.1 |
| | oliv | 0.4 | 0.6 | 0.1 |
| M8 | garnet | 0.06 | $<0.05$ | 0.03 |
| | cpx | 7.8 | 1.1 | 0.9 |
| | opx | 1.2 | 0.6 | 0.05 |
| | oliv | 0.4 | 0.3 | 0.05 |
| M11 | garnet | 0.06 | $<0.05$ | 0.07 |
| | cpx | 0.4 | 0.2 | 0.1 |
| | opx | 1.4 | 0.1 | 0.06 |
| | oliv | 0.2 | $<0.02$ | 0.08 |
| M13 | garnet | 0.1 | 0.6 | 0.07 |
| | cpx | 0.4 | 1.2 | 0.08 |
| | opx | 1.0 | 0.4 | 0.2 |
| | oliv | 0.9 | 0.3 | 0.08 |

**B. Mineral separates of Bushveld rocks (Gijbels et al., 1974). Sample W-53-10-343 is a chromite bronzitite from the upper Critical Zone. W-53-8 is a mixture of chromitite (A), bronzitite (B) and anorthosite (C), apparently collected as a single sample.**

| | | Os | Ru | Ir |
|---|---|---|---|---|
| W-53-10-343 | opx | 22.6 | 235 | 36 |
| W-53-10-343 | chromite | 51.5 | 370 | 33.7 |
| W-53-8 A | chromite | 103.7 | 680 | 89 |
| W-53-8 B | chromite | 114.4 | 680 | 105 |
| W-53-8 B | opx | 57 | 360 | 84 |
| W-53-8 C | plag | 1.86 | 11.2 | 2.6 |

**C. Chromite separates from harzburgite and dunite of the Thetford Mine ophiolite before and after acid leaching (Oshin and Crocket, 1982). The chromites were washed in HF followed by hot concentrated $H_2SO_4$.**

| | Ir | Pt | Pd | Au |
|---|---|---|---|---|
| Before leaching | | | | |
| tectonized harzburgite | 446 | 301 | 107 | 1.5 |
| dunite cumulate | 158 | 22.5 | 6.9 | 3.3 |
| After leaching | | | | |
| tectonized harzburgite | 3.5 | 22 | 18 | 1.5 |
| dunite cumulate | 110 | 13 | 7.7 | 5.7 |

diopside–anorthite system using proton induced x-ray emission (PIXE). [PIXE is a microbeam technique analogous to electron probe microanalysis. A characteristic x-ray spectrum is generated by excitation with high-energy protons produced in a particle accelerator and focused to a microbeam. The *bremsstrahlung*, and thus the background, of the spectrum generated by proton excitation is lower than that generated by electron excitation. Consequently, the former has the advantage of better detection limits, which are typically in the range 1–10 ppm. See Cabri et al. (1984) for applications of PIXE to sulfide analysis.] The data are summarized in Table 4.3. Malvin et al. explicitly warned that their data are preliminary and not quantitative, so they should be used with caution.

TABLE 4.3—Concentrations (ppm) and crystal-melt $K_D$ for gold and some PGE in phases in the forsterite-diopside-anorthite system at 1300°C. Data of Malvin et al. (1986).

| | Os | Ir | Pt | Au |
|---|---|---|---|---|
| **Forsterite-melt experiment** | | | | |
| Forsterite | 2.4 | 3.6 | 2.6 | nd |
| Melt | <2 | <0.4 | 12 | 7.5 |
| forsterite/melt | >1 | >7 | 0.22 | <1 |
| **Reconnaissance experiment for other phases** | | | | |
| diopside/melt | | 1.4–2.0 | | |
| spinel/melt | | 25 | 3 | |

nd = not determined

A second set of reconnaissance experiments are those of Watson et al. (1987). They determined approximate clinopyroxene/liquid $K_D$ for osmium and several other trace elements at 1250°C and 1 atm in an iron-free synthetic "basalt." Their estimated $K_D$ for osmium is 0.08; that for rhenium it is 0.04.

## PGE PATTERNS OF ROCKS

It has become customary to present relative abundances of the PGE and gold normalized to their abundances in C1 chondrites and plotted in order of increasing melting point (Naldrett and Duke, 1980). The resulting PGE patterns (Fig. 4.1) are characteristic for different rock types. Gold, however, does not appear to behave in any systematic fashion and must be subject to processes that do not affect the PGE. As expected, spinel lherzolite xenoliths (Mitchell and Keays, 1981; Morgan et al., 1981) and komatiitic rocks (Ross and Keays, 1979; Keays, 1982; Crocket and MacRae, 1985), which most closely reflect bulk mantle compositions, are characterized by relatively flat patterns (Fig. 4.1a). General patterns of other rocks can be interpreted from the following additional considerations:

(a) Iridium and osmium must be more compatible than platinum and palladium because fractional crystallization of mafic magma results in residual melts depleted in the former and enriched in the latter elements, i.e., Pd/Ir ratios increase. This is most clearly illustrated by comparison of compositions of the basal ultramafic cumulate and the spinifex-textured top of a single komatiite flow (Fig. 4.1a).

(b) Steep PGE patterns characterize sulfide ores of mafic and ultramafic rocks (e.g., Naldrett, 1981; Barnes and Naldrett, 1987), including the sulfide-dominated Bushveld Merensky Reef and Stillwater Howland Reef (Fig. 4.1b). Assuming that the original magmas were characterized by more or less chondritic PGE patterns, then sulfide melt–silicate melt $K_D$ of platinum and palladium are higher than those of iridium and osmium. The $K_D$ of gold also appears to be greater than that of iridium (Fig. 4.1b).

(c) MORB, which are thought to be generated from depleted mantle, are characterized by positive PGE patterns (Fig. 4.1a). Assuming that their sources are represented by depleted spinel lherzolites such as those from Kilbourne Hole (Mitchell and Keays, 1981), then MORB evidently possess higher Pd/Ir ratios than their source rocks. The PGE signature of MORB must be controlled by a combination silicate and sulfide fractionation, but this is not yet precisely understood. See Hamlyn et al. (1985) for a more detailed discussion of PGE patterns in MORB and other lava types.

(d) Chromitite horizons in layered intrusions, e.g., the UG–2 (Fig. 4.1b), are also characterized by high Pd/Ir ratios (e.g., McLaren and DeVilliers, 1982; Gain, 1985). However, unlike the other stratabound PGE-rich horizons, the chromitites are not sulfide-rich, and other processes appear to have influenced them. One possibility, as noted above, is that the chromitites retain the characteristic sulfide PGE pattern because magmatic sulfide was originally concentrated in them, but that sulfur was subsequently lost because sulfide decomposed during subsolidus recrystallization (von Gruenewaldt et al., 1986; Naldrett and Lehmann, 1987). This hypothesis is developed in more detail by Naldrett (this volume). The geochemical evolution of the chromitites is not well understood.

(e) Podiform chromitites in ophiolites exhibit distinctive negative PGE patterns (Fig. 4.1b), just opposite those of the layered intrusion chromitites. The former are also distinct from the chondritic patterns of spinel *separates* from spinel lherzolite xenoliths (Table 4.1a and Fig. 4.1b). Spinel lherzolites are generally considered to be residues of partial melting, so it appears that partial melting does not yield residual chromite enriched in iridium. Therefore, the podiform chromitite patterns cannot have been generated simply from partial melting. Also, assuming equilibrium among all the phases in the spinel lherzolites and in the ophiolitic chromitites, it seems unlikely that the distinctly different PGE patterns of chromite in the two different rocks could both be due to equilibria between chromite and the crystalline silicate assemblage. Possibly the ophiolite patterns reflect chromite-sulfide or chromite-fluid equilibria.

An unknown in interpretation of these patterns is how PGE are influenced by fluids. It is possible that for the important fluid-mineral reactions PGE patterns are imposed primarily by the specific minerals involved, in which case evidence for the involvement of fluids will obviously not be found in distinctive patterns.

## PGE TRANSPORT BY FLUIDS

The only systematic attempt to evaluate the ability of fluids to transport PGE in the environment of a crystallizing layered intrusion is that of Wood (1987), who used available thermodynamic data collected at low temperature to compute the volatilities of PGE metals, oxides and chlorides at high temperature. Wood showed that most of the relevant equilibria, and thus metal solubilities in fluids, are strongly dependent on oxygen fugacity ($f_{O_2}$) and temperature, with solubilities generally decreasing with decreasing $f_{O_2}$ and increasing temperature. In the metal-oxide systems, the calculations indicate that only palladium exhibits any significant solubility in aqueous vapor under geologic conditions, but even then its maximum solubility in fluid in equilibrium with the metal at QFM (the quartz–fayalite–magnetite oxygen buffer) and 1400°K is only ~300 ppt.

One of the suprising results of the calculations is that the halogens enhance PGE solubilities only marginally. For

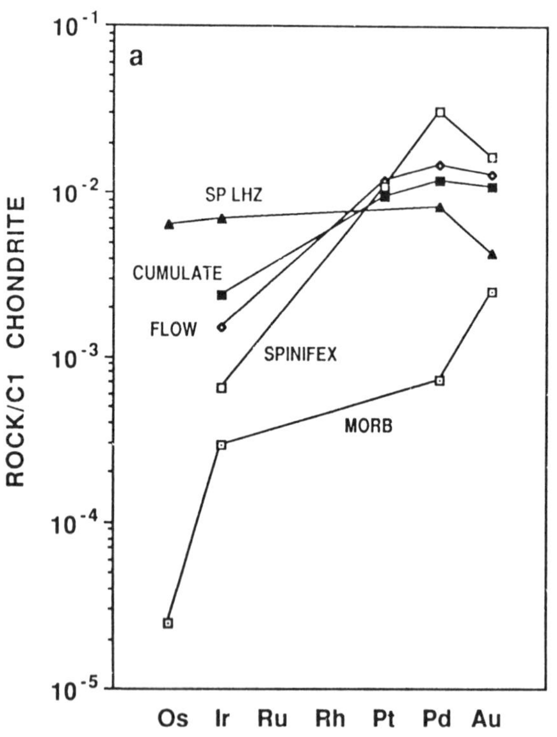


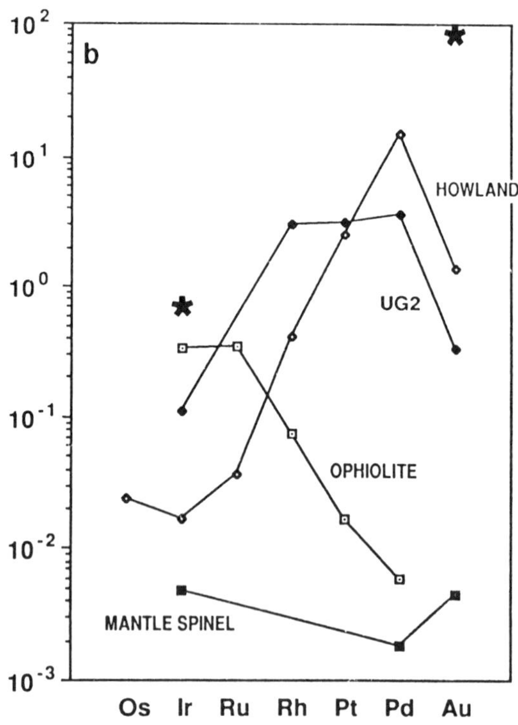


FIGURE 4.1—Chondrite normalized PGE patterns for various rock types. **A.** SP LHZ = average of 14 spinel lherzolites (Morgan et al., 1981); SPINIFEX, FLOW and CUMULATE = bulk compositions of spinifex top, cumulate base and total of "Fred's" komatiite flow, Munro Township, Ontario (Crocket and MacRae, 1986); MORB = composition of FAMOUS glass 529–4 (Hertogen et al., 1980). **B.** HOWLAND = average ÷ 10 of 36 analyses of sulfide-rich rocks from the Howland Reef (Barnes and Naldrett, 1985, table 1); UG–2 = average UG–2 (McLaren and DeVilliers, 1982); OPHIOLITE = podiform chromitite, New Caledonia (Page et al., 1982); MANTLE SPINEL = average of three spinel separates from Kilbourne Hole spinel lherzolite (Mitchell and Keays, 1981); stars = composition of sulfide separated from FAMOUS glass 526–1 (Peach and Mathez, 1987). The normalizing chondrite composition (Naldrett and Duke, 1980) is (ppb): Os (514); Ir (540); Ru (690); Rh (200); Pt (1020); Pd (545); Au (152).

example, $IrF_6$(s) was found to be much less volatile than $IrO_3$(s) at magmatic temperature, suggesting that the iridium enrichment of Kilauea gases is not related to their fluorine contents. Similarly, the few thermodynamic data available indicate relatively low solubilities for PGE chlorides, except in extremely HCl-rich fluids.

As an example of the nature of these calculations, consider the equilibrium between $RuS_2$ and an $H_2O$–HCl fluid. The reaction, which is probably a reasonable approximation of nature, may be written as (Wood, 1987)

$$RuS_2(s) + 3HCl(g) + {}^3/_4O_2(g) = RuCl_3(g) + S_2(g) + {}^3/_2H_2O(g), \quad [1]$$

where (g) and (s) refer to the vapor and solids, respectively. An illustration of the magnitude of $f_{RuCl_3}$ for various fluid compositions and temperatures is presented in Fig. 4.2. The calculations were performed at log $f_{O_2}$ of 2.5 orders of magnitude below QFM and log $f_{S_2}$ of $-2.64$, conditions which probably approximate those attendant on the partially molten Stillwater and Bushveld cumulates (Mathez, this volume). It can be seen that the solubility of $RuCl_3$ is dependent on temperature and on fluid composition only at low values of $f_{HCl}/f_{H_2O}$.

The solubilities of PGE chlorides are enhanced if $f_{S_2}$ is sufficiently low that the metals rather than the sulfides are stable. Under such conditions, solubility is controlled by an equilibrium analogous to (1):

$$Ru(s) + 3HCl(g) + {}^3/_4O_2(g) = RuCl_3(g) + {}^3/_2H_2O(g). \quad [2]$$

$RuCl_3$ (and $PdCl_2$) concentrations in aqueous fluids obviously depend on $f_{HCl}$ and $f_{H_2O}$ in the same way as for equilibrium (1) and may reach 10s to 100s of ppb. For example, Wood calculates that for a ($CO_2$+CO)-rich fluid at $P_{total}$ = 1 kbar in which $f_{H_2O}$ = 1 bar and $f_{HCl}$ = 100 bars, the solubilities of

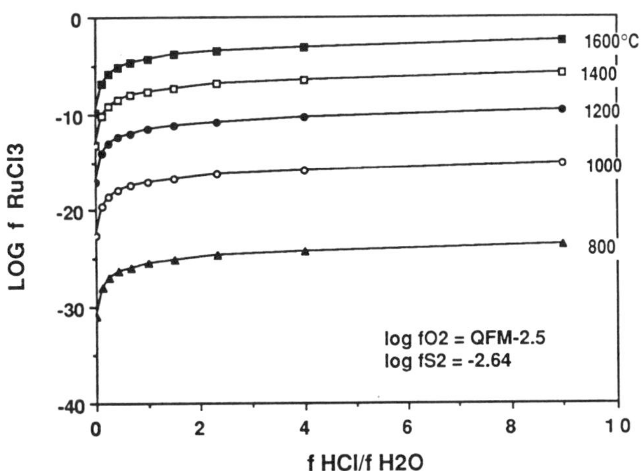


FIGURE 4.2—Variation of $f_{RuCl_3}$ as a function of fluid composition for various temperatures according to reaction (1) (see text).

palladium and ruthenium are 500 and 600 ppb, respectively, at 1400°K. Circumstances may be imagined, therefore, where fluid transport of the PGE is important. However, Wood also pointed out that aqueous fluid should dissolve orders of magnitude more chlorides of iron, nickel and copper than of PGE. He notes, in addition, that metals and alkalis will decrease the amount of PGE dissolved in the fluid by competing for available chlorine (i.e., by effectively reducing $f_{HCl}$) and that metal transport by and subsequent deposition from fluids does not, at least within the compositional contraints of the computed system, account for enrichment of PGE relative to the transition metals.

Several points should be added. The fact that solubilities of PGE and transition metals are widely different leaves open the possibility of enrichment of one over the other. There are at least two ways this could happen. Sulfides may become enriched in PGE by preferential loss of the transition metals to a transient chlorine-bearing fluid. Alternatively, precipitation of the PGE would occur first from fluid carrying both PGE and transition metals in response to changing conditions. Finally, it should be remembered, as emphasized by Wood, that the thermodynamic data base is extremely limited, and no data exist that would allow evaluating the possibility of PGE dissolution as sulfur or more complex species.

Existing thermochemical data on the PGE appropriate for high-temperature calculations have been collected and tabulated in Appendix II. In general, experimental determinations of enthalpies, from which heat capacities are derived, have been made only at room-temperature or slightly higher. Extrapolations of heat capacities to higher temperatures have been accomplished according to well-defined but empirical laws (Kubaschewski and Alcock, 1979) that, although inexact, probably do not yield errors of sufficient magnitude to invalidate Wood's basic conclusions. This point and the thermodynamic data base are discussed in more detail in Appendix II.

## CONCLUSION

It should be obvious that our knowledge of the high-temperature geochemistry of the PGE is extremely limited. Although the elements are obviously chalcophile, the only study containing even approximate data on sulfide melt/silicate melt $K_D$ is that of Peach and Mathez (1987) on submarine basalts. It indicates that the $K_D$(Ir) and $K_D$(Au) are 2 $\times 10^4$ and 3 $\times 10^4$, respectively. Similarly, few data exist on the distribution behavior of PGE among silicate melts and crystalline phases. The experimental data are limited to several preliminary measurements (Malvin et al., 1986), which in general suggest a moderate compatibility of iridium and incompatibility of platinum during fractionation of spinel and olivine from mafic melt. Detailed geochemical studies of komatiitic flows and other mafic rocks are in general accord with this conclusion. Finally, ample evidence has accumulated that PGE distribution, at least in the Stillwater and Bushveld complexes, is influenced in part by fluids. The limited thermodynamic data on PGE volatility at high temperature give little indication of how this might occur, however.

ACKNOWLEDGEMENTS—A review of this paper by A. J. Naldrett is gratefully acknowledged. This work was supported by NSF grant EAR8720982.

## REFERENCES

Barin, I., Knacke, O., and Kubaschewski, O., 1973, Thermochemical Properties of Inorganic Substances: Springer–Verlag, New York, 921 pp.

Barin, I., Knacke, O., and Kubaschewski, O., 1977, Thermochemical Properties of Inorganic Substances—Supplement: Springer–Verlag, New York, 861 pp.

Barnes, S.-J., and Naldrett, A. J., 1987, Fractionation of the platinum-group elements and gold in some komatiites of the Abitibi greenstone belt, northern Ontario: Econ. Geol., v. 82, pp. 165–183.

Barnes, S. J., and Naldrett, A. J., 1985, Geochemistry of the J–M (Howland) Reef of the Stillwater Complex, Minneapolis adit area. 1. Sulfide chemistry and sulfide-olivine equilibrium: Econ. Geol., v. 80, pp. 627–645.

Boudreau, A. E., Mathez, E. A., and McCallum, I. S., 1986, Halogen geochemistry of the Stillwater and Bushveld Complexes: Evidence for transport of the platinum-group elements by Cl-rich fluids: J. Petrol., v. 27, pp. 967–986.

Brügmann, G. E., Arndt, N. T., Hofmann, A. W., and Tobschall, H. J., 1987, Noble metal abundances in komatiite suites from Alexo, Ontario, and Gorgona Island, Columbia: Geochim. Cosmochim. Acta, v. 51, pp. 2159–2169.

Cabri, L. J., Blank, H., el Goresy, A., Laflamme, J. H. G., Nobiling, R., Sizgoric, M. B., and Traxel, K., 1984, Quantitative trace-element analyses of sulfides from Sudbury and Stillwater by proton microprobe: Canadian Mineral., v. 22, pp. 521–542.

Campbell, I. H., and Barnes, S. J., 1984, A model for the geochemistry of the platinum group elements in magmatic sulfide deposits: Can. Mineral., v. 22, pp. 151–160.

Campbell, I. H., Naldrett, A. J., and Barnes, S. J., 1983, A model for the origin of the platinum rich sulfide horizons in the Bushveld and Stillwater Complexes: J. Petrol., v. 24, pp. 133–165.

Crocket, J. H., and MacRae, W. E., 1986, Platinum-group element distribution in komatiitic and tholeiitic volcanic rocks from Munro Township, Ontario: Econ. Geol., v. 81, pp. 1242–1251.

Gain, S. B., 1985, The geologic setting of the platiniferous UG–2 chromitite layer on the farm Maandagshoek, eastern Bushveld Complex: Econ. Geol., v. 80, pp. 924–943.

Gijbels, R., Millard, H. T., Jr., Desborough, G. A., and Bartel, A. J., 1974, Osmium, ruthenium, iridium and uranium in silicates and chromite from the eastern Bushveld complex, South Africa: Geochim. Cosmochim. Acta, v. 38, pp. 319–337.

Greenland, L. P., 1971, Variation of iridium in a differentiated tholeiitic diabase. Geochim. Cosmochim. Acta, v. 35, pp. 319–322.

Greenwood, N. N., and Earnshaw, A., 1984, Chemistry of the elements: Pergamon Press, Oxford, 1542 pp.

Hamlyn, P. R., Keays, R. R., Cameron, W. E., Crawford, A. J., and Waldron, M. H., 1985, Precious metals in magnesian low-Ti lavas: Implications for metallogenesis and sulfur saturation in primary magmas: Geochim. Cosmochim. Acta, v. 49, pp. 1797–1811.

Hertogen, J., Janssens, M.-J., and Palme, H., 1980, Trace elements in ocean ridge basalt glasses: implications for fractionations during mantle evolution and petrogenesis: Geochim. Cosmochim. Acta, v. 44, pp. 2125–2143.

Howland, A. L., Peoples, J. W., and Sampson, E., 1936, The Stillwater igneous complex and associated occurrences of nickel and platinum group metals: State of Montana Bureau of Mines and Geology, Miscellaneous Contribution 7, 15 pp.

Irvine, T. N., Keith, D. W., and Todd, S. G., 1983, The J–M platinum–palladium reef of the Stillwater Complex, Montana: II. Origin by double-diffusive convective magma mixing and implications for the Bushveld Complex: Econ. Geol., v. 78, pp. 1287–1334.

Keays, R. R., 1982, Palladium and iridium in komatiites and associated rocks: application to petrogenetic problems; *in* Arndt, N. T., and Nisbet, E. G. (eds.), Komatiites: George Allen, London, pp. 435–357.

Keays, R. R., and Campbell, I. H., 1981, Precious metals in the Jimberlana Intrusion, Western Australia: Implications for the genesis of platiniferous ores in layered intrusions: Econ. Geol., v. 76, pp. 1118–1141.

Keays, R. R., Ross, J. R., and Woolrich, P., 1981, Precious metals in volcanic peridotite-associated nickel sulfide deposits in Western Australia. II: Distribution within the ores and host rocks at Kambalda: Econ. Geol., v. 76, pp. 1645–1674.

Keays, R. R., and Davison, R. M., 1976, Palladium, iridium and gold in the ores and host rocks of nickel sulfide deposits in Western Australia: Econ. Geol., v. 71, pp. 1214–1228.

Kruger, F. J., and Marsh, S. J., 1985, The mineralogy, petrology and origin of the Merensky cyclic unit in the eastern Bushveld Complex: Econ. Geol., v. 80, pp. 958–974.

Kubaschewski, O., and Alcock, C. B., 1979, Metallurgical thermochemistry: Pergammon Press, New York, 449 pp.

Lee, C. A., and Fesq, H. W., 1986, Au, Ir, Ni and Co in some chromitites of the eastern Bushveld Complex, South Africa: Chem. Geol., v. 62, pp. 227–237.

Lee, C. A., and Tredoux, M., 1985, PGE abundances in the Lower and Lower Critical Zones of the eastern Bushveld Complex: Econ. Geol., v. 81, pp. 1087–1095.

Malvin, D. J., Drake, M. J., Benjamin, T. M., Duffy, C. J., Hollander, M., and Rogers, P. S. Z., 1986, Experimental partitioning studies of siderophile elements amongst lithophile phases; Preliminary results using PIXE microprobe analysis (abs): Lunar Planet. Sci. XVII, pp. 49–50.

Mathez, E. A., 1976, Sulfur solubility and magmatic sulfides in submarine basalt glass: J. Geophys. Res., v. 81, pp. 4269–4276.

McLaren, C. H., and DeVilliers, J. P. R., 1982, The platinum-group chemistry and mineralogy of the UG–2 chomitite layer of the Bushveld Complex: Econ. Geol., v. 77, pp. 1348–1366.

Mitchell, R. H., and Keays, R. R., 1981, Abundance and distribution of gold, palladium and iridium in some spinel and garnet lherzolites: Implications for the nature and origin of precious metal-rich intergranular components in the upper mantle: Geochim. Cosmochim. Acta, v. 45, pp. 2425–2442.

Moore, J. G., and Schilling, J.–G., 1973, Vesicles, water and sulfur in Reykjanes Ridge basalts: Contrib. Mineral. Petrol., v. 41, pp. 105–118.

Morgan, J. W., Ganapathy, R., Higuchi, H., and Krahenbuhl, U., 1976, Volatile and siderophile trace elements in anorthositic rocks from Fiskenaesset, West Greenland: Comparison with lunar and meteoritic analogues: Geochim. Cosmochim. Acta, v. 40, pp. 861–887.

Morgan, J. W., Wandless, G. A., Petrie, R. K., and Irving, A. J., 1981, Composition of the Earth's upper mantle—I. Siderophile trace elements in ultramafic nodules: Tectonophysics, v. 75, pp. 47–67.

Naldrett, A. J., 1981, Platinum group element deposits—a review; *in* Cabri, L. J. (ed.), Platinum group elements: mineralogy, geology, recovery: Canadian Institute of Mining and Metallurgy, Spec. Vol. 23, pp. 197–231.

Naldrett, A. J. and Duke, J. M., 1980, Platinum metals in magmatic sulfide ores: Science, v. 208, pp. 1417–1424.

Naldrett, A. J., Gasparrini, E. C., Barnes, S. J., von Gruenewaldt, G. and Sharpe, M. R., 1986, The upper Critical Zone of the Bushveld Complex and the origin of the Merensky-type ores: Econ. Geol., v. 81, pp. 1105–1117.

Naldrett, A. J., and Lehmann, J., 1987, Spinel non-stoichiometry as the explanation for Ni-, Cu-, and PGE-enriched sulphides in chromitites (abs.): Geoplatinum 87 Symposium, Open University, Paper T10.

Olmez, I., Finnegan, D. L., and Zoller, W. H., 1986, Iridium emissions from Kilauea volcano: J. Geophys. Res., v. 91, pp. 653–663.

Oshin, I. O., and Crocket, J. H., 1982, Noble metals in Thetford Mines ophiolites, Quebec, Canada—Part I: Distribution of gold, iridium, platinum and palladium in the ultramafic and gabbroic rocks: Econ. Geol., v. 77, pp. 1556–1570.

Pankratz, L. B., Stuve, J. M., and Gokcen, N. A., 1984, Thermodynamic data for mineral technology: U.S. Bureau of Mines, Bull. No. 677, 355 pp.

Peach, C. L., and Mathez, E. A., 1987, Gold and iridium in sulphides from submarine basalt glasses (abs): Geo-Platinum 87 Symosium, Open University, P12.

Prichard, H. M., Potts, P. J., and Neary, C. R., 1985, Platinum-group minerals in ophiolite complexes: An example from Shetland (abs): Canadian Mineral., v. 23, p. 311.

Ross, J. R., and Keays, R. R., 1979, Precious metals in volcanic-type nickel sulfide deposits in Western Australia—I: Relationship with the composition of the ores and their host rocks: Canadian Mineral., v. 17, pp. 417–435.

Watson, E. B., Othman, D. B., Luck, J.–M., and Hofmann, A. W., 1987, Partitioning of U, Pb, Cs, Yb, Hf, Re, and Os between chromian diopsidic pyroxene and haplobasaltic liquid: Chem. Geol., v. 62, pp. 191–208.

Wendlandt, R. F., 1982, Sulfide saturation of basalt and andesite melts at high pressures and temperatures: Am. Mineral., v. 67, pp. 877–885.

Wood, S. A., 1987, Thermodynamic calculations of the volatility of the platinum group elements (PGE): The PGE content of fluids at magmatic temperatures: Geochim. Cosmochim. Acta, v. 51, pp. 3041–3050.

von Gruenewaldt, G., 1979, A review of some recent concepts of the Busvheld Complex, with particular reference to sulfide mineralization: Canadian Mineral., v. 17, pp. 233–256.

von Gruenewaldt, G., Hatton, C. J., Merkle, R. K. W., and Gain, S. B., 1986, Platinum-group element-chromitite associations in the Bushveld Complex: Econ. Geol., v. 81, pp. 1067–1079.

## APPENDIX I. ETYMOLOGY*

Ruthenium. After *Ruthenia,* the Latin of Russia, first discovered in ores from Urals. By K. Klaus, 1844.

Rhodium. From the Greek *rhodon,* rose, for the color of an aqueous solution of its salts. By W. H. Wollaston, 1803.

Palladium. After the asteroid, Pallas, itself named for the Greek goddess of wisdom, Pallas Athena. By W. H. Wollaston, 1803.

Osmium. From the Greek *osme,* odour, because of characteristic and pungent smell of volatile oxide, $OsO_4$. By S. Tennant, 1803.

Iridium. After the Greek goddess Iris, whose sign was that of the rainbow, *iris,* because its compounds exhibit a variety of colors. By S. Tennant, 1803.

Platinum. After *platina,* Spanish for little silver. By Spanish astronomer and naval officer, A. de Ulloa, in 1736, who observed a brittle white metal in gold mines of what is now Columbia. (Impurities of Cu, Fe made the platinum brittle.)
*(from Greenwood and Earnshaw, 1984)

## APPENDIX II

Compilations of thermodynamic data for the PGE include those of Barin et al. (1973, 1977), Kubachewski and Alcock (1979) and Pankratz et al. (1984). None of these sources contain all the available data, so they are tabulated here. The data are presented as $\Delta H^\circ_{f,298.15}$, $S^\circ_{298.15}$, and $C_p$, from which other quantities may be derived. In addition, compilations of $\Delta G_{f,T}$ may be found in the original references. The reference state is the pure substance at 298.15°K and 1 atmosphere. Where more than one set of values exists for a single substance, all the data are presented.

Estimations of $C_p$ are based in part on theory (knowledge of the molecular properties of a substance and how they control its ability to absorb heat) and in part on empirical observation (values for cationic and anionic contributions to the $C_p$ at 298.15°K). In a few instances, values for $S^\circ_{298.15}$ and $\Delta H^\circ_{f,298.15}$ are also estimated for solid phases. A detailed discussion of methods for estimating the various quantities is given by Kubaschewski and Alcock (1979, pp. 179–209). These methods have been found to be fairly accurate when compared to available experimental data. All estimated values or equations are indicated in the following tabulation.

### Notes

1. $\Delta H^\circ_{f,298.15}$ in kJ/mole; $S^\circ_{298.15}$ in J/mole.
2. (s) and (g) refer to the solid and gaseous state of the substance.
3. Constants a, b, c and d are the coefficients for the $C_p$ equation in J/mole/K:

$$C_p = a + b\times10^{-3}T + c\times10^{5}T^{-2} + d\times10^{-6}T^2.$$

4. All values are given with the same number of decimal places as appear in the literature.
5. Errors are listed when known.
6. ACC: Quality of measurement as defined in Barin et al. (1977), where
   A. "Key value," well-established.
   B. "Good value," no immediate need to redetermine.
   C. "Moderately good value," redetermination desirable.
   D. "Better than nothing."
7. References: a. Kubaschewski and Alcock (1979); b. Barin et al. (1973); c. Barin et al. (1977); d. Pankratz et al. (1984).

*estimated value
**extrapolated from $C_p$ at 298.15.

| | $\Delta H^\circ_{f,298.15}$ | $S^\circ_{298.15}$ | $C_p$ (T) a | b | c | d | T °K | Ref. | ACC. |
|---|---|---|---|---|---|---|---|---|---|
| Ir (s) | | 35.53 ± 0.21 | 23.266 | 5.942 | | | 298–1800 | a | |
| | | 35.51 | 23.366 | 5.36 | | | 298–600 | c | |
| | | | 21.337 | 8.005 | 1.661 | −0.151 | 600–2716 | c | |
| | | 35.489 | 22.797 | 6.185 | 0.402 | | 298–2000 | d | |
| Ir (g) | 669.5 ± 6.3 | 193.53 | | | | | | a | |
| | 665.336 | 193.495 | 17.47 | 5.892 | 1.389 | | 298–2000 | d | |

| | $\Delta H^\circ_{f,298.15}$ | | $S^\circ_{298.15}$ | | $C_p$ (T) a | b | c | d | T °K | Ref. | ACC. |
|---|---|---|---|---|---|---|---|---|---|---|---|
| $IrF_6$ (g) | −544 | ± 83.7 | 357.8 | ± 1.7 | | | | | | a | |
| $IrCl_3$ (s) | −254.4 | ± 16.7 | 114.86 | ± 16.74 | 84.9 | 18.8 | −4.2 | | 298–1080 | a ** | B–C |
| $IrBr_3$ (s) | −138.5 | | −176* | | 99.26 | 20.34 | | | 298–800 | c | B–C |
| IrI (s) | −46 | ± 13 | 109 | ± 21 | 48.5 | 12.6 | | | 298–800 | c ** | D |
| $IrI_2$ (s) | −84 | ± 13 | 159 | ± 21 | 74.5 | 16.7 | | | 298–800 | c ** | D |
| $IrO_2$ (s) | −241 | ± 13 | 56 | ± 6 | 72.5 | 8.1 | −15.4 | | 298–1400 | a ** | B–C |
| $IrO_3$ (g) | 13.4 | ± 16.7 | 290.4 | ± 8.4 | 82.18 | 1.76 | −5.52 | | 298–1600 | c ** | B–C |
| $Ir_2S_3$ (s) | −244.4 | ± 7.5 | 97.1 | | 108.8 | 40.51 | −10.54 | | 298–1200 | a ** | |
| | −244.4 | | 97.1 | | 110.3 | 32.97 | −9.62 | | 298–1400 | c | B–C |
| $IrS_2$ (s) | −144.8 | ± 6.3 | 61.5 | | 68.58 | 15.78 | −6.57 | | 298–1400 | c | B–C |
| Os (s) | | | 32.6 | ± 2.1 | 23.56 | 3.85 | | | 298–2000 | b | |
| | | | 32.639 | | 23.584 | 3.8 | | | 298–2000 | d | |
| Os (g) | 788.4 | ± 9.6 | 192.5 | | | | | | | a | |
| | 790.871 | | 192.487 | | 16.675 | 6.486 | 1.937 | | 298–2000 | d | |
| $OsO_2$ (s) | −294.6* | ± 9.6 | 59.9* | ± 10.9 | 69.96 | 10.38 | −14.19 | | 298–1200 | a | B |
| $OsO_4$ (s) | −393.8 | ± 8.4 | 136.8 | ± 8.4 | 151.5 | | | | 298–314 | c ** | B |
| $OsO_4$ (g) | −336.22 | ± 8.79 | 293.67 | ± 0.33 | 85.99 | 20.42 | −15.98 | | 298–1000 | a | |
| $OsS_2$ (s) | −100.4 | ± 10.5 | 84.1 | ± 10.5 | 68.54 | 11.84 | −8.8 | | 298–1400 | c | B |
| $OsSe_2$ | −120.1 | | 81.6 | | 73.65 | −11.09 | −4 | | 298–1200 | c | C |
| $OsP_2$ (s) | −131.8* | ± 18.8 | 82.0* | ± 13 | 66.5 | 15.5 | | | 298–1400 | c ** | B–C |
| Pd (s) | | | 37.87 | ± 0.21 | | | | | | a | |
| | | | 37.83 | | 23.781 | 7.394 | | | 298–400 | c | D |
| | | | | | 24.576 | 5.331 | | | 400–1400 | c | D |
| | | | | | 3.519 | 15.198 | 142.449 | | 1400–1825 | c | D |
| | | | 37.715 | | 24.571 | 5.448 | −0.272 | | 298–1827 | d | |
| | | | | | 37.912 | | | | 1827–2500 | d | |
| Pd (g) | 377 | ± 2.9 | 167.05 | | | | | | | a | |
| | 378.279 | | 166.97 | | 10.997 | 11.315 | 5.703 | | 298–2500 | d | |
| $PdF_2$ (s) | −469 | | 88.7 | | 59.8 | 20.5 | | | 298–1100 | c ** | |
| $PdCl_2$ (s) | −173.2 | | 103.8 | | 69 | 20.9 | | | 298–952 | c ** | C |
| $PdCl_2$ (g) | 121.56 | | 313.63 | | | | | | | a | |
| $PdI_2$ (s) | −63.6 | ± 16.7 | 151 | ± 8.4 | 68.2 | 23 | | | 298–833 | c ** | C |
| | | | | | 68.2 | 23 | | | 833–900 | c ** | |
| PdO (s) | −112.6* | | 39.3* | | 45.318 | 7.03 | −1.255 | 0.38 | 298–1200 | a ** | |
| $Pd_4S$ (s) | −69 | ± 2.1 | 180.69 | ± 0.42 | 100.4 | 48.79 | | | 298–1034 | c | B |
| PdS (s) | −70.7 | ± 6.3 | 56.5 | ± 6.3 | 41.72 | 17.2 | −3.05 | | 298–1243 | c | C |
| $PdS_2$ (s) | −78.3 | ± 12.6 | 88 | ± 13 | 68.58 | 15.78 | −6.57 | | 298–1245 | c | C |
| PdTe (s) | −38 | ± 17 | 89.63 | ± 0.42 | 47.45 | 12.93 | | | 298–993 | c | C |
| Pt (s) | | | 41.64 | ± 0.21 | | | | | | a | |
| | | | 41.64 | | 24.253 | 5.377 | | | 298–2043 | b | |
| | | | 41.636 | | 24.379 | 5.272 | −0.084 | | 298–2042 | d | |
| | | | | | 34.731 | | | | 2042–2200 | d | |
| Pt (g) | 564.9 | ± 1.7 | 192.32 | | | | | | | a | |
| | 565.326 | | 192.32 | | 29.497 | −4.662 | −2.285 | | 298–2200 | d | |
| $PtCl_2$ (s) | −110.9 | ± 12.6 | 130* | | 64.4 | 36.8 | | | 298–700 | c ** | C |
| $PtCl_3$ (s) | −174.1 | ± 12.6 | 148.5* | | 121.4 | | | | 298–500 | c ** | C |
| $PtCl_4$ (s) | −236.8 | ± 16.7 | 205* | ± 21 | 150.6 | | | | 298–400 | c ** | C |
| $PtBr_2$ (s) | −80.3 | ± 12.6 | 155* | | 69 | 23 | | | 298–889 | c ** | C |
| $PtBr_3$ (s) | −127.6 | ± 16.7 | 180* | | 92.9 | 25.1 | | | 298–1000 | c ** | C |
| $PtBr_4$ (s) | −140.6 | ± 16.7 | 251* | | 146.5 | | | | 298–500 | c ** | C |
| $PtI_4$ (s) | −72.8 | ± 16.7 | 281.2* | ± 25.1 | 148.5 | | | | 298–700 | c ** | D |
| $PtO_2$ (g) | 168.6 | ± 10.5 | 255.88 | ± 10.46 | 55.44 | 2.1 | −11.51 | | 298–2000 | a ** | B |
| PtS (s) | −83.1 | ± 2.51 | 55.07 | ± 0.21 | 50 | 10.88 | −8.87 | | 298–1100 | a ** | |
| | −83.1 | | 55.07 | | 41.72 | 17.2 | −3.05 | | 298–1500 | c | B |
| $PtS_2$ (s) | −111.7 | ± 2.9 | 74.69 | ± 0.42 | 58.88 | 29.63 | −1.67 | | 298–1000 | a ** | |
| | −110.5 | | 74.69 | | 68.58 | 15.78 | −6.57 | | 298–1500 | c | B |

| | $\Delta H^{\circ}_{f,298.15}$ | $S^{\circ}_{298.15}$ | $C_p$ (T) a | b | c | d | T °K | Ref. | ACC. |
|---|---|---|---|---|---|---|---|---|---|
| $Pt_5Se_4$ (s) | −265.3 ± 25.1 | 327.2 ± 21.8 | | | | | | a | |
| | −243 | 336.9 | 188.3 | 78.46 | | | 298–1100 | c | C |
| $PtSb_2$ (s) | | | 63.9 | 21.17 | | | 298–900 | a | |
| Rh (s) | | 31.51 ± 0.21 | 20.814 | 13.449 | 0.339 | −2.268 | 298–2233 | c | |
| | | 31.559 | 21.115 | 11.106 | 0.427 | | 298–2237 | d | |
| | | | 50.591 | | | | 2237–2500 | d | |
| Rh (g) | 553.2 ± 6.3 | 185.7 | 23 | 8.6 | | | 298–1900 | a | |
| | 556.957 | 185.737 | 21.508 | 4.235 | −1.565 | | 298–2000 | d | |
| $RhCl_3$ (s) | −275.3 ± 25.1 | 126.8 ± 14.6 | 105.4 | 27.6 | −19.08 | | 298–1200 | c ** | B–C |
| $RhBr_3$ (s) | −209.2 ± 16.7 | 188.3 ± 29.3 | | | | | | a | |
| | −210.9 | 188 | 93.3 | 26.4 | | | 298–1000 | c ** | C |
| $Rh_2O_3$ (s) | −382.9 ± 16.7 | 92* ± 15 | 86.79 | 57.7 | | | 298–1300 | c | B |
| $RhO_2$ (g) | −195.8 ± 8.4 | 264 ± 10 | 58.6 | | −13.8 | | 298–1500 | c ** | B |
| RhO (s) | −90.8 | 54.02 | 41.18 | 22.3 | | | 293–1023 | b | |
| $Rh_2O$ (s) | −95 | 114.7 | 65.24 | 27.07 | | | 292–973 | b | |
| Ru (s) | | 28.54* ± 0.42 | 22.09 | 4.6 | | 1.3 | 298–2000 | a | |
| | | 28.54 | 21.964 | 4.929 | 0.456 | 1.113 | 298–1600 | c | |
| | | | 15.562 | 10.775 | | | 1600–2523 | c | |
| | | 28.618 | 19.696 | 10.076 | 1.138 | | 298–2607 | d | |
| | | | 52 | | | | 2607–3000 | d | |
| Ru (g) | 651.5 ± 8.4 | 186.42 | | | | | | a | |
| | 642.739 | 186.419 | 24.35 | | −2.967 | | 298–3000 | d | |
| $RuF_5$ (s) | −892.97 ± 10.46 | 161.1* ± 14.6 | 163.2 | | | | 298–374 | c ** | C |
| $RuCl_3$ (s) | −253.2 ± 12.6 | 127.6 ± 10.5 | 115.1 | | | | 298–1000 | c ** | B–C |
| $RuCl_3$ (g) | 56.1 ± 16.7 | 398 ± 21 | 56.91 | 7.66 | −3.01 | | 298–1500 | c ** | B–C |
| $RuCl_4$ (g) | −93.3 ± 16.7 | 374.5 ± 20.9 | 95.8 | | −10.5 | | 298–1500 | c ** | C |
| $RuO_2$ (s) | −304.6 ± 3.3 | 60.7 ± 4.6 | 69.9 | 10.5 | −14.85 | | 298–1300 | c ** | B–C |
| | −305.05 | 58.165 | 76.639 | 10.821 | −24.124 | | 298–1200 | d | |
| $RuO_3$ (g) | −78.3 ± 11.7 | 276 ± 13 | 78.7 | 2.5 | −17.78 | | 298–1900 | c ** | C |
| $RuO_4$ (g) | −184 ± 8 | 290.7 ± 0.42 | 101.81 | 3.05 | −24.1 | | 298–2000 | c ** | B–C |
| $RuS_2$ (s) | 205.9 ± 20.9 | 54 | 68.54 | 11.84 | −8.79 | | 298–1500 | c | B–C |
| $RuSe_2$ (s) | −161.5 | 81.6 | 73.65 | 11.09 | −4.2 | | 298–1500 | c | |

## Chapter 5

# KOMATIITE-ASSOCIATED NICKEL SULFIDE DEPOSITS

C. M. Lesher

### INTRODUCTION

Nickel sulfide ores associated with komatiites in Archean greenstone belts are the most important examples of magmatic mineralization in volcanic rocks. They contain about 25 percent of the world total identified nickel resource in deposits with ⩾0.8 percent Ni, as well as significant amounts of copper and platinum-group elements (PGE). Experimental studies suggest that komatiites erupted at very high temperatures, consequently, these deposits provide critical information concerning sulfide segregation processes in high temperature volcanic systems.

The general geological characteristics of these deposits, compositions of sulfide ores, and resource information have been discussed in detail by Groves and Hudson (1981), Marston et al. (1981), Naldrett (1981), and Ross and Travis (1981). The emphasis here will be on the physical volcanology of the deposits and on some of the more contentious problems of sulfide genesis and ore localization.

### NOMENCLATURE AND CLASSIFICATION

Following the recommendation of Arndt and Nisbet (1982), *komatiites* are defined here as ultramafic volcanic rocks with volatile-free magnesium contents ranging between 18 and 32% MgO. Petrogenetically related basalts and olivine cumulate rocks are identified with the prefix *komatiitic* (Table 5.1). All Archean komatiitic rocks are metamorphosed to some degree, but pseudomorphous and palimpsest textures are commonly well preserved, so igneous nomenclature will be used throughout this chapter.

To simplify discussion, komatiite-associated nickel sulfide deposits are divided into four classes, comprising two lithological groups and two ore distribution types:

I) **Komatiitic Peridotite-Hosted Deposits**
   A) *Stratiform Deposits:* small (0.5–5 × $10^6$ tonnes), high grade (2–4% Ni) deposits of massive/matrix/disseminated sulfides at the base of komatiitic peridotites
   B) *Strata-Bound Deposits:* medium-sized (5–30 × $10^6$ tonnes), low grade (<1% Ni) deposits of disseminated/blebby sulfides within komatiitic peridotites

II) **Komatiitic Dunite-Hosted Deposits**
   A) *Stratiform Deposits:* small to medium-sized (1–40 × $10^6$ tonnes), high grade (1.5–3.5% Ni) deposits of massive/matrix/disseminated sulfides at the base of komatiitic dunites
   B) *Strata-Bound Deposits:* large (up to 300 × $10^6$ tonnes), low grade (<1% Ni) deposits of fine disseminated sulfides within komatiitic dunites

This scheme facilitates consideration of the deposits in terms of either host rock (group I vs. group II) or ore distribution (type A vs. type B) without implying any particular volcanic setting. Groups I and II are analogous to the "volcanic peridotite" and "intrusive dunite" associations of Marston et al. (1981) and Ross and Travis (1981), and classes IA, IIA, and IIB are analogous to the three groups defined by Naldrett (1981).

Of particular significance is the reclassification of Agnew, previously considered to be an anomalously high grade (2.05% Ni) group II deposit (Marston et al., 1981; Billington, 1984), as a giant (45 × $10^6$ tonnes) combination IA/IIB deposit. Barnes et al. (1988a) demonstrate that the deposit is hosted by a komatiitic peridotite flow that predates emplacement of a transgressive komatiitic dunite flow.

The characteristics of the host rocks for these deposits are summarized in Table 5.2. A list of the better characterized deposits, tentative classifications, and most important geological references is given in Table 5.3. Unless noted otherwise, descriptive data for individual deposits are taken from those references.

### AGE AND DISTRIBUTION

Komatiite-associated nickel sulfide deposits occur in Archean greenstone belts in Canada, Western Australia, and Zimbabwe. Most occur in "younger" Archean (3.0–2.7 Ga) terrains: the Norseman–Wiluna and Southern Cross–Forrestania greenstone belts in Western Australia (Marston et al., 1981; Groves and Lesher, 1982; Marston, 1984), the Abitibi greenstone belt in Ontario (Naldrett and Gasparrini, 1971; Coad, 1979), and the Zimbabwe Province in Southern Africa (Williams, 1979; Hammerbeck, 1984). No significant deposits have been reported in "older" Archean (3.5–3.3 Ga) terrains (e.g. Barberton, South Africa; east Pilbara, Western Australia; Sebakwian, Zimbabwe), and none have been described in greenstone belts in the Brazilian, Indian, or Baltic Shields. Analogous deposits are associated with komatiitic basalts in lower Proterozoic mobile belts in Canada (Kilburn et al., 1969; Peredery et al., 1982).

Komatiite-associated nickel sulfide deposits in Western Australia exhibit a markedly heterogeneous distribution on at least three scales within the Yilgarn Block (Figs. 5.1 and 5.2):

1) All deposits are concentrated in the Eastern Goldfields Province, particularly within the Norseman–Wiluna greenstone belt, whereas significant deposits are absent in the Murchison Province and northern part of the Southern Cross Province, reflecting somewhat different tectonic settings for those provinces (Groves et al., 1984).

TABLE 5.1—Nomenclature for komatiitic rocks.

| Rock | Composition[1] | Mesostasis texture[2] | Phenocryst/cumulus texture[2,3] |
|---|---|---|---|
| **Aphyric** | | | |
| komatiitic basalt | 10–18% MgO | glassy with random acicular or radiating-plumose clinopyroxene | |
| komatiite | 18–32% MgO | glassy with random acicular clinopyroxene, random platy olivine, and dendritic chromite | |
| **Porphyritic** | | | |
| porphyritic komatiitic basalt | 10–18% MgO | glassy | polyhedral-hopper olivine or acicular clinopyroxene |
| pyroxene spinifex-textured komatiitic basalt | 15–18% MgO | glassy | parallel acicular clinopyroxene |
| olivine spinifex-textured komatiite | 18–32% MgO | glassy with random acicular clinopyroxene and dendritic chromite | random or parallel platy olivine |
| olivine porphyritic komatiite | 28–38% MgO, 25–40% olivine | glassy with random acicular clinopyroxene and intercumulus chromite | polyhedral or hopper olivine |
| **Cumulate** | | | |
| komatiitic peridotite | 34–48% MgO, 40–90% olivine | glassy with random acicular clinopyroxene and euhedral-subhedral intercumulus chromite | polyhedral (orthocumulate or mesocumulate) or branching (crescumulate) olivine |
| komatiitic dunite | 44–50% MgO, >90% olivine | lobate intercumulus chromite or clinopyroxene | polyhedral-anhedral (adcumulate) olivine |

[1]volatile free; [2]pseudomorphed by metamorphic phases; [3]terminology after C. H. Donaldson (1982).

TABLE 5.2—General characteristics of komatiitic peridotite (Group I) and komatiitic dunite (Group II) hosts to nickel sulfide mineralization in Archean greenstone belts.

| | Komatiitic peridotite | Komatiitic dunite |
|---|---|---|
| **Internal Structure** | | |
| Form | both: lenticular cross section, highly elongate | |
| Thickness | thick, generally 30–100 m | very thick, commonly 300–1000 m |
| Layering | both: fine-scale cryptic-rhythmic (olivine texture, olivine composition, sulfide content) | |
| Texture | orthocumulate to mesocumulate, some crescumulate | mesocumulate to adcumulate |
| Margin(s) | spinifex-textured or aphyric | poikilitic or recrystallized |
| Grain size | fine-medium, generally 0.5–5 mm | medium-coarse, generally 1–10 mm |
| **Composition** | | |
| Whole rock | 38–45% MgO* | 45–50% MgO* |
| Modal olivine | 50–75% | 75–95%, normally >90% |
| Olivine | both: 90–95 mole% forsterite | |
| Parent | both: 28–32% MgO* | |
| Chilled margins | 16–26% MgO* | 20% MgO* |

*Volatile-free.
Compositional data for chilled margins are limited.

2) Class IA deposits cluster in the southern part of the Norseman–Wiluna belt, class IIA deposits cluster in the southern part of the Southern Cross–Forrestania belt, and class IIB deposits cluster in the northern part of the Norseman–Wiluna belt, reflecting somewhat different volcanic settings for those domains.
3) Class IIA and IIB deposits occur as isolated pods, exposed discontinuously along narrow fold belts, and class IA deposits occur in groups around discrete structural highs, reflecting different styles of deformation and different levels of exposure.

A similarly heterogeneous distribution of deposits is observed in Canada, where class IA deposits are concentrated in the southwestern part of the Abitibi belt, mainly around the flanks of granite-cored domes (Coad, 1979; Green and Naldrett, 1981), a few class IA and IIB deposits occur in the central part of the Abitibi belt (Naldrett and Gasparrini, 1971; Duke, 1986a), and analogs of class IIA deposits are concentrated in the Thompson and Cape Smith–Wakeham Bay fold belts (Kilburn et al., 1969; Peredery et al., 1982).

The komatiite-associated nickel sulfide deposits in Zimbabwe are more widely distributed (Williams, 1979; Hammerbeck, 1984); most appear to be group I deposits.

## TECTONIC SETTING

The tectonic settings of Archean greenstone belts are uncertain, but they are most likely intracratonic rift zones (e.g. Groves et al., 1984) or extensional zones within wrench fault systems (e.g. Ludden et al., 1986). The presence of multiple horizons of interflow sediments, multiple cycles of

TABLE 5.3—Classification and tentative volcanic settings of Archean komatiite-associated nickel sulfide deposits and Proterozoic analogs.

| District/Deposit | Class | Setting | Important geological references |
|---|---|---|---|
| **Pilbara Block, Western Australia** | | | |
| Ruth Well | IA | ? | Tomich (1974); Nisbet & Chinner (1981) |
| **Eastern Goldfields Province, Yilgarn Block, Western Australia** | | | |
| *Norseman–Wiluna Greenstone Belt, Western Australia* | | | |
| Honeymoon Well | IIB | D | Donaldson & Bromley (1981) |
| Betheno | IIB | D | Groves & Keays (1979) |
| Mt. Keith | IIB | D | Burt & Sheppy (1975); Groves & Keays (1979) |
| Six Mile–Goliath | IIB | D | Naldrett & Turner (1977); Hill (1982); Hill et al. (1987) |
| Black Swan | IIB | D | Groves et al. (1974) |
| Agnew | IA/IIB | D | Billington (1984); Hill et al. (1987); Barnes et al. (1988a–c) |
| Scotia | IA | P | Stolz & Nesbitt (1981); Page & Schmulian (1981) |
| Nepean | IA | D | Barrett et al. (1976); Sanders (1982) |
| Kambalda district | IA | D | Woodall & Travis (1969); Ewers & Hudson (1972); Keele & Nickel (1974); Ross & Hopkins (1975); Marston & Kay (1980); Gresham & Loftus-Hills (1981); Bavinton (1981); Lesher et al. (1984); Paterson et al. (1984); Gresham (1986); Groves et al. (1986); Cowden (1988); Evans et al. (1988); Frost & Groves (1988) |
| Widgiemooltha district | IA | D | McQueen (1981a, b) |
| Windarra district | IA | D | Schmulian (1982, 1984) |
| Forrestania district | IIA/IIB | D | Porter & McKay (1981) |
| General references | | | Marston et al. (1981); Groves & Lesher (1982); Marston (1984) |
| **Superior Province, Canada** | | | |
| *Abitibi Greenstone Belt* | | | |
| Alexo | IA | D | Naldrett (1966); Barnes (1985); Barnes & Naldrett (1987) |
| Dundonald | IA | D | Muir & Comba (1979); Barnes (1985); Barnes & Naldrett (1987) |
| Langmuir | IA | P | Green & Naldrett (1981) |
| Hart | IA | D | Barnes (1985); Barnes & Naldrett (1987) |
| Redstone | IA | D | Robinson & Hutchinson (1982) |
| Texmont | IA | D | Barnes (1985); Barnes & Naldrett (1987) |
| Marbridge | IA | D | Naldrett & Gasparrini (1971) |
| Dumont | IIB | S | Eckstrand (1975); Duke (1986) |
| General references | | | Naldrett & Gasparrini (1971); Coad (1979) |
| **Zimbabwe Province, Zimbabwe** | | | |
| Damba district | IB | P | Viljoen & Bernasconi (1979) |
| Shangani | IA | C | Viljoen et al. (1976); Viljoen & Bernasconi (1979) |
| Trojan | IA | D | Chimimba (1984) |
| General references | | | Williams (1979); Hammerbeck (1984) |
| **Cape Smith–Wakeham Bay Fold Belt, Quebec, Canada** | | | |
| Katiniq | IIA | D | Kilburn et al. (1969); Barnes et al. (1982) |
| **Thompson–Wabowden Fold Belt, Manitoba, Canada** | | | |
| Thompson district | (IIA) | ? | Peredery (1979); Peredery et al. (1982) |

I = komatiitic peridotite-hosted, II = komatiitic dunite-hosted, A = stratiform, B = strata-bound, S = subvolcanic, C = central volcanic, P = proximal volcanic, D = distal volcanic.

ultramafic/mafic/felsic volcanism and clastic sedimentation, and absence of sheeted dikes indicates that they do not represent oceanic crust.

Groves and Batt (1984) subdivided the greenstone belts in Western Australia into two types that differ in terms of tectonic setting and degree of mineralization: *platform-phase greenstones* are characterized by equi-spaced granitoid batholiths with intervening stellate greenstone belts, a coherent volcanic stratigraphy, abundant komatiitic basalts, and shallow water volcaniclastic rocks or oxide facies iron-formation. They are interpreted to have formed in relatively shallow water under conditions of low crustal extension. In contrast, *rift-phase greenstones* are characterized by elongate granitoid domes, linear tectonic patterns, complex volcanic stratigraphy, abundant komatiites (including komatiitic peridotites and komatiitic dunites), and sulfidic shales or cherts. They are interpreted to have formed in relatively deep water under conditions of high crustal extension, and appear to be superimposed on older platform-phase greenstones. Most class IA and IIB deposits in Western Australia (Fig. 5.1) occur within rift-phase greenstones (e.g. Kambalda–Widgiemooltha and Agnew–Mt. Keith districts), but some occur in plat-

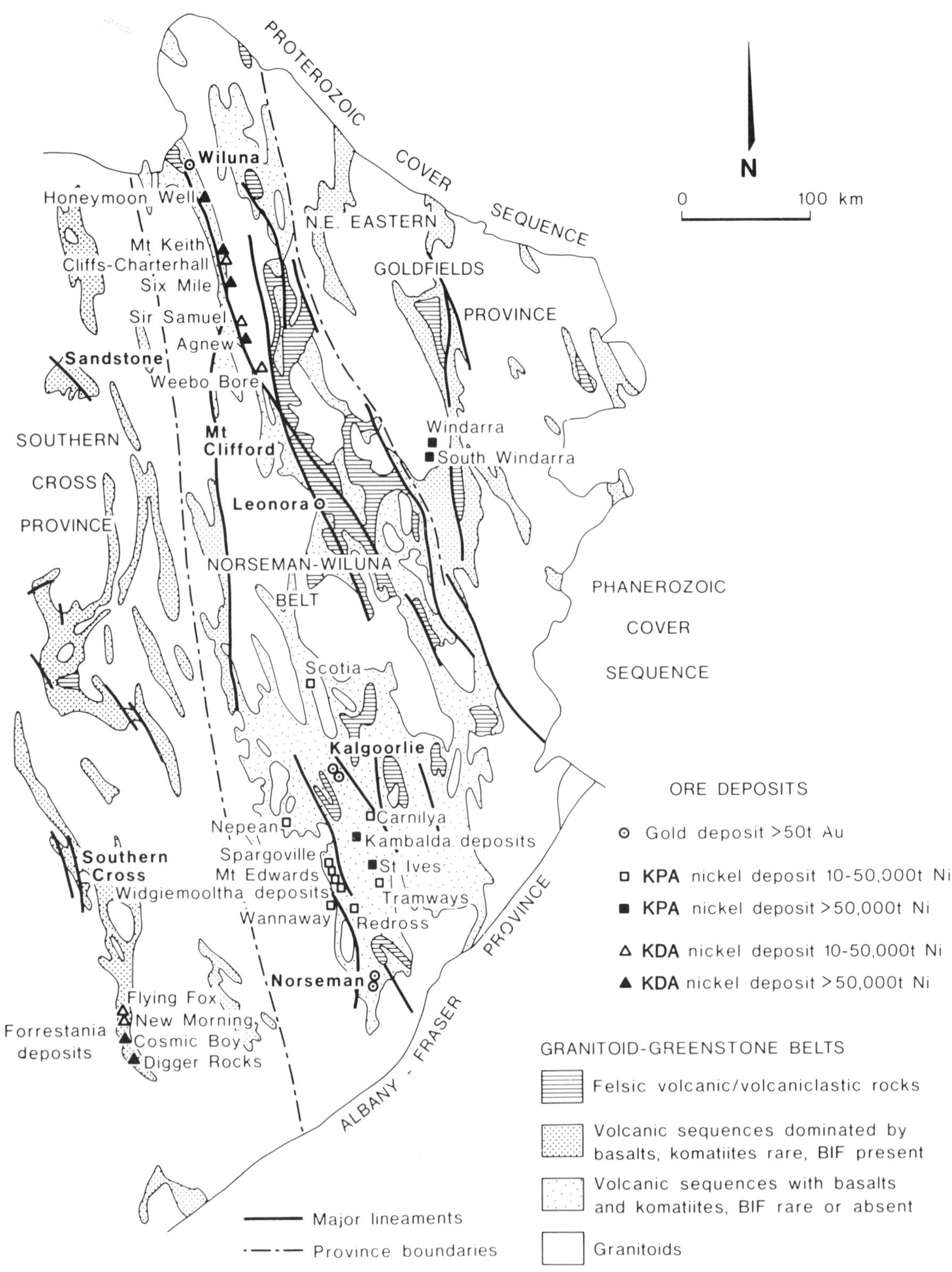


FIGURE 5.1—Solid rock geology of eastern part of Yilgarn Block, Western Australia, illustrating setting of major nickel deposits in terms of lithofacies, structure, and major subdivisions of granitoid-greenstone terrains (after Groves et al., 1984).

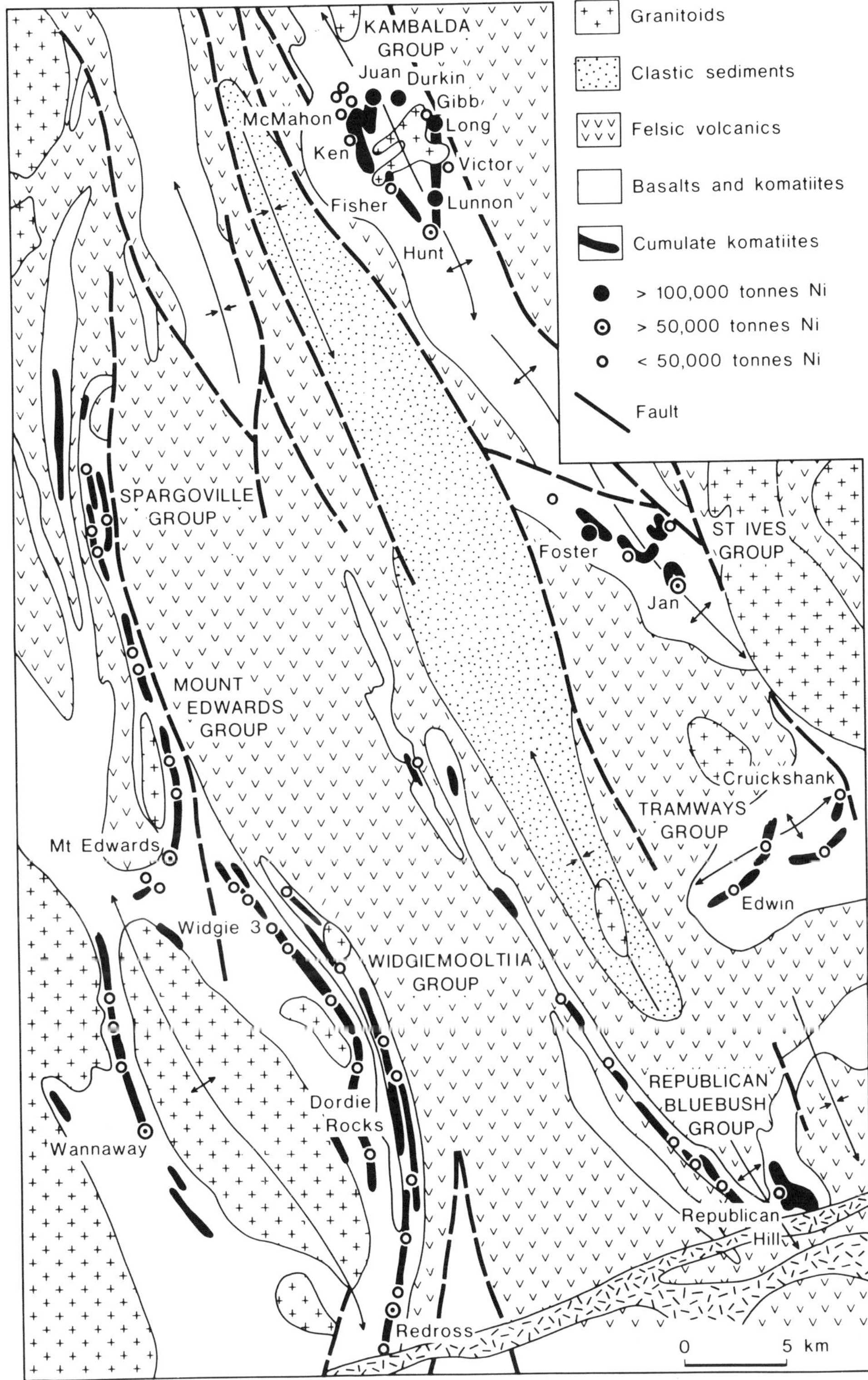


FIGURE 5.2—Solid rock geology of Kambalda–St. Ives–Tramways–Widgiemooltha area in the southern part of the Norseman–Wiluna greenstone belt, Western Australia, showing lithological associations and distribution of komatiitic peridotite-hosted nickel deposits (after Groves et al., 1984).

form-phase greenstones (e.g. Windarra and Forrestania districts).

Analogous subdivisions have been defined in the Abitibi greenstone belt by Ludden et al. (1986); komatiite-associated nickel sulfide deposits and large volcanic-associated massive Cu–Zn sulfide deposits (Franklin et al., 1981) are confined to a southern volcanic zone, interpreted as a series of fault-bound rift basins superimposed on a volcanic arc complex.

Class IA and IB deposits in Zimbabwe are also interpreted to have formed in an intracratonic rift zone, superimposed on an older volcanic cycle (Wilson, 1979; Hammerbeck, 1984).

## STRATIGRAPHIC SETTING

Stratigraphic studies are hampered by poor exposure, metamorphism, and complex polyphase deformation. Only at Kambalda (Figs. 5.3 and 5.4), Agnew (Fig. 5.5), and Scotia (Fig. 5.6) have systematic diamond drilling, extensive underground mapping, and detailed petrological research established the nature of the stratigraphic sequence in any detail. Most other deposits are characterized only by reconnaissance surface mapping, limited diamond drilling, and restricted underground mapping.

### Regional stratigraphy

Despite structural complications and probable facies variations discussed below, mineralized komatiite sequences appear to be correlative over distances of hundreds of kilometers, possibly over entire greenstone provinces. The class IA deposits in the southern part of the Norseman–Wiluna belt (Fig. 5.2), for example, are all interpreted to occur at the same stratigraphic level (Gemuts and Theron; 1975; Gresham and Loftus–Hills, 1981). This correlation can probably also be extended to Nepean and Scotia (Fig. 5.1). The lithologically identical, albeit only weakly mineralized, Yilmia–Bluebush sequence (Fig. 5.2) was considered to be a younger cycle by Gemuts and Theron (1975) and Gresham and Loftus–Hills (1981), but is more likely a faulted equivalent of the Kambalda–Widgiemooltha sequences (Martyn, 1987).

The temporal and stratigraphic relationships between these sequences and those that host class IIB deposits in the northern part of the Norseman Wiluna belt (Fig. 5.1) is uncertain. They overlie different footwall lithologies and have been interpreted to have been emplaced at different stages of evolution along the greenstone belt (Marston and Groves, 1981) or contemporaneously into different tectonic settings within or across the rift zone (Groves et al., 1984), but they may also be distal facies equivalents.

Similar stratigraphic correlations have been made in Canada and Zimbabwe. Class IA deposits in the Timmins region of the Abitibi greenstone belt, for example, are all assigned to the lowermost, ultramafic unit of the Upper Supergroup (Green and Naldrett, 1981; Jensen and Pyke, 1982), whereas komatiites at the base of the Lower Supergroup are apparently barren. Note that the classic komatiite locality in Munro Township (Arndt et al., 1977) contains no significant nickel sulfide mineralization, but is correlative with the mineralized sequences.

Group I deposits in Zimbabwe occur at, or close to, the base of the Upper Greenstones of the Bulawayan Group (Hammerbeck, 1984). Another classic komatiite locality at Belingwe (Nisbet et al., 1977, 1982) is also barren, but correlative with the mineralized sequences.

### Local stratigraphy

Many mineralized komatiite sequences appear to grade systematically upwards from thick, komatiitic peridotites or komatiitic dunites at the base through thinner, aphyric and spinifex-textured komatiite flows to massive or pillowed komatiitic basalt flows at the top (Figs. 5.3–14). Komatiitic peridotite flows in Western Australia are commonly separated by interflow sediments, except near ore zones. Thus, the thickness of komatiite flows, time between eruptions, degree of olivine enrichment, and magnesium content of liquids all decrease upwards. This probably reflects an evolutionary change from voluminous, episodic eruptions to smaller, more continuous eruptions.

There are also variations in the composition and texture of komatiitic lavas across mineralized volcanic piles. Komatiitic peridotites grade laterally into porphyritic komatiites, komatiitic dunites grade laterally into komatiitic peridotites, and barren komatiites overlying ore zones may contain more cumulate flows and may be better compositionally and texturally differentiated than those further from mineralization. At Kambalda, these stratigraphic variations define elongate prisms of rock parallel to and overlying linear ore shoots (Figs. 5.3 and 5.4), indicating a strong volcanic control on ore localization. This has been attributed previously to ponding of lavas adjacent to proximal feeding fissures Ross and Hopkins, 1975; Gresham and Loftus–Hills, 1981; Gresham, 1986), but the absence of feeders and other petrographic and geochemical constraints discussed below suggest that it represents channelization of flows from a distal eruptive site (Lesher et al., 1981, 1984; Cowden, 1988).

In contrast, barren komatiite sequences in Barberton, South Africa (Viljoen and Viljoen, 1969; 1982), Belingwe, Zimbabwe (Nisbet et al., 1977; 1982), and Munro Township, Ontario (Arndt et al., 1977; Jensen and Pyke, 1982) contain few komatiitic peridotites or dunites, and commonly comprise interlayered komatiite and basaltic komatiite. They may represent less voluminous eruptions, or more distal portions of komatiite lava piles.

### Komatiitic peridotites and dunites

Mineralized komatiitic peridotites are anomalously thick, highly magnesian components of komatiite sequences, but are texturally and compositionally gradational with overlying and adjacent orthocumulate, porphyritic, and spinifex-textured komatiites. Some host units are highly elongate, and confined to a large degree within footwall embayments (e.g. Figs. 5.6, 5.12 and 5.13). Others appear to be more sheet-like in form (e.g Figs. 5.7, 5.9 and 5.14). Some correlate with a single flow unit along strike (Fig. 5.13); others appear to correlate with multiple flows along strike (Fig. 5.12). The host units have been interpreted previously as i) phenocryst-enriched lava flows (Ross and Hopkins, 1975; Gresham and Loftus–Hills, 1981; Gresham, 1986), ii) overflowing lava ponds (Naldrett, 1973), and iii) ponded lava flows (Ussleman et al., 1979; Stolz and Nesbitt, 1981; Naldrett and Campbell, 1982), but petrographic and geochemical constraints dis-

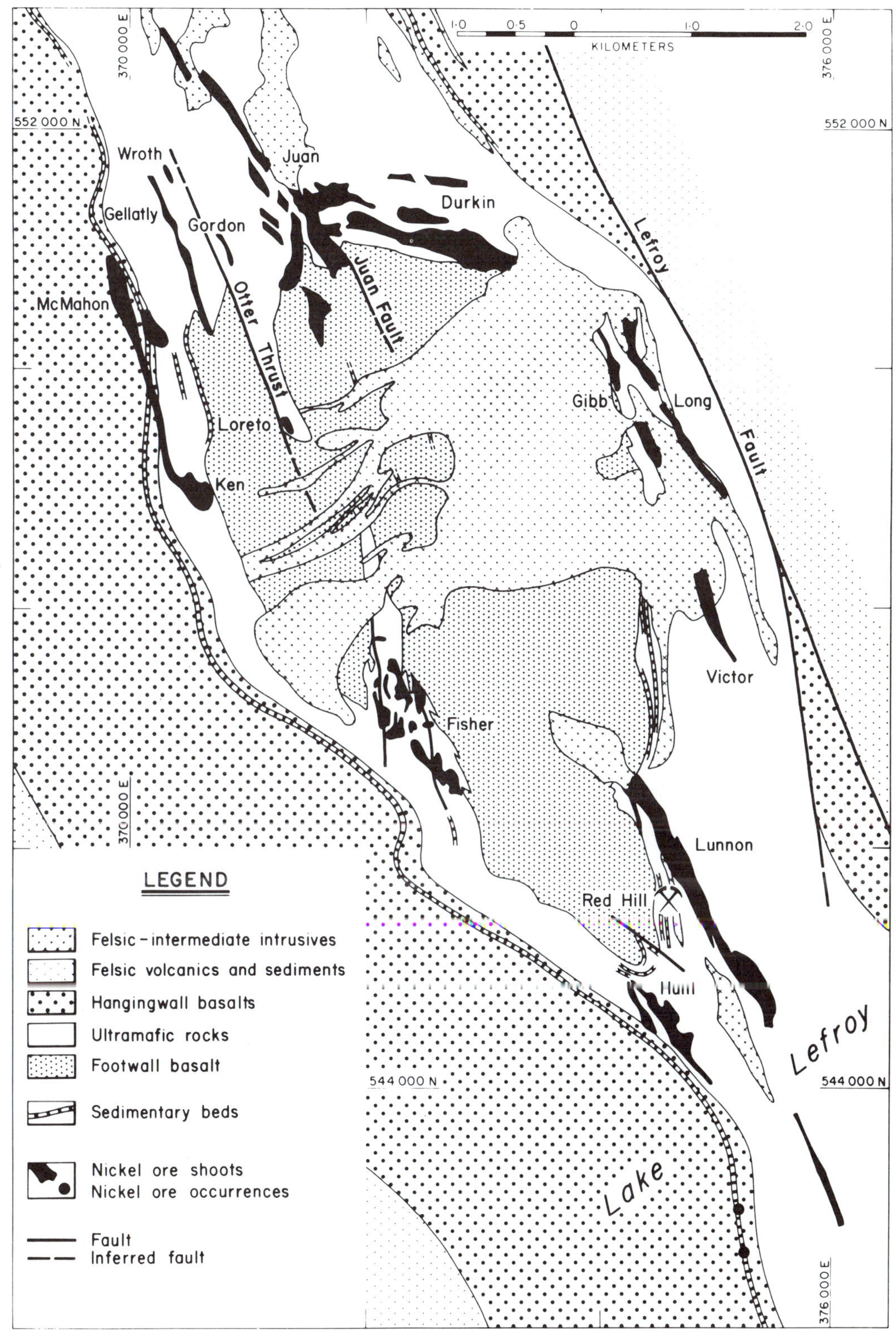


FIGURE 5.3—Solid rock geology of Kambalda dome, Western Australia, showing ore shoots in horizontal projection (after Gresham and Loftus-Hills, 1981).

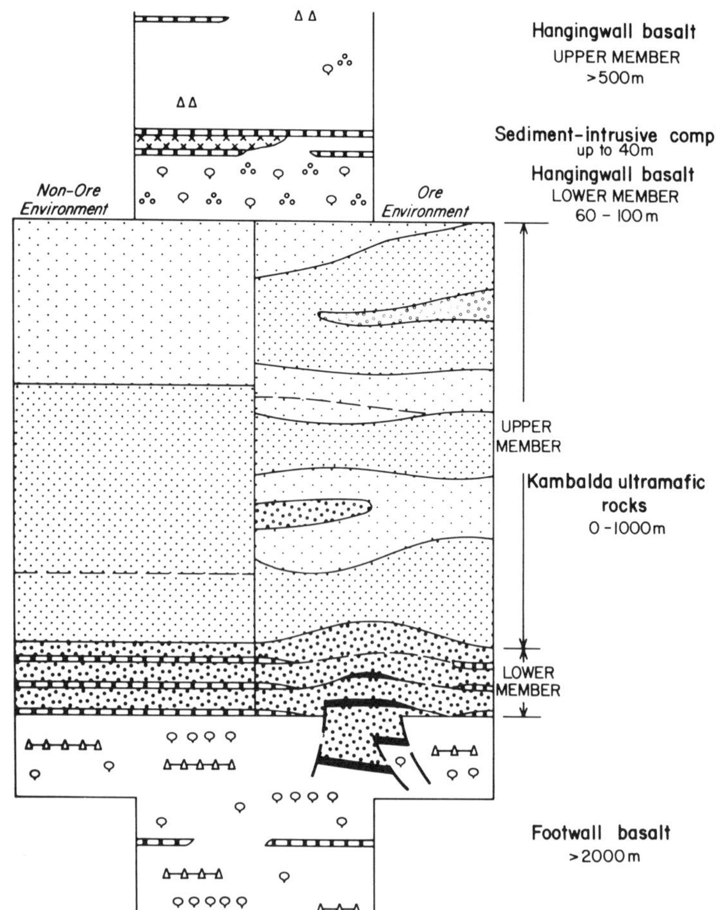


FIGURE 5.4—Diagramatic stratigraphic column of volcanic sequence at the Kambalda dome illustrating differences between ore and non-ore environments (after Gresham and Loftus-Hills, 1981). Ornamentation used in all subsequent figures.

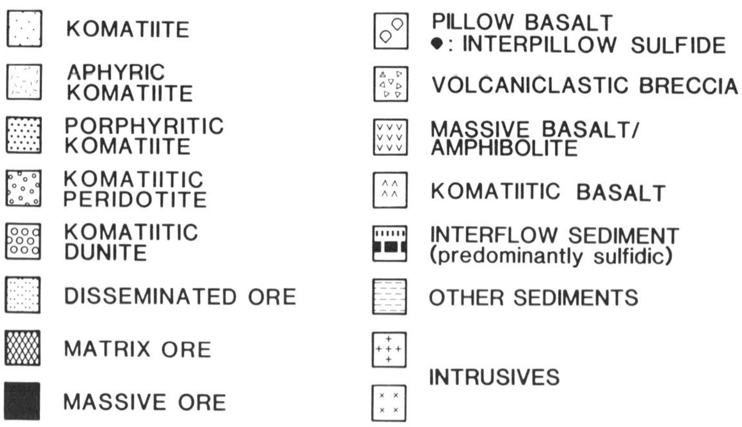

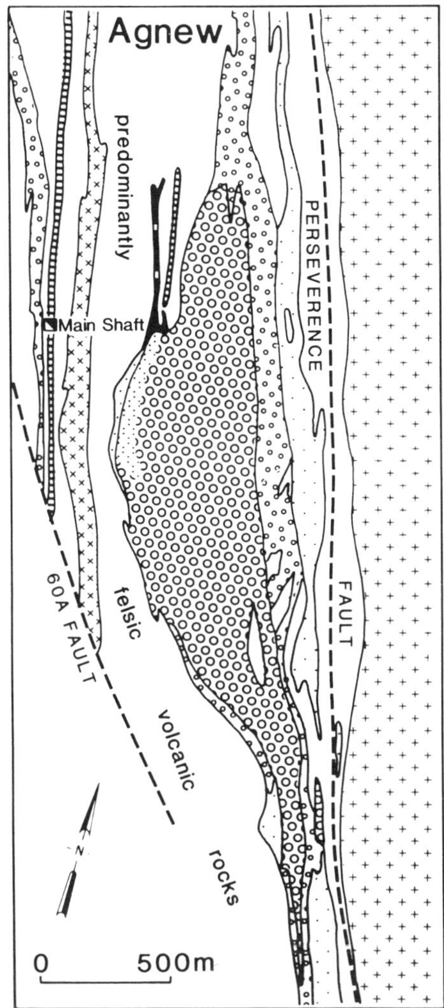


FIGURE 5.5—Solid rock geology of Agnew mine area, Western Australia (after Barnes et al., 1988a). Ornamentation as for Figure 5.4.

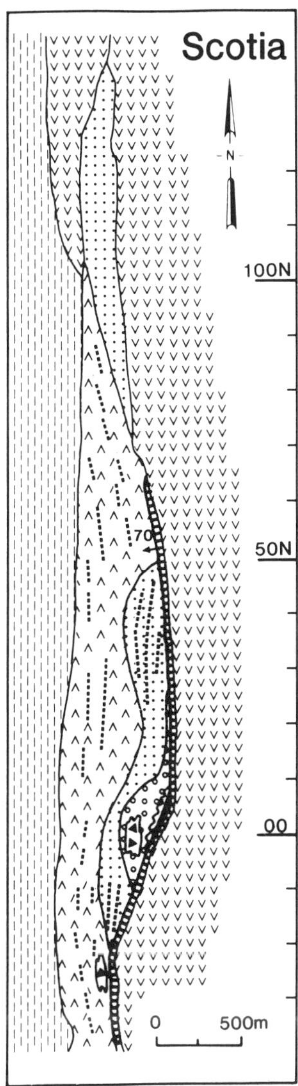


FIGURE 5.6—Solid rock geology of Scotia mine area, Western Australia (after Page and Schmulian, 1981). Ornamentation as for Figure 5.4.

cussed below indicate that they are probably dynamic lava channels (Lesher et al., 1984; Lesher and Groves, 1986; Barnes and Naldrett, 1987; Cowden, 1988; see also Barnes et al., 1983; Barnes, 1985).

Mineralized komatiitic dunites in Western Australia are lenticular in cross section; thicker and more magnesian in central parts, thinner and less magnesian at margins (e.g. Figs. 5.5 and 5.8). They occupy thickened zones at or very near the base of komatiite sequences, grade along strike into komatiitic peridotites and komatiites, and are conformable with overlying komatiites, komatiitic basalts, and tholeiitic basalts. They have previously been interpreted as dikes (Binns et al., 1977) or subvolcanic sills (Naldrett and Turner, 1977), but the above stratigraphic relationships and similarities with komatiitic peridotites suggest that many are dynamic lava channels (Donaldson et al., 1986; Hill et al., 1987; Barnes et al., 1988c).

The Katiniq deposit has been interpreted previously as a subvolcanic sill (Barnes et al., 1982), but stratigraphic relationships with overlying komatiitic basalts suggest that it is probably also extrusive (CML., unpub. data). Spinifex-texture has been reported along the upper margin (S.J. Barnes and S.J. Barnes, pers. comm., 1988).

### Footwall rocks

The principal footwall lithology varies from district to district, including i) tholeiitic basalt in the Kambalda–Widgiemooltha district, ii) mafic-felsic volcaniclastic rock in the Mt. Keith–Agnew district, iii) sulfide facies iron-formation in the Windarra district, iv) oxide facies iron-formation in the Forrestania district, v) andesitic volcanic rock in the Timmins

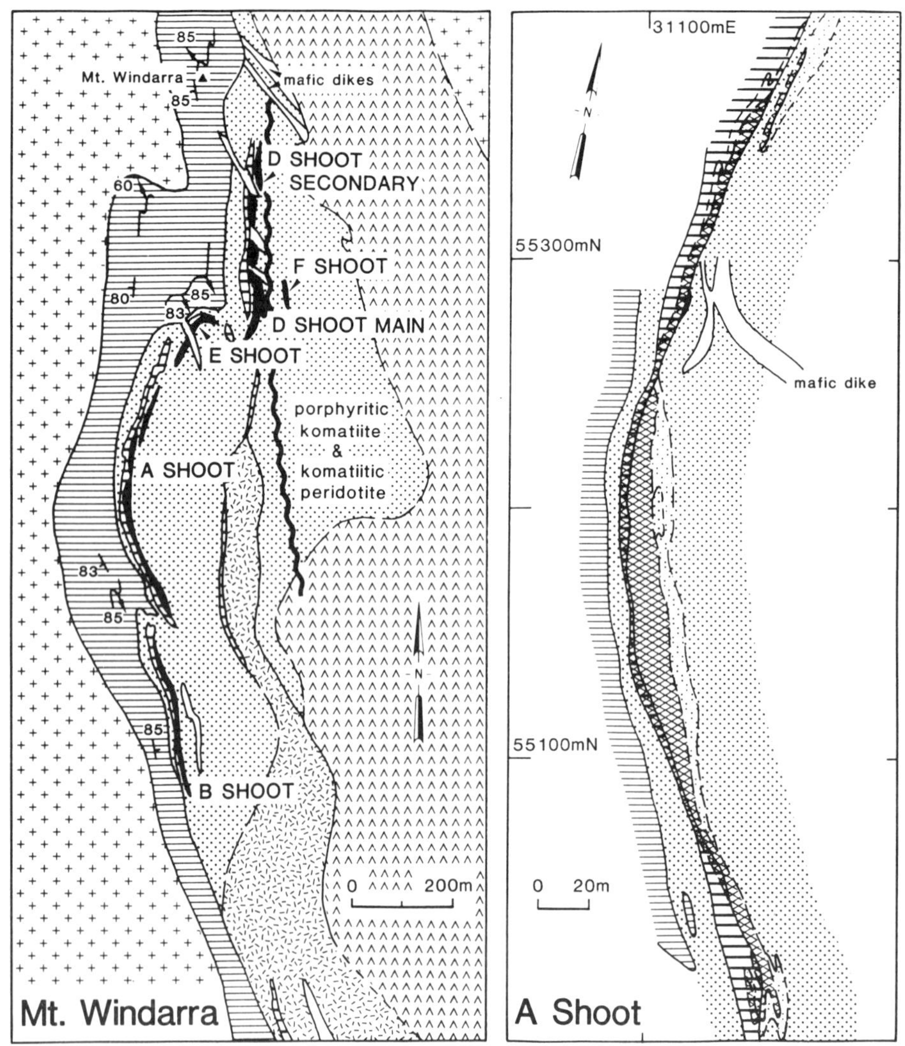


FIGURE 5.7—Solid rock geology of Mt. Windarra mine area (**A**) and A shoot (**B**), Windarra district, Western Australia (after Marston, 1984). Ornamentation as for Figure 5.4.

district, and vi) quartzo-feldspathic sediment in the Damba–Shangani region. This indicates that these deposits formed in somewhat different tectonic settings, at somewhat different stages in the evolution of greenstone belts, or in somewhat different environments within the rift zones.

### Interflow sediments

Most class IA and IIA deposits overlie or are stratigraphically correlative with sulfidic sediments. Class IB and IIB deposits also occur in sequences that contain sulfidic sediments, but stratigraphic relationships are more obscure. The nature and composition of the sediments varies from siliceous and carbonaceous, sulfidic shales in the Norseman–Wiluna belt to sulfidic cherts in the Forrestania, Windarra, Timmins, and Trojan districts. Although oxide facies iron-formation forms the footwall to several deposits in the Forrestania district, komatiites are intercalated with sulfidic cherts.

Ore-sediment relationships vary considerably, even between deposits in the same district. Ores may i) grade laterally into sediments (e.g. Redstone; parts of Juan, Lunnon, and Jan shoots, Kambalda: e.g. Fig. 5.13), ii) directly overlie or lap onto sediments which may be altered, thinned, or discontinuous beneath the ore zone (e.g. Langmuir; Scotia: Fig. 5.6; Mt. Edwards; Wannaway; Windarra: Fig. 5.7; Trojan; Foster, Ken, McMahon, and Cruickshank shoots, Kambalda; Digger Rocks and New Morning, Forrestania: Figs. 5.8a–b), or iii) occur along strike from sediments, separated by a zone of barren contact (e.g. Nepean; Miriam and Widgiemooltha 3; most shoots at Kambalda: e.g. Figs. 5.4, 5.12, and 5.14).

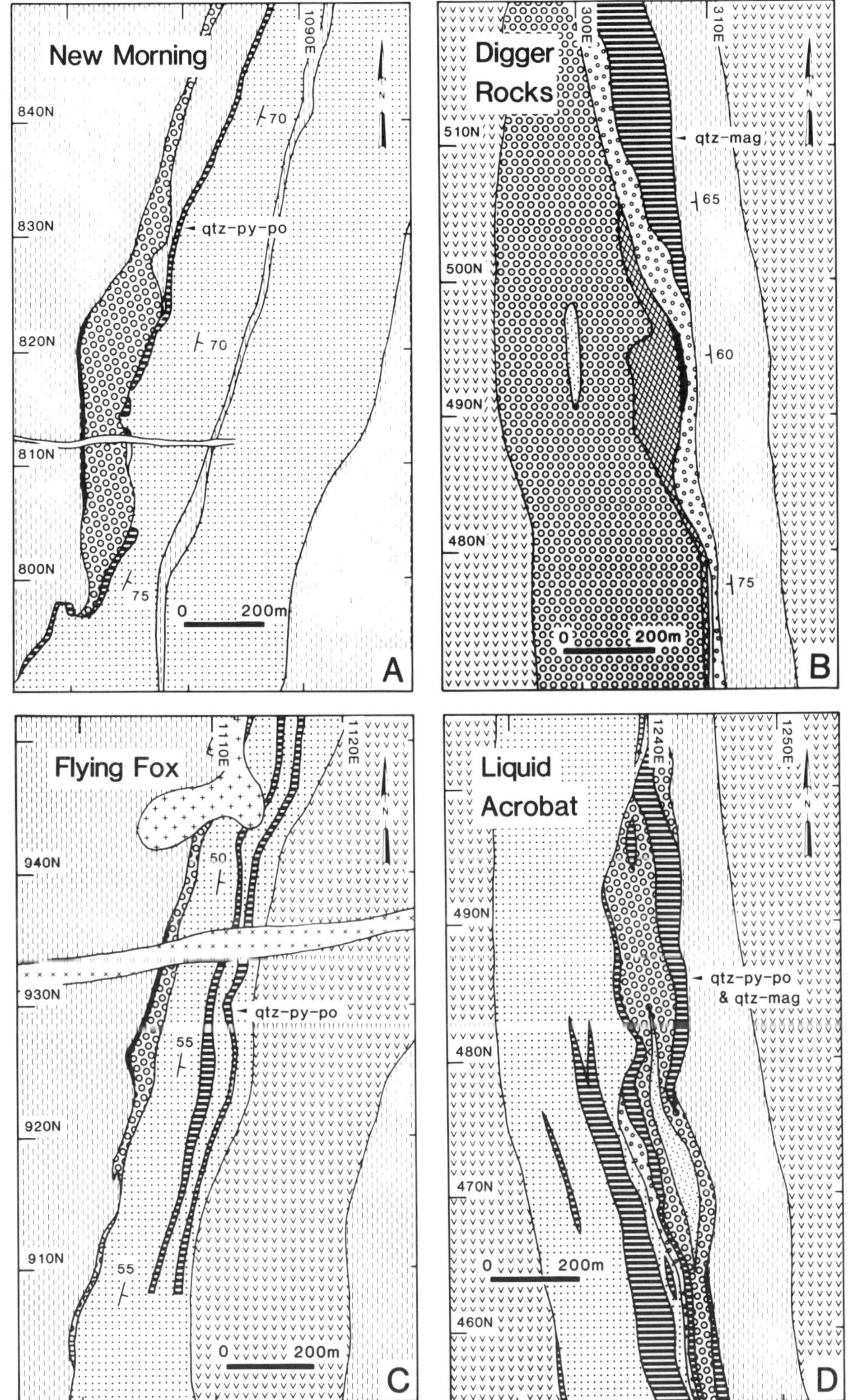


FIGURE 5.8—Solid rock geology of New Morning (**A**), Digger Rocks (**B**), Flying Fox (**C**), and Liquid Acrobat (**D**) deposits, Forrestania district, Western Australia (after Porter and McKay, 1981). Ornamentation as for Figure 5.4.

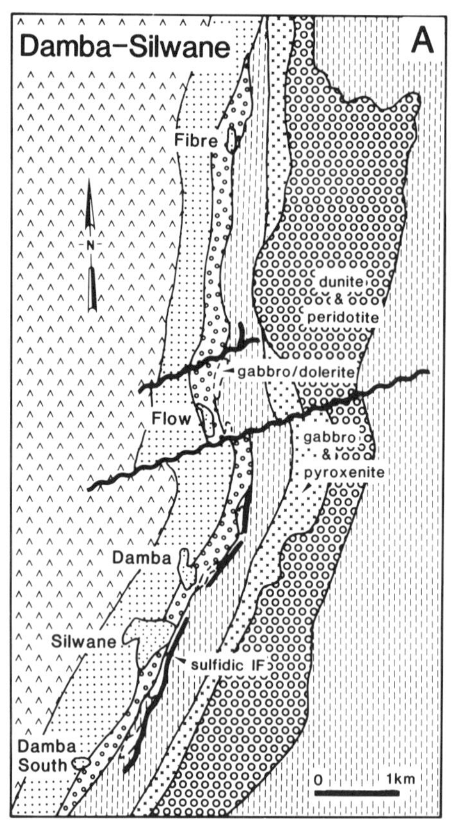


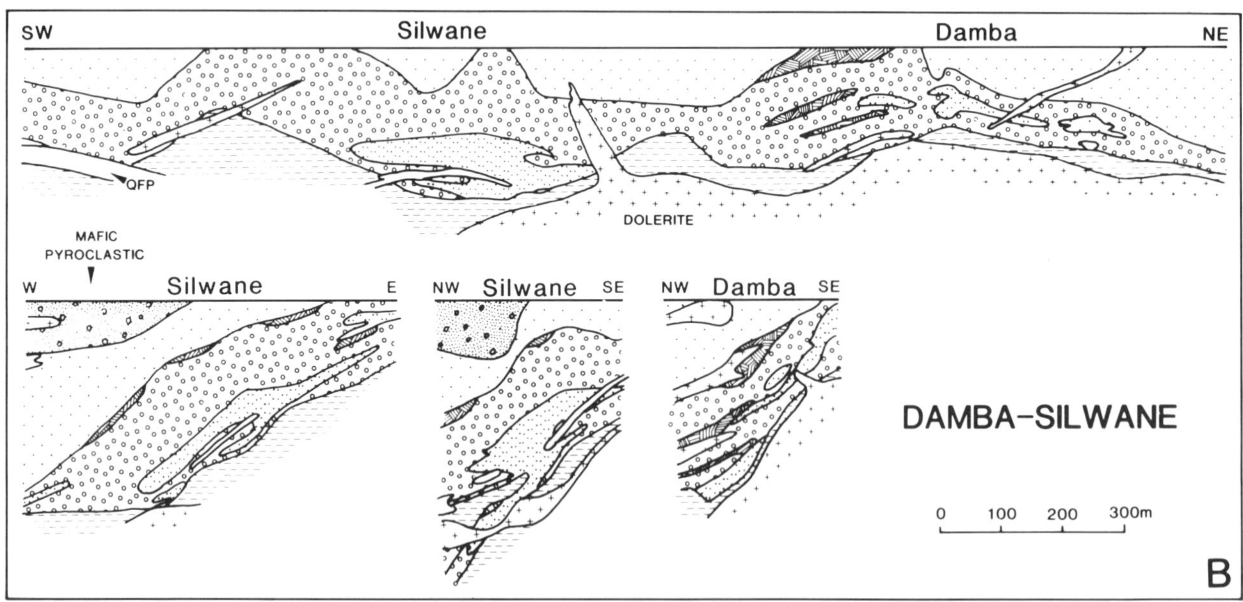


FIGURE 5.9—Solid rock geology (**A**: after Viljoen and Bernasconi, 1979) and cross sections (**B**: after Williams, 1979) of Damba–Silwane deposits, Zimbabwe. Ornamentation as for Figure 5.4.

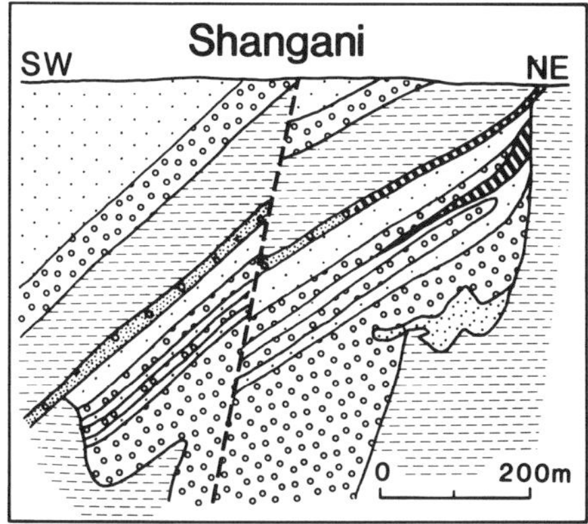


FIGURE 5.10—Cross section of Shangani deposit, Zimbabwe (after Viljoen and Viljoen, 1976). Ornamentation as for Figure 5.4.

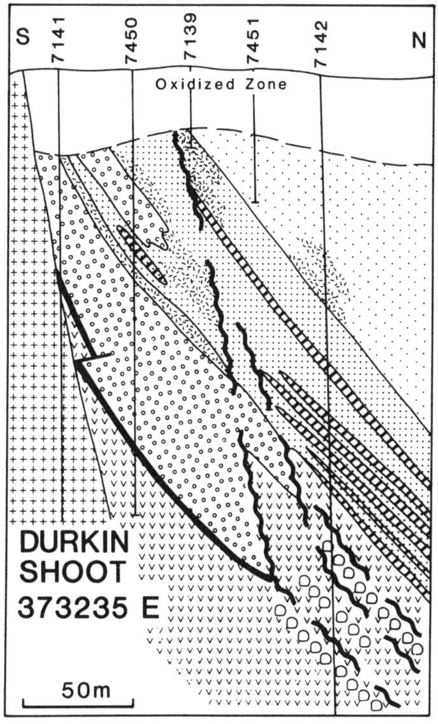


FIGURE 5.12—Cross section of Durkin shoot, Kambalda (modified from Hayden, 1976; cf. Marston et al., 1981). Ornamentation as for Figure 5.4.

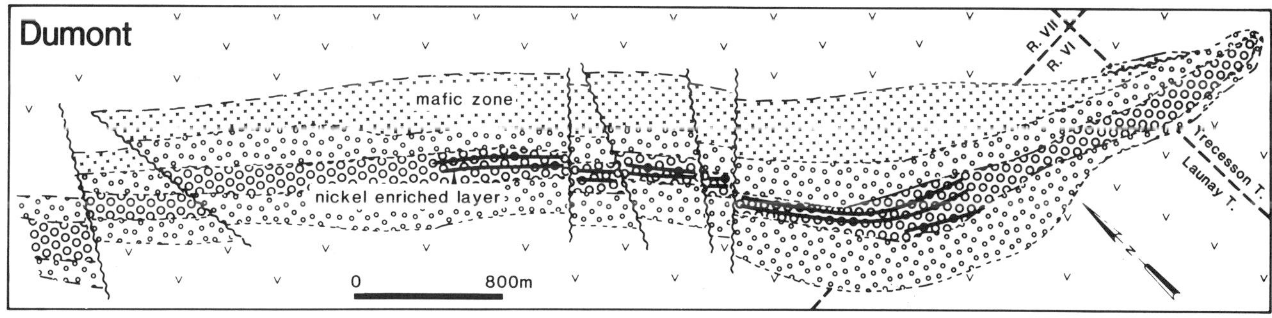


FIGURE 5.11—Solid rock geology of Dumont sill, Quebec (after Duke, 1986). Random V pattern = gabbro; other ornamentation as for Figure 5.4.

## Footwall embayments

Komatiite-associated nickel deposits are generally localized in or over embayments in the footwall. The only exceptions appear to be extremely deformed deposits (e.g. Nepean, Redross, Trojan) and intrusive bodies (e.g. Dumont). The geometry varies considerably from deposit to deposit, including i) broad, shallow depressions (e.g. Damba: Fig. 5.9b, Forrestania: Fig. 5.8, Scotia: Fig. 5.6, Wannaway, Alexo, Katiniq), ii) discontinuous, re-entrant troughs (e.g. most Kambalda shoots: Figs. 5.12–14, Langmuir), and iii) deep trenches that are transgressive to the footwall stratigraphy (e.g. Agnew). Some embayments are highly elongate, parallel to the length of the ore shoots (e.g. most Kambalda shoots), others are more irregular (e.g. Langmuir), and some are elliptical (parts of the Ken and Juan complexes: e.g. Fig. 5.15).

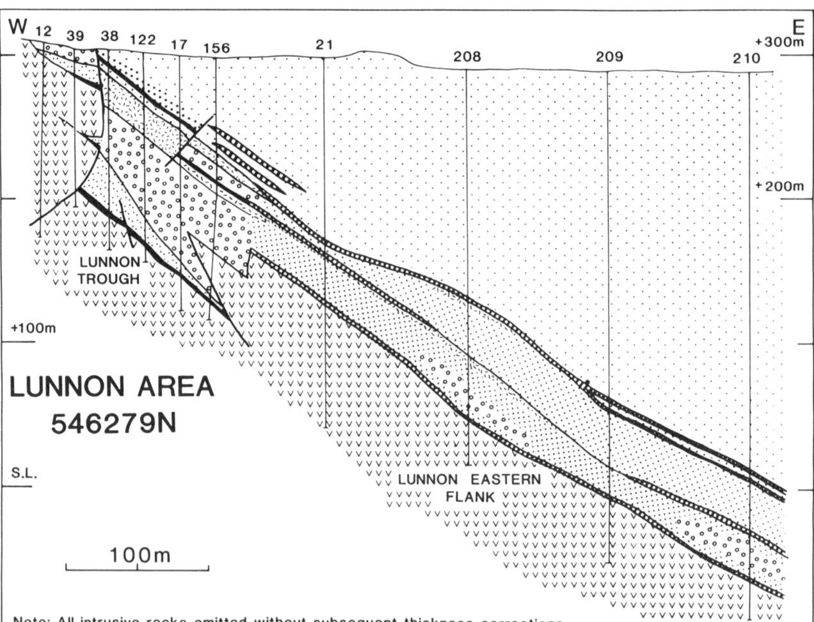


FIGURE 5.13—Cross section of Lunnon shoot, Kambalda (modified from Western Mining Corporation, unpubl. data). Ornamentation as for Figure 5.4.

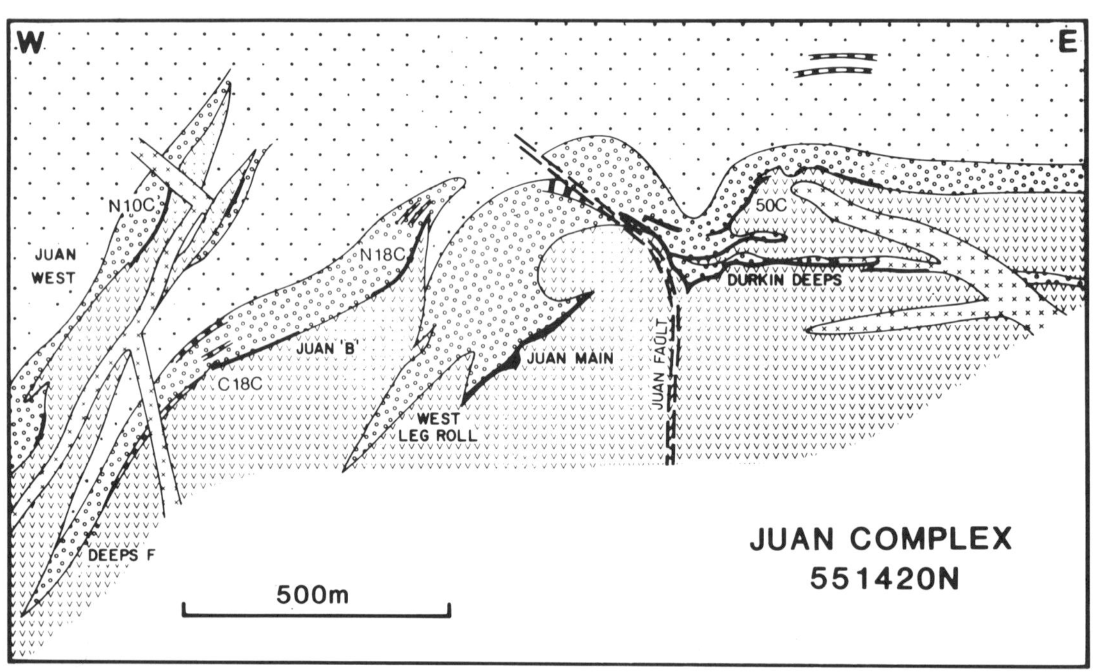


FIGURE 5.14—Cross section of Juan Complex, Kambalda (after Gresham, 1982). Ornamentation as for Figure 5.4.

The embayments at Kambalda have been interpreted previously as syn-volcanic grabens (Ross and Hopkins, 1975; Gresham and Loftus-Hills, 1981; Gresham, 1986), but more likely they have been produced by a combination of structural deformation (Cowden and Archibald, 1987; Cowden, 1988), thermal erosion (Huppert et al., 1984; Evans et al., 1988), and volcanic topography (Green and Naldrett, 1981; Lesher et al., 1984).

## HOST UNITS

The host rocks in many ore deposits play a passive role in ore genesis, by contributing ore metals or a favorable environment for accumulation, but the rocks that host komatiite-associated nickel sulfide ores played an active role in the genesis of sulfide ores. They exhibit a number of characteristics that distinguish them from unmineralized komatiites and provide important constraints on ore genetic models.

### Internal structure and composition

Most komatiitic peridotite host units are characterized by thick lower cumulate zones and thin upper aphyric and spinifex-textured zones. Several textural/mineralogical/compositional zones can normally be distinguished (Fig. 5.16a), but their thicknesses vary considerably and not all zones are present in all units. Stratigraphically equivalent, barren flanking units are thinner and less magnesian (Figs. 5.16b and c). Parts of the chilled upper margins of many basal host units at Kambalda are missing (cf. Figs. 5.16a and 5.17) and may have have been thermally eroded by overlying flows. The basal host unit at Scotia lacks any liquid-rich upper division (Page and Schmulian, 1981; Stolz and Nesbitt, 1981), but it is overlain by a shear zone (Christie, 1975) and may have been tectonically beheaded. Cumulate zones typically contain granular or polyhedral ortho-mesocumulate olivine, but some contain in situ-crystallized branching crescumulate olivine, in the center and near their bases (e.g. Victor and Durkin shoots, Kambalda: Lesher, 1983). This precludes models involving emplacement of olivine-liquid mushes or ponding/riffling of intratelluric olivine.

Komatiitic dunite host units are typically adcumulate and undifferentiated. Barren analogs at Marshall Pool and Mt. Clifford exhibit thin chilled margins (Donaldson et al., 1986), but most mineralized bodies simply grade to olivine orthocumulate at their margins (Hill et al., 1987). The Dumont dunite is significantly different from other mineralized komatiitic dunites in that it is overlain by Fe-rich tholeiitic basalts, not komatiites, and contains a thick upper zone of gabbroic differentiates (Duke, 1986a).

Excluding Dumont, both types of host units are texturally and compositionally gradational with overlying komatiite flows and are interpreted to be integral parts of the extrusive lava sequence. Most that have been studied in detail exhibit systematic variations in olivine texture, olivine composition (see below), and/or whole-rock composition with stratigraphic height (Ross and Hopkins, 1975; Naldrett and Turner, 1977; Donaldson and Bromley, 1981; Stolz and Nesbitt, 1981; Hill, 1982; Lesher et al., 1984; Donaldson et al., 1986; Hill et al., 1987; Cowden, 1988).

### Mineralogy

Komatiitic peridotites and dunites originally comprised olivine–clinopyroxene–glass–chromite and olivine–chromite–clinopyroxene ± glass assemblages, respectively. However, all have been metamorphosed to some degree and most have been variably hydrated or carbonated. The present mineralogy of these rocks is therefore a function of metamorphic grade, bulk composition, and degree of hydration and carbonation (Binns et al., 1977; Jolly, 1982; Gole et al., 1987).

Komatiitic peridotites now comprise lizardite–chlorite–chromite–magnetite–clay ± albite ± clinopyroxene (very low grade), antigorite–tremolite–chlorite–ferrochromite–magnetite ± clinopyroxene (low-medium grade, hydrated), talc–magnesite–chlorite–magnetite (low-medium grade, carbonated), metamorphic olivine–talc–tremolite–chlorite–ferrochromite ± olivine (medium-high grade, hydrated), and metamorphic olivine–chlorite–magnetite ± anthophyllite ± enstatite (medium-high grade, carbonated) assemblages.

Komatiitic dunites now comprise lizardite–brucite–magnetite–chlorite–chromite (very low grade), antigorite–brucite–ferrochromite–magnetite–tremolite–chlorite ± clinopyroxene (low-medium grade, hydrated), talc–magnesite–chromite–magnetite–chlorite (low-medium grade, carbonated), metamorphic olivine–talc–ferrochromite–tremolite–chlorite ± olivine (medium-high grade, hydrated), and metamorphic olivine ± anthophyllite ± enstatite (medium-high grade, carbonated) assemblages.

Only three relict phases have been identified in these rocks: olivine, chromite, and clinopyroxene. The first two are important petrogenetic indicators and potentially useful exploration tools (Groves et al., 1977; Duke and Naldrett, 1978; Lesher and Groves, 1984; Naldrett et al., 1984).

#### Olivine

Relict igneous olivine is rarely preserved in mineralized komatiitic peridotites at Wannaway (McQueen, 1981b), Scotia (Stolz and Nesbitt, 1981), and Kambalda (Donaldson, 1983; Lesher, 1983). Forsterite contents in basal host units vary with stratigraphic height (e.g. Fig. 5.17) and are systematically higher than that in stratigraphically equivalent barren flanking units (Fig. 5.16b). Igneous olivine is typically brownish in color, owing to microscopic inclusions of chromite that exsolved during metamorphism, and contains 0.10–0.29% CaO, 0.05–0.23% MnO, 0.11–0.24% $Cr_2O_3$, and 0.36–0.57% NiO (Table 5.4), similar to olivine in unmineralized komatiites. The lower values in these ranges are attributed to loss during metamorphism; altered igneous olivines and olivines regenerated during metamorphism are colorless and contain negligible amounts of the same minor elements. The high Cr, Ca, and Mn contents of the igneous olivines are attributable to the high temperature and low viscosity of komatiitic liquids, together with little opportunity for subsolidus exsolution during rapid cooling. The high Ni contents indicate that olivine did not equilibrate with large amounts of sulfide liquid (Duke and Naldrett, 1978).

Olivine at Victor shoot, Kambalda also contains 0.09–1.1 ppb Pd, 1.5–8.4 ppb Ir, and 0.10–0.46 ppb Au (Keays et al., 1981), as well as $0.011 \pm 0.004\%$ $V_2O_3$, $0.022 \pm 0.006\%$ CoO, $<0.019\%$ Cu, $<0.019\%$ Zn, and $<0.013\%$ Pb (Table 5.4; 3 $\sigma$ errors). Incorporation of Ir in olivine has been attributed to i) crystallographic substitution (Brügmann et al., 1987), ii) nucleation of olivine on Ir-rich alloys (Keays, 1982; Campbell and Barnes, 1984; Barnes et al., 1985), or iii) preferential non-structural incorporation of labile Ir during rapid crystallization (Sarah-Jane Barnes, pers. comm., 1988).

Relict igneous olivine ($Fo_{95-86}$) is preserved in mineralized komatiitic dunite at Agnew (Barnes et al., 1988c), Betheno (Donaldson, 1983), and Dumont (Duke, 1986a), and exhibits similar variations in composition with stratigraphic height. It typically has lower minor element contents than that in komatiitic peridotites, probably reflecting slower cooling rates (Donaldson et al., 1986). Ni contents of olivines in the mineralized zones at Dumont are systematically lower (min. 1200 ppm Ni) than those in unmineralized zones (min. 2700 ppm Ni) at equivalent forsterite contents ($Fo_{91-93.5}$), probably reflecting in situ separation and equilibration with sulfides (Duke, 1986a).

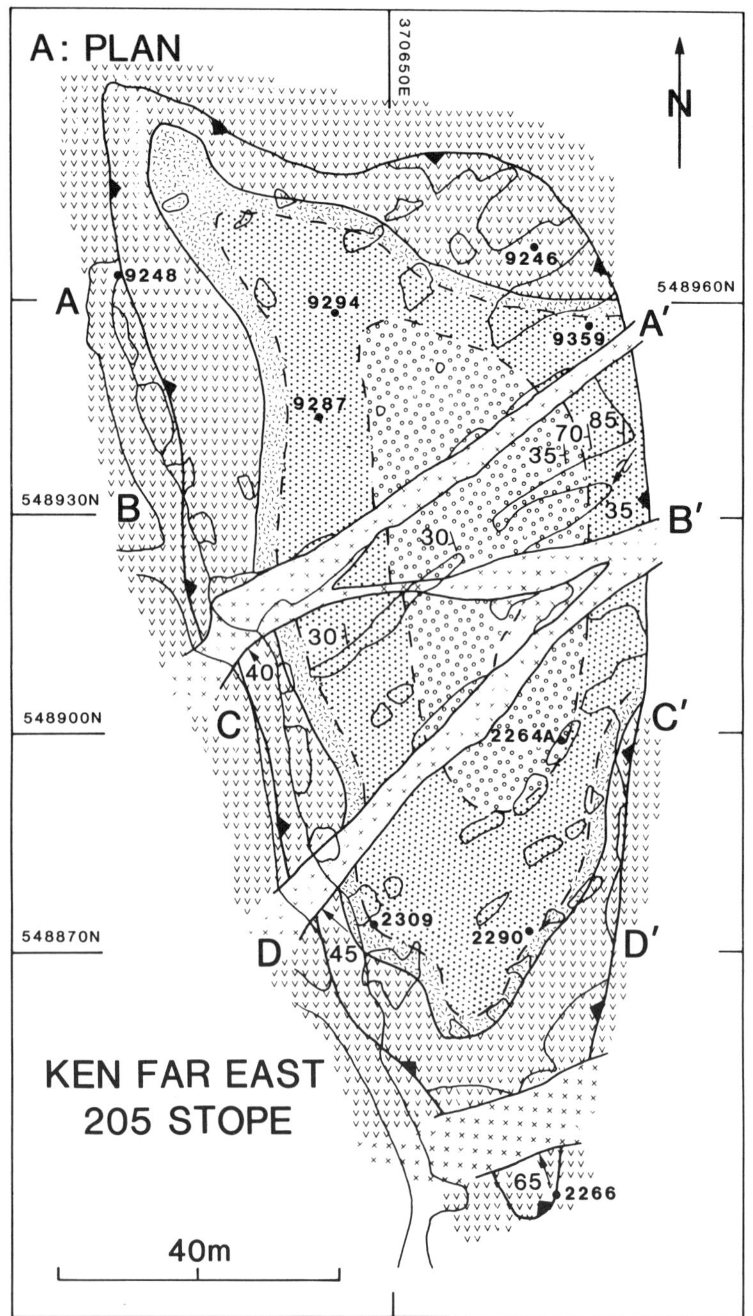


FIGURE 5.15—Plan (**A**) and serial cross sections (**B**) of Ken Far East shoot, Kambalda (mapping by C.M.L. except for data from sand-filled NW corner of stope compiled from previous mapping by J.S. Reeve and G. Evans). Ornamentation as for Figure 5.4.

## Chromite

Chromite is an ubiquitous cumulus-intercumulus accessory phase in komatiites and komatiitic dunites. Most grains exhibit Cr-magnetite rims that increase in width with increasing metamorphic grade (Donaldson, 1983). Chromites in mineralized komatiite sequences contain 71–74 mole % $R^{+2}Cr_2O_4$, 21–24 mole % $R^{+2}Al_2O_4$, and 4–5 mole % $R^{+2}Fe_2O_4$, and 0.6–2.2 atomic % Zn (Table 5.5). Chromites in mineralized komatiitic dunites and unmineralized komatiite sequences exhibit similar, albeit more variable, $Cr^{+3}$–$Al^{+3}$–$Fe^{+3}$ contents, but contain less than 0.6 atomic % Zn. Chromites in mineralized komatiite sequences have high Zn contents, irrespective of the stratigraphic position of the ultramafic flow in which they occur (Groves et al., 1977; Lesher and Groves, 1984), although there is evidence that Zn in chromites varies with stratigraphic height within individual units (Donaldson, 1983; Lesher, 1983).

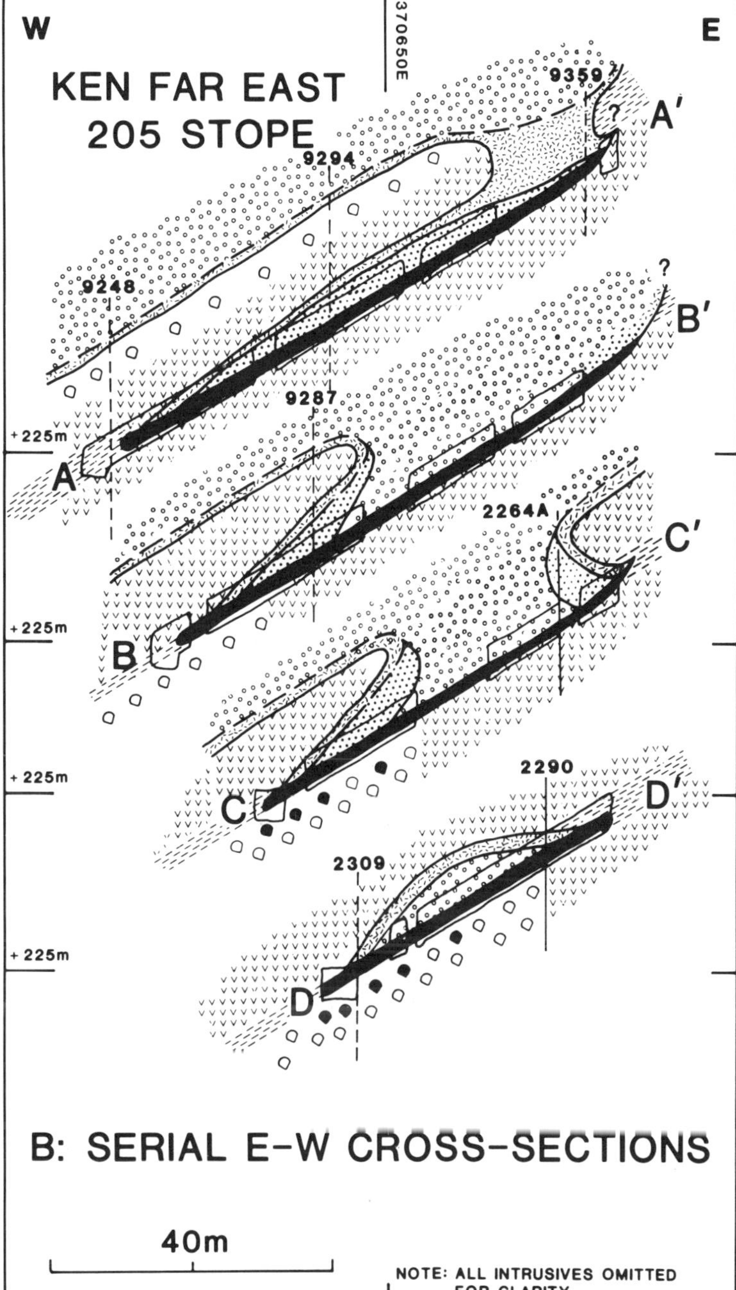


FIGURE 5.15—Plan (**A**) and serial cross sections (**B**) of Ken Far East shoot, Kambalda (continued).

The different Zn contents of chromites from mineralized komatiitic peridotites and komatiitic dunites may reflect early sulfide saturation in the former and late sulfide saturation in the latter (Lesher and Groves, 1984), but higher Zn in komatiitic peridotites may also result from assimilation of a proportionately larger amount of Zn-rich sediments during emplacement (see below).

### Whole-rock geochemistry

Regional metamorphism, including serpentinization and commonly talc-carbonation, has chemically modified most komatiites. Alkalic (Cs, Rb, K, Na) and calc-alkalic elements (Ba, Sr, Ca, $Eu^{+2}$) were most mobile, especially in cumulate rocks where there are few metamorphic phases capable of housing those elements. Si, Al, Mg, trivalent transition metals (Sc, Ti, V, Cr), some divalent transition metals (Fe, Co,

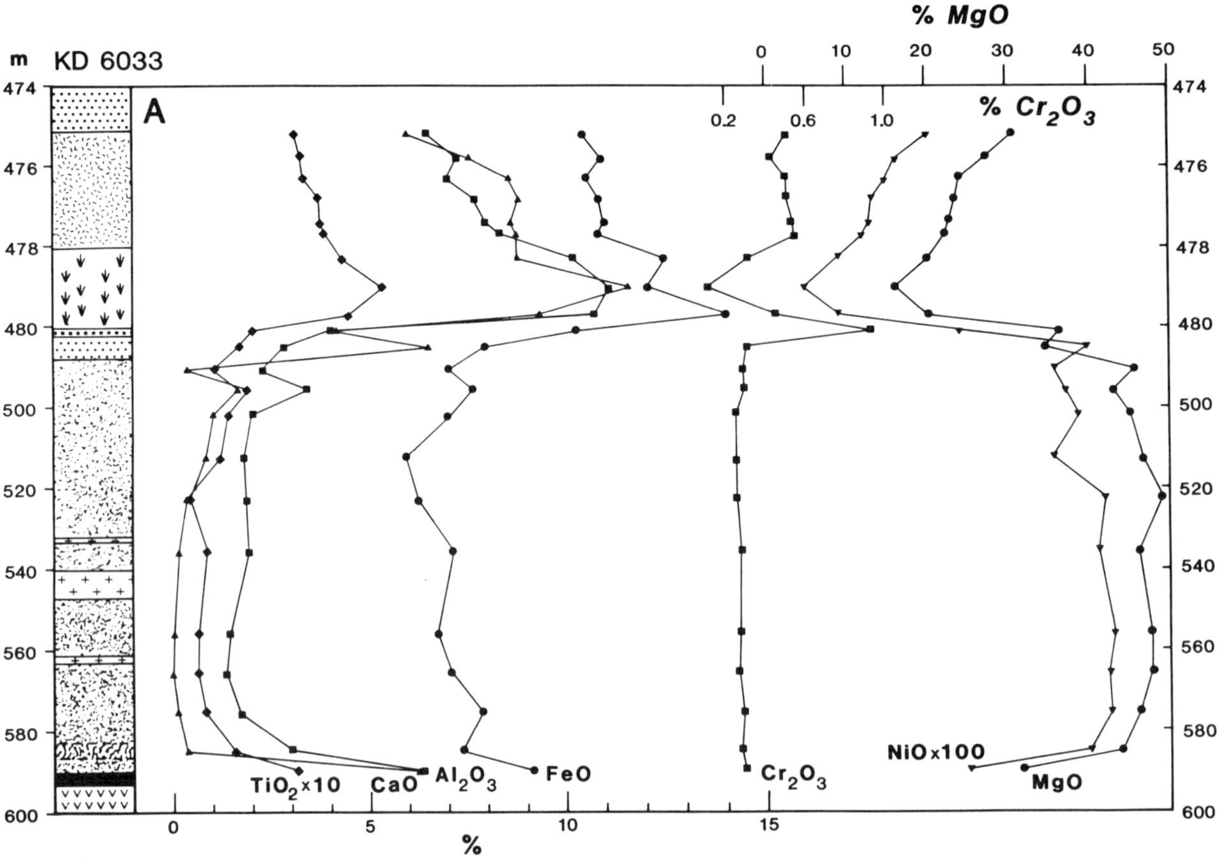

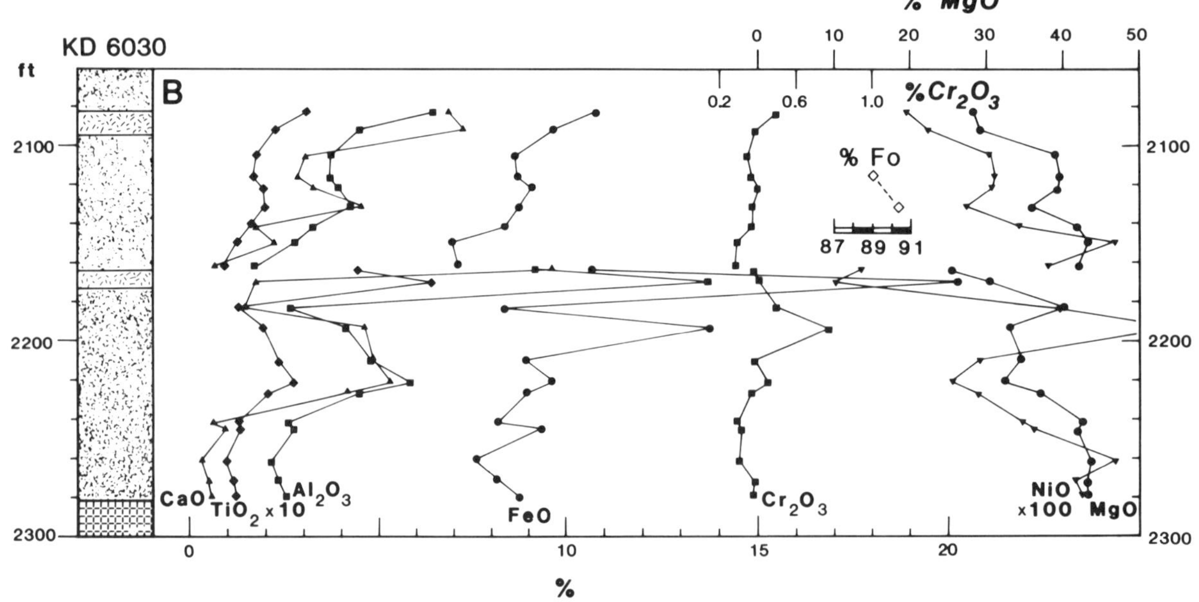


FIGURE 5.16—Drill log and whole-rock geochemical profiles through basal host unit (**A**), stratigraphically adjacent flanking units (**B**), and stratigraphically equivalent flanking units (**C**), Victor shoot, Kambalda (data from Lesher, 1983).

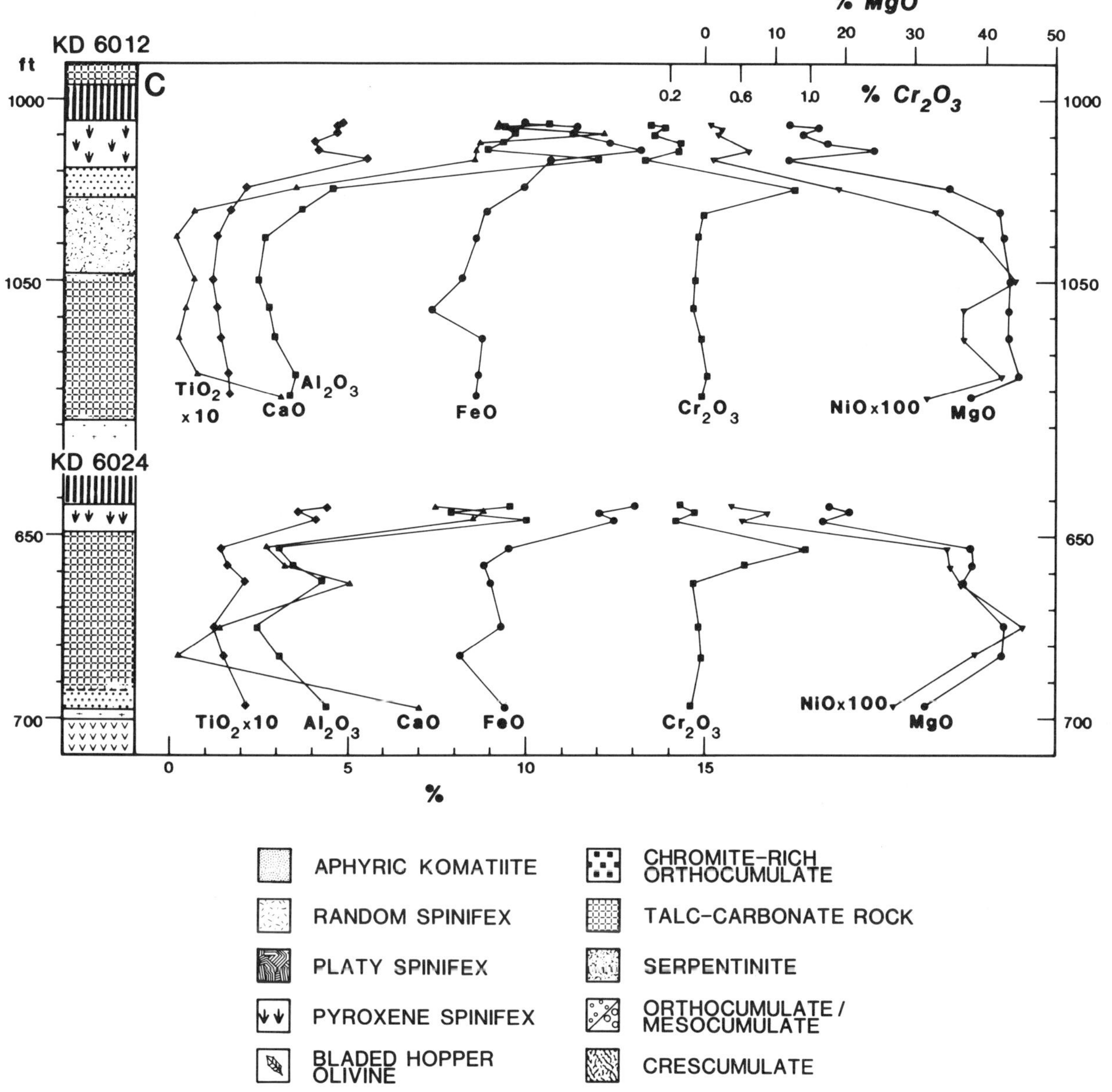


FIGURE 5.16—Drill log and whole-rock geochemical profiles (continued).

Ni), and most rare-earth elements (REE) and high field-strength elements (Y, Zr, Nb) remained relatively immobile.

### Aphyric and spinifex-textured rocks

Major element geochemical variations within aphyric and spinifex-textured komatiites at Kambalda, both within the komatiite sequence (Fig. 5.18) and within individual basal host units (Fig. 5.16), and at most other deposits are consistent with fractional crystallization of olivine (Fig. 5.19).

The compositions of initial liquids may be deduced from the compositions of upper chilled margins or random spinifex-textured zones of the host units, calculated from the compositions of cumulus olivine using experimentally determined partition coefficients, or inferred from the compositions of conformably overlying lavas (Table 5.6). The initial liquids for these deposits appear to range from an apparent maximum of around 30% MgO for most Archean deposits, to a low of around 18% MgO for Proterozoic analogs, and include both aluminum-undepleted ($Al_2O_3/TiO_2$ = 18–22) and aluminum-depleted ($Al_2O_3/TiO_2$ = 9–13) types (Nesbitt et al., 1979). Note that these are not necessarily the parental magmas, nor the lavas which equilibrated with the sulfides (see below).

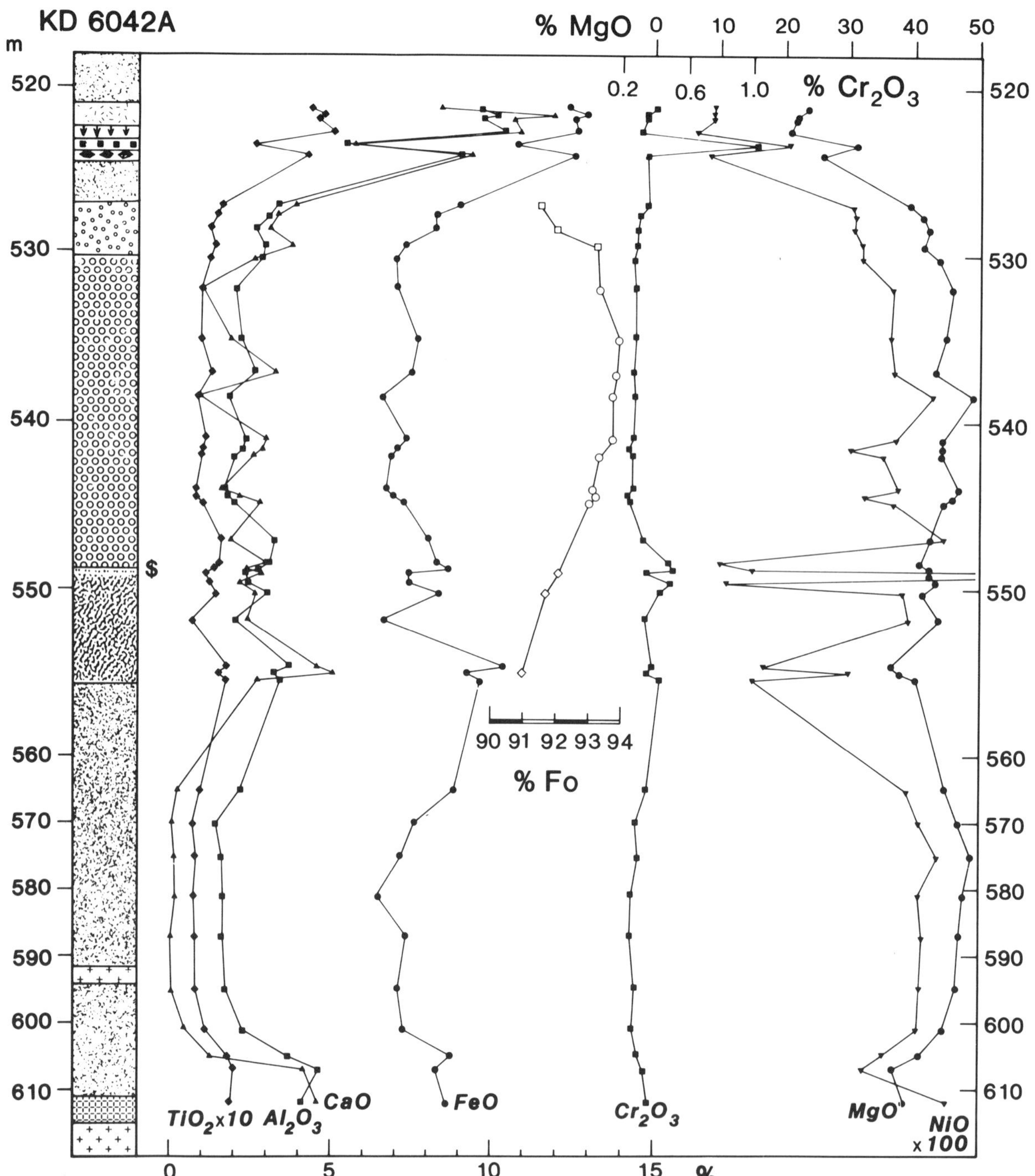


FIGURE 5.17—Drill log and whole-rock geochemical profiles through basal host unit, Victor shoot, Kambalda (data from Lesher, 1983). Note systematic textural and compositional variations in relict igneous olivine; symbols: squares = orthocumulates, circles = ortho-mesocumulates, diamonds = crescumulates. Ornamentation as for Figure 5.16.

TABLE 5.4—Electron probe microanalyses of olivine in komatiitic peridotites and dunites.

| | 1 Victor | 2 Victor | 3 Victor | 4 Victor | 5 Victor | 6 Victor | 7 Scotia | 8 Scotia | 9 Scotia | 10 Wannaway | 11 Wannaway | 12 Betheno | 13 Betheno | 14 Mt. Goode | 15 Mt. Goode |
|---|---|---|---|---|---|---|---|---|---|---|---|---|---|---|---|
| $SiO_2$ | 40.04 | 41.25 | 40.50 | 40.92 | 40.88 | 40.42 | 41.78 | 40.55 | 41.24 | 38.6 | 40.1 | 41.64 | 42.11 | 41.5 | 41.2 |
| $TiO_2$ | 0.01 | 0.01 | 0.01 | <0.01 | <0.01 | <0.01 | – | – | 0.04 | – | – | – | – | – | – |
| $Al_2O_3$ | 0.05 | 0.05 | 0.03 | 0.05 | 0.03 | 0.03 | 0.09 | 0.08 | <0.01 | – | – | 0.01 | <0.01 | 0.01 | <0.01 |
| $Cr_2O_3$ | 0.16 | 0.16 | 0.15 | 0.13 | 0.18 | 0.13 | 0.28 | 0.17 | 0.11 | 0.16 | 0.20 | 0.04 | 0.01 | 0.02 | 0.01 |
| $FeO_t$ | 8.32 | 5.79 | 8.45 | 7.63 | 10.58 | 8.97 | 6.64 | 6.96 | 6.85 | 19.1 | 9.1 | 7.39 | 5.92 | 8.28 | 7.51 |
| MnO | 0.18 | 0.10 | 0.15 | 0.21 | 0.38 | 0.17 | <0.05 | 0.13 | 0.12 | 0.33 | 0.17 | 0.11 | 0.14 | 0.16 | 0.13 |
| MgO | 50.01 | 51.68 | 49.24 | 51.21 | 47.97 | 49.62 | 51.01 | 50.92 | 51.10 | 41.8 | 49.8 | 51.31 | 52.22 | 50.3 | 51.0 |
| NiO | 0.40 | 0.39 | 0.38 | 0.45 | 0.35 | 0.40 | 0.40 | 0.48 | 0.54 | 0.22 | 0.40 | 0.46 | 0.22 | 0.38 | 0.55 |
| CaO | 0.24 | 0.23 | 0.23 | 0.27 | 0.22 | 0.23 | 0.16 | 0.20 | 0.11 | 0.02 | 0.03 | 0.05 | 0.02 | 0.02 | 0.02 |
| Total | 99.41 | 99.66 | 99.14 | 100.87 | 100.59 | 99.97 | 100.36 | 99.49 | 100.11 | 100.23 | 99.80 | 101.01 | 100.64 | 100.67 | 100.42 |
| Mg/Fe + Mg | 91.5 | 94.1 | 91.2 | 92.3 | 89.0 | 90.8 | 93.2 | 92.9 | 93.0 | 79.6 | 90.7 | 92.5 | 94.0 | 91.5 | 92.4 |

– not determined
1 Average of 8 orthocumulate olivines, basal host unit, KD6042A/527.0m (Lesher, 1983)
2 Average of 5 mesocumulate olivines, basal host unit, KD6042A/537.1m (Lesher, 1983)
3 Average of 9 crescumulate olivines, basal host unit, KD6042A/555.0m (Lesher, 1983) also contains $0.011 \pm 0.004\%$ $V_2O_3$, $0.022 \pm 0.006\%$ CoO, <0.019% Cu, <0.019% Zn, and <0.013% Pb (3 $\sigma$ errors)
4 Average of 9 crescumulate olivines, lower part of basal host unit, KD6033/585.0m (Lesher, 1983)
5 Average of 7 crescumulate olivines, barren flanking unit, KD6030/2116.0′ (Lesher, 1983)
6 Average of 7 orthocumulate olivines, barren flanking unit, KD6030/1829.0′ (Lesher, 1983)
7 Spinifex olivine, barren unit, SD4/193′ (Stolz and Nesbitt, 1981)
8 Skeletal olivine, barren unit, SD48/1132′ (Stolz and Nesbitt, 1981)
9 Crescumulate olivine, barren unit, SD48/1116′ (Stolz and Nesbitt, 1981)
10 Average of 5 bladed olivines, upper part, basal host unit, WAD12a/506′ (McQueen, 1981b); note low Ca
11 Average of 4 cumulate olivines, lower part, basal host unit, WAD13b/810.2′ (McQueen, 1981b); note low Ca
12 Average of 7 adcumulate olivines, distant from pentlandite, MKD52/1175′ (Donaldson, 1983); note low Cr and Ca
13 Average of 3 adcumulate olivines, distant from pentlandite, MKD52/1761.5′ (Donaldson, 1983); note low Cr and Ca
14 Average of 4 fragments in lizardite serpentinite, MGD22D/282.7m (Donaldson, 1983); note low Cr and Ca
15 Average of 4 fragments in lizardite serpentinite, MGD106A/344.7m (Donaldson, 1983); note low Cr and Ca

A more detailed discussion of the geochemistry of these rocks is beyond the scope of this chapter, but it is worth mentioning that there are systematic differences in Ti–Al–Ca–Sc–V–REE ratios between i) barren komatiites (e.g. Barberton), ii) barren komatiites that are correlative with mineralized sequences (e.g. Belingwe, Munro), and iii) komatiites that are mineralized (e.g. Kambalda) or overlie mineralized komatiitic dunites (e.g. Yakabindie) that appear to reflect differences in their petrogenesis (Beswick, 1982; Ludden and Gélinas, 1982; Smith and Erlank, 1982). The possible relevance of this to the genesis of nickel sulfide deposits is discussed further below.

## Cumulates

Geochemical variations within the cumulate rocks (Fig. 5.19) are controlled primarily by accumulation of olivine and minor chromite into komatiitic liquids. The compositions of some komatiitic peridotite and komatiitic dunite host rocks are given in Tables 5.7 and 5.8. Most are depleted in Ca, Na, and K as a consequence of alteration.

Within mineralized komatiite sequences, host units are typically more magnesian than stratigraphically equivalent units flanking the ore zone (Gresham and Loftus-Hills, 1981; Hill et al., 1987). At Kambalda, for example, the basal host units are significantly enriched in Mg and Ni, and depleted in Ti, Al, Cr, Fe and Zn relative to adjacent barren flanking units. These variations were studied by Lesher and Groves (1984) who analyzed chilled margins, cumulate rocks, and relict olivine, and modeled the geochemical variations in the cumulate rocks (Fig. 5.20). They concluded that the basal host units crystallized from more magnesian liquids (28–31% MgO) and more forsteritic olivine ($Fo_{94-91}$) than the adjacent units (<26% MgO and $Fo_{91-89}$, respectively) (Fig. 5.21). Similar variations are observed at Agnew (Barnes et al., 1988c). These variations preclude models involving accumulation of intratelluric olivine.

## Lower/lateral chilled margins

Some host units exhibit lower chilled margins, preserved beneath ore zones (e.g. Alexo, Wannaway) or along lateral margins of embayments (e.g. most Kambalda shoots). Analogous contact zones occur along the margins of some komatiitic dunites (e.g. Mt. Clifford and Marshall Pool). Those at Kambalda are typically less magnesian than the upper margins, but are compositionally distinct from metasomatic reaction zones produced during metamorphism (cf. Tables 5.6 and 9), suggesting that they are hybridized komatiites rather than metasomatic reaction zones (Lesher, 1985). Trace element geochemical studies are in progress to test this possibility.

## Chalcophile element depletion

Chalcophile elements such as Ni, Co, Cu, and PGE partition preferentially into the sulfide phase relative to olivine or silicate magma. Thus, the abundances of these elements in derivative silicate liquids should be strongly influenced by the proportions of fractionated olivine and sulfide that equilibrate in a magma (Duke and Naldrett, 1978; Campbell and Naldrett, 1979; Duke, 1979; Campbell and Barnes, 1984) and should be a sensitive test of magmas that have equilibrated with sulfides (Lesher et al., 1981; Naldrett et al., 1984).

TABLE 5.5—Electron probe microanalyses of chromite in komatiitic peridotites and dunites, and ferrochromites in sulfide ores.

| | 1<br>Victor | 2<br>Victor | 3<br>Victor | 4<br>Durkin | 5<br>Scotia | 6<br>Wannaway | 7<br>Windarra | 8<br>Betheno |
|---|---|---|---|---|---|---|---|---|
| $TiO_2$ | 0.28 | 0.27 | 0.26 | – | 0.24 | 0.36 | 0.33 | 1.55 |
| $Al_2O_3$ | 12.39 | 12.29 | 12.22 | 12.25 | 10.66 | 8.76 | 8.64 | 17.87 |
| $Cr_2O_3$ | 51.67 | 51.37 | 52.06 | 54.71 | 54.52 | 52.00 | 53.25 | 46.75 |
| $Fe_2O_3$* | 4.25 | 4.55 | 4.14 | 3.40 | 3.86 | 5.92 | 4.56 | 2.23 |
| FeO* | 20.05 | 20.89 | 20.95 | 16.69 | 24.51 | 26.55 | 27.30 | 22.64 |
| MnO | 1.24 | 1.18 | 1.22 | 1.24 | 0.49 | 0.76 | 0.54 | 0.55 |
| MgO | 7.33 | 6.59 | 6.53 | 9.22 | 6.21 | 2.35 | 2.02 | 8.18 |
| NiO | 0.07 | 0.06 | 0.06 | $<0.25$ | $<0.25$ | – | 0.09 | 0.06 |
| ZnO | 1.28 | 1.82 | 1.89 | 1.84 | 0.53 | 2.92 | 2.59 | 0.71 |
| $V_2O_5$ | – | – | – | – | – | – | – | – |
| Total | 98.56 | 99.02 | 99.33 | 99.35 | 101.02 | 99.62 | 99.32 | 100.54 |
| $RCr_2O_4$ | 69.7 | 69.4 | 70.1 | 71.8 | 73.6 | 73.6 | 75.6 | 61.9 |
| $RAl_2O_4$ | 24.9 | 24.8 | 24.5 | 24.0 | 21.4 | 18.5 | 18.3 | 35.3 |
| $RFe_2O_4$ | 5.5 | 5.9 | 5.3 | 4.2 | 5.0 | 8.0 | 6.2 | 2.8 |
| 100Mg/R | 36.7 | 33.1 | 32.7 | 45.6 | 30.3 | 12.3 | 10.6 | 37.9 |

– not determined

1 Average of 8 chromites in orthocumulate, basal host unit, KD6042A/529.4m (Lesher, 1983)

2 Average of 3 chromites in mesocumulate, basal host unit, KD6042A/537.1m (Lesher, 1983)

3 Average of 5 chromites in crescumulate, basal host unit, KD6042A/555.0m (Lesher, 1983) also contains $0.042 \pm 0.004\%$ $V_2O_3$, $0.125 \pm 0.008\%$ CoO, $<0.025\%$ Cu, and $<0.018\%$ Pb ($3\ \sigma$ errors)

4 Chromite in partially serpentinized peridotite, sample 79378 (Groves et al., 1977)

5 Chromite in partially serpentinized peridotite, sample 66923 (Groves et al., 1977)

6 Chromite in disseminated ore in serpentinite, sample 79834 (Groves et al., 1977)

7 Chromite in talc-carbonate rock above ore zone, sample 79375 (Groves et al., 1977)

8 Average of 3 chromites in serpentinized dunite, MKD52/630.0′ (Donaldson, 1983)

*$Fe^{+2}$ and $Fe^{+3}$ calculated from total Fe assuming spinel stoichiometry after subtracting Ti and a proportionate amount of Fe to form ülvospinel; $R = Fe^{+2} + Mn + Mg + Zn$

| | 9<br>Betheno | 10<br>Mt. Goode | 11<br>Mt. Goode | 12<br>Mt. Clifford | 13<br>Lunnon | 14<br>KPA | 15<br>Zimbabwe | 16<br>KDA |
|---|---|---|---|---|---|---|---|---|
| $TiO_2$ | 0.19 | 0.21 | 0.22 | 0.34 | 0.84 | 0.29 | 2.23 | 0.34 |
| $Al_2O_3$ | 17.27 | 9.26 | 9.43 | 11.39 | 0.66 | 0.65 | 1.45 | 1.82 |
| $Cr_2O_3$ | 51.25 | 49.19 | 51.86 | 54.05 | 51.93 | 52.59 | 43.23 | 49.81 |
| $Fe_2O_3$* | 1.55 | 10.38 | 6.94 | 5.99 | 12.74 | 11.64 | 16.24 | 16.75 |
| FeO* | 18.86 | 20.99 | 25.09 | 14.60 | 28.10 | 26.17 | 28.68 | 28.97 |
| MnO | 0.60 | 0.88 | 0.67 | 0.65 | 1.64 | 1.88 | 1.08 | 1.30 |
| MgO | 9.56 | 6.07 | 4.48 | 12.04 | 0.46 | 0.24 | 0.06 | 1.24 |
| NiO | 0.05 | 0.11 | 0.08 | 0.10 | 0.04 | 0.01 | 0.00 | 0.01 |
| ZnO | 1.00 | 2.28 | 0.96 | 0.09 | 1.86 | 2.88 | 3.59 | 0.53 |
| $V_2O_5$ | – | – | – | – | – | – | 1.24 | – |
| Total | 100.33 | 99.37 | 99.73 | 99.25 | 98.27 | 96.35 | 97.80 | 100.77 |
| $RCr_2O_4$ | 65.3 | 67.5 | 71.5 | 70.4 | 79.8 | 81.4 | 71.0 | 72.8 |
| $RAl_2O_4$ | 32.8 | 18.9 | 19.4 | 22.1 | 1.5 | 1.5 | 3.6 | 4.0 |
| $RFe_2O_4$ | 1.9 | 13.6 | 9.1 | 7.4 | 18.6 | 17.1 | 25.4 | 23.3 |
| 100Mg/R | 45.5 | 31.1 | 23.0 | 58.2 | 2.5 | 1.4 | 0.3 | 6.7 |

–not determined

9 Average of 2 chromites in partially serpentinized dunite, MKD52/1761.5′ (Donaldson, 1983)

10 Average of 2 chromites in serpentinized dunite, MGD22C/578.5′ (Donaldson, 1983)

11 Average of 2 chromites in serpentinized dunite, MGD22D/333.2′ (Donaldson, 1983)

12 Chromite in barren serpentinized peridotite (Barnes et al., 1974); note low Zn

13 Average of 42 ferrochromites in massive and matrix ores (Groves et al., 1981)

14 Average of 17 ferrochromites in sulfide ores from Redross, Scotia, Wannaway, and Windarra (Groves et al., 1981)

15 Average of 31 ferrochromites in sulfide ores from Trojan and Perseverance (Groves et al., 1981)

16 Average of 9 ferrochromites in sulfide ores from Agnew and Digger Rocks (Groves et al., 1981)

*$Fe^{+2}$ and $Fe^{+3}$ calculated from total Fe assuming spinel stoichiometry after subtracting Ti and a proportionate amount of Fe to form ülvospinel; $R = Fe^{+2} + Mn + Mg + Zn$

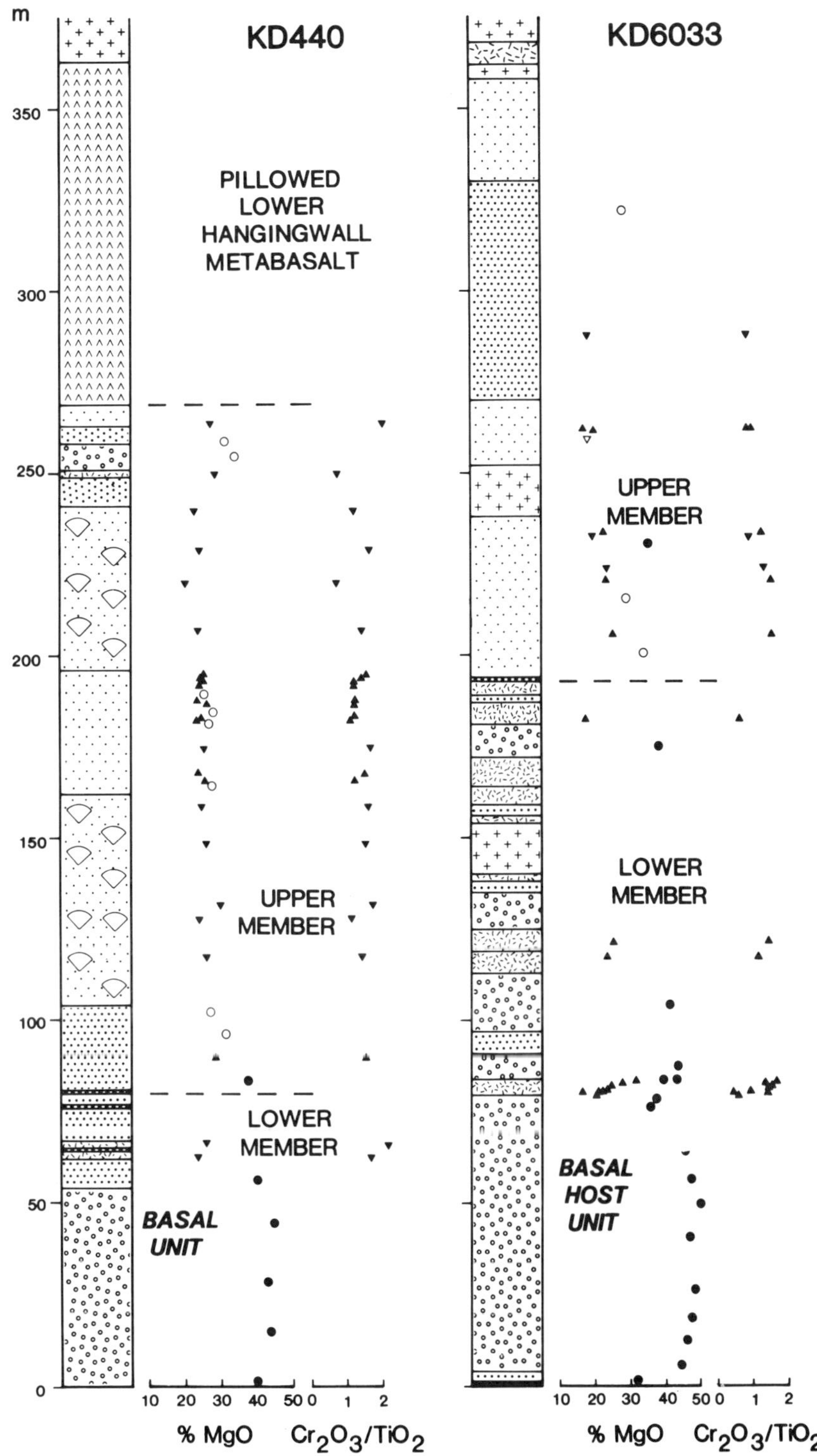


FIGURE 5.18—Drill logs and whole-rock geochemical profiles through unmineralized (**A**) and mineralized (**B**) sections of komatiite sequence near Hunt and Victor shoots (see Figure 5.3), Kambalda (data from Lesher, 1983). Vertical scale is true height above footwall basalt contact. Symbols: circles = porphyritic and cumulate rocks, triangles = aphyric and spinifex-textured rocks, inverted triangles = recrystallized aphyric rocks, solid symbols = XRF data, open symbols = AAS data. Ornamentation as for Figure 5.4.

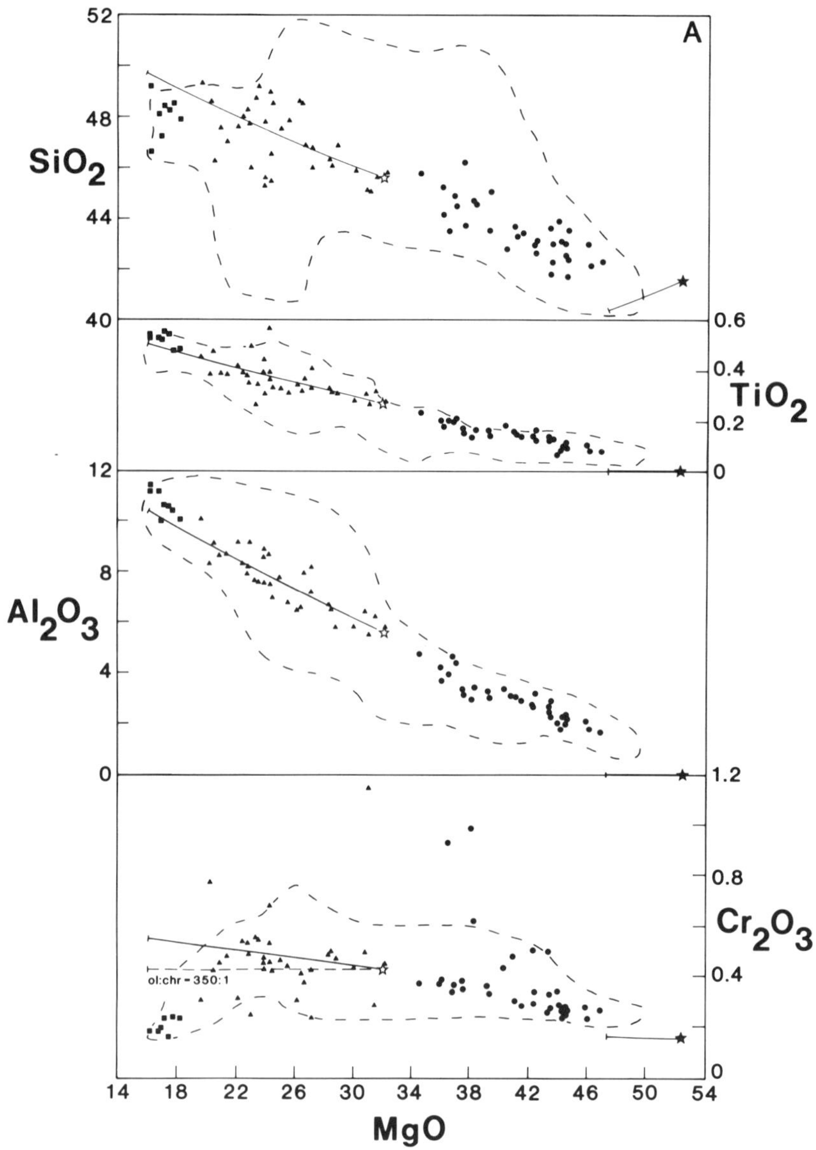


FIGURE 5.19—Major element variations in Kambalda komatiites (data from Lesher, 1983). Data points = 36 porphyritic and cumulate komatiites, 31 aphyric and spinifex-textured komatiites, and 8 pyroxenitic komatiites selected as least altered and most representative (open star = model komatiite, solid star = equilibrium olivine; other symbols as in Figure 5.18). Dashed field encloses 325 analyzed samples. Solid lines are calculated liquid and olivine fractionation curves ($K_D^{Fe-Mg}$ = 0.33).

Spinifex-textured komatiites from barren sequences (e.g. Barberton), barren sequences that correlate with mineralized sequences (e.g. Belingwe, Munro), barren flows overlying class IIB deposits (e.g. Yakabindie), or flows that are associated with only weak disseminated mineralization (e.g. Mt. Clifford: see Lesher et al., 1981) plot along a trend that corresponds closely with that of a sulfide-unsaturated model (Figs. 5.22a and b). In contrast, spinifex-textured komatiites from mineralized sequences at Kambalda (Fig. 5.22c), Scotia (Fig. 5.22d), and possibly Widgiemooltha (see McQueen, 1981a) are slightly, but significantly depleted in Ni (and Co, not shown) relative to barren komatiites and to sulfide-undersaturated models. Spinifex-textured rocks from thin flow units above Lunnon and Long shoots (Keays et al., 1981) are similarly depleted. Cu (and Zn) scatter owing to mobility during alteration and metamorphism.

Lesher et al. (1981) found that there were no systematic variations between degree of Ni depletion and type of alteration (serpentinization vs. carbonation), proximity to sulfide mineralization (mineralized vs. unmineralized flows), or stratigraphic position (lower vs. upper member), and attributed the depletion at Kambalda to scavenging of chalcophile elements by sulfides at the magmatic stage, prior to eruption. They noted, however, that dispersion in the data required involvement of more than one petrogenetic process. This process is now believed to be variable degrees of equilibration with sulfides during emplacement. Coarse platy spinifex-textured rocks at Lunnon shoot (Lunnon "metapicrite": Keays et al., 1981) are not as depleted in Ni and Co, which may be attributed to crystallization from replenished liquids that did not equilibrate with sulfides.

PGE have much higher sulfide/silicate partition coeffi-

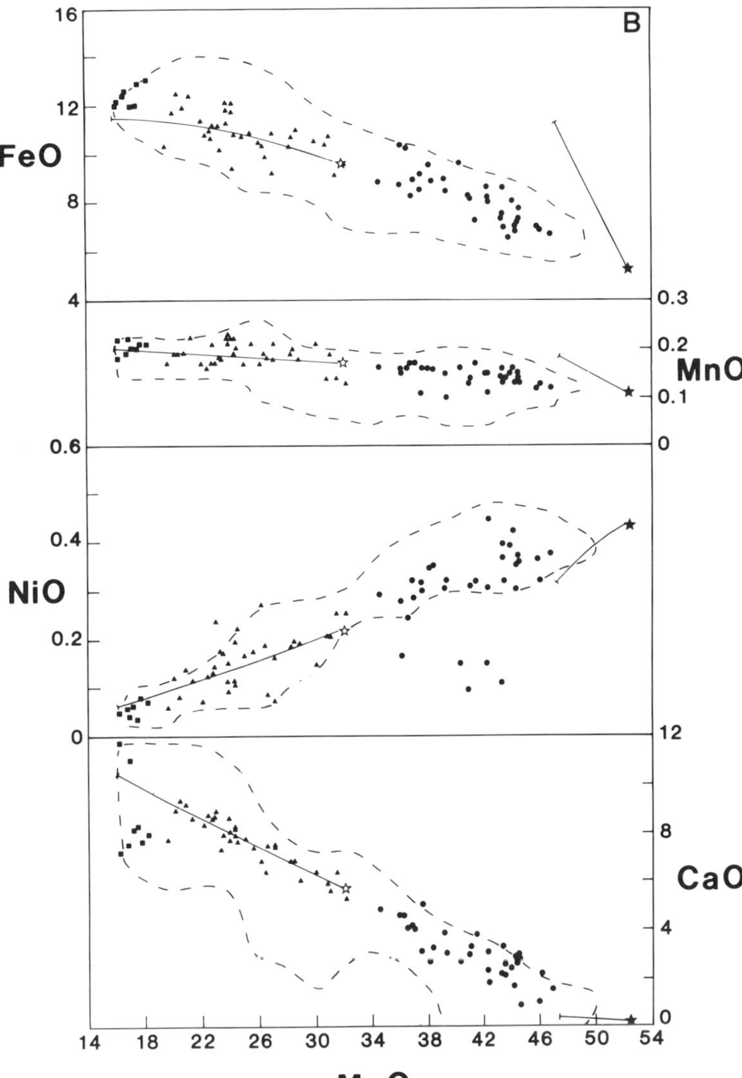


FIGURE 5.19—Major element variations in Kambalda komatiites (continued).

cients, of the order of $10^3$ (Barnes et al., 1985) to $10^5$ (Campbell and Barnes, 1984), and should be more sensitive to sulfide separation than Ni, Cu, or Co, but are more difficult to sample and analyze because of their low abundances in silicate rocks. Comparison of data from Alexo (Brügmann et al., 1987), Long, Lunnon, and Victor shoots, Kambalda (Keays et al., 1981), Pyke Hill, Munro Township (Crocket and MacRae, 1986), and Mt. Clifford (Keays, 1982) indicates that PGE variations are much more complex. Although Pd–MgO variations are too disperse to infer any differences attributable to sulfide separation (Fig. 5.23a), Kambalda samples are not depleted in Pd relative to the other suites and most of the variation is broadly consistent with olivine fractionation (Pd is relatively incompatible in olivine). Why Kambalda lavas appear to be depleted in Ni, but not Pd is an unresolved problem.

Ir is higher in spinifex-textured rocks at Alexo and Mt. Clifford than at Kambalda or Munro (Fig. 5.23b), but interpretation is complicated by the fact that Ir appears to be variably compatible in olivine. Ir contents in olivine separates from Victor shoot, for example, vary between 1.5 and 8.4 ppb (4 fractions: Keays et al., 1981), and Ir contents in orthocumulate, mesocumulate, and crescumulate whole rocks vary between 3.3 and 10.2 ppb (9 samples: Keays et al., 1981; C.M.L. and R.R. Keays, unpubl. data). Similar variations are evident in samples from Lunnon. Keays (1982) contended that the high Ir content of olivine in mesocumulate rocks at Victor shoot is diagnostic of an intratelluric origin, and the low Ir content of olivine in orthocumulate rocks at Mt. Clifford and Munro ($<1$ ppb) calculated by mass balance) is diagnostic of crystallization in situ, but the high Ir contents of crescumulates, which must crystallize in situ,

TABLE 5.6—Whole-rock geochemical analyses of komatiites and calculated parental liquids. Analyses recalculated to 100% volatile-free.

| | 1<br>Forrestania | 2<br>Yakabindie | 3<br>Agnew | 4<br>Nepean | 5<br>Scotia | 6<br>Lunnon | 7<br>Victor | 8<br>Durkin | 9<br>Juan W. | 10<br>Foster |
|---|---|---|---|---|---|---|---|---|---|---|
| $SiO_2$ | 48.56 | 44.16 | 48.83 | 46.36 | 45.40 | 45.81 | 45.11 | 44.48 | 43.52 | 47.43 |
| $TiO_2$ | 0.36 | 0.27 | 0.25 | 0.35 | 0.31 | 0.281 | 0.306 | 0.353 | 0.476 | 0.344 |
| $Al_2O_3$ | 4.57 | 5.26 | 5.28 | 7.20 | 6.19 | 5.84 | 6.43 | 8.03 | 10.55 | 7.70 |
| $Cr_2O_3$ | 0.072 | 0.37 | 0.24 | 0.44 | 0.49 | 0.440 | 0.500 | 0.502 | 0.600 | 0.408 |
| $FeO_t$ | 11.54 | 10.99 | 7.12 | 10.23 | 9.95 | 10.59 | 10.44 | 11.91 | 12.63 | 10.73 |
| MnO | 0.17 | 0.22 | 0.15 | 0.18 | 0.14 | 0.21 | 0.14 | 0.16 | 0.11 | 0.22 |
| MgO | 26.45 | 32.67 | 32.07 | 29.57 | 30.51 | 30.02 | 30.79 | 27.97 | 29.69 | 24.05 |
| NiO | 0.189 | 0.20 | – | 0.20 | 0.192 | 0.147 | 0.205 | 0.125 | 0.146 | 0.120 |
| CaO | 7.66 | 5.50 | 6.28 | 5.22 | 6.19 | 6.37 | 5.92 | 6.29 | 2.11 | 8.91 |
| $Na_2O$ | 0.30 | 0.32 | 0.05 | 0.18 | 0.46 | 0.15 | 0.11 | 0.14 | 0.09 | 0.04 |
| $K_2O$ | 0.08 | 0.04 | 0.03 | 0.03 | 0.13 | <0.01 | 0.01 | 0.01 | <0.01 | 0.02 |
| $P_2O_5$ | 0.03 | – | – | 0.03 | 0.03 | 0.03 | 0.04 | 0.02 | 0.07 | 0.02 |
| S | 0.09 | – | – | <0.1 | – | – | 0.20 | 0.04 | <0.01 | 0.07 |
| Volatiles | 5.81 | 7.37 | – | 5.40 | 6.42 | 5.58 | 7.49 | 6.76 | 10.26 | 15.62 |
| $Al_2O_3/TiO_2$ | 12.7 | 19.8 | 21.1 | 20.7 | 19.8 | 20.8 | 21.0 | 22.7 | 22.2 | 22.4 |
| $CaO/Al_2O_3$ | 1.68 | 1.05 | 1.19 | 0.73 | 1.00 | 1.09 | 0.92 | 0.78 | 0.20 | 1.16 |
| Fe + Mn + Mg + Ni/Si | 1.02 | 1.32 | 1.10 | 1.14 | 1.19 | 1.17 | 1.22 | 1.17 | 1.26 | 0.95 |
| Mg/Fe + Mg | 80.3 | 84.1 | 88.9 | 83.7 | 84.5 | 83.5 | 84.0 | 80.7 | 80.7 | 80.0 |

– not determined
1 Average of 10 komatiites (Porter and McKay, 1981)
2 Average of 2 spinifex-textured komatiites, layered series, samples 106 and 108 (Naldrett and Turner, 1977; Ni from Duke and Naldrett, 1978)
3 Average of 18 tremolite-chlorite-cummingtonite schists, main mineralized flow (Barnes et al., 1988a)
4 Tremolite-chlorite-olivine rock with relict spinifex texture, unit 1, sample 73834 (Barrett et al., 1976)
5 Average of 3 "fresh" spinifex-textured rocks (Stolz and Nesbitt, 1981)
6 Spinifex-textured komatiite, basal host unit, KD134/482.5′ (Lesher, 1983)
7 Spinifex-textured komatiite, basal host unit, sample 90699 (Lesher, 1983)
8 Spinifex-textures komatiite, basal host unit, sample 90282 (Lesher, 1983)
9 Spinifex-textured komatiite, basal host unit, sample 90297 (C.M.L., unpubl. data)
10 Spinifex-textured komatiite, basal host unit, sample 90305 (C.M.L., unpubl. data)

| | 11<br>Wannaway | 12<br>Windarra | 13<br>Alexo | 14<br>Dundonald | 15<br>Hart | 16<br>Texmont | 17<br>Langmuir | 18<br>Dumont | 19<br>Katiniq | 20<br>Shangani |
|---|---|---|---|---|---|---|---|---|---|---|
| $SiO_2$ | 46.4 | 51.22 | 45.27 | 45.41 | 46.33 | 47.17 | 47.9 | 47.6 | 46.9 | 50.0 |
| $TiO_2$ | 0.38 | 0.261 | 0.42 | 0.35 | 0.45 | 0.41 | 0.35 | 0.32 | 0.61 | 0.32 |
| $Al_2O_3$ | 8.8 | 4.12 | 8.02 | 6.97 | 8.08 | 8.62 | 6.50 | 6.52 | 9.81 | 6.30 |
| $Cr_2O_3$ | 0.19 | 0.532 | 0.437 | 0.356 | 0.434 | 0.453 | – | 0.3 | – | 0.742 |
| $FeO_t$ | 7.6 | 9.14 | 11.85 | 10.74 | 11.16 | 11.37 | 10.7 | 11.1 | 14.4 | 9.91 |
| MnO | 0.15 | 0.16 | 0.19 | 0.20 | 0.19 | 0.19 | 0.21 | 0.18 | 0.3 | 0.25 |
| MgO | 28.2 | 30.03 | 25.94 | 28.77 | 24.96 | 23.21 | 28.3 | 27.6 | 18.9 | 26.44 |
| NiO | 0.258 | 0.224 | 0.180 | 0.219 | 0.145 | 0.143 | – | 0.178 | – | 0.152 |
| CaO | 7.62 | 4.22 | 7.58 | 6.88 | 8.14 | 8.15 | 6.00 | 6.02 | 8.6 | 5.79 |
| $Na_2O$ | 0.12 | 0.06 | 0.04 | 0.04 | 0.04 | 0.18 | – | 0.15 | 0.3 | 0.01 |
| $K_2O$ | 0.18 | 0.03 | 0.04 | 0.04 | 0.04 | 0.04 | – | – | 0.05 | 0.01 |
| $P_2O_5$ | 0.09 | 0.00 | 0.03 | 0.03 | 0.04 | 0.06 | – | – | 0.15 | 0.06 |
| S | 0.20 | – | – | – | – | 0.09 | – | – | – | 0.09 |
| Volatiles | 7.12 | – | 9.43 | 8.84 | 7.86 | 5.93 | 10.12 | – | – | 6.34 |
| $Al_2O_3/TiO_2$ | 23.1 | 15.8 | 19.2 | 20.0 | 18.1 | 21.1 | 18.6 | 20.3 | 16.1 | 19.7 |
| $CaO/Al_2O_3$ | 0.87 | 1.02 | 0.95 | 0.99 | 1.01 | 0.95 | 0.92 | 0.92 | 0.88 | 0.92 |
| Fe + Mn + Mg + Ni/Si | 1.05 | 1.03 | 1.08 | 1.15 | 1.01 | 0.94 | 1.07 | 1.06 | 0.86 | 0.96 |
| Mg/Fe + Mg | 86.8 | 85.4 | 79.6 | 82.7 | 80.0 | 78.4 | 82.5 | 81.5 | 70.1 | 82.6 |

– not determined
11 Aphric komatiite, basal host unit, sample 84515 (McQueen, 1981a)
12 Average of 5 tremolite-chlorite-carbonate-serpentine rocks, upper zones of lower thick units (Groves and Hudson, 1981)
13 Upper chilled margin, mineralized flow, small pit, AX57 (Barnes, 1985)
14 Spinifex-textured komatiite, sample DU22 (Barnes, 1985)
15 Spinifex-textured komatiite, sample HT22 (Barnes, 1985)
16 Spinifex-textured komatiite, sample TX23 (Barnes, 1985)
17 Spinifex-textured flow top (5′ composite), sample L7653 (Green and Naldrett, 1981)
18 Model composition of parental magma (Duke, 1986)
19 Model composition of parental magma (Barnes et al., 1982)
20 Average of 4 metapyroxenites, upper selvedge of mineralized body, samples V25, V25A, V107, and V107B (Viljoen and Bernasconi, 1979)

TABLE 5.7—Whole-rock geochemical analyses of komatiitic peridotites. Analyses recalculated to 100% volatile-free.

| | 1<br>Scotia | 2<br>Nepean | 3<br>Lunnon | 4<br>Victor | 5<br>Agnew | 6<br>Agnew | 7<br>Wannaway | 8<br>Windarra | 9<br>Alexo | 10<br>Shangani | 11<br>Damba |
|---|---|---|---|---|---|---|---|---|---|---|---|
| $SiO_2$ | 42.92 | 44.26 | 44.88 | 42.89 | 43.1 | 44.1 | 43.2 | 47.2 | 45.54 | 42.96 | 45.0 |
| $TiO_2$ | 0.09 | 0.14 | 0.113 | 0.122 | 0.07 | 0.12 | 0.12 | 0.13 | 0.261 | 0.09 | 0.02 |
| $Al_2O_3$ | 1.66 | 2.76 | 2.30 | 2.55 | 1.6 | 2.3 | 2.3 | 2.7 | 5.22 | 2.39 | 2.65 |
| $Cr_2O_3$ | – | 0.35 | 0.287 | 0.331 | 0.18 | 0.24 | 0.27 | 0.69 | 0.090 | 0.62 | 0.32 |
| $FeO_t$ | 8.53 | 6.38 | 8.12 | 7.61 | 7.1 | 7.9 | 8.0 | 8.6 | 8.91 | 10.15 | 9.24 |
| MnO | 0.14 | 0.05 | 0.10 | 0.14 | 0.1 | 0.1 | 0.13 | 0.12 | 0.23 | 0.13 | 0.14 |
| MgO | 46.11 | 45.61 | 43.35 | 43.25 | 46.8 | 44.1 | 45.1 | 38.7 | 34.92 | 42.64 | 41.6 |
| NiO | – | 0.38 | 0.372 | 0.354 | – | – | 0.401 | 0.317 | (0.874) | 0.453 | 0.48 |
| CaO | 0.52 | 0.03 | 0.21 | 2.62 | 0.9 | 1.2 | 0.51 | 1.50 | 4.81 | 0.53 | 0.42 |
| $Na_2O$ | – | 0.01 | 0.27 | 0.13 | 0.0 | 0.0 | <0.01 | 0.05 | 0.00 | 0.01 | 0.02 |
| $K_2O$ | 0.02 | 0.03 | <0.01 | 0.01 | 0.0 | 0.0 | 0.02 | 0.04 | 0.01 | 0.01 | 0.02 |
| $P_2O_5$ | 0.01 | 0.01 | <0.01 | 0.02 | – | – | 0.01 | 0.02 | 0.00 | 0.01 | 0.02 |
| S | – | <0.1 | – | 0.08 | 0.8 | 0.3 | 0.30 | – | 2.31 | <0.01 | – |
| Volatiles | 11.62 | 12.44 | 21.16 | 4.36 | – | – | 12.60 | – | 11.20 | 14.28 | 10.65 |
| $Al_2O_3/TiO_2$ | 18.1 | 19.9 | 20.4 | 20.9 | 23 | 19 | 20.0 | 20.8 | 20.0 | 25.3 | – |
| $CaO/Al_2O_3$ | 0.31 | 0.01 | 0.09 | 1.03 | 0.56 | 0.52 | 0.22 | 0.56 | 0.92 | 0.22 | 0.16 |
| Fe + Mn + Mg + Ni/Si | 1.77 | 1.66 | 1.60 | 1.66 | 1.8 | 1.6 | 1.72 | 1.38 | 1.30 | 1.69 | 1.56 |
| Mg/Fe + Mg | 90.6 | 92.7 | 90.5 | 91.0 | 92.0 | 91.0 | 91.0 | 88.9 | 87.5 | 88.2 | 88.9 |

1 Average of 48 komatiitic peridotites, ore-bearing ultramafic (Stolz and Nesbitt, 1981)
2 Average of 3 komatiitic peridotites, host units 2 and 3, samples 73736, 73905, and 73906 (Barrett et al., 1976)
3 Average of 15 talc-carbonated komatiitic peridotites, basal host unit, KD29 (Lesher, 1983)
4 Average of 20 komatiitic peridotites with relict igneous olivine, basal host unit, KD6042A (Lesher, 1983)
5 Average of 23 komatiitic peridotites from Perseverence Ultramafic north of dunite lens (Barnes et al., 1988c)
6 Average of 40 komatiitic peridotites from Perseverence Ultramafic south of dunite lens (Barnes et al., 1988c)
7 Average of 2 serpentinized komatiitic peridotites, basal host unit, samples 84520 and 84521 (McQueen, 1981a)
8 Average of 7 serpentinites and talc-carbonate rocks (Groves and Hudson, 1981)
9 Average of 2 komatiitic peridotites, mineralized flow, large pit, AX12 and AX18 (Barnes, 1985)
10 Average of 4 komatiitic peridotites, eastern lobe and stem, mineralized body, samples V42, V43, V69, and V71 (Viljoen and Bernasconi, 1979)
11 Average of 3 komatiitic peridotites, Inyati belt near Damba, samples D7, DAM2, and D4 (Viljoen and Bernasconi, 1979)

TABLE 5.8—Whole-rock geochemical analyses of komatiitic dunites. Analyses recalculated to 100% volatile-free.

| | 1<br>Forrestania | 2<br>Agnew | 3<br>Betheno | 4<br>Honeymoon Well | 5<br>Mt. Goode | 6<br>Six Mile | 7<br>Goliath N. | 8<br>Dumont | 9<br>Katiniq |
|---|---|---|---|---|---|---|---|---|---|
| $SiO_2$ | 41.34 | 41.7 | 41.06 | 42.0 | 42.71 | 40.36 | 41.02 | 40.8 | 43.95 |
| $TiO_2$ | 0.04 | 0.01 | 0.01 | 0.01 | 0.03 | 0.02 | 0.01 | 0.004 | 0.31 |
| $Al_2O_3$ | 0.29 | 0.3 | 0.15 | 0.13 | 0.56 | 0.29 | 0.43 | 0.24 | 4.73 |
| $Cr_2O_3$ | 0.045 | 0.23 | 0.25 | 0.20 | 0.22 | 0.21 | 0.34 | 0.20 | 0.66 |
| $FeO_t$ | 8.33 | 5.6 | 6.33 | 6.21 | 7.34 | 7.13 | 8.82 | 7.89 | 11.52 |
| MnO | 0.10 | 0.1 | 0.13 | 0.11 | 0.12 | 0.09 | 0.15 | 0.12 | 0.18 |
| MgO | 49.09 | 51.9 | 51.48 | 50.5 | 48.14 | 51.86 | 49.00 | 50.1 | 34.62 |
| NiO | 0.429 | – | 0.46 | 0.43 | 0.48 | – | – | 0.412 | 0.74 |
| CaO | 0.26 | 0.1 | 0.10 | 0.13 | 0.08 | 0.04 | 0.21 | 0.30 | 2.92 |
| $Na_2O$ | 0.03 | 0.0 | 0.01 | 0.24 | 0.30 | <0.01 | 0.01 | <0.01 | 0.19 |
| $K_2O$ | 0.02 | 0.0 | 0.01 | 0.01 | 0.01 | <0.01 | <0.01 | <0.01 | 0.02 |
| $P_2O_5$ | 0.02 | – | 0.01 | 0.02 | 0.02 | – | – | – | 0.17 |
| S | – | 1.0 | 0.04 | 0.07 | 0.31 | – | – | – | – |
| Volatiles | – | – | 4.47 | – | – | 15.37 | 16.48 | – | – |
| $Al_2O_3/TiO_2$ | 7.3 | 30.0 | 14.0 | 13.0 | 22.4 | 14.5 | 32.7 | 60.0 | 15.2 |
| $CaO/Al_2O_3$ | 0.90 | 0.33 | 0.71 | 1.00 | 0.14 | 0.12 | 0.48 | 1.25 | 0.62 |
| Fe + Mn + Mg + Ni/Si | 1.95 | 1.97 | 2.01 | 1.93 | 1.84 | 2.07 | 1.96 | 2.00 | 1.41 |
| Mg/Fe + Mg | 91.3 | 94.3 | 93.6 | 93.5 | 92.1 | 92.8 | 90.8 | 91.9 | 84.3 |

1 Average of 4 komatiitic dunites (Porter and McKay, 1981)
2 Average of 12 komatiitic dunites from central part of dunite lens, Perseverence Ultramafic (Barnes et al., 1988c)
3 Average of 2 least-altered komatiitic dunites, samples MKD52/1699' and MKD52/1465' (Donaldson, 1983)
4 Average of 2 lizardite serpentinites, samples 87191 and 87181 (Donaldson, 1983)
5 Average of 2 lizardite serpentinites, samples 87239 and 87256 (Donaldson, 1983)
6 Average of 3 komatiitic dunites, samples 234, 253, and 272 (Naldrett and Turner, 1977)
7 Average of 3 komatiitic dunites, samples 375, 376, and 377 (Naldrett and Turner, 1977)
8 Average of dunite subzone (Duke, 1986)
9 Average of 3 dunites, samples 480, 516, and 648 (Barnes et al., 1982)

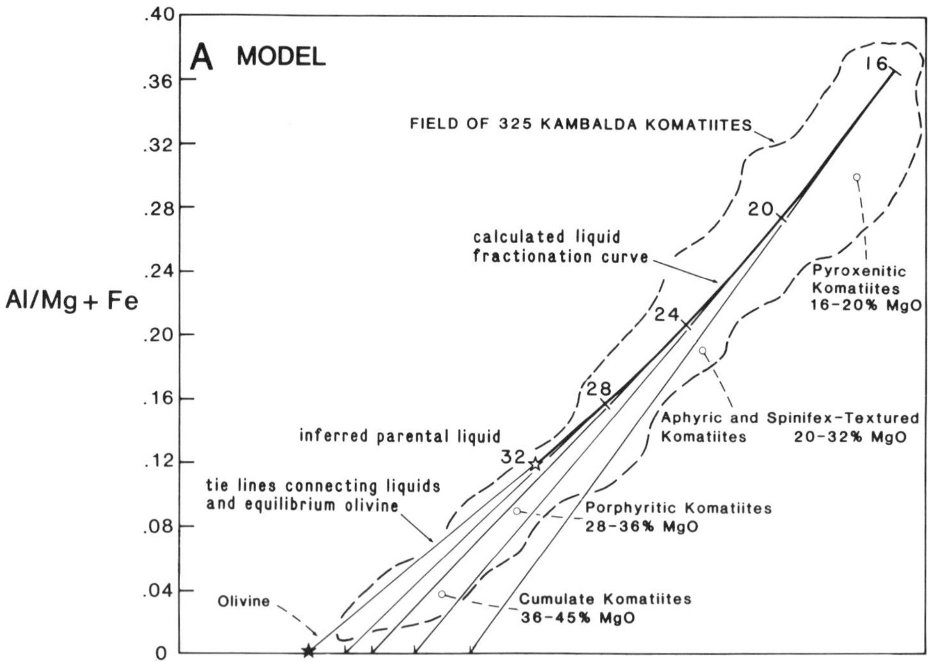


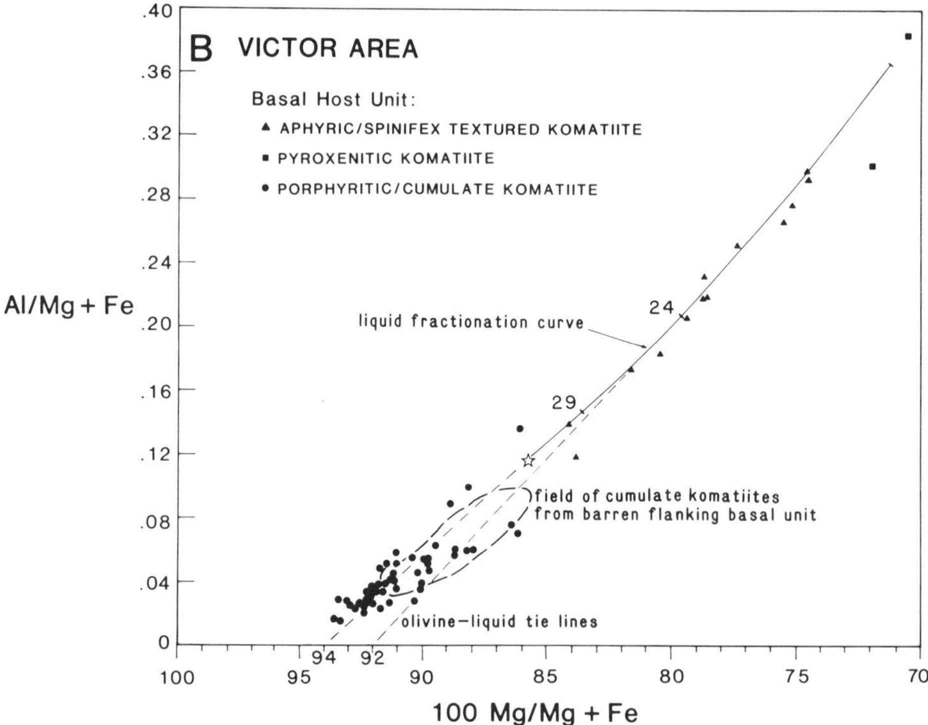


FIGURE 5.20—Reference diagram (**A**) and fractional crystallization model (**B**) for Kambalda komatiites at Victor shoot, Kambalda (after Lesher and Groves, 1984, as adapted from Irvine, 1977b; data from Lesher, 1983). Cumulate tie-lines are tangential to liquid curve in equilibrium with olivine which plots on horizontal axis. Trend of cumulates is oblique to that of liquids and fields from basal host unit and barren flanking units span different ranges of tie-lines (cf. Figure 5.21).

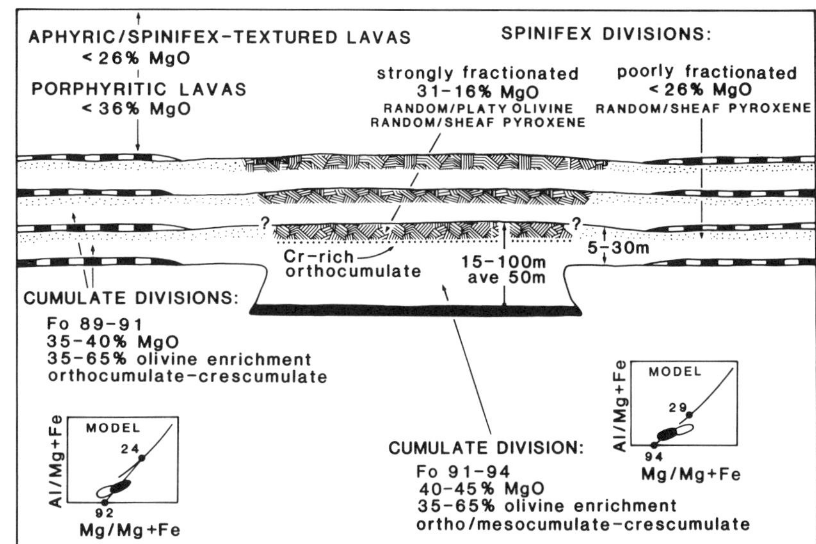


FIGURE 5.21—Sketch summarizing stratigraphic, textural, and geochemical relationships at a typical Kambalda ore shoot (modified from Lesher and Groves, 1984). Ornamentation as for Figure 5.4.

TABLE 5.9—Whole-rock geochemical analyses of lower marginal zones of komatiitic dunites, lateral/lower margins of komatiitic peridotites, and metasomatized komatiitic peridotites. Analyses recalculated to 100% volatile-free.

| | 1<br>Mt. Clifford | 2<br>Marshall Pool | 3<br>Lunnon | 4<br>Lunnon | 5<br>Victor | 6<br>Gibb | 7<br>Wannaway | 8<br>Alexo | 9<br>Alexo | 10<br>Lunnon | 11<br>Lunnon | 12<br>Durkin |
|---|---|---|---|---|---|---|---|---|---|---|---|---|
| $SiO_2$ | 47.87 | 50.38 | 49.11 | 49.68 | 46.04 | 48.52 | 43.94 | 47.54 | 45.70 | 42.50 | 56.26 | 53.33 |
| $TiO_2$ | 0.42 | 0.43 | 0.37 | 0.39 | 0.49 | 0.39 | 0.31 | 0.39 | 0.37 | 0.130 | 0.072 | 0.137 |
| $Al_2O_3$ | 8.08 | 7.85 | 7.51 | 7.89 | 9.17 | 7.61 | 6.46 | 8.51 | 8.35 | 2.77 | 1.84 | 4.05 |
| $Cr_2O_3$ | 0.39 | 0.46 | 0.24 | – | 0.25 | 0.31 | 0.42 | 0.45 | 0.45 | 0.270 | 0.136 | 0.187 |
| $FeO_t$ | 11.30 | 8.25 | 10.61 | 9.93 | 11.22 | 9.79 | 15.34 | 8.43 | 10.95 | 8.11 | 8.09 | 7.53 |
| MnO | 0.19 | 0.16 | 0.19 | 0.24 | 0.17 | 0.19 | 0.18 | 0.19 | 0.24 | 0.22 | 0.24 | 0.20 |
| MgO | 20.37 | 20.66 | 23.03 | 24.36 | 22.96 | 25.61 | 25.20 | 24.01 | 23.99 | 31.37 | 20.08 | 23.72 |
| NiO | 0.10 | 0.08 | – | – | 0.24 | – | 0.15 | 0.53 | 0.63 | – | – | – |
| CaO | 9.50 | 11.42 | 8.67 | 7.24 | 8.92 | 7.29 | 7.68 | 9.68 | 9.20 | 14.49 | 12.95 | 10.70 |
| $Na_2O$ | 0.57 | 0.13 | 0.23 | 0.16 | 0.34 | 0.22 | 0.26 | 0.13 | 0.03 | 0.11 | 0.14 | 0.12 |
| $K_2O$ | 1.11 | 0.08 | 0.01 | 0.02 | 0.10 | 0.02 | 0.03 | 0.05 | 0.05 | 0.03 | 0.20 | 0.01 |
| $P_2O_5$ | 0.10 | 0.11 | 0.04 | 0.09 | 0.12 | 0.03 | 0.03 | 0.08 | 0.02 | 0.01 | <0.01 | 0.02 |
| S | 1.71 | 0.51 | – | – | – | – | 3.30 | 5.38 | 1.81 | – | – | – |
| Volatiles | – | – | 4.71 | 7.47 | 5.05 | 5.29 | 4.80 | 4.00 | 6.95 | 20.50 | 1.87 | 4.28 |
| $Al_2O_3/TiO_2$ | 19.3 | 18.4 | 20.5 | 20.1 | 18.6 | 19.6 | 20.7 | 21.6 | 22.3 | 21.3 | 25.6 | 29.6 |
| $CaO/Al_2O_3$ | 1.18 | 1.46 | 1.15 | 0.92 | 0.97 | 0.96 | 1.19 | 1.14 | 1.10 | 5.23 | 7.04 | 2.64 |
| Fe + Mn + Mg + Ni/Si | 0.84 | 0.75 | 0.88 | 0.90 | 0.95 | 0.96 | 1.15 | 0.91 | 1.00 | 1.26 | 0.66 | 0.78 |
| Mg/Fe + Mg | 76.3 | 81.7 | 79.5 | 81.4 | 78.5 | 82.3 | 74.6 | 83.5 | 79.6 | 87.3 | 81.6 | 84.9 |
| footwall/adj. rock | sed | sed | basalt | basalt | basalt | basalt | basalt | andesite | andesite | basalt | basalt | basalt |

1 Contact facies (cf. analysis 2, Table 5.6), sample 85350 (Donaldson et al., 1986)
2 Contact facies (cf. analysis 2, Table 5.6), sample 85442 (Donaldson et al., 1986)
3 Lateral margin (cf. analysis 6, Table 5.6), Lunnon TSW1004, sample 91974 (Lesher, 1983)
4 Lateral margin (cf. analysis 6, Table 5.6), Lunnon UR704/8S, sample 91819 (Lesher, 1983)
5 Lateral margin (cf. analysis 7, Table 5.6), VF1, sample 91844 (Lesher, 1983)
6 Lateral margin, Gibb 3LS, sample 91789 (Lesher, 1983)
7 Lower chilled margin (cf. analysis 11, Table 5.6), basal host unit, sample 84529 (McQueen, 1981a)
8 Lower chilled margin, mineralized flow, main pit, sample AX26 (Barnes et al., 1985)
9 Lower chilled margin (cf. analysis 13, Table 5.6), mineralized flow, small pit, sample AX55 (Barnes et al., 1985)
10 Metasomatized komatiitic peridotite adjacent to Lunnon Main Shear (LF/1: Fig. 30) (Lesher, 1983); note high Ca, low Ti, Al, and Cr
11 Metasomatized komatiitic peridotite adjacent to Lunnon No. 1 East Thrust (1/ET: Fig. 30) (Lesher, 1983); note high Ca, low Ti, Al, and Cr
12 Metasomatized komatiitic peridotite adjacent to Durkin 503 Thrust (Lesher, 1983); note high Ca, low Ti, Al, and Cr

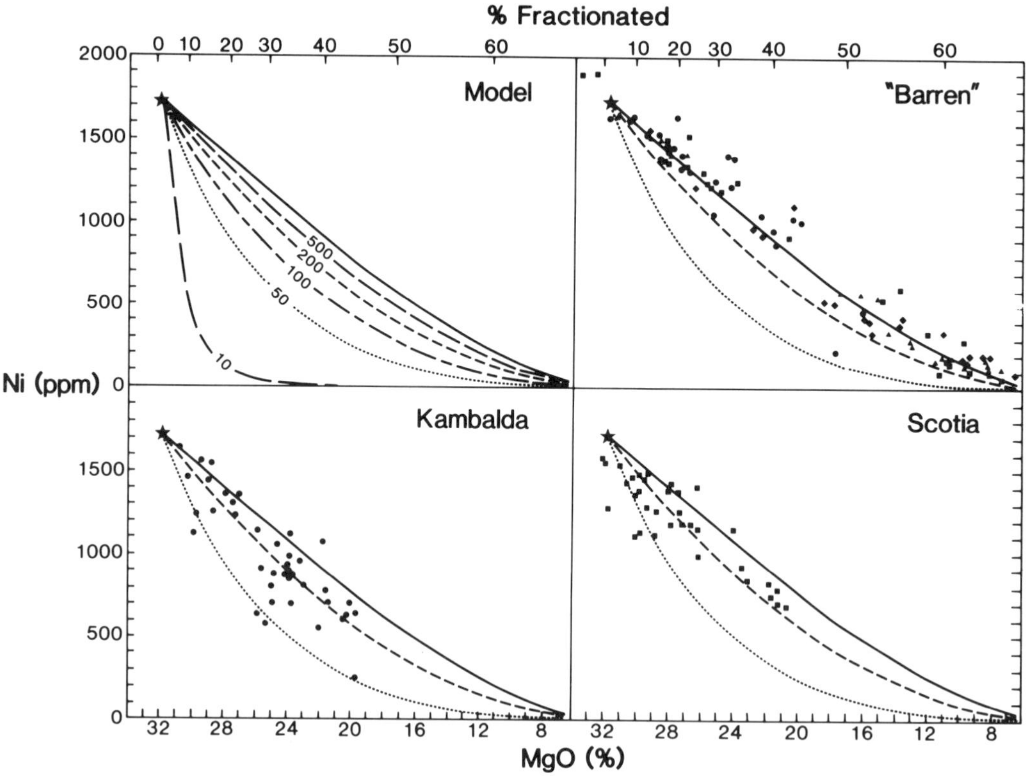


FIGURE 5.22—Fractional crystallization model of Ni–MgO relationships in aphyric and spinifex-textured komatiites (after Naldrett et al., 1984). **A**: Reference diagram showing different magma: sulfide ratios. **B**: "Barren" komatiites from Barberton (diamonds), Belingwe (circles), Yakabindie (triangles), and Munro (squares). **C** and **D**: Mineralized komatiites from Kambalda and Scotia. Data from Herrmann et al. (1976), Nisbet et al. (1977), Naldrett and Turner (1977), J.M. Duke (unpubl. data), Lesher et al. (1981), and Stolz (1981).

and mesocumulates, in which some cumulus growth must occur, are inconsistent with that interpretation.

## MINERALIZATION

### Distribution

Mineralization in komatiite-associated nickel sulfide deposits varies from stratiform, massive/matrix/disseminated sulfides in classes IA and IIA through strata-bound, disseminated/blebby sulfides in class IB to strata-bound, fine disseminated sulfides in class IIB. Stratiform mineralization occurs at or near the base of the basal host unit and is referred to as *contact ore*. Stratiform mineralization at the base of overlying flow units is referred to as *hanging wall ore*. Strata-bound, coarse disseminated spheroidal sulfides, termed *blebby ore*, occur centrally disposed within some komatiitic peridotite host units, isolated from contact ores (e.g. Otter shoot: Keele and Nickel, 1974). Isolated zones of fine disseminated sulfides are often termed *cloud sulfides*.

A typical contact ore profile comprises a thin, discontinuous layer of massive sulfides (<20% gangue) directly overlying footwall rock and overlain successively by a thick, more continuous layer of *matrix (net-textured)* sulfides (20–60% gangue), *disseminated sulfides* (60–90% gangue), and barren host rocks. The proportion of massive ore appears to increase with shoot size (Marston et al., 1981), but massive ore thicknesses are influenced to a very large degree by deformation (see below). Although some host units contain very thick zones of low-grade disseminated mineralization overlying the contact ores (e.g. Lunnon shoot: Ross and Hopkins, 1975), most contain negligible mineralization outside of the ore zones.

Class IA ore shoots vary considerably in size and grade both within and between major districts. They are generally less than $5 \times 10^6$ tonnes, normally less than $2 \times 10^6$ tonnes, and contain between 2 and 4 percent Ni. The shoots are typically ribbon-like in form with dimensions varying from 100 m to 2.5 km, or more, in length, and from 50 to 250 m in width. Thicknesses vary from 1 to 5 m in those containing more than 2 percent Ni, to 5 to 20 m in lower grade shoots (Marston et al., 1981). Class IB deposits are not well described, and most appear to be subeconomic unless associated with stratiform mineralization (e.g. Keele and Nickel, 1974).

Class IIA ores are generally fairly high grade (1.5–3.5% Ni), but only moderate tonnage ($<5 \times 10^6$ tonnes, up to 40 $\times 10^6$ tonnes in Proterozoic analogs). Ore zones in the Forrestania belt, for example, have dimensions of the order of 800 × 400 × 4 m. Class IIB ores are typically low grade (<1% Ni), but large tonnage (up to $250 \times 10^6$ tonnes). Disseminated zones in the Agnew–Wiluna belt are 100–300 m wide and 1–2 km long, several of which have been drilled

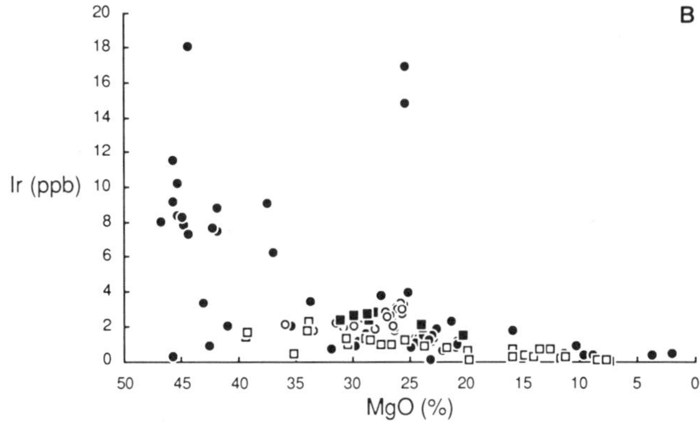


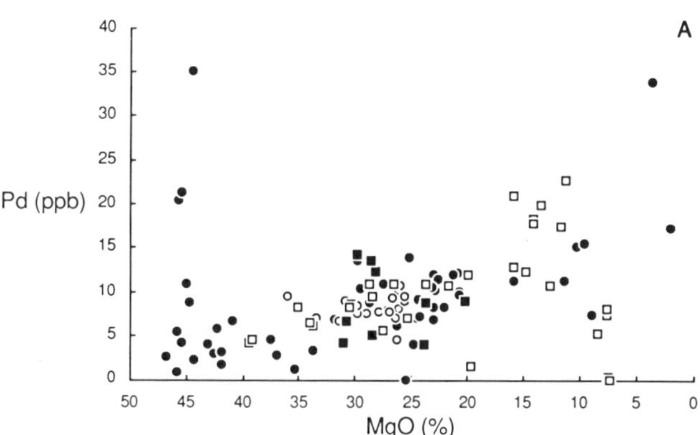


FIGURE 5.23—Pd–MgO (**A**) and Ir–MgO (**B**) variation diagrams (data from Brügmann et al., 1987, Keays et al., 1981; Keays, 1982; and Crockett and MacRae, 1986). Symbols: solid circles = Kambalda, open circles = Mt. Clifford, solid squares = Alexo, and open squares = Munro.

to over 1 km depth (Lesher et al., 1982). At some Forrestania deposits (e.g. New Morning: Porter and McKay, 1981) there are multiple ore zones, both stratiform (contact) and strata-bound (internal). Sulfides at Six Mile are layered on a centimeter scale and abrupt transitions to barren dunite coincide with grain size and textural changes in the dunite (Hill, 1982). In other areas, mineralized dunites grade into sulfide-free dunite, so ore boundaries are arbitrarily defined by cutoff grades.

### Ore mineralogy

Primary (i.e. non-supergene) stratiform ores have a relatively simple mineralogy dominated by pyrrhotite–pentlandite–pyrite, lesser pentlandite–pyrite, and rarer pentlandite–pyrite–millerite sulfide assemblages. Chalcopyrite, magnetite, and ferrochromite are ubiquitous minor phases. Magnetite is more abundant in matrix and disseminated ores, whereas pyrite is more abundant in massive ores (e.g. Cowden and Woolrich, 1987). Groves and Hudson (1981) and Hudson and Donaldson (1984) provide comprehensive lists of rarer native elements, oxides, sulfides, tellurides, arsenides, and antimonides at Kambalda.

Ferrochromites are normally concentrated at the margins of massive ore layers, and in rare, possibly magmatically intrusive ore pinchouts at Kambalda (Fig. 5.24). They are normally coarse-grained (0.5–2.0 mm), euhedral, and most exhibit narrow Cr-magnetite rims; those in the above ore pinchouts are unzoned and contain skeletal inclusions of sulfides (po-pn) or actinolite (after clinopyroxene) (Lesher, 1983). Compositionally, they are Al-poor Fe-chromites with 20–30 mole % $R^{+2}Fe_2O_4$ (Table 5.5; Fig. 5.25a), similar to spinels crystallized experimentally by Ewers et al. (1976) and Woolrich et al. (1981). They are thought to have crystallized at the magmatic stage on interfaces between layers of different sulfide content in response to gradients in $f_{O_2}$ (Marston and Kay, 1980; Woolrich et al., 1981).

Finely disseminated strata-bound sulfides are mineralogically more variable and generally more Ni-rich, as a result of reactions with host rocks during metamorphism (Eckstrand, 1975; Groves et al., 1974; Groves and Keays, 1979; Donaldson, 1981). Typical sulfide assemblages for progressively altered dunites are: pentlandite (dunite), pentlandite–heazlewoodite–magnetite (partially serpentinized dunite), pyrrhotite–pentlandite ± pyrite (black lizardite serpentinite), pyrrhotite–pentlandite or pentlandite–millerite (green antigorite serpentinite), pyrrhotite–pyrite–pentlandite or pyrite–millerite–polydymite (talc-carbonate rock). Chromite and chalcopyrite are ubiquitous accessory phases; additional phases include godlevskite, vaesite, awaruite, bravoite, cobaltite, nickeliferous linnaeite, and cubanite.

### Deformation

Owing to the greater ductility of sulfides than silicate rocks, massive ore layers have been a locus for penetrative deformation resulting in mobilization along fault planes and reforming in response to differential wall rock movements. Contacts between massive ores and more disseminated ores or wall rocks are generally tectonic boundaries. Massive ores are commonly thickened in fold hinges, attenuated along faults, and tectonically displaced into wall rocks (Barrett et al., 1977; Marston and Kay, 1980; Maiden et al., 1986). Repeated deformation of the ores is indicated by cross cutting relationships with late-metamorphic, largely post-tectonic dikes and hydrothermal quartz-carbonate veins.

As a consequence, footwall rocks are often mineralized to varying degrees; tectonic breccia ores are common at Windarra (Barrett et al., 1977; Schmulian, 1984), footwall stringers rich in chalcopyrite and PGE are present beneath many shoots at Kambalda (Ross and Keays, 1979), and interstitial spaces in pillow basalts and broken pillow breccias may be mineralized above and below ore horizons at Kambalda (Lesher and Keays, 1984).

### Metamorphism

Metamorphism ranges from prehnite-pumpellyite facies at Alexo and Dundonald (Barnes, 1985) through lower amphibolite facies at Kambalda (Cowden and Archibald, 1987) to upper amphibolite facies at Nepean and Forrestania (Barrett et al., 1977). Petrofabric (Cowden and Woolrich, 1987) and experimental (McQueen, 1979; Hill, 1984; J. Mac-

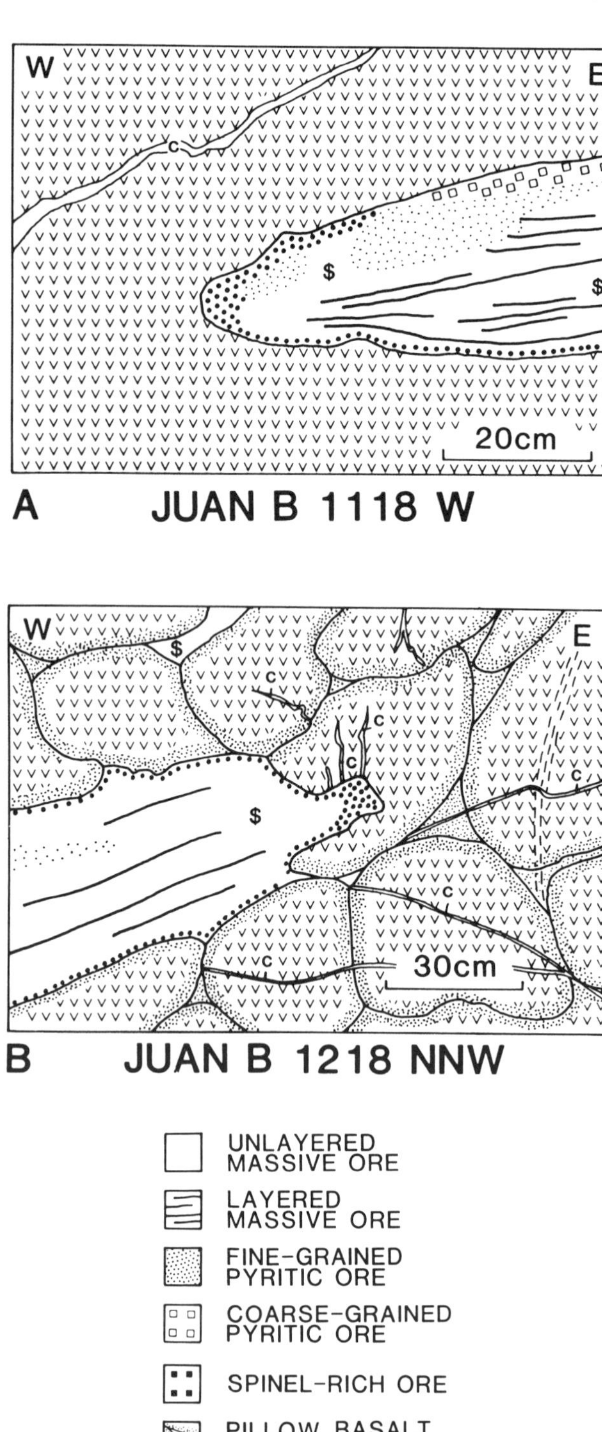


FIGURE 5.24—Face sketches of possible magmatic intrusive ore pinchouts into massive unfoliated basalt at Juan B 1118W (**A**: N18C in Figure 5.14) and pillow basalt at Juan B 1218 NNW (**B**: C18C in Figure 5.14). Both pinchouts are bordered by skeletal ferrochromites (see discussion in text).

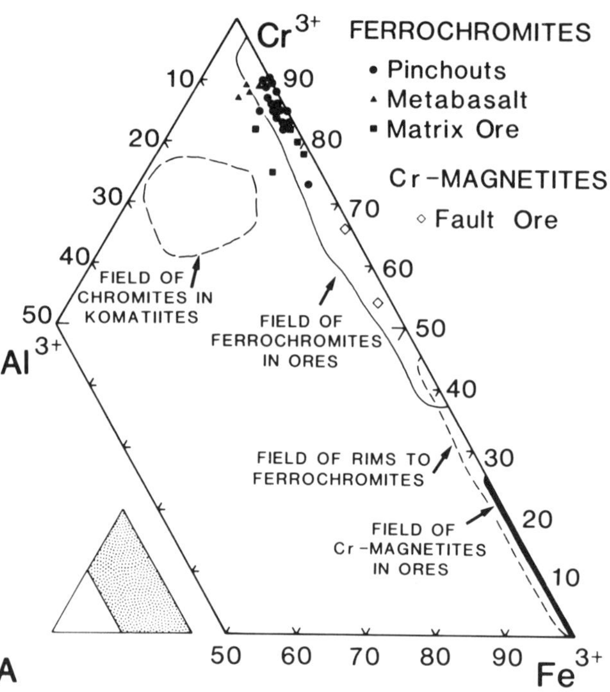


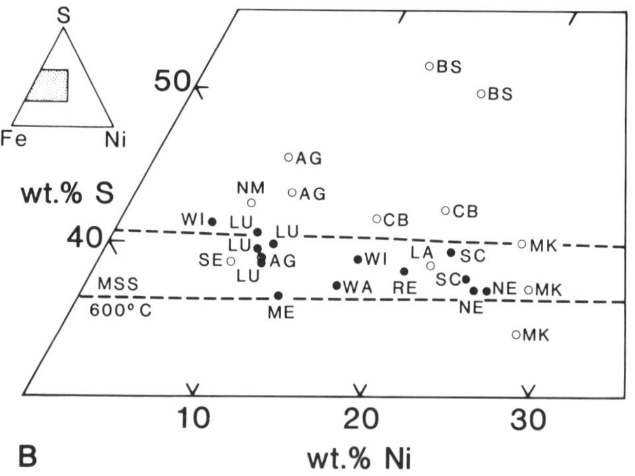


FIGURE 5.25—**A**: $Al^{3+}$–$Cr^{3+}$–$Fe^{3+}$ plot of ferrochromites and Cr-magnetites in ores at Kambalda (after Lesher, 1983; compositional fields from Groves et al., 1981). **B**: Fe–Ni–S plot of ore compositions for some komatiite-associated nickel sulfide deposits (after Marston et al., 1981, who provide sources of data and ore type). AG = Agnew, BS = Black Swan, CB = Cosmic Boy, LA = Liquid Acrobat, LU = Lunnon shoot, ME = Mt. Edward, MK = Mt. Keith, NE = Nepean, NM = New Morning, RE = Redross, SC = Scotia, SE = South Endeavor, WA = Wannaway, WI = Windarra.

donald and M. Paterson, reported in Cowden and Woolrich, 1987) studies indicate that the ores reverted to mixtures of Fe- and/or Ni-rich monosulfide solid solution (MSS), Cu–Fe intermediate solid solution (ISS), pentlandite and/or pyrite, and relict spinels during the metamorphic climax. The precise assemblage depends on temperature and composition, not only Fe–Ni–S, but also Cu content.

As a consequence of deformation and metamorphism, most massive ores exhibit metamorphic phase layering and tectonite fabrics. Ore fabrics at Kambalda mimic tectonite fabrics in adjacent silicate assemblages and preserve the total sequence of deformation (Cowden and Woolrich, 1987), indicating that the ores predate metamorphism and the earliest recognizable deformation. In higher grade, more complexly deformed areas, ore fabrics record only deformation or annealing events at lower metamorphic temperatures following unmixing of Fe–Ni–Cu sulfides from MSS (Barrett et al., 1977). Stress-induced diffusion of Cu and PGE accompanied the deformation of massive ores (Barrett et al., 1977; Keays et al., 1981), and sulfur diffusion via the vapor phase was common during waning metamorphism in some ores Seccombe et al., 1981). Philpotts (1961) and Dillon–Leitch et al. (1986) describe the generation of "reverse net-textured sulfides," almost complete pseudomorphous replacement of serpentinized olivine by sulfides.

In disseminated ores, oxidation, sulfidation, and carbonation reactions between sulfides, igneous silicates, and metamorphic fluids have substantially modified original sulfide compositions (Groves et al., 1974; Eckstrand, 1975; Groves and Keays, 1979; Donaldson, 1981) and possibly also sulfur isotopic compositions (Seccombe et al., 1981). Note that all ore metals, except possibly Ir, are potentially mobile during metamorphism. At Kambalda, for example, Fe, Ni, Cu, Co, Cr, Zn, and especially Pd and Au have been locally mobilized into post-tectonic hydrothermal quartz-sulfide ± carbonate veins in the footwall (Lesher and Keays, 1984).

### Ore chemistry

Because of the desirability of comparing analyses of whole-rock ore samples containing different amounts of gangue phases, it is customary to i) selectively leach sulfides and oxides (e.g. Davis, 1972); ii) recalculate analyses to 100% sulfides (e.g. Ross and Keays, 1979; Naldrett et al., 1979); or iii) recalculate analyses to 100% sulfides plus oxides (Cowden and Woolrich, 1987). Each method involves assumptions regarding the siting of Fe, Ni, and PGE in the rock, and each introduces systematic errors into the analysis. As different workers have used different analytical/recalculation methods, care is required when comparing data from different studies.

Average ore compositions from a number of these deposits have been compiled by Ross and Keays (1979), Marston et al. (1981), and Naldrett (1981). When plotted within the system Fe–Ni–S, most fall within the field of MSS at 600°C (Fig. 5.25b), consistent with a magmatic origin. Individual samples, however, exhibit considerable scatter owing to variable degrees of magmatic fractionation and metamorphic alteration.

Komatiite-associated ores are characterized by high Ni/Cu and low Pd/Ir ratios that distinguish them from most other magmatic sulfide ores (Naldrett, 1981; Barnes et al., 1988) and from hydrothermal and volcanic-exhalative nickel sulfides (Keays et al., 1982). Sulfides hosted by high-Mg komatiites (e.g. Kambalda, Dundonald, Langmuir, Trojan) contain near-chondritic abundances of PGE and exhibit slightly fractionated chondrite-normalized PGE distribution patterns (Pd/Ir = ca. 10), whereas those associated with low-Mg komatiites (e.g. Katiniq), or believed to have equilibrated with fractionated komatiites (e.g. Alexo, possibly Shangani), exhibit more fractionated PGE patterns (Pd/Ir = ca. 30) (Barnes et al., 1985; Barnes and Naldrett, 1987). High-Mg and low-Mg komatiites exhibit complementary patterns at 0.001–0.01x chondritic abundances, consistent with high degree partial melting of a mantle source, minor retention of Ir-group PGE (Os>Ir>Ru) in restite, and strong partitioning of all PGE into sulfide (Barnes et al., 1985).

Most chalcophile elements (Ni, Cu, Co, Pt, Pd, Os, Ir, Rh, and Ru) correlate negatively with Fe, and chalcophile element ratios vary systematically with Ni content (Ross and Keays, 1979; Keays et al., 1981; Barnes and Naldrett, 1986; Cowden et al., 1986; Cowden and Woolrich, 1987). Metal ratios are relatively constant within single ore horizons, but vary considerably from deposit to deposit, among different ore shoots within a single deposit, and across individual ore profiles. The ranges at Kambalda are given in Table 5.10. Variations across ore profiles have been attributed to metamorphic mobilization and/or magmatic fractionation (cf. Keays and Davison, 1976; Keays et al., 1981; Barnes et al., 1985; Cowden et al., 1986). Variations between deposits are interpreted to represent original magmatic variations (Ross and

TABLE 5.10—Compositional variation within Kambalda nickel ores.[1]

| Element | Range | Ratio | Range |
|---|---|---|---|
| Fe | 54–38% | Ni/Fe | 0.15–0.57 |
| Ni | 8–22% | Ni/S | 0.24–0.59 |
| S | 39–35% | Ni/Cu | 9–24 |
| Cu | 0.6–1.3% | Ni/Co | 11–16 |
| Co | 0.2–0.3% | Cu/Co | 3.6–4.5 |
| Cr | 0.18–0.40% | Cr/Zn | 9–10 |
| Zn | 0.02–0.04% | Metal/S | 1.6–2.0 |
| As | 5–20 ppm | | |
| Pb | 15–25 ppm | | |
| Bi | 5–10 ppm | | |
| Ag | 2–4 ppm | | |
| Os | 270–980 ppb | Pd/Ir | 5–14 |
| Ir | 160–350 ppb | Pd/Pt | 0.7–2.6 |
| Ru | 710–1560 ppb | Pt + Pd/Ru + Ir + Os | 1.2–3.9 |
| Rh | 210–490 ppb | | |
| Pt | 1000–2430 ppb | | |
| Pd | 1510–3460 ppb | | |
| Au | 260–700 ppb (up to 7000 ppb at Fisher, Hunt, and Lunnon) | | |
| $\delta^{34}S$ | −0.5 to +3.8‰ | S/Se | 7,600–11,300 |

[1]Recalculated to 100% sulfides; major and minor element data from Cowden and Woolrich (1987), precious metal data from Cowden et al. (1986), sulfur isotopic data from Seccombe et al. (1981), sulfur/selenium data from Groves et al. (1979).

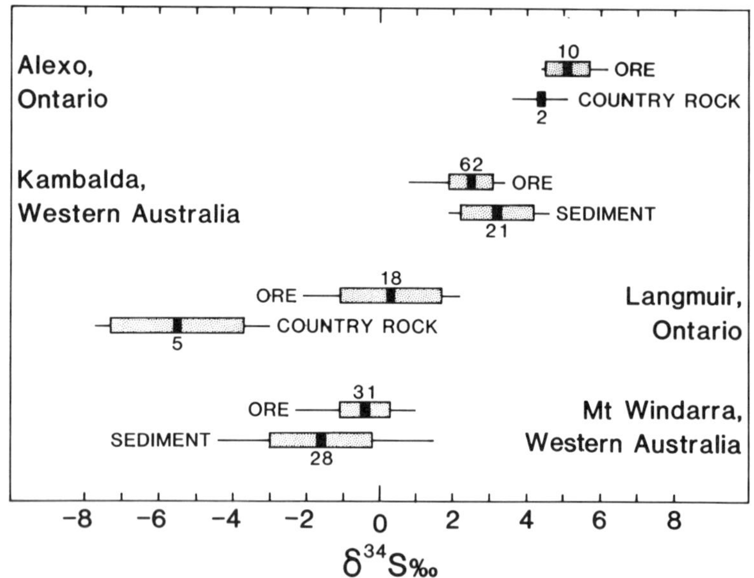


FIGURE 5.26—Sulfur isotopic variations in ores and sulfidic sediments at Alexo, Kambalda, Langmuir, and Mt. Windarra (after Lesher and Groves, 1986). Data from Naldrett (1966), Seccombe et al. (1978, 1981), and Green and Naldrett (1981).

Keays, 1979; Woolrich et al., 1981; Barnes and Naldrett, 1987; Cowden and Woolrich, 1987).

Footwall stringers and hydrothermal sulfide veins are enriched in Cu and PGE (Ross and Keays, 1979; Lesher and Keays, 1984).

### Sulfur isotopes and S/Se ratios

Detailed sulfur isotopic data for ores and sulfidic country rocks in class IA deposits are available only for Alexo, Kambalda, Langmuir #2, and Windarra (Fig. 5.26). Reconnaissance data for ores and sediments at other deposits in the southern part of the Norseman–Wiluna belt (Miriam, Nepean, Scotia, Redross, and Wannaway) are within the range of Kambalda samples (Groves et al., 1979; Donnelly et al., 1978). These data indicate that sulfur isotopic ratios of sulfides in ores and country rocks within an individual district are generally very similar (Langmuir is an exception), but systematically different from other districts. Where significantly different, ores are typically closer to the chondritic value (i.e. zero per mil $\delta^{34}S$) than associated sediments. At Kambalda, sulfur isotopic compositions of pyrrhotites from massive and matrix ores are identical and there is no apparent variation in isotopic composition with stratigraphic height through an ore profile (Seccombe et al., 1981).

Reconnaissance sulfur isotopic values for komatiitic dunite-hosted ores in the northern part of the Norseman–Wiluna belt (Donnelly et al., 1978) are more negative: −2.2 at Agnew (ave. 3 samples) and −2.3 at Mt. Keith (ave. 2 analyses), probably reflecting modification during alteration of the host rocks.

S/Se ratios in ores and sediments exhibit complementary variations (Groves et al., 1979; Green and Naldrett, 1981; Eckstrand and Hulbert, 1987; R. Eckstrand, pers. comm., 1988): 5–10 × $10^3$ (i.e. near-chondritic) at Mt. Edwards, Nepean, Forrestania, Dumont, and Mt. Keith; 5–30 × $10^3$ at Kambalda and Wannaway; and 10–55 × $10^3$ at Langmuir and Windarra.

S/Se ratios of hydrothermal (2–9 × $10^3$) and metamorphically mobilized (6–12 x $10^3$) sulfides are slightly, but significantly lower than normal sulfide ores (10–20 × $10^3$) at Kambalda (Lesher and Keays, 1984), suggesting that S and Se have decoupled slightly during deformation and metamorphism.

## PHYSICAL VOLCANOLOGY OF HOST KOMATIITES

Experimental and theoretical studies suggest that komatiite lavas had very low viscosities (1–10 g/cm-sec), high densities (2.7–2.8 g/cm³), high liquidus temperatures (1500 to 1650°C), a large interval between the liquidus and solidus (350–450°C), and a high heat content (ca. 200 cal/g). The influence of these properties on the physical volcanology of komatiites has been discussed by Nisbet (1982), Huppert et al. (1984), Lesher et al. (1984), Huppert and Sparks (1985a, b), Arndt (1986), Lesher and Groves (1986), Turner et al. (1986), and Hill et al. (1987), and is summarized below.

### Magma generation

High magnesium contents, low abundances of incompatible elements, and near-chondritic ratios for many incompatible elements indicate that komatiites are derived by partial melting of the mantle. Various physical, geochemical, petrogenetic, and thermodynamic constraints have been used to propose a number of different models for their generation, including: i) high degree partial melting (e.g. Nesbitt et al., 1979), ii) multi-stage melting (e.g. Arndt, 1977; Malyuk, 1985), iii) low degree partial melting at high pressure (e.g. Takahashi and Scarfe, 1985), iv) polybaric assimilation (e.g. Bickle et al., 1977), v) two-stage melting-mixing (e.g. Smith and Erlank, 1982), and vi) derivation from a laterally extensive mantle melt layer (e.g. Nisbet and Walker, 1982). Each of these models appears internally consistent, but none individually accounts for the geochemical variations between suites (i.e. Barberton vs. Belingwe vs. Abitibi vs. Norseman–Wiluna), so some degree of mantle heterogeneity has been inferred (Beswick, 1982; Ludden and Glinas, 1982; Smith and Erlank, 1982).

The degree of partial melting required is especially important because it influences the volume of magma produced and therefore i) the discharge rate and ii) the sulfur content of the melt and the likelihood of it being saturated in sulfide (see below). As many komatiites, especially those at Kambalda, exhibit near-chondritic incompatible element ratios, high degrees of melting are likely.

### Ascent and eruption

The eruption rate of magma varies inversely with viscosity, so komatiites should ascend rapidly, have high discharge rates, and form very extensive flows, consistent with their broad regional extent. Huppert and Sparks (1985b) modeled the vertical ascent of komatiite magma and concluded that high temperature, low viscosity komatiites should ascend turbulently, may melt and assimilate wall rocks, and could become contaminated by up to 30% of crustal materials. High MgO contents, LREE-depletion, low abundances of incompatible elements, and near-chondritic incompatible element ratios indicate that the degree of crustal contamination of the komatiites at Kambalda is very small (Arndt and Jenner, 1986; Lesher and Arndt, in prep.), but geochemical and isotopic studies by Chauvel et al. (1985), Arndt and Jenner (1986), and Compston et al. (1986) do suggest that similar komatiites may have assimilated up to 8% of granitic crust and fractionated to generate geochemically anomalous, variolitic komatiitic basalts containing zircon xenocrysts.

This, plus the upwards change from thick, olivine-enriched to thin, unenriched flows in most mineralized komatiite sequences, suggests an evolutionary change in the rate of magma eruption and degree of crustal contamination. The earliest komatiitic magmas (i.e. komatiites) probably erupted very rapidly through unheated conduits and consequently assimilated volumetrically negligible amounts of crustal material. Later magmas (i.e. komatiitic basalts) probably erupted less rapidly through heated conduits and consequently assimilated volumetrically more crustal material. As the least contaminated lavas are mineralized and the most-contaminated lavas are barren, crustal contamination during eruption is not considered to have been important in the genesis of sulfide ores.

### Fractional crystallization

The magnesium contents of lavas decreases upwards through the komatiite sequence at Kambalda (Fig. 5.18) and many other deposits. Part of this variation is attributable to decreasing amounts of olivine accumulation, but aphyric lavas in the upper part of the pile are less magnesian than the upper chilled margins of cumulate flows in the lower part. If the lavas had erupted from a proximal eruptive site, these geochemical variations would require fractionation in a high level magma chamber prior to eruption, but if the lavas are distal, as inferred for many deposits, fractionation may have occurred also during emplacement.

MgO contents of olivines at Agnew (Hill et al., 1987; Barnes et al., 1988c), Mt. Clifford (M.J. Donaldson, 1982), and Dumont (Duke, 1986a) increase upwards through the host units. This has been attributed to decreasing amounts of olivine fractionation prior to emplacement, but may also reflect higher discharge rates in the upper parts, resulting in less fractionation during fractional accumulation (see below), and/or more contamination in the lower parts.

### Lava emplacement

High temperature, low viscosity, high density komatiite magma reaching the sea-floor should form highly mobile flows which, unless ponded, should travel great distances. Because of the high heat content and large interval between the liquidus and solidus, komatiite lava should solidify relatively slowly, resulting in very extensive flows.

Lava flows with high discharge rates should become channelized, especially within topographic depressions in the substrate, with most lava transport being confined to central channels between more slowly flowing margins (Lesher et al., 1984). Fluctuations in discharge rate may cause lava channels to overflow, producing compound flanking units (see also Cowden, 1988). If komatiites erupted into crustal rift zones, as inferred above, then emplacement of lava piles may have been controlled by topographic gradients within the rift zones (Lesher et al., 1981; Groves et al., 1984). Such piles would comprise multiple overlapping lava channels, separated by massive sheet flows and rare lava ponds, grading down the pitchline into compound lava flows.

Observed facies variations (Table 5.11; Fig. 5.27), reflecting variable degrees of olivine enrichment and internal differentiation, probably result from variations in effusion rate (decreasing with time and distance from vent), degree of channelization (variable across strike), and rate of cooling and crystallization (increasing with distance from vent). Pyroclastic komatiites are rare, probably because of rapid exsolution of volatiles (if any) from low viscosity lava.

Huppert et al. (1984) and Huppert and Sparks (1985a) modeled the emplacement of komatiite lava and concluded that high temperature, low viscosity komatiite lava should flow turbulently, may melt and assimilate footwall rocks, and could be contaminated by up to 10% of sedimentary and volcanic rocks. This has been confirmed at Kambalda, where field, geochemical, and Pb-isotopic studies by Lesher et al. (1984), Groves et al. (1986), Evans et al. (1988), Frost and Groves (1988), and McNaughton et al. (1988) indicate that komatiite lava channels have thermally eroded i) tops of underlying komatiite flows, producing spinifex-textured ores and domes of silicate liquid (Fig. 5.28), ii) sulfidic sediments, producing rare ocellar komatiites (xenomelts) that accumulated along levees adjacent to lava channels (Fig. 5.28), and iii) footwall basalt, enhancing the re-entrant geometry of the embayments (Fig. 5.24).

Thermal erosion has also been proposed at Agnew (Hill et al., 1987; Barnes et al., 1988a, c).

### Olivine enrichment

Previous models for olivine enrichment in these units include: i) eruption of magma-olivine liquid-crystal mushes (Ross and Hopkins, 1975; Naldrett and Turner, 1977; Groves et al., 1979; Marston et al., 1981; Gresham, 1986), ii) accumulation of intratelluric olivine in overflowing lava ponds (Naldrett, 1973), and iii) riffling of intratelluric olivine into

TABLE 5.11—Physical volcanology of komatiites.[1]

**Massive Komatiite Flows**

compound lava flows formed at lowest effusion rates, levees to differentiated komatiite flows, or distal facies of komatiitic peridotite flows; 1–10 m thick; massive, less commonly pillowed, rarely volcaniclastic; aphyric to porphyritic; compositionally uniform; most common facies in unmineralized sequences and in barren upper parts of mineralized sequences

**Spinifex-textured Komatiite Flows**

lava tubes, compound lava flows formed at low effusion rates, or distal facies of komatiitic peridotite flows; 3–30 m thick; moderately fractionated upper spinifex-textured divisions, lower olivine porphyritic-orthocumulate divisions; bulk composition similar to chilled margins; less common facies in unmineralized sequences and in barren upper parts of mineralized sequences

**Differentiated Komatiite Flows**

lava ponds formed at low effusion rates; 15–100 m thick; strongly fractionated upper spinifex-textured and differentiated gabbroic divisions, lower olivine porphyritic-orthocumulate divisions; bulk composition similar to chilled margins; uncommon facies in unmineralized sequences; not reported in mineralized sequences

**Cumulate Sheet Flows**

*Komatiitic Peridotite*

sheet flows formed at low-intermediate effusion rates, levees to komatiitic peridotite flows, or distal parts of komatiitic dunite flows; 10–30 m thick; massive; thin aphyric upper/lower chilled margins, thick central olivine orthocumulate or crescumulate zones; bulk composition 20–50% more magnesian than chilled margins; minor facies in unmineralized sequences; common facies in barren lower parts of mineralized sequences

*Komatiitic Dunite*

sheet flows formed at high effusion rates, or levees to lava channels; 100–1000 m thick; massive; upper/lower zones of olivine orthocumulate or crescumulate, central zone of olivine mesocumulate and adcumulate; bulk composition 50–70% more magnesian than chilled margins (or estimated liquid composition); not reported in most unmineralized sequences; common facies in barren parts of mineralized sequences

**Cumulate Lava Channels**

*Komatiitic Peridotite*

dynamic lava channels formed at intermediate-high effusion rates; 30–100 m thick; thin, fractionated upper spinifex-textured divisions, anomalously thick lower olivine orthocumulate or crescumulate divisions; bulk composition up to 20–50% more magnesian than chilled margins; localized in embayments; rare in unmineralized sequences; *host units in mineralized sequences*

*Komatiitic Dunite*

dynamic lava channels formed at highest effusion rates; 300–1000 m thick; upper/lower zones of olivine orthocumulate or crescumulate, central zone of olivine mesocumulate and adcumulate; bulk composition 50–70% more magnesian than chilled margins (or estimated liquid composition); localized in embayments; rare in unmineralized sequences; *host units in mineralized sequences*

[1]adapted from Arndt et al. (1979), Lesher et al. (1984), and Hill et al. (1987).

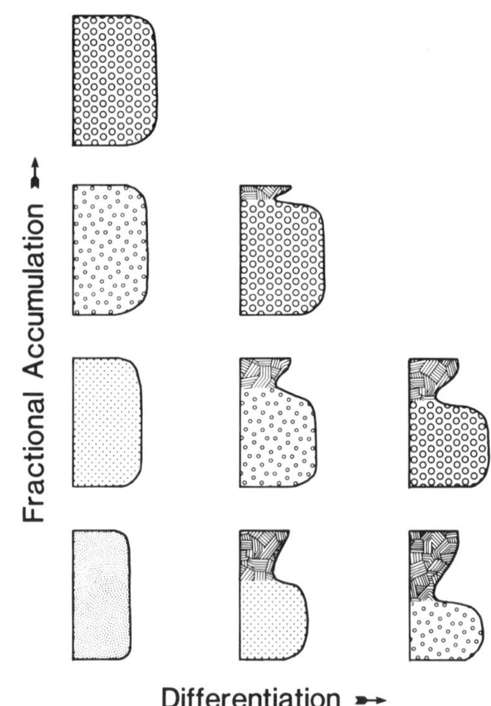


FIGURE 5.27—Facies variations in komatiitic lavas as a function of olivine enrichment (fractional accumulation) and in situ differentiation (fractional crystallization) (modified from Lesher et al., 1984). Aphyric lava is unornamented, other ornamentation as in Figure 5.4. Width of profiles is proportional to MgO content.

footwall embayments (Stolz and Nesbitt, 1981; Naldrett and Campbell, 1982). However, mesocumulate-adcumulate textures in komatiitic dunites, textural and compositional variations of olivine with stratigraphic location within komatiitic peridotites and komatiitic dunites, the absence of phenocrysts in upper chilled margins of komatiitic peridotites, the presence of crescumulate olivine within and at the base of some komatiitic peridotites, and geochemical models of komatiitic peridotites and dunites indicate that most olivine crystallized in situ.

The best interpretation of these units is that they represent dynamic lava channels (Lesher et al., 1984; Lesher and Groves, 1986; Barnes and Naldrett, 1987; Cowden, 1988; see also Barnes et al., 1983; Barnes, 1985) in which olivine crystallized during lava flow; i.e. fractional accumulation. Textural and compositional variations (e.g. Fig. 5.27) reflect varying flow conditions. Continuous or periodic flow of lava within the channel would replenish crystallizing olivine components without producing excessive fractionation of the silicate liquid. The degree of olivine enrichment in the host units at Kambalda (35–65%, calculated by mass balance) could be produced with negligible fractionation of the lava (1–2%) by only 5–10 times the volume of the host unit. The degree of olivine enrichment of dunitic host units would be proportionately larger and would require proportionately larger volumes of lava.

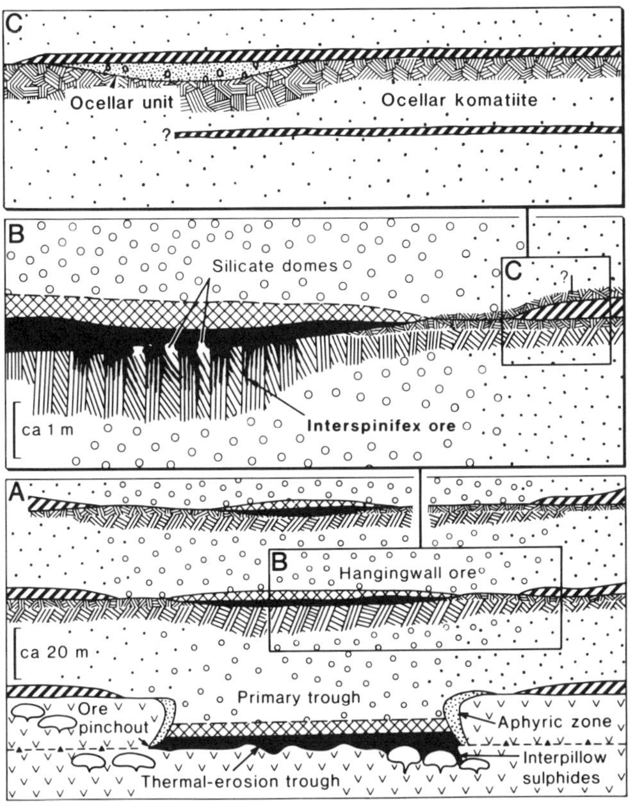


FIGURE 5.28—Sketches summarizing stratigraphic relationships at Kambalda: contact ores (**A**: modified from Lesher et al., 1981), stratiform hanging wall "interspinifex" ores (**B**: after Groves et al., 1986), and ocellar komatiites (**C**: after McNaughton et al., 1988). Ornamentation as in Figure 5.4.

Proximal parts of lava channels process more lava than distal parts, so lava facies should grade from komatiitic dunite through komatiitic peridotite to komatiite, and possibly komatiitic basalt, along the length of the flow (Fig. 5.29). Gresham (1986) pointed out that differentiated lavas have not been observed along strike from the deposits at Kambalda; but initial liquid and cumulate compositions at the Kambalda Dome (Tables 5.6 and 5.7) are slightly more magnesian than those at St. Ives, 20 km to the south (D. Evans, pers. comm., 1988). However, as lavas with large flow rates could extend for hundreds of kilometers (Huppert and Sparks, 1985a), the lava channels (and associated ore deposits) are probably more extensive than the structural blocks in which they are exposed. Note that spinifex-textured rocks associated with komatiitic dunites in the northern part of the Norseman–Wiluna belt are more magnesian (max. 33% MgO: Naldrett and Turner, 1977; Barnes et al., 1988c) than those associated with komatiitic peridotites at Kambalda (max. 31% MgO: Lesher et al., 1981), and that the southernmost part of the belt (Fig. 5.1) is dominated by komatiitic basalt (Gemuts and Theron, 1975).

### Crystallization

Because of their high temperature and low viscosity, komatiites should convect vigorously and cool relatively rapidly (Huppert et al., 1984). The abundance of porphyritic and cumulate olivine textures reflects the large temperature interval over which olivine is the only crystallizing phase (Arndt, 1976). Textural variations are produced by variations in the degree of supercooling and rates of nucleation and crystallization within the lava flow profile (C.H. Donaldson, 1982; Arndt, 1986; Turner et al., 1986). Cooling rates are initially extremely rapid, producing glassy margins at the seawater interface. As the lava crust thickens, cooling rate and degree of supercooling decline and produce in succession: dendritic ("random spinifex"), platy (plate spinifex), branching (crescumulate), hopper (skeletal or embayed), and polyhedral (orthocumulate) olivine textures.

Assimilation of footwall rocks beneath lava channels may depress the liquidus, resulting in sustained crystallization of polyhedral or crescumulate olivine. Away from the central channel, at the terminal flow front, or in associated thinner flows, lavas would cool more rapidly, fractionate more, and crystallize as porphyritic and aphyric flows. The preservation of glassy mesostases in many cumulate rocks indicates that they crystallized rapidly once flow ceased, prior to reaching the clinopyroxene-plagioclase cotectic (ca. 1230°C: Arndt, 1976). This may explain the apparent "gap" in lava compositions between 16 and 20% MgO reported in many areas (see Arndt and Nisbet, 1982) as a frequency minimum between i) the limited range of lava compositions produced by surface fractional crystallization of high-Mg parents and ii) the wider range of lava compositions produced by variable degrees of partial melting, crustal assimilation, and fractional crystallization prior to eruption.

As discussed by Hill et al. (1987), formation of adcumulates requires a dynamic regime in which temperatures never dropped very far below the liquidus. Very low degrees of supercooling favor growth of existing crystals, rather than nucleation of new ones, and continued crystal growth produces a thin boundary layer of olivine-depleted liquid around the crystal. Removal of the boundary layer during vigorous convection or turbulent lava flow inhibits nucleation of new crystals.

### Sediment distribution

The absence of sediments in the ore zones at most Kambalda shoots (Figs. 5.4, 5.12–14) continues to be attributed by some workers to non-deposition (Ross and Hopkins, 1975; Bavinton, 1981; Gresham and Loftus-Hills, 1981; Gresham, 1986; Cowden, 1988). This requires i) a local eruptive site which inhibited sedimentation owing to turbulence in the overlying seawater (if the contact sediment predates emplacement of the basal host unit), or ii) a local topographic high on which sediments did not accumulate (if the contact sediment post-dates emplacement of the host unit.

There is no direct evidence, whatsoever, for a local eruptive site at Kambalda. Despite virtually complete exposure of the footwall during development and mining operations, no feeders have been observed beneath any of the shoots. Systematic grid diamond drilling (hundreds of thousands of meters) has also failed to intersect any transgressive ultramafic units in the area. It is unlikely that all of the feeders

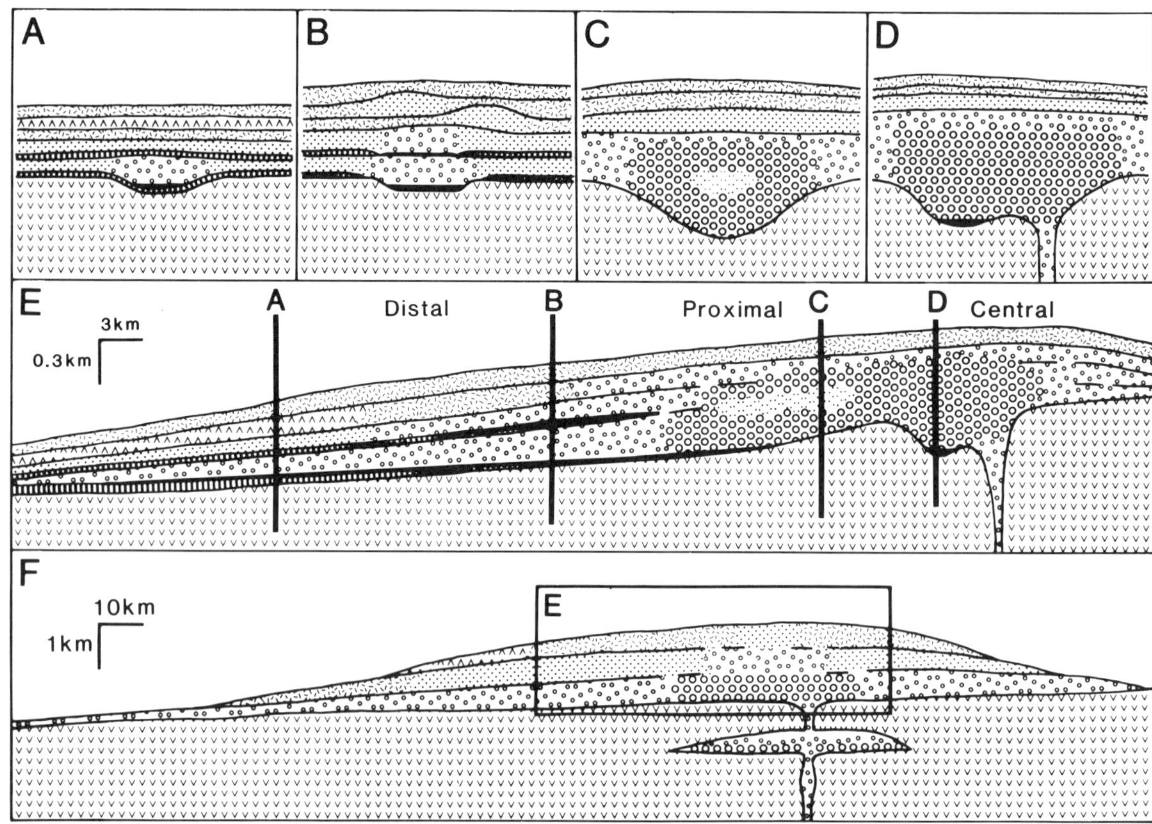


FIGURE 5.29—Schematic cross section through an Archean komatiite volcano showing possible volcanic settings (**A–D**) and regional-scale stratigraphic relationships (**E–F**) of komatiite-associated nickel sulfide deposits (modified from Lesher and Groves, 1986). Ornamentation as in Figure 5.4.

have been obliterated by faulting and dike intrusion (cf. Gresham, 1986).

The second possibility creates a geological dilemma: if the basal host unit was a topographic high that prevented sediment accumulation in the ore zone, then how can the same unit overlie the contact sediment away from the ore zones? In many areas the basal host units are stratigraphically correlative and contiguous with the basal flanking units (Figs. 5.12–14). Cowden's (1988) model implies that the sediments accumulated between the time of emplacement of a central lava channel and lateral spillage onto the flanks. It is very unlikely that 1–10 meters of fine-grained detrital-exhalative sediments (Bavinton, 1981) could accumulate in this time interval, or that the host unit could form such a narrow tube at such low viscosities and high discharge rates.

As noted above, however, there is direct physical evidence that sediments have been thermally eroded and assimilated by the komatiites (Groves et al., 1986) and a variety of other theoretical, stratigraphic, geochemical, and isotopic data to support this interpretation (Huppert et al., 1984; Lesher et al., 1984; McNaughton et al., 1988; Evans et al., 1988; Frost and Groves, 1988).

### Footwall embayments

Sulfide ores, irrespective of whether stratiform or strata-bound, are normally localized in or over embayments in the footwall (Figs. 5.4–15). Resolution of their origin has been hampered by incomplete exposure and a paucity of marker horizons. Many different models have been proposed over the years, but only four persist at the present time: 1) syn-volcanic faulting (Ross and Hopkins, 1975; Gresham, 1986), 2) post-volcanic structural deformation (Cowden and Archibald, 1987; Cowden, 1988), 3) thermal erosion (Huppert et al., 1984; Huppert and Sparks, 1985a; Barnes et al., 1988a, c; Evans et al., 1988), and 4) volcanic topography (Green and Naldrett, 1981; Lesher et al., 1984). None of these are mutually exclusive and it is possible that all have influenced the present geometry of the embayments to one degree or another. The discussion below is aimed mainly at pointing out the difficulties in their interpretation.

### Syn-volcanic faulting

This model is based on the Lunnon shoot, which is bounded by a series of sub-vertical faults (Figs. 5.30a–d) that have

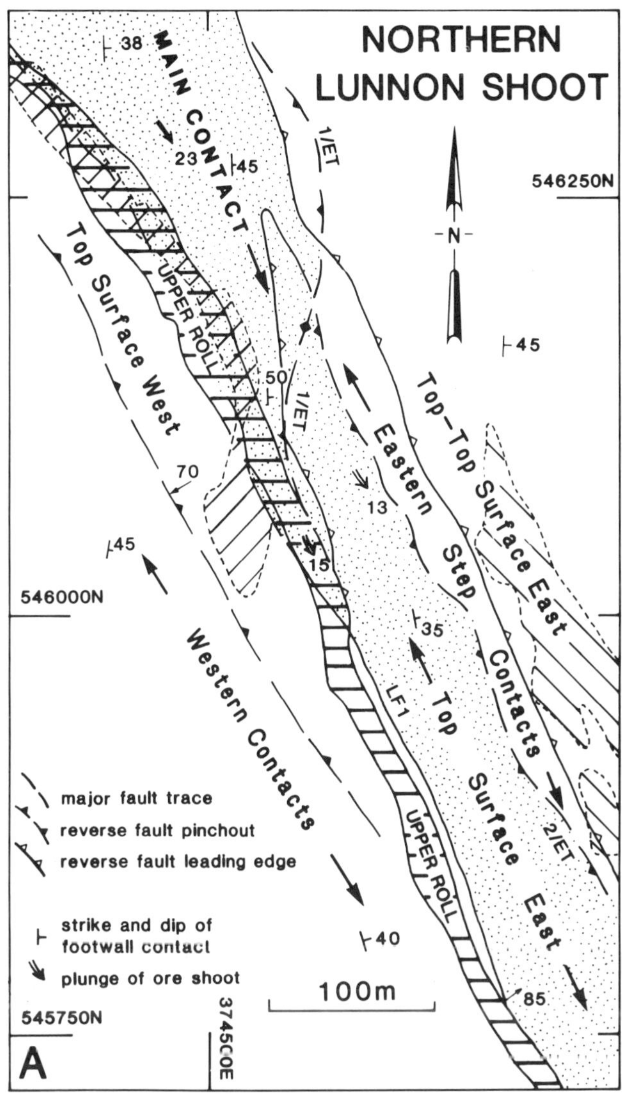


FIGURE 5.30—Map of northern part of Lunnon Shoot, Kambalda showing ore surfaces and fault traces in horizontal projection (**A**: from Middleton, 1980) and cross sections (**B–D**: modified from Middleton, 1980).

been interpreted as syn-volcanic on the basis of apparent differential displacements between the komatiite/basalt contact and marker horizons in the komatiite sequence (Ross and Hopkins, 1975; Gresham and Loftus-Hills, 1981; Gresham, 1986).

As discussed by Lesher (1983; see also Cowden, 1988), however, the Lunnon fault system comprises a series of faults with different senses of displacement that merge and bifurcate along the plunge of the shoot, producing net displacements on marker horizons in the komatiite sequence that appear smaller than those along the komatiite/basalt contact (cf. LF2 and 1/ET: Figs. 5.30b–d). This eliminates the necessity of syn-volcanic movement on most faults at Lunnon, but still leaves a component of pre-tectonic relief between the embayment contact (UR–MC–TSE) and outside flanking contacts (TSW, ES–TTSE).

The remaining "dislocations" (e.g. LF2: Fig. 5.30b) separate contacts that are underlain by different lava facies (cf. TSW and UR: Fig. 5.30b) and/or correlate with cooling unit boundaries in the footwall basalt (e.g. ES: Figs. 5.30c–d). These appear to be primary volcanic discontinuities and may be *modified* by folding and faulting, but could not be *generated* entirely by deformation. Cowden (1988) contended that the embayment at Lunnon was generated from a roughly planar contact by strike-slip faults cutting dipping contacts, but this cannot explain the volcanological and stratigraphic relationships. Some syn-volcanic faulting is possible in a rift zone, but these areas are not analogous to mid-ocean ridges (cf. Gresham, 1986) and it is unlikely that many of the embayments have been generated by syn-volcanic faulting.

### Post-volcanic deformation

Most embayments at Kambalda have been modified to some degree by superimposed polyphase deformation; the question is how much. Lesher et al. (1984) attributed their present re-entrant geometries to deformation, but maintained the presence of 2–30 m deep pre-tectonic embayments. Cowden (1988) suggested that most have been generated entirely by deformation of a near-planar contact. Good cases have been made for both interpretations.

For example, the embayments at Juan (Fig. 5.14) and Hunt (Fig. 5.31) shoots are defined by complex fold-thrust couples developed during $D_2$ deformation. At Hunt, $S_1$ foliations are folded around $F_2$ folds, pillow basalts in the footwall are flattened into the plane of the $S_2$ foliation, parallel to the pitchline of the ore shoot, and there is no evidence of a primary embayment (Cowden, 1988). Similar fold-thrust features are probably present at Ken shoot (Fig. 5.15b). As noted by Cowden (1988) the embayment at Durkin shoot is defined by thrust faults (Fig. 5.12; cf. Marston et al., 1981: fig. 5.10a).

Nevertheless, detailed underground mapping of stratigraphy in the footwall basalt in other areas at Kambalda (Figs. 5.32–35) supports the presence of pre-tectonic embayments. Cooling unit boundaries defined by variations in lava facies (massive vs. pillowed vs. broken-pillow breccia) are unfolded, pillows and flow-top breccias are apparently undeformed, and pillow facings are right-way-up. In many cases, embayment contacts correlate with cooling unit boundaries in the basalt (Figs. 5.32a–b, 33). Regardless of the style of deformation (folding or faulting), differences in lava facies between the embayment contact and flanking basal contact (Figs. 5.34, 35a–b) preclude structural equivalence.

### Thermal erosion

Huppert et al. (1984) and Huppert and Sparks (1985b) suggested that the embayments at Kambalda may have been generated by thermal erosion beneath linear lava channels. This interpretation is consistent with direct field evidence for thermal erosion of sediment and basalt at Kambalda (Lesher, 1983; 1985; Groves et al., 1986; Evans et al., 1988; Frost and Groves, 1988), and the undercut geometry of

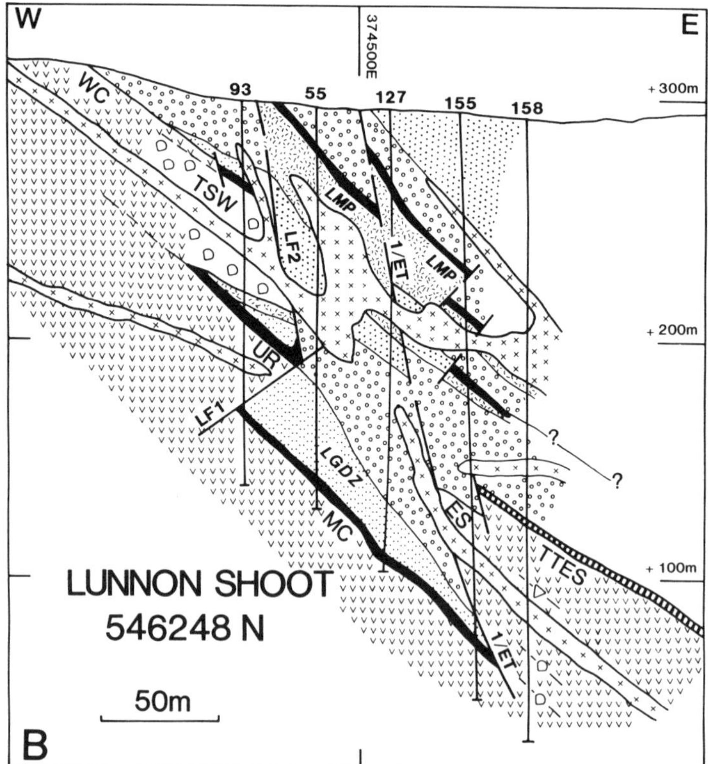


FIGURE 5.30—Cross sections of northern part of Lunnon Shoot, Kambalda (continued).

experimental channels mimics that of many shoots at Kambalda.

Thermal erosion also explains the general correlation between depths of embayments and ore-sediment relationships at class IA deposits. Most embayments at Kambalda, for example, are relatively deep (Figs. 5.12–15) and highly elongate (Fig. 5.3), do not contain sediments, and are flanked by wide zones of barren contact that are occupied by sediments along strike (Fig. 5.4). In contrast, the embayments at Wannaway (McQueen, 1981b), Scotia (Fig. 5.6), and Windarra (Fig. 5.7) are considerably shallower and less elongate, and the ores directly overlie thinned or partially preserved sediments that are contiguous along strike. These variations between deposits are consistent with variable degrees of thermal erosion of sediments and footwall rocks to generate embayments.

The embayments in class IB, IIA, and IIB deposits are also strongly deformed, but most are probably thermal erosional (Lesher and Groves, 1986; Hill et al., 1987). The deep embayment at Agnew, for example, transgresses a considerable amount of the footwall as well as a class IA ore deposit (Fig. 5.5).

There are other field relationships, however, which suggest that some embayments may not have been generated entirely by thermal erosion: 1) embayments at Scotia (Fig. 5.6) and Wannaway (McQueen, 1981b) contain preserved or partially preserved sediments that predate emplacement of the host unit; 2) the bases of most embayments at Kambalda are parallel to the stratigraphy in the footwall, many correlate with cooling unit boundaries in the adjacent basalts, and therefore appear to be stratigraphically conformable within the footwall sequence (Figs. 5.32–35), and 3) some embayments at Kambalda are elliptical (Fig. 5.15) and are not likely to have formed by thermal erosion.

The embayment in the central part of the Foster shoot (Fig. 5.35b) is highly deformed, but there is clear evidence of thermal erosion (Evans et al., 1988). The presence of two different lava facies (massive to the northwest and pillowed to the southeast) suggests that it may have been localized on a topographic feature.

### Volcanic topography

Based on the above stratigraphic and structural relationships, Lesher et al. (1984) proposed that the embayments represent kipukas, inliers between non-overlapping basalt flows, which have been variably modified by thermal erosion and superimposed structural deformation. Volcanic topography would also explain the general correlation between depths of embayments and nature of the substrate that is evident in class IA deposits; embayments in the southern part of the Norseman–Wiluna belt and Timmins area of the Abitibi belt, for example, overlie volcanic rocks and are more pronounced than those in the Windarra district or Zimbabwe Province that overlie sedimentary rocks.

It seems almost certain that there were original topo-

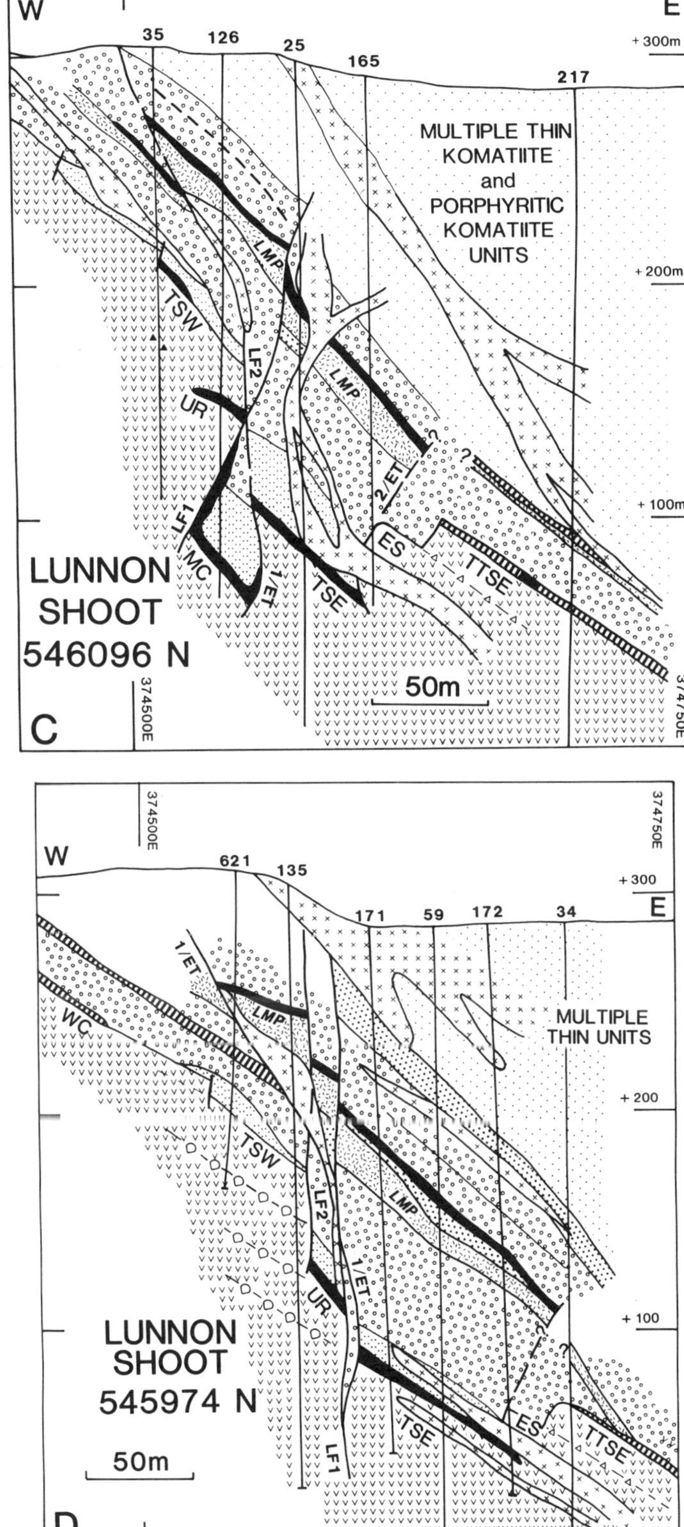


FIGURE 5.30—Cross sections of northern part of Lunnon Shoot, Kambalda (continued).

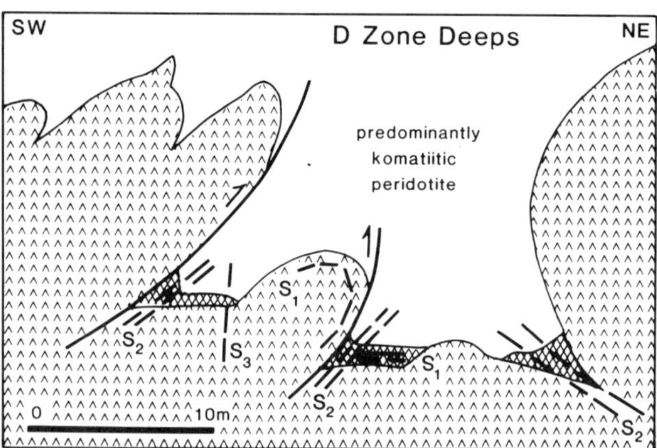


FIGURE 5.31—Simplified cross section of D Zone Deeps area, Hunt shoot, Kambalda showing principal structural elements (after Cowden and Archibald, 1987; based on mapping by B. Thompson). Ornamentation as in Figure 5.4.

graphic variations in volcanic terrains and it seems reasonable that these would focus and channelize lava flow, just as they do in modern volcanic environments. The relative significance of pre-existing topography and subsequent thermal erosion and deformation would vary with the nature of the substrate (mafic vs. felsic vs. sedimentary), the therlaminated sulfidic shales, sulfidic cherts, or sulfide-facies iron-formation, suggesting emplacement into a relatively deep subaqueous environment.

Proximity to eruptive sites may be inferred from proximity to volcanic feeders, physical volcanology of lavas, and stratigraphic relationships with associated volcanic and sedimentary rocks. Although there is some controversy regarding the settings of individual deposits, resulting from different interpretations of volcanological features and stratigraphic relationships, interdeposit comparisons suggest a continuum of volcanic environments ranging from subvolcanic through central and proximal volcanic to distal volcanic (Fig. 5.29). Tentative volcanic settings are given in Table 5.3, and some examples are discussed below.

The Dumont dunite (Fig. 5.11) appears to be derived from a komatiitic parental magma (Table 5.6), but it is significantly different from other class IIB deposits in two important respects (Duke, 1986a): 1) it is intrusive into a sequence of tholeiitic basalts and 2) it contains a thick upper zone of gabbroic differentiates. Thus, it intruded into petrogenetically unrelated rocks and evolved by fractional crystallization without significant magma replenishment. Dumont is probably the only unequivocal example of an intrusive komatiite-associated nickel sulfide deposit. The Shangani and Epoch deposits in Zimbabwe are associated with barren differentiated komatiitic sills (Williams, 1979).

The Shangani deposit (Fig. 5.10) appears to occur directly within a volcanic vent, is associated with a thick diffrenmal activity of the channel (discharge rate, duration of eruption, proximity to eruptive site, etc.), and the style and degree of deformation.

### Volcanic setting

Most host komatiite sequences are underlain and/or overlain by pillow basalts and contain intercalations of planetiated komatiitic intrusive complex, and occurs within the central, thickest part of the ultramafic lava sequence (Viljoen et al., 1976; Viljoen and Bernasconi, 1979; Williams, 1979). Part of the mineralization at the stratigraphically equivalent Silwane deposit (Fig. 5.9) also occurs within a possible footwall feeder (Williams, 1979). These appear to be the only examples of deposits in central volcanic settings.

The Damba (Fig. 5.9) and Langmuir #2 (Green and Naldrett, 1981) deposits occur near possible feeders, and footwall andesites at Langmuir contain barren komatiitic intrusives. Both deposits have been interpreted to occur in proximal volcanic settings (Williams, 1979; Green and Naldrett, 1981). The komatiite sequence at Scotia (Fig. 5.6) contains ultramafic fragmental rocks interpreted as pyroclastic and a thick massive peridotite interpreted as intrusive which have been used to infer a proximal volcanic setting (Page and Schmulian, 1981). However, the breccias contain fragments of all rocks types (spinifex-textured komatiite, porphyritic komatiite, and ortho-mesocumulate komatiite) and may be epiclastic (see Gélinas et al., 1977), and the peridotite may represent the beheaded upper part of the basal host unit (see above).

The majority of these deposits, including all of those in the southern and northern parts of the Norseman–Wiluna Belt, do not occur near volcanic feeders and are not associated with pyroclastic rocks or intrusives; they are therefore interpreted to occur in distal volcanic settings (Lesher et al., 1981; 1984; Donaldson et al., 1986; Lesher and Groves, 1986; Hill et al., 1987). The concentration of class IA deposits in the southern part of the belt and class IIB deposits in the northern part may simply reflect different eruption rates (high vs. very high) in the two areas, komatiitic dunites may have been ponded to a greater degree, or komatiitic peridotites may represent a distal facies of komatiitic dunites (Fig. 5.29).

## ORE GENESIS

A variety of different models have been proposed for the genesis of komatiite-associated nickel sulfide deposits, most involving segregation of immiscible sulfide-oxide liquids from komatiitic magmas (Table 5.12). Groves et al. (1976, 1979), Naldrett (1979), Green and Naldrett (1981), and Naldrett and Campbell (1982) have argued convincingly against alternative models. An ultimate magmatic origin for the nickel sulfide ores in these deposits is supported by the following field, geochemical, and experimental evidence:

1) The sulfide ores are associated exclusively with the most magnesian, lowermost units in the host komatiite sequence. Class IA deposits are spatially related to variations in the internal structure, physical volcanology, and stratigraphy of the overlying komatiite sequence that indicate a strong volcanic control on ore localization.
2) Strata-bound disseminated sulfides and stratiform massive sulfides predate deformation, metamorphism, and alteration of host rocks. Although sulfides have been physically mobilized into veins and stringers during deformation of some komatiitic peridotite-hosted depos-

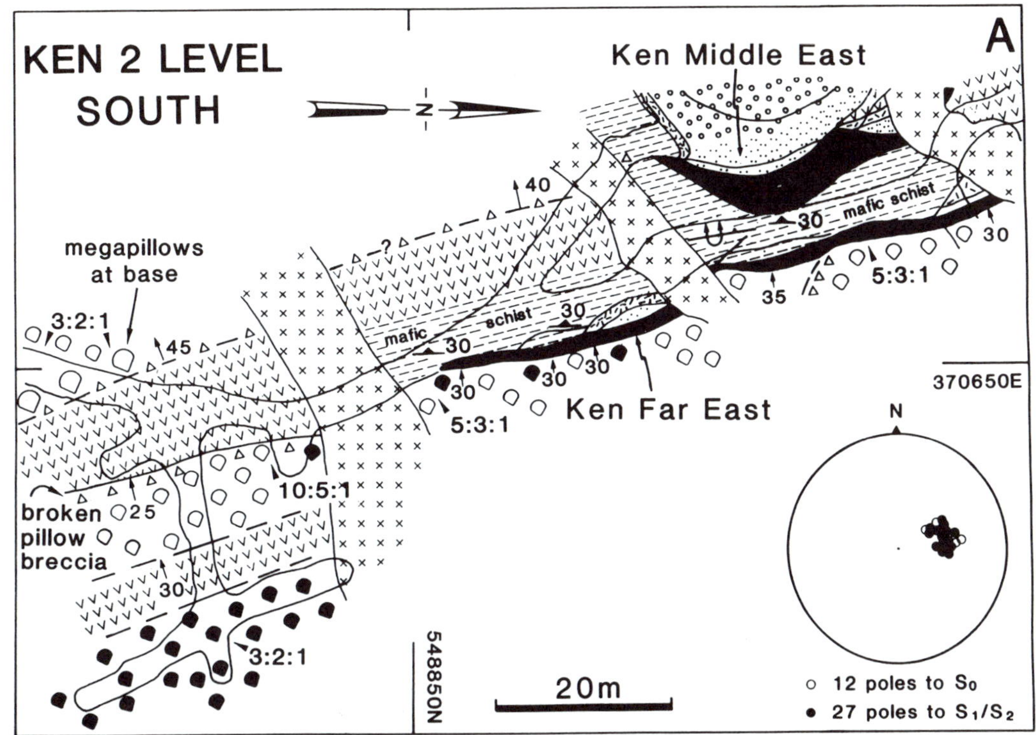


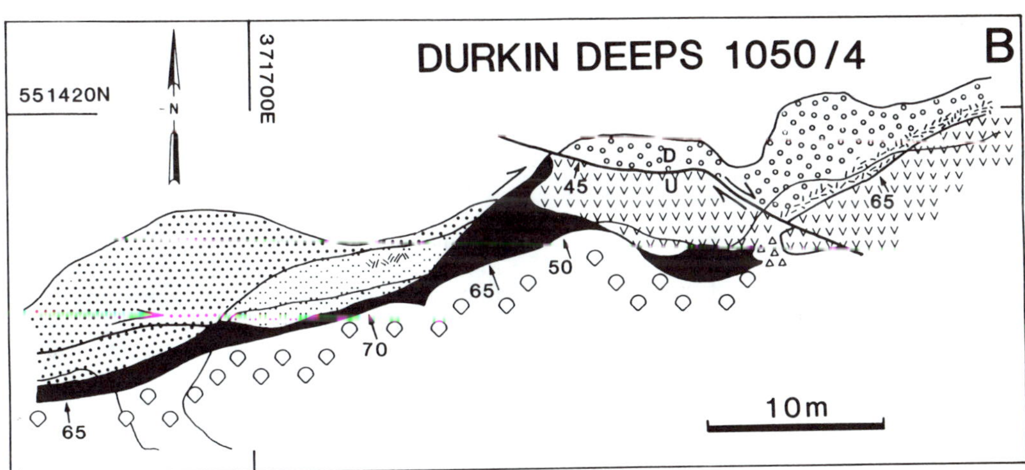


FIGURE 5.32—Maps showing stratigraphic relationships in footwall basalt adjacent to ore pinchouts at Kambalda. **A**: South end of 2 level, Ken shoot (mapping by C.M.L. except for data in mined-out Ken Middle East compiled from Western Mining Corporation level plan; cf. Figure 5.15). Note conformability of ore horizons within basalt sequence. **B**: Durkin Deeps 1050/4 cut and fill stope (mapping by C.M.L. and H.L. Paterson; cf. Figures 5.14 and 5.37). Note correlation between ore horizon and basalt flow front breccia (see also Lesher and Keays, 1984: fig. 5). Ornamentation as in Figure 5.4.

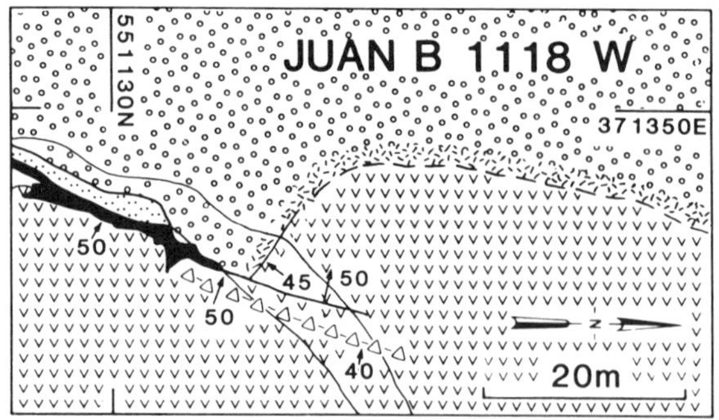


FIGURE 5.33—Map showing stratigraphic relationships in footwall basalt adjacent to ore pinchout at Juan B 1118 W (N10C in Figure 5.14), Kambalda (mapping by B.D. Kay and D.N. Harley). Note correlation between ore horizon and broken pillow flow-top breccia. Ornamentation as in Figure 5.4.

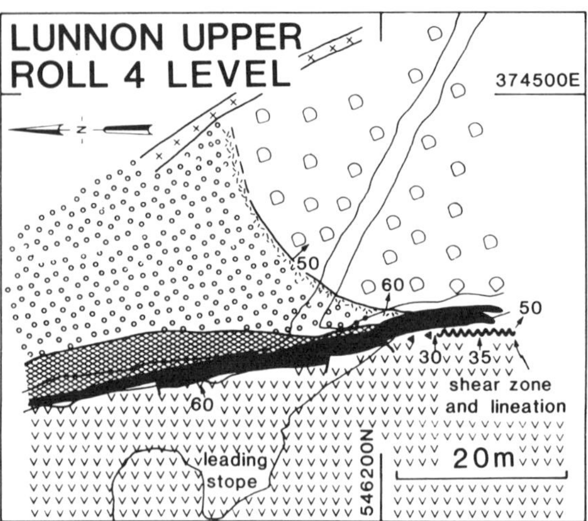


FIGURE 5.34—Map showing stratigraphic relationships in footwall basalt adjacent to ore pinchout on 4 level, Lunnon Upper Roll, Kambalda (mapping by C.M.L. except for data in leading stope compiled from Western Mining Corporation level plan; cf. Figure 5.30B). Note that basalt above ore horizon is pillowed (faces SE) whereas that below is massive; they are not structurally equivalent. Ornamentation as in Figure 5.4.

its, and some footwall rocks exhibit evidence of local volcanic or metamorphic interaction with the ores and host rocks, footwall rocks are otherwise unaltered and barren.

3) Stratiform sulfides exhibit a graded massive/disseminated/matrix ore segregation profile, indicative of gravitational settling or flow segregation. The ores contain distinctive chalcophile ferrochromites, similar to those crystallized experimentally from sulfide-oxide liquids.
4) The high Ni and PGE contents, high Ni/Cu and Pd/Ir ratios, and low Zn contents of the ores are consistent with known and inferred metal partitioning behavior in mafic-ultramafic magmatic systems. There are unresolved conflicts in experimental data for olivine-sulfide systems, but olivine-silicate and silicate-sulfide data are internally consistent and support a magmatic origin for the sulfides.

The major uncertainties are i) the timing of sulfide separation and ii) the source of the sulfur. Each of these is discussed below.

### Timing of sulfide separation

Sulfide separation occurs in a silicate melt when it becomes saturated in $S^{-2}$. As discussed by Naldrett in Chapter 2, sulfur solubility in silicate melts has been shown to decrease with: i) decreasing temperature, ii) decreasing $a_{FeO}$ or increasing $a_{SiO2}$, iii) decreasing $f_{S_2}$ or increasing $f_{O_2}$, and/or iv) increasing pressure.

#### Strata-bound deposits

The disseminated nature of strata-bound sulfide ores, their concentration in central parts of host units, and the presence of multiple ore horizons in some deposits indicate that sulfide saturation occurred relatively late in their crystallization history. The amount of sulfide (1–3%) in class IB and IIB deposits is greater than that capable of being dissolved in the small amount of interstitial liquid, indicating that sulfides accumulated during olivine crystallization. The proportions of olivine and sulfide (60:1) at Dumont, for example, are roughly consistent with crystallization along the intersection of the olivine liquidus and the sulfide-silicate liquid solvus in a komatiite magma (Duke, 1986a). However, there is not any simple relationship between degree of accumulation and degree of mineralization; Hill (1982), for example, reported that the mineralization at Six Mile is most abundant in mesocumulate layers and less abundant in adcumulate layers. Duke (1986a, b) suggested that sulfide saturation occurred when fractionated, sulfur-enriched intercumulus liquid was expelled upward by filter-pressing of partially molten underlying cumulates and mixed with less evolved liquid at the temporary floor of the magma chamber. If the sulfide-olivine cotectic is curved (MacLean, 1969), then mixing of a sulfide-undersaturated primitive magma with a sulfide-undersaturated evolved magma could induce sulfide saturation (see Irvine, 1977a).

In this model the proportions of olivine and sulfide which separate from the melt are constrained by the thermodynamic properties of the melt and there is a limit to the amount of sulfides that may precipitate (Duke, 1986b), consistent with the restricted range of grades for deposits of this type ($\leq$0.8 % Ni). This would not be the case for deposits formed by accumulation of immiscible sulfides.

#### Stratiform deposits

In contrast, the segregation of stratiform ores at the base of the host unit indicates that sulfide saturation occurred relatively early in their crystallization history. The question has been whether sulfides i) were transported directly from the mantle, ii) separated in a subvolcanic magma chamber, iii) separated during ascent, or iv) separated during eruption and emplacement.

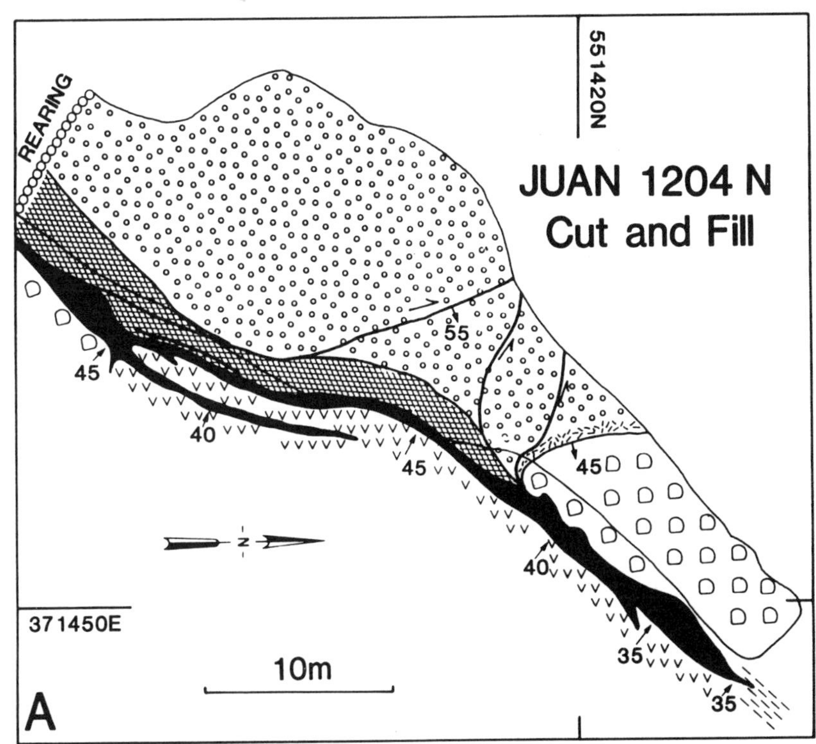


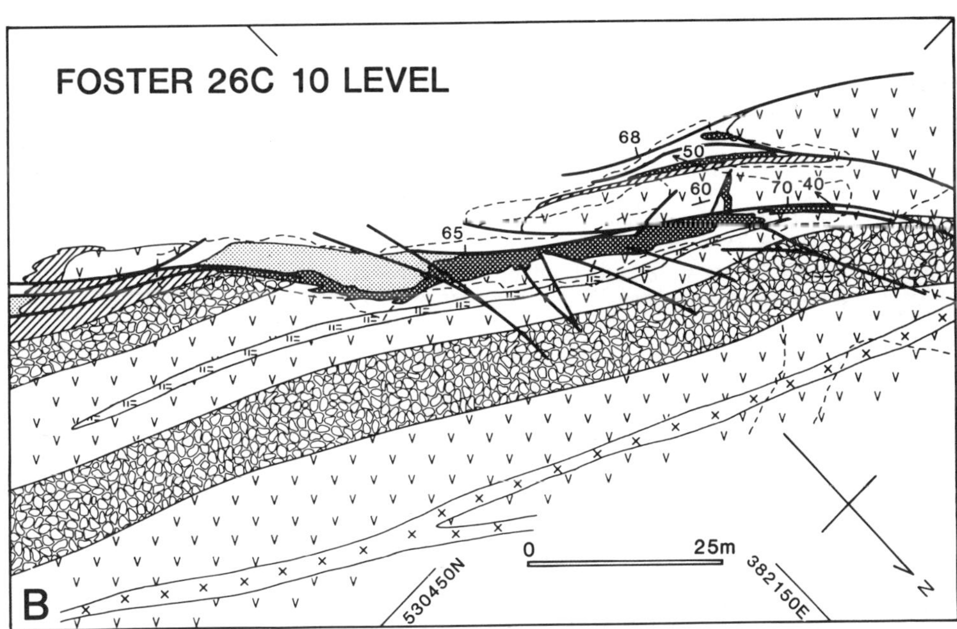


FIGURE 5.35—Maps showing stratigraphic relationships in footwall basalt adjacent to ore pinchouts at Kambalda and St. Ives. **A**: Juan 1204N cut and fill stope (mapping by C.M.L. and H.L. Paterson; cf. Figures 5.14 and 5.37). Note that basalt above ore horizon is pillowed (faces NW) whereas that below is massive; they are not structurally equivalent. **B**: Foster shoot 10 level (after Evans et al., 1988; cf. Figure 5.2). Note that footwall basalt to south (beneath sediment) is pillowed, whereas that beneath and to north of ore horizon is massive; also note that ore is transgressive to footwall basalt (cf. Figure 5.28). Ornamentation as in Figure 5.4.

TABLE 5.12—Magmatic models for class IA komatiite-associated nickel sulfide deposits.

**Timing of Sulfide Separation**

A) Magmatic sulfur
  1) Derivation of sulfides directly from the mantle (Nesbitt, 1971; Naldrett, 1973; Naldrett and Cabri, 1976)
  2) Separation of sulfides in a sub-volcanic magma chamber (Groves and Hudson, 1981)
  3) Exsolution of sulfides during ascent (Lesher et al., 1981)
  4) Exsolution of sulfides during emplacement (Hudson, 1972)

B) Crustal sedimentary sulfur
  1) Assimilation and separation during ascent (Groves et al., 1979; Naldrett and Green, 1981)
  2) Assimilation and separation during emplacement (Lesher et al., 1984; see also Naldrett, 1966 and Hopwood, 1981 for intrusive analogs)

**Mechanism of Segregation**

  1) Static gravity segregation (Nesbitt, 1971; Naldrett, 1973; Ussleman et al., 1979; Naldrett and Campbell, 1982)
  2) Separate emplacement of massive and matrix/disseminated ores (Ross and Hopkins, 1975)
  3) Dynamic flow segregation (Hudson, 1972; Groves et al., 1979; Marston et al., 1981; Lesher et al., 1984)

## Sulfur source

### Mantle sulfur

Nesbitt (1971), Naldrett (1973), Naldrett and Turner (1977), and Malyuk (1984) proposed that sulfides were transported and erupted directly from the mantle, whereas Ross and Hopkins (1975) and Groves and Hudson (1981) suggested that magmatic sulfides may have separated and segregated in a subsurface magma chamber prior to eruption. Fluid dynamic calculations by Lesher and Groves (1986) indicate that finely dispersed sulfide droplets could be supported during rapid ascent of turbulent, low-viscosity komatiite magmas, but they noted several petrogenetic, thermodynamic, geochemical, and geological constraints that militate against eruption of sulfide-saturated komatiites. These and several others are discussed below.

**Petrogenetic Constraints**—Whether or not melts are saturated in sulfide depends on the amount of sulfur in the source and the degree of partial melting. Depending on the mechanism of magma generation, there are several possible scenarios: 1) Komatiites derived from a laterally extensive mantle melt layer (Nisbet and Walker, 1982) should not be saturated in sulfide, because of the large volume of melt. 2) Komatiitic melts generated by multistage melting or mixing processes (e.g. Arndt, 1977; Bickle et al., 1977; Smith and Erlank, 1982) are unlikely to be saturated in sulfide. Sulfur would partition into early melts and/or be physically lost as dense sulfide during melt segregation, and residual melts will be driven off the sulfur saturation surface by continued partial melting/mixing. 3) Komatiitic melts generated by single-stage partial melting (e.g. Nesbitt et al., 1979; Takahashi and Scarfe, 1985), may or may not be sulfur-saturated, depending on the degree of melting and the abundance of sulfur in the source area.

Achieving concentrations of 0.16–0.27% S (Shima and Naldrett, 1975) at 50–70% partial melting, for example, would require a source containing 800–1900 ppm S, but at 10–30% melting only 160–810 ppm S. Current estimates of S abundance in the Archean mantle (350–1000 ppm: Sun, 1982) span this range, but assume that mantle melts are not saturated in sulfide and that 50% partial melting is required to produce a komatiitic magma. S abundances in mantle nodules are low (Jagoutz et al., 1979), but may have experienced prior melt extraction.

**Thermodynamic Constraints**—Experimental studies by Helz (1977), Huang and Williams (1980), and Wentlandt (1982) have shown i) there is a strong negative pressure dependence on sulfur solubility in basaltic melts and ii) sulfur saturation isopleths are parallel to anhydrous silicate liquidi. If sulfur-saturated at depth, magmas which ascend along liquidus paths (see discussion by Naldrett, Chapter 2) should be sulfur-saturated during ascent (Fig. 5.36) and may therefore exsolve an immiscible sulfide liquid during and after eruption. But there is no evidence of significant intratelluric phenocryst contents in komatiites, and magmas of such low viscosity should ascend rapidly and lose little heat. Komatiites should ascend along steeper P–T trajectories, nearer to an adiabatic gradient (Fig. 5.36), and should arrive at the surface superheated and sulfur-undersaturated.

**Geochemical Constraints**—Nickel depletion in spinifex-textured komatiites and Zn enrichment of chromites throughout mineralized komatiite sequences in Western Australia have been used to infer that the precursor magmas were sulfur-saturated prior to eruption (Lesher et al., 1981; Lesher and Groves, 1984; Naldrett et al., 1984), but this does not necessarily restrict ore formation to the pre-eruptive stage. The structure and dimensions of mineralized volcanic piles are not known and is quite possible that the barren upper parts of a mineralized sequence in one area of the volcanic belt may be the facies equivalent of mineralized parts of a sequence in another area of the belt (see Fig. 5.29). Thus, chalcophile element depletion and Zn enrichment may be attributed to assimilation of Zn-rich sulfidic sediments and precipitation of sulfides during emplacement.

Because sulfide/melt partition coefficients are so high (Campbell and Barnes, 1984; Barnes et al., 1985), the similar chalcophile element abundances in komatiites from Alexo, Munro, Mt. Clifford, and Kambalda require that i) all were unsaturated in sulfide in the source region or ii) all were saturated in sulfide in the source region, formed at similar degrees of partial melting (see Barnes et al., 1985), and exsolved similar amounts of sulfides during ascent and emplacement. The latter possibility is rejected as too fortuitous. The former possibility only requires that the sulfides at Alexo and Kambalda separated after emplacement of the host units, as discussed below.

The sulfur isotopic differences between districts and significant deviations from chondritic values (Fig. 5.26) are not consistent with derivation from a homogeneous mantle source. It is unlikely that there is any oxidized sulfur in the mantle, and at such high temperatures significant fractionation of sulfur isotopes between oxidized and reduced species would be unlikely anyway (Ohmoto, 1986).

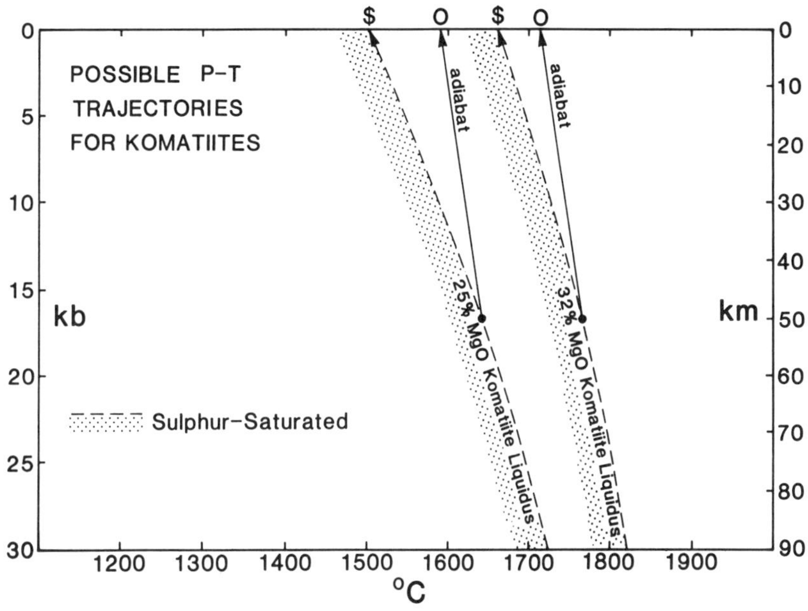

FIGURE 5.36—Sulfide saturation model showing possible P–T trajectories of komatiites (after Lesher and Groves, 1986). Dashed curves are liquidus ascent paths (Arndt, 1976; Bickle et al., 1977; Bickle, 1978); solid curves are adiabatic ascent paths (dT/dP = 3°C/kb). $ = sulfide-saturated lava, O = sulfide-undersaturated lava.

**Geological Constraints**—Although the concentration of stratiform sulfides in these deposits is too great to have once been dissolved in the overlying host unit (cf. Nesbitt, 1971; Naldrett and Cabri, 1976), the present distribution of sulfides is not necessarily representative of their abundance in the original magma, because the host units are laterally more extensive than the ores. Mass balance calculations provide only minimum values of magma/sulfide ratios.

Possibly the strongest argument that the komatiites were not erupted carrying sulfides is the virtual absence of mineralization outside of the ore zones. If magmas were erupted carrying sulfides, regardless of whether mantle-derived, exsolved during ascent, or produced by assimilation of crustal sulfur, then the entire host unit should contain ubiquitous disseminated mineralization: i) sulfides should have been trapped in the chilled margins of host units, ii) sulfides should have continued to exsolve throughout host units as the lavas cooled, oxidized, and crystallized after emplacement; and iii) fractional accumulation of sulfur-saturated komatiitic lavas should have resulted in significant enrichment of sulfide as well as olivine.

### Crustal sulfur

Sedimentary sulfur sources were proposed by Naldrett (1966), Prider (1970), Hudson (1972), Lusk (1976), and Hopwood (1981). Various aspects of these models were retracted or shown to be invalid (Naldrett, 1973; Groves et al., 1979), but all of these workers recognized deficiencies in the magmatic model. More recently, assimilation of crustal sulfur has been proposed to explain the anomalous S/Se and sulfur isotopic variations between deposits; Groves et al. (1979) and Green and Naldrett (1981) have proposed assimilation of sulfur from sediments during ascent through the crust, whereas Lesher et al. (1984), and Lesher and Groves (1986) have proposed assimilation of sulfur from sediments during lava emplacement.

Stratigraphic, geochemical, and Pb-isotopic studies by Lesher et al. (1984), Groves et al. (1986), Evans et al. (1988), Frost and Groves (1988), and McNaughton et al. (1988) indicate that komatiite lava channels at Kambalda have thermally eroded sulfidic sediments, producing rare ocellar komatiites (xenomelts) that accumulated along levees adjacent to the lava channels. Ocellar komatiites at Kambalda are depleted in chalcophile elements, interpreted to reflect scavenging by sulfide liquids, and exhibit high U/Pb, low Th/U, and high Pb isotopic ratios, resulting from partitioning of U and Pb between xenomelt and komatiite (McNaughton et al., 1988). Aphyric flow tops of flanking basalt units at Victor shoot (Fig. 5.16c) are anomalously evolved (down to 11% MgO; up to 3.5% $Na_2O$, 3.9% $K_2O$, 0.10% $P_2O_5$) and strongly depleted in Ni (down to 0.009% NiO) relative to the initial liquid in the adjacent basal host unit (31% MgO, 0.196–0.205% NiO), reflecting contamination and fractionation during emplacement relative to the basal host unit (Lesher, 1983).

Xenoliths of sediment are only rarely preserved in sulfide ores or basal host units (Bavinton, 1981), owing to the dynamics of the lava channel and dilution during lava replenishment. Pb–Pb studies of iron sulfides from nickel ores and adjacent sulfidic sediments suggest mixing of mantle and older crustal lead (Parker, 1984). Biotite-rich zones in the Six Mile dunite are also interpreted to represent partially digested sediments (Naldrett and Turner, 1977).

The restriction of sulfide ores to embayments in the footwall, in areas where sulfidic sediments are thinned or locally absent, suggests that they were thermally eroded beneath thermally active lava conduits, melting sulfide which scavenged chalcophile elements from the komatiite lava to form Fe–Ni–Cu sulfide ores. The absence of sediments and mineralization in areas adjacent to the ore zones may indicate assimilation of sulfide into sulfide undersaturated lava, or physical mobilization into the central part of the channel.

Geochemical and isotopic data provide only slight, somewhat equivocal, evidence for contamination of the basal host units (Arndt and Jenner, 1986; Lesher and Arndt, in prep.), consistent with the inferred dynamic nature of the lava channel; most of the unit would have crystallized from replenished lava.

The sulfur isotopic variations in these deposits are best attributed to fractionation of sulfur isotopes in a hydrothermal-exhalative-sedimentary environment with subsequent incorporation in and mixing with magmatic sulfur in komatiitic magmas. Windarra and Langmuir occur in volcanic sequences that contain oxide facies iron-formations, indicating a more oxidizing environment than the sulfidic sediments at Kambalda and Trojan, consistent with their higher S/Se ratios. Mixing of crustal and magmatic sulfur would explain the minor, but systematic differences between sulfur isotopic compositions of ores and associated sulfidic country rocks.

## Ore tenor variations

Variations in bulk ore compositions between deposits are most apparent at Kambalda where they appear to define parallel belts of differing ore tenor (Fig. 5.37). These belts occur within the same tectonic and metamorphic setting, are hosted by similar lithologies, and appear to have formed contemporaneously, suggesting a magmatic origin for the variations (Marston and Kay, 1980; Woolrich et al., 1981). Possible causes include: i) variations in the compositions of host silicate magmas, ii) variations in $f_{O_2}$ and/or $f_{S_2}$ among different magma batches, and/or iii) variations in the silicate magma to sulfide ratio. The limitations of experimentally determined and thermodynamic modeled sulfide-silicate partition relationships are discussed by Naldrett in Chapter 2, but these data do provide some general constraints on the problem. It appears that all three factors may have contributed to some degree in generating ore tenor variations.

### Magma composition

Rajamini and Naldrett (1978) demonstrated that the composition of the silicate melt strongly influences the partitioning behavior of chalcophile elements between immiscible sulfide liquid and silicate magma. Their equilibrium crystallization model, however, requires over 50% fractionation to produce a basaltic derivative liquid in order to produce the range of Kambalda ore compositions (Fig. 5.38). Fractional crystallization models by Duke and Naldrett (1978) and Duke (1979) indicate that the range of Kambalda ore compositions could be accounted for by 40% fractionation of olivine and sulfide in a ratio of 100:1 from a 32% MgO parental komatiite to produce a 22% MgO derivative liquid (Fig. 5.38).

Lesher and Groves (1986) and Cowden and Woolrich (1987) argued that the limited 28–32% MgO range of initial liquid compositions for the host units (Table 5.6) indicates that ore tenor variations between different ore shoots could not be attributed entirely to variations in the composition of the komatiite magma. However, this interpretation is valid only if komatiites were erupted carrying immiscible sulfide droplets. Although the absence of a chilled margin of komatiite beneath the ore zones at most Kambalda shoots has been used previously to infer that sulfides were emplaced early (Ross and Hopkins, 1975; Groves et al., 1979; Marston et al., 1981), chilled margins were probably thermally eroded (see Huppert and Sparks, 1985a). If sulfide separation and equilibration occurred after formation of the chilled margin, then sulfides may have equilibrated with more fractionated/hybridized lavas. Compositions within the spinifex-textured zones at Victor and Lunnon shoot range between 31 and 16% MgO; calculated magnesium contents of liquids in equilibrium with relict olivines at Victor shoot range between 29 and 22% MgO (Lesher, 1983). Thus, variable degrees of fractionation and hybridization during emplacement in a dynamic lava conduit could account for the range of ore compositions within the different shoots at Kambalda.

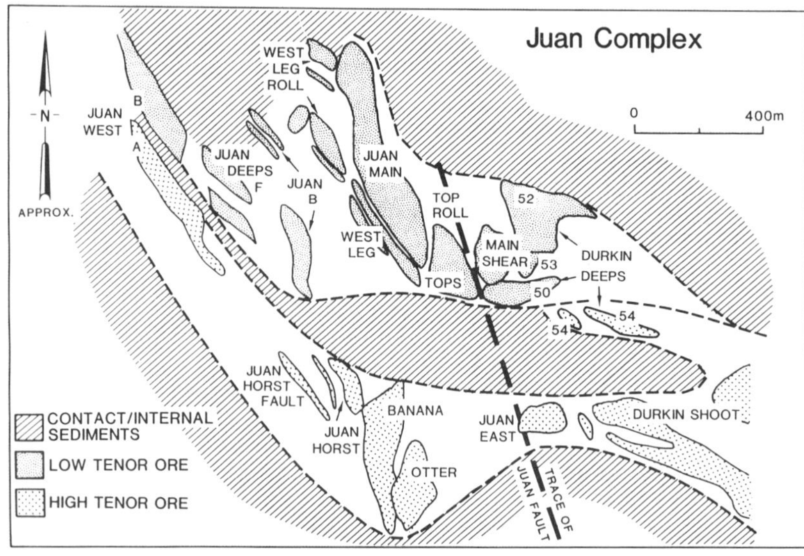


FIGURE 5.37—Map of Juan Complex showing distribution of ore shoots within two broad belts of differing ore tenor (after Marston and Kay, 1980). Note disparate distribution of sediments and ores (cf. Figure 5.4).

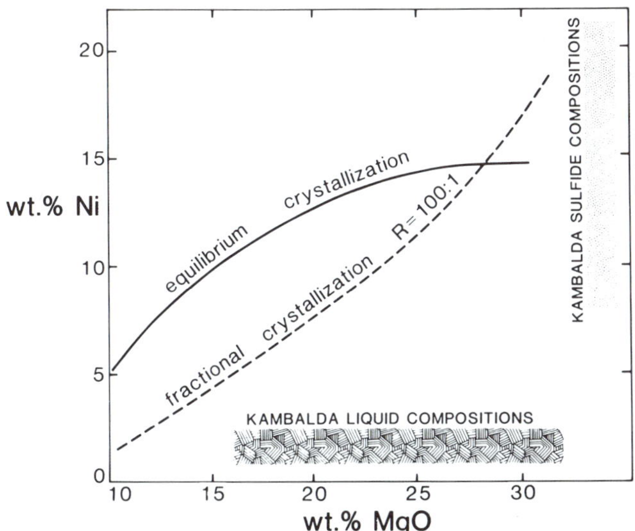


FIGURE 5.38—Equilibrium and fractional crystallization models of Ni–MgO relationships between sulfide ores and komatiite magmas (modified from Cowden and Woolrich, 1987). Equilibrium crystallization after Rajamini and Naldrett (1978); fractional crystallization after (Duke and Naldrett, 1978); Kambalda sulfide compositions after Cowden and Woolrich (1987); Kambalda liquid compositions after Lesher (1983; see also Figures 5.16A, 5.18, 5.19).

Barnes and Naldrett (1986) noted that the ore composition at Alexo is consistent with equilibration with a low-Mg liquid, not the high-Mg liquid represented by the chilled margins, and proposed that sulfides separated after the flow had begun to crystallized and fractionate. This is consistent with the preservation of a chilled margin beneath the ore zone at Alexo.

### Variations in $f_{O_2}$

Cowden and Woolrich (1987) note that oxidation of a silicate magma will result in a reduction of the number of octahedral sites available to metals. As the octahedral site preference energy (OSPE) for Ni is greater than that for Fe, they theorize that Ni/Fe in sulfide should increase with increasing $f_{O_2}$ and calculate that the entire range of Ni abundances in Kambalda ores could be accounted for by increasing $f_{O_2}$ by only one log unit. They suggest that high tenor ores reflect equilibration during emplacement, whereas low- and medium-tenor ores reflect disequilibrium compositions inherited from equilibrium at depth. In contrast, Doyle and Naldrett (1987) suggest that the Ni–Fe distribution constant between sulfide and olivine decreases with increasing O/O+S in the magma, suggesting that Ni/Fe in sulfide should decrease with increasing $f_{O_2}$ in the magma. The relative magnitudes of these two effects are difficult to evaluate (see discussion by Naldrett in Chapter 2), and they may offset each other to some degree.

The Cowden and Woolrich (1987; see also Woolrich et al., 1981) model requires sulfides to be erupted from depth and requires different ore shoots to be emplaced separately as different eruptive episodes. This is inconsistent with the above evidence against early sulfide separation, and with observed stratigraphic relationships: the ore tenor belts are only a few hundred meters apart, closer in parts of the Juan shoot (Fig. 5.37), and the basal host unit at the Juan Complex (Fig. 5.14) is laterally extensive, not confined to the embayments. Low viscosity komatiites should form sheet-like flows; the different ore shoots (lava channels) were probably emplaced at or near the same time.

Woolrich et al. (1981) suggested that variable oxidation also could have occurred in the cooling lava; small volumes of sulfides (high tenor shoots) may have oxidized more rapidly than large volumes of sulfides (low tenor shoots). Another possibility is that oxidation occurred during lava emplacement by assimilation of hydrous sediment.

### Magma:sulfide ratio

Campbell and Naldrett (1979) noted that the partitioning behavior of metals between sulfide and silicate liquids should be strongly influenced by the mass ratio (R) of the two phases. Magmas that equilibrate with large amounts of sulfide (small R) should be rapidly depleted in chalcophile elements, resulting in lower tenor ores, whereas magmas that equilibrate with small amounts of sulfide (large R) should retain normal abundances of chalcophile elements, resulting in higher tenor ores. Lesher and Groves (1986) suggested that the lower tenor of the thicker ore shoots at Kambalda confirms a relationship between ore tenor and the amount of sulfide.

Cowden and Woolrich (1987) modeled the abundances of Co, Ni, Co, and Pt at different values of R and sulfide-silicate partition coefficients (D), and contended that there is no unique value of R that can account for the range of metal values observed at Kambalda (Fig. 5.39). A problem with their approach is that the data used to estimate the composition of the initial liquid were from samples that had already equilibrated with sulfide and are depleted in chalcophile elements (Lesher et al., 1981). Based on experimentally determined and calculated partition coefficients, the degree of depletion can be predicted to be Pt > Cu > Ni > Co, which is precisely the sense required to eliminate the mismatch of different R values calculated by Cowden and Woolrich (1987).

Given the uncertainties in partition coefficients, initial metal abundances in the magma, and the relative volumes of silicate and sulfide involved, the ranges are in broad agreement (Naldrett, 1981), and variations in magma:sulfide ratios should be expected in dynamic lava conduits.

## DISTAL VOLCANIC-ASSIMILATION MODEL

The genetic model that best explains these deposits (Lesher et al., 1984; Groves and Lesher, 1986; Barnes et al., 1988a, b, c) involves voluminous eruption of komatiites into rift-phase greenstone belts, forming large (group I) and very large (group II) channelized flows that assimilated variable amounts of sulfidic footwall rocks during emplacement (Figs. 5.29 and 40). The distribution of mineralization was influenced by: i) the volcanic setting and mode of emplacement of the host unit; ii) the composition of the substrate and degree of assimilation; iii) the volume of lava and its capacity to dissolve sulfur/silica; and iv) the timing of sulfide saturation and opportunity for sulfide segregation. Flows that assimilated proportionately large amounts of footwall rocks, especially sulfidic sediments, achieved sulfide saturation early in their crystallization history. Sulfides segregated rapidly, settled to the base of the flow, and formed stratiform massive/matrix/disseminated sulfide ores (type A). Flows that assimilated proportionately small amounts material achieved sulfide saturation later in their crystallization history and formed strata-bound disseminated sulfide ores (type B).

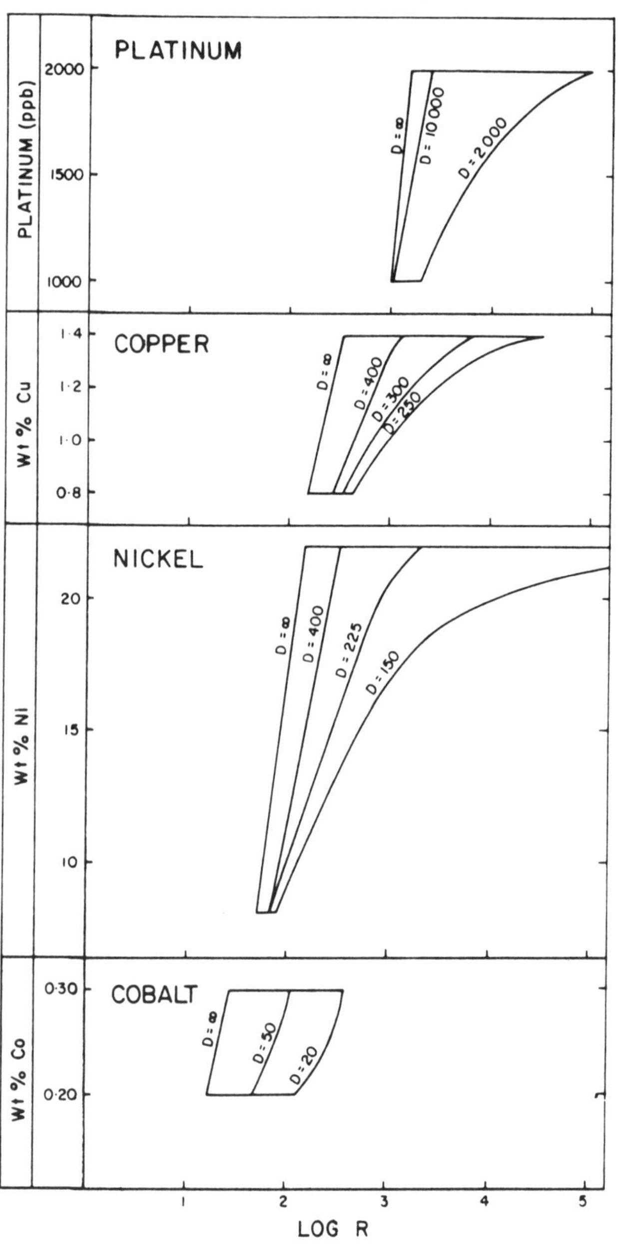


FIGURE 5.39—Magma:sulfide equilibration models of Pt–Cu–Ni–Co variations in Kambalda sulfide ores (after Cowden and Woolrich, 1987). R = magma:sulfide ratio (Campbell and Naldrett, 1979; Naldrett, 1981).

Although the greatest degree of thermal erosion and assimilation would occur beneath the lava channel, minor amounts may occur beneath flanking parts of the flow, especially near the channel. This is consistent with the well-defined margins of the ore shoots, but absence of sediments along flanking contacts and minor mineralization in flanking areas at Kambalda (cf. Gresham, 1986). Hanging wall ores at Kambalda occur at the base of overlying units emplaced along the same volcanic pitchline. The presence of thermally highly conductive sulfides (ca. 0.05 cal/cm-s-°C: Williams et al., 1972 vs. 0.003–0.006 cal/cm-s-°C for komatiite: Bickle, 1982) at the base of the flow may have enhanced thermal erosion (Groves et al., 1986), but accumulation of sulfides in areas with the greatest degree of thermal erosion is attributable primarily to these being topographic lows.

Assimilation may occur along the entire length of the lava channel, but sulfide saturation may not occur until some intermediate point along the length of the channel. All ores would have been above their liquidus (1120–1160°C: see Chapter 2) at temperatures above the komatiite solidus (ca. 1200°C: Arndt, 1976), so low viscosity sulfides (0.027–0.030 g/cm-sec: Dobrovinski et al., 1969) may have been mobilized subsequent to segregation. As a consequence, distal parts of flows may be more mineralized that proximal parts. The intensity of mineralization at any location would be influenced by the topography of the footwall rocks which would trap dense sulfides and pond komatiite flows.

The observed massive/matrix/disseminated ore profile in class 1A deposits can probably be attributed to a combination of dynamic flow segregation (Hudson, 1972) and static buoyancy ("billiard ball model:" Naldrett, 1973). The interpretation that ferrochromites preferentially crystallized at the margins of individual ore layers obviates models involving multiple stage emplacement (cf. Ross and Hopkins, 1975; Marston and Kay, 1980; Groves et al., 1979). Variations in the mode of emplacement of the host unit influenced the degree of assimilation and probably account for variations in ore tenor, intensity of mineralization, and stratigraphic relationships with sediments that are observed in different deposits.

This model explains the absence of deposits in areas that contain sulfidic sediments, but no lava channels, or lava channels, but no sulfur source (cf. Claoué–Long and Nesbitt, 1985; Nesbitt, 1986). Of major consequence is that mineralization will be localized in specific parts of the volcanic pile, i.e lava channels. Barren cumulate sheet flows in one part of an area may be lateral facies equivalents of cumulate lava channels in another part.

## EXPLORATION GUIDES

Although class IIB deposits represent a very large Ni resource, most are presently subeconomic because of their low grade and the difficulty of extracting fine-grained disseminated sulfides from the host rock. Despite their high grade, isolated class IA deposits may also be subeconomic. The best exploration targets are clusters of class IA deposits. Some of the features that may be used as guides in exploration for komatiite-associated nickel sulfide deposit are summarized in Table 5.13. Geochemical and geophysical exploration methods used to further define prospective areas are described by Smith (1984) and Pridmore et al. (1984).

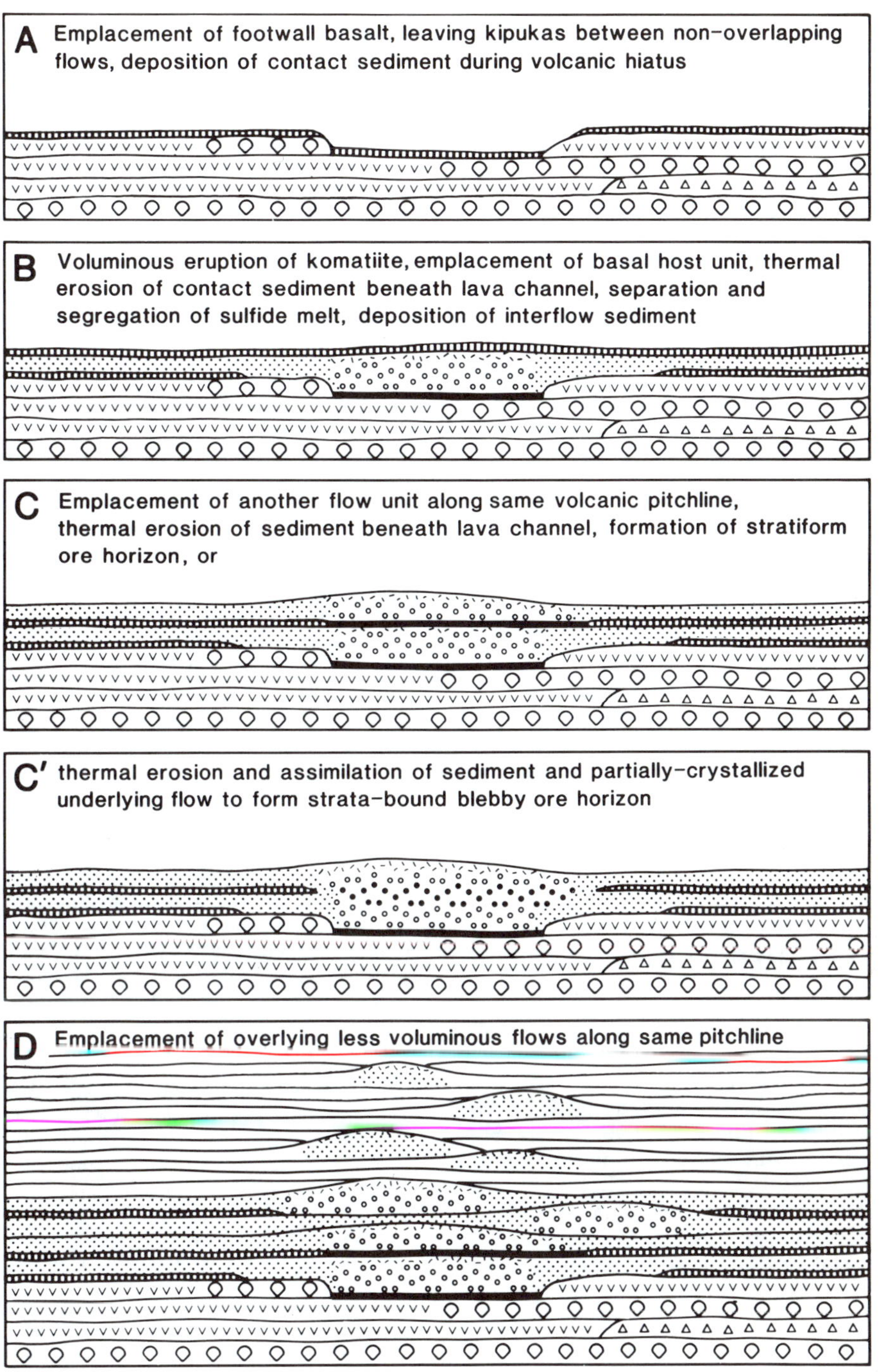


FIGURE 5.40. Schematic interpretive cross section through a class IA komatiite-associated nickel sulfide deposit (adapted from Lesher et al., 1984); based on Kambalda, but applicable with modification to most other deposits. **A**: deposition of footwall rocks, leaving irregular topographic surface, deposition of contact sediment during volcanic hiatus; **B**: emplacement of basal host unit, channelization within embayment, thermal erosion of footwall rocks beneath channel, separation of sulfide liquid to form stratiform contact ore, fractional accumulation of olivine in channel, deposition of hanging wall sediment during volcanic hiatus; **C**: emplacement of second unit, thermal erosion of hanging wall sediment, separation of sulfide liquid to form stratiform hanging wall ore horizon *or* **C'** thermal erosion of sediment and underlying partially consolidated flow to form strata-bound hanging wall ore horizon; **D**: emplacement of remainder of komatiite sequence, channelization along same volcanic pitchline to form observed variations in physical volcanology (cf. Figures 5.4, 5.12–5.14).

TABLE 5.13—Exploration guidelines for komatiite-associated nickel sulfide deposits.[1]

**Metallogenic Province Selection**

1) Younger Archean (2.7–3.0 Ga) greenstone belts or lower Proterozoic fold belts
2) Rift-phase greenstones (elongate granitoid domes; linear tectonic patterns; complex volcanic stratigraphy; abundant komatiites, including komatiitic peridotites and komatiitic dunites; sulfidic shales and cherts)

**Intra-Province Area Selection**

1) Komatiite sequences with komatiitic peridotites or komatiitic dunites in lower part, typically not interlayered with komatiitic basalts
2) Komatiites containing Zn-rich ferrochromites
3) Komatiites exhibiting chalcophile element depletion
4) Structural highs

**Selection of Local Ore Environments**

1) Thickened areas of komatiite sequences
2) More "disordered" stratigraphic sequence (sporadic komatiitic peridotites in upper parts, poor lateral continuity of flow units)
3) Locally better textural and compositional differentiation in overlying komatiites
4) Locally absent sulfidic sediments
5) Anomalously thick, highly magnesian basal host units (depleted in Ti, Al, Cr, and Zn relative to flanking units; more magnesian olivine than flanking units; more magnesian chilled margins)
6) Footwall embayments

[1]adapted from Lesher et al. (1982), Gresham and Loftus-Hills (1981), Lesher and Groves (1984)

## ACKNOWLEDGEMENTS

Many of the ideas in this chapter have developed as a result of collaborative research with David Groves, Mike Donaldson, and Nick Arndt. I have benefitted immeasurably from discussions with many Western Mining Corporation geologists, especially Jeff Gresham and Alistair Cowden. I am very grateful to Nick Arndt, Sarah-Jane Barnes, Steve Barnes, Alistair Cowden, Murray Duke, Roger Eckstrand, Dave Evans, David Groves, Rob Hill, and Herbert Huppert for beneficial discussions and for providing preprints or reprints of their most recent research, to Sheree Miller and Jan Cook for typing the initial draft of the manuscript, and to Tony Naldrett, Jim Whitney, and Jamie Robertson for their editorial efforts.

## REFERENCES

Arndt, N. T., 1976, Melting relations of ultramafic lavas (komatiites) at 1 atm and high pressure: Carnegie Institute of Washington Yearbook, v. 75, pp. 555–562.

Arndt, N. T., 1977, Ultrabasic magmas and high degree melting of the mantle: Contributions to Mineralogy and Petrology, v. 64, pp. 205–221.

Arndt, N. T., 1986, Differentiation of komatiite flows: Journal of Petrology, v. 27, pp. 279–301.

Arndt, N. T., and Jenner, G. A., 1986, Crustally contaminated komatiites and basalts from Kambalda, Western Australia: Chemical Geology, v. 56, pp. 229–255.

Arndt, N. T., and Nisbet, E. G. (eds.), 1982, Komatiites: Allen & Unwin, London, 526 pp.

Arndt, N. T., Francis, D., and Hynes, A. J., 1979, The field characteristics and petrology of Archean and Proterozoic komatiites: Canadian Mineralogist, v. 17, pp. 147–163.

Arndt, N. T., Naldrett, A. J., and Pyke, D. R., 1977, Komatiitic and iron-rich tholeiitic lavas of Munro township, northeast Ontario: Journal of Petrology, v. 18, pp. 319–369.

Barnes, R. G., Lewis, J. D., and Gee, R. D., 1974, Archaean ultramafic lavas from Mount Clifford: Geologic Society of Western Australia, Annual Report for 1973, pp. 59–70.

Barnes, Sarah-Jane, 1985, The petrology and geochemistry of komatiite flows from the Abitibi greenstone belt, Canada, and a model for their formation: Lithos, v. 18, pp. 241–270.

Barnes, Sarah-Jane, and Naldrett, A. J., 1986, Variations in platinum group element concentrations in the Alexo mine komatiite, Abitibi greenstone belt, northern Ontario: Geological Magazine, v. 123, pp. 515–524.

Barnes, Sarah-Jane, and Naldrett, A. J., 1987, Fractionation of platinum group elements and gold in some komatiites of the Abitibi Greenstone Belt, northern Ontario: Economic Geology, v. 82, pp. 165–183.

Barnes, Sarah-Jane, Boyd, R., Korneliussen, A., Nilsson, L.-P., Often, M., Pedersen, R. B., and Robins, B., 1988, The use of mantle normalization and metal ratios in discriminating between the effects of partial melting, crystal fractionation and sulphide segregation on platinum group elements, gold, nickel and copper: Examples from Norway; *in* Prichard, H. M., Potts, P. J., Bowles, J. F. W., and Cribb, S. J. (eds.), Geo-Platinum 87: Elsevier, Essex, pp. 113–141.

Barnes, Sarah-Jane, Gorton, M. P., and Naldrett, A. J., 1983, A comparative study of olivine and clinopyroxene spinifex flows from Alexo, Abitibi greenstone belt, Ontario, Canada: Contributions to Mineralogy and Petrology, v. 83, pp. 293–308.

Barnes, Sarah-Jane, Naldrett, A. J., and Gorton, M. P., 1985, The origin of the fractionation of platinum-group elements in terrestrial magmas: Chemical Geology, v. 53, pp. 303–323.

Barnes, Stephen J., Coates, C. J. A., and Naldrett, A. J., 1982, Petrogenesis of a Proterozoic nickel sulfide-komatiite association: The Katiniq sill, Ungava, Quebec: Economic Geology, v. 77, pp. 413–429.

Barnes, Stephen J., Gole, M. J., and Hill, R. E. T., 1988a, The Agnew nickel deposit, Western Australia: Part I. Structure and stratigraphy: Economic Geology, v. 83, pp. 524–537.

Barnes, Stephen J., Gole, M. J., and Hill, R. E. T., 1988b, The Agnew nickel deposit, Western Australia: Part II. Sulfide geochemistry, with emphasis on the platinum-group elements: Economic Geology, v. 83, pp. 537–550.

Barnes, Stephen J., Hill, R. E. T., and Gole, M. J., 1988c, The Perseverance ultramafic complex, Western Australia: The product of a komatiite lava river: Journal of Petrology, v. 29, pp. 305–331.

Barrett, F. M., Binns, R. A., Groves, D. I., Marston, R. J., and McQueen, K. G., 1977, Structural history and metamorphic modification of Archean volcanic-type nickel deposits, Yilgarn block, Western Australia: Economic Geology, v. 72, pp. 1195–1223.

Barrett, F.M., Groves, D.I., and Binns, R.A., 1976, Importance of metamorphic processes at the Nepean nickel deposit, Western Australia: Institution of Mining and Metallurgy, Transactions, Section B, v. 85, pp. 251–274.

Bavinton, O. A., 1981, The nature of sulfidic sediments at Kambalda and their broad relationships with associated ultramafic rocks and nickel ores: Economic Geology, v. 76: pp. 1606–1628.

Beswick, A. E., 1982, Some geochemical aspects of alteration and genetic relations in komatiitic suites; *in* Arndt, N. T., and Nisbet, E. G. (eds.), Komatiites: George Allen and Unwin, London, pp. 283–307.

Bickle, M. J., 1978, Melting experiments on peridotitic komatiites: Progress in Experimental Petrology, Natural Environmental Research Council, 4th Progress Report, series D, no. 11, pp. 187–195.

Bickle, M. J., 1982, The magnesium contents of komatiitic liquids: *in* Arndt, N. T., and Nisbet, E. G. (eds.), Komatiites: George Allen and Unwin, London, pp. 479–493.

Bickle, M. J., Ford, C. E., and Nisbet, E. G., 1977, The petrogenesis of peridotitic komatiites: Evidence from high-pressure melting experiments: Earth and Planetary Science Letters, v. 37, pp. 97–106.

Billington, L. G., 1984, Geological review of the Agnew nickel deposit, Western Australia; *in* Buchanan, D. L., and Jones, M. J. (eds.), Sulphide deposits in mafic and ultramafic rocks, Proceedings of International Geological Correlation Program Projects 161 and 91, Third Nickel Sulphide Field Conference, Perth, Western Australia: Institution of Mining and Metallurgy, London, pp. 43–54.

Binns, R. A., Groves, D. I., and Gunthorpe, R. J., 1977, Nickel sulphides in Archaean ultramafic rocks of Western Australia; *in* Sidorenko, A. V (ed.), Correlation of the Precambrian, v. 2: Nauka, Moscow, pp. 349–380.

Brügmann, G. E., Arndt, N. T., Hofmann, A. W., and Tobschall, H. J., 1987, Noble metal abundances in komatiite suites from Alexo, Ontario, and Gorgona Island, Colombia: Geochimica et Cosmochimica Acta, v. 51, pp. 2159–2169.

Burt, D. R. L., and Sheppy, N. R., 1975, Mount Keith nickel sulphide deposit; *in* Knight, C. L. (ed.), Economic Geology of Australia and Papua–New Guinea, I. Metals, Monograph 5: Australasian Institute of Mining and Metallurgy, Parkville, pp. 159–168.

Campbell, I. H., and Barnes, S. J., 1984, A model for the Geochemistry of the platinum-group elements in magmatic sulfide deposits: Canadian Mineralogist, v. 22, pp. 151–160.

Campbell, I. H., and Naldrett, A. J., 1979, The influence of silicate: sulfide ratios on the geochemistry of magmatic sulfides: Economic Geology, v. 74, pp. 1503–1505.

Chauvel, C., Dupre, B., and George A. Jenner, 1985, The Sm–Nd age of Kambalda volcanics is 500 Ma too old!: Earth and Planetary Science Letters, v. 74, pp. 315–324.

Chimimba, L. R., 1984, Geology and mineralogy at Trojan Mine, Zimbabwe; *in* Buchanan, D. L., and Jones, M. J. (eds.), Sulphide deposits in mafic and ultramafic rocks, Proceedings of International Geological Correlation Program Projects 161 and 91, Third Nickel Sulphide Field Conference, Perth, Western Australia: Institution of Mining and Metallurgy, London, pp. 147–155.

Christie, D., 1975, Scotia nickel deposit; *in* Knight, C. L. (ed.), Economic Geology of Australia and Papua–New Guinea, I. Metals, Monograph 5. Australasian Institute of Mining and Metallurgy, Parkville, pp. 121–125.

Claoué-Long, J. C., and Nesbitt, R. W., 1985, Contaminated komatiites: Nature, v. 313, pp. 247.

Coad, P. R., 1979, Nickel sulphide deposits associated with ultramafic rocks of the Abitibi Belt and economic potential of mafic-ultramafic intrusions. Ontario Geological Survey Study 20, 84 pp.

Compston, W., Williams, I. S., Campbell, I. H., and Gresham, J. J. 1986, Zircon xenocrysts from the Kambalda volcanics: age constraints and direct evidence for older continental crust below the Kambalda–Norseman greenstones: Earth and Planetary Science Letters, v. 76, pp. 299–311.

Cowden, A., 1988, Emplacement of komatiite lava flows and associated nickel sulfides at Kambalda, Western Australia: Economic Geology, v. 83, pp. 436–442.

Cowden, A., and Archibald, N. T., 1987, Massive-sulfide fabrics at Kambalda and their relevance to the inferred stability of monosulfide solid-solution: Canadian Mineralogist, v. 25, pp. 37–50.

Cowden, A., and Woolrich, P., 1987, Geochemistry of the Kambalda iron-nickel sulfides: Implications for models of sulfide-silicate partitioning: Canadian Mineralogist, v. 25, pp. 21–36.

Cowden, A., Donaldson, M. J., Naldrett, A. J., and Campbell, I. H., 1986, Platinum-group elements and gold in the komatiite-hosted Fe–Ni–Cu sulfide deposits at Kambalda, Western Australia: Economic Geology, v. 81, pp. 1226–1235.

Crocket, J. H., and MacRae, W. E., 1986, Platinum-group element distribution in komatiitic and tholeiitic volcanic rocks from Munro township, Ontario: Economic Geology, v. 81, pp. 1242–1251.

Davis, C. E. S., 1972, Analytical methods used in the study of an ore intersection from Lunnon Shoot, Kambalda, Western Australia; *appendix to* W. E. Ewers and D. R. Hudson, An interpretive study of a nickel-iron sulfide ore intersection, Lunnon Shoot, Kambalda, Western Australia: Economic Geology, v. 67, pp. 1075–1092.

Dillon–Leitch, H. C. H., Watkinson, D. H., and Coats, C. J. A., 1986, Distribution of platinum-group elements in the Donaldson West deposit, Cape Smith Belt, Quebec: Economic Geology, v., 81, pp. 1147–1158.

Dobrovinski, I. E., Esin, O. A., Barmin, L. N., and Chuchmarev, S. K., 1969, Physicochemical properties of sulphide melts: Russian Journal of Physical Chemistry, v. 43, pp. 1769–1771.

Donaldson, C. H., 1982, Spinifex-textured komatiites: a review of textures, mineral compositions and layering; *in* Arndt, N. T., and Nisbet, E. G. (eds.), Komatiites: George Allen and Unwin, London, pp. 213–242.

Donaldson, M. J., 1981, Redistribution of ore elements during serpentinization and talc-carbonate alteration of some Archaean dunites, Western Australia: Economic Geology, v. 76, pp. 1698–1713.

Donaldson, M. J., 1982, Mount Clifford–Marshall Pool Area; *in* Groves, D. I., and Lesher, C. M. (eds.), Regional geology and nickel deposits of the Norseman–Wiluna Greenstone Belt, Western Australia, Publication 7: University of Western Australia Department of Geology and Extension Service, Perth, part C, pp. 53–67

Donaldson, M. J., 1983. Progressive alteration of barren and weakly mineralized Archaean dunites: A petrological, mineralogical and geochemical study of some intrusive dunites from Western Australia: Unpublished Ph.D. thesis, University of Western Australia, Perth, 345 pp.

Donaldson, M. J., and Bromley, G. J., 1982, The Honeymoon Well nickel sulfide deposits, Western Australia: Economic Geology, v. 76, pp. 1550–1564.

Donaldson, M. J., Lesher, C. M., Groves, D. I., and Gresham, J. J, 1986, Comparison of Archaean dunites and komatiites associated with nickel mineralization in Western Australia: Implications for dunite genesis: Mineralium Deposita, v. 21, pp. 296–305.

Donnelly, T. H., Lambert, I. B., Oehler, D. Z., Hallberg, J. A., Hudson, D. R., Smith, J. W., Bavinton, O. A., and Golding, L. Y., 1978, A reconnaissance study of stable isotope ratios in Archaean rocks from the Yilgarn Block, Western Australia: Journal of the Geological Society of Australia, v. 24, pp. 409–420.

Doyle, C. D., and Naldrett, A. J., 1987, The oxygen content of "sulphide" magma and its effect on the partitioning of nickel between coexisting olivine and molten ores: Economic Geology, v. 82, pp. 208–211.

Duke, J. M., 1979, Computer simulation of the fractionation of olivine and sulfide from mafic and ultramafic magmas: Canadian Mineralogist, v. 76, pp. 507–514.

Duke, J. M., 1986a, Petrology and economic geology of the Dumont sill: An Archean intrusion of komatiitic affinity in Northwestern Quebec, Economic Geology Report 35: Geological Survey of Canada, Ottawa, 56 pp.

Duke, J. M., 1986b, The Dumont nickel deposit: A genetic model for disseminated magmatic sulphide deposits of komatiitic affinity; *in* Gallagher, M. J., Ixer, R. A., Neary, C. R., and Prichard, H. M. (eds.), Metallogeny of Basic and Ultrabasic Rocks: Institution of Mining and Metallurgy, London, pp. 151–160.

Duke, J. M., and Naldrett, A. J., 1978, A numerical model of the fractionation of olivine and molten sulfide from komatiite magma: Earth and Planetary Science Letters, v. 39, pp. 255–266.

Eckstrand, O. R., 1975, The Dumont serpentinite: A model for control of nickeliferous opaque assemblages by alternation products in ultramafic rocks: Economic Geology, v. 70, pp. 83–201.

Eckstrand, O. R., and Hulbert, L. J., 1987, Selenium and the source of sulphur in magmatic nickel and platinum deposits: Geological Association of Canada, Mineralogical association of Canada, Program with Abstracts, v. 12, p. 40.

Evans, D. M., Cowden, A., and Barratt, R. M., 1988, Deformation and thermal erosion at the Foster nickel deposit, Kambalda–St. Ives, Western Australia; *in* Proceedings of the Fifth Magmatic Sulphide Field Conference, Harare, Zimbabwe: Institution of Mining and Metallurgy, London.

Ewers, W. E., and Hudson, D. R., 1972, An interpretive study of a nickel-iron sulfide ore intersection, Lunnon shoot, Kambalda, Western Australia: Economic Geology, v. 67, pp.1075–1092.

Ewers, W. E., Graham, J., Hudson, D. R., and Rolls, J. M., 1976, Crystallization of chromite from nickel-iron sulphide melts: Contributions to Mineralogy and Petrology, v. 54, pp. 61–64.

Franklin, J. M., Lydon, J. W., and Sangster, D. F., 1981, Volcanic-associated massive sulfide deposits; *in* Skinner, B. J. (ed.), Seventy-Fifth Anniversary Volume: Economic Geology Publishing Company, El Paso, pp. 485–627.

Frost, K. M., and Groves, D. I., 1988, Ocellar units at Kambalda: Evidence for sediment assimilation by komatiite lavas; *in* Proceedings of the Fifth Magmatic Sulphide Field Conference, Harare, Zimbabwe: Institution of Mining and Metallurgy, London.

Gemuts, I., and Theron, A. C., 1975, The Archaean between Coolgardie and Norseman—stratigraphy and mineralization; *in* Knight, C. L. (ed.), Economic Geology of Australia and Papua–New Guinea, I. Metals, Monograph 5: Australasian Institute of Mining and Metallurgy, Parkville, pp. 66–74.

Gélinas, L., Lajoie, J., and Brooks, C., 1977, The origin and significance of Archean ultramafic volcaniclastics from Spinifex Ridge, La Motte township, Quebec; *in* Baragar, W. R. A., Coleman, L. C., and Hall, J. M. (eds.), Volcanic Regimes in Canada, Special Paper 16: Geological Association of Canada, Waterloo, pp. 297–309.

Gole, M.J., Barnes, S. J., and Hill, R. E. T, 1987, The role of fluids in the metamorphism of komatiites, Agnew nickel deposit, Western Australia: Contributions to Mineralogy and Petrology, v. 96, pp. 151–162.

Green, A. H., and Naldrett, A. J., 1981, The Langmuir volcanic peridotite-associated nickel sulphide deposits: Canadian equivalents of the Western Australian occurrences: Economic Geology, v. 76, pp. 1503–1523.

Gresham, J. J., 1986, Depositional environments of volcanic peridotite-associated nickel sulphide deposits with special reference to the Kambalda dome; *in* Friedrich, G. H., Genkin, A., Naldrett, A. J., Ridge, J. D., Sillitoe, R. H., and Vokes, F. M. (eds.), Geology and Metallogeny of Copper Deposits, Proceedings of the Twenty-Seventh International Geological Congress, Moscow: Springer–Verlag, Berlin, pp. 63–90.

Gresham, J. J., and Loftus–Hills, G. D., 1981, The geology of the Kambalda nickel field, Western Australia: Economic Geology, v. 76, pp. 1373–1416.

Groves, D. I., and Batt, W. D., 1984, Spatial and temporal variations of Archaean metallogenic associations in terms of evolution of granitoid-greenstone terrains with particular emphasis on the Western Australian Shield; *in* Kroner, A., Hanson, G. N., and Goodwin, A. M. (eds.), Archean geochemistry: Springer–Verlag, Berlin, pp. 73–98.

Groves, D. I., and Hudson, D. R., 1981, The nature and origin of Archaean strata-bound volcanic-associated nickel-iron-copper sulfide deposits; *in* K. H. Wolf (ed.), Handbook of Strata-Bound and Stratiform Ore Deposits, v. 9: Elsevier, Amsterdam, pp. 305–410.

Groves, D. I., and Keays, R. R., 1979, Mobilization of ore-forming elements during alteration of dunites, Mt. Keith–Betheno, Western Australia: Canadian Mineralogist, v. 17, pp. 373–389.

Groves, D. I., and Lesher, C. M. (eds.), 1982, Regional geology and nickel deposits of the Norseman–Wiluna Greenstone Belt, Western Australia, Publication 7: University of Western Australia Department of Geology and Extension Service, Perth, 234 pp.

Groves, D.I., Barrett, F.M., and Brotherton, R.H., 1981, Exploration significance of chrome-spinels in mineralized ultramafic rocks and Ni–Cu ores; *in* de Villiers, J. R. R., and Cawthorn, G. A. (eds.), ICAM 81—Proceedings of the First International Conference on Applied Mineralogy: Geological Survey of South Aftrica Special Publication No. 7, Johannesburg, South Africa, pp. 21–30.

Groves, D. I., Barrett, F. M., and McQueen, K. G., 1976, A possible volcanic-exhalative origin for lenticular nickel sulfide deposits of volcanic association, with special reference to those in Western Australia: Discussion: Canadian Journal of Earth Sciences, v. 13, pp. 1646–1650.

Groves, D. I., Barrett, F. M., and McQueen, K. G., 1979, The relative roles of magmatic segregation, volcanic exhalation and regional metamorphism in the generation of volcanic-associated nickel ores of Western Australia: Canadian Mineralogist, v. 17, pp. 319–336.

Groves, D. I., Barrett, F. M., Binns, R. A., and McQueen, K. G., 1977, Spinel phases associated with metamorphosed volcanic-type iron-nickel sulfide ores from Western Australia: Economic Geology, v. 72, pp. 1224–1244.

Groves, D. I., Hudson, D. R., and Hack, T. B. C., 1974, Modification of iron-nickel sulfides during serpentinization and talc-carbonate alteration at Black Swan, Western Australia Economic Geology, v. 69, pp. 1265–1281.

Groves, D. I., Korkiakoski, E. A., McNaughton, N. J., Lesher, C. M., and Cowden, A., 1986, Thermal erosion by komatiites at Kambalda and genesis of nickel ores: Nature, v. 319, pp. 136–139.

Groves, D. I., Lesher, C. M., and Gee, D. G., 1984, Tectonic setting of the sulphide nickel deposits of Western Australia; *in* Buchanan, D. L., and Jones, M. J. (eds.), Sulphide deposits in mafic and ultramafic rocks, Proceedings of International Geological Correlation Program Projects 161 and 91, Third Nickel Sulphide Field Conference, Perth, Western Australia: Institution of Mining and Metallurgy, London, pp. 1–13.

Hammerbeck, E. C. I., 1984, Aspects of nickel metallogeny of Southern Africa; *in* Buchanan, D. L., and Jones, M. J. (eds.), Sulphide deposits in mafic and ultramafic rocks, Proceedings of International Geological Correlation Program Projects 161 and 91, Third Nickel Sulphide Field Conference, Perth, Western Australia: Institution of Mining and Metallurgy, London, pp. 130–140.

Helz, R. T., 1977, Determination of the P–T dependence of the first appearance of FeS-rich liquid in natural basalts to 20 kb: EOS, v. 58, pp. 523.

Herrmann, A. G., Blanchard, D. P., Haskin, L. A., Jacobs, J. W., Knake, D., Korotev, R. L., and Brannon, J. C., 1976, Major, minor, and trace element compositions of peridotitic and basaltic komatiites from the Precambrian crust of Southern Africa: Contributions to Mineralogy and Petrology, v. 59, pp. 1–12.

Hill, R. E. T., 1982, The six mile well nickel deposit; *in* Groves, D. I., and Lesher, C. M. (eds.), Regional geology and nickel deposits of the Norseman–Wiluna Greenstone Belt, Western Australia, Publication 7: University of Western Australia Department of Geology and Extension Service, Perth, part C, pp. 101–109.

Hill, R. E. T., Gole, M. J., and Barnes, S. J., 1987, Physical volcanology of komatiites, A field guide to the komatiites between Kalgoorlie and Wiluna, Eastern Goldfields Province, Yilgarn Block, Western Australia, Excursion Guide Book No. 1: Geological Society of Australia, Western Australian Division, Perth, 74 pp.

Hopwood, T., 1981, The significance of pyritic black shales in the genesis of Archaean nickel sulphide deposits; *in* Wolf, K. H. (ed.), Handbook of Strata-Bound and Stratiform Ore Deposits, v. 9: Elsevier, Amsterdam, pp. 412–467.

Huang, W.–L., and Williams, R. J., 1980, Melting relations of the system Fe–S–Si–O to 32 kb with implications as to the nature of the mantle-core boundary: Lunar Planetary Science Conference XI, Lunar and Planetary Science Institute, Houston, pp. 486–488.

Hudson, D. R., 1972, Evaluation of genetic models for Australian sulphide nickel deposits: Australasian Institute of Mining and Metallurgy Conference, Newcastle, pp. 59–68.

Hudson, D. R., and Donaldson, M. J., 1982, Mineralogy of platinum group elements in the Kambalda nickel deposits, Western Australia; *in* Buchanan, D. L., and Jones, M. J. (eds.), Sulphide deposits in mafic and ultramafic rocks, Proceedings of International Geological Correlation Program Projects 161 and 91, Third Nickel Sulphide Field Conference, Perth, Western Australia: Institution of Mining and Metallurgy, London, pp. 55–61.

Huppert, H. E., Sparks, R. S. J., Turner, J. S., and Arndt, N. T., 1984, Emplacement and cooling of komatiite lavas: Nature, v. 309, pp. 19–22.

Huppert, H. H., and Sparks, S. J., 1985a, Komatiites I: Eruption and flow: Journal of Petrology, v. 26, pp. 694–725.

Huppert, H. H., and Sparks, S. J., 1985b, Cooling and contamination of mafic and ultramafic magmas during ascent through continental crust: Earth and Planetary Science Letters, v. 74, pp. 371–386.

Irvine, T. N., 1977a, Origin of chromitite layers in the Muskox intrusion and other stratiform intrusions: A new interpretation: Geology, v. 5, pp. 273–277.

Irvine, T. N., 1977b, Definition of primitive liquid compositions for basic magmas: Carnegie Institution of Washington, Yearbook, v. 76, pp. 454–461.

Jagoutz, E., Palme, H., Baddenhausen, H., Blum, K., Cendales, M., Dreibus, G., Spettel, B., Lorenz, V., and Wänke, H., 1979, The abundance of major, minor, and trace elements in the Earth's mantle as derived from primitive ultramafic nodules: Proceedings of the Lunar Science Conference, v. 10, pp. 2031–2050.

Jensen, L. S., and Pyke, D. R., 1982, Komatiites in the Ontario portion of the Abitibi belt; *in* Arndt, N. T., and Nisbet, E. G. (eds.), Komatiites: George Allen and Unwin, London, pp. 147–156.

Jolly, W. T., 1982, Progressive metamorphism of komatiites and related Archaean lavas of the Abitibi area, Canada; *in* Arndt, N. T., and Nisbet, E. G. (eds.), Komatiites: George Allen and Unwin, London, pp. 247–265.

Keays, R. R., 1982, Palladium and iridium in komatiites and associated rocks: Application to petrogenetic problems; *in* Arndt, N. T., and Nisbet, E. G. (eds.), Komatiites: George Allen and Unwin, London, pp. 435–457.

Keays, R. R., and Davison, R. M., 1976, Palladium, iridium and gold in the ores and host rocks of nickel sulfide deposits in Western Australia: Economic Geology, v. 71, pp. 1214–1220.

Keays, R. R., Ross, J. R., and Woolrich P., 1981, Precious metals in volcanic peridotite-associated nickel sulfide deposits in Western Australia, II: Distribution within ores and host rocks at Kambalda: Economic Geology, v. 76, pp. 1645–1674.

Keele, R.A., and Nickel, E.H., 1974, The geology of a primary millerite-bearing sulfide assemblage and supergene alteration at the Otter shoot, Kambalda, Western Australia: Economic Geology, v.69, pp.1102–1117.

Kilburn, L. C., Wilson, H. D. B., Graham, A. R., Ogura, Y., Coats, C. J. A., and Scoates, R. F. J., 1969, Nickel sulfide ores related to ultrabasic intrusions in Canada; *in* Wilson, H. D. B. (ed.), Magmatic Ore Deposits, Monograph 4: Economic Geology Publishing Company, Lancaster, pp. 267–293.

Lesher, C. M., 1983, Localization and genesis of komatiite-hosted Fe–Ni–Cu sulphide mineralization at Kambalda, Western Australia: Unpublished Ph.D. dissertation, University of Western Australia, Perth, 318 pp.

Lesher, C. M., 1985, Evidence for thermal erosion of basalt and hybridization of komatiite: Geological Society of America Annual Meeting, Program with Abstracts, v. 17, no. 7, p. 642.

Lesher, C. M., and Groves, D. I., 1984, Geochemical and mineralogical criteria for the identification of mineralized komatiites in Archaean greenstone belts of Australia; *in* Petrology: Igneous and Metamorphic Rocks, Proceedings of the Twenty-Seventh International Geological Congress, Moscow, Volume 9: VNU Science Press, Utrecht, pp. 283–302.

Lesher, C. M., and Groves, D. I., 1986, Controls on the formation of komatiite-associated nickel-copper sulfide deposits; *in* Friedrich, G. H., Genkin, A., Naldrett, A. J., Ridge, J. D., Sillitoe, R. H., and Vokes, F. M. (eds.), Geology and Metallogeny of Copper Deposits, Proceedings of the Twenty-Seventh International Geological Congress, Moscow: Springer–Verlag, Berlin, pp. 43–62.

Lesher, C. M., and Keays, R. R., 1984, Metamorphically and hydrothermally mobilized Fe–Ni–Cu sulphides at Kambalda, Western Australia; *in* Buchanan, D. L., and Jones, M. J. (eds.), Sulphide deposits in mafic and ultramafic rocks, Proceedings of International Geological Correlation Program Projects 161 and 91, Third Nickel Sulphide Field Conference, Perth, Western Australia: Institution of Mining and Metallurgy, London, pp. 62–69.

Lesher, C. M., Arndt, N. T., and Groves, D. I., 1984, Genesis of komatiite-associated nickel sulphide deposits at Kambalda, Western Australia: A distal volcanic model; *in* Buchanan, D. L., and Jones, M. J. (eds.), Sulphide deposits in mafic and ultramafic rocks, Proceedings of International Geological Correlation Program Projects 161 and 91, Third Nickel Sulphide Field Conference, Perth, Western Australia: Institution of Mining and Metallurgy, London, pp. 70–80.

Lesher, C. M., Donaldson, M. J., and Groves, D. I., 1982, Nickel deposits and their host rocks in the Norseman–Wiluna Belt; *in* Groves, D. I., and Lesher, C. M. (eds.), Regional geology and nickel deposits of the Norseman–Wiluna Greenstone Belt, Western Australia, Publication 7: University of Western Australia Department of Geology and Extension Service, Perth, part B, pp. 1–63

Lesher, C. M., Lee, R. F., Groves, D. I., Bickle, M. J., and Donaldson, M. J., 1981, Geochemistry of komatiites at Kambalda: I. Chalcophile element depletion—a consequence of sulfide liquid separation from komatiitic magmas: Economic Geology, v. 76, pp. 1714–1728

Ludden, J., Hubert, C., and Gariepy, C., 1986, The tectonic evolution of the Abitibi greenstone belt of Canada: Geological Magazine, v. 123, pp. 153–166.

Ludden, J. N., and Gélinas, L., 1982, Trace element characteristics of komatiites and komatiitic basalts from the Abitibi metavolcanic belt of Quebec; *in* Arndt, N. T., and Nisbet, E. G. (eds.), Komatiites: George Allen and Unwin, London, pp. 331–345.

Lusk, J., 1976, A possible volcanic-exhalative origin for lenticular nickel sulfide deposits of volcanic association with special reference to those in Western Australia: Canadian Journal of Earth Sciences, v.13, pp. 451–458.

MacLean, W. H., 1969, Liquidus phase relations in the FeS–FeO–$Fe_2O_3$–$SiO_2$ system and their application in geology: Economic Geology, v. 64, pp. 865–884.

Maiden, K. J., Chimimba, L. R., and Smalley, T. J., 1986, Cuspate ore-wall rock interfaces, piercement structures, and the localization of some sulfide ores in deformed sulfide deposits: Economic Geology, v. 81, pp. 1464–1472.

Malyuk, B. I., 1984, A juvenile-magmatic model of the formation of stratiform nickel-sulfide ore deposits in komatiite host rocks: Geologiya i Geofizika, v. 25, pp. 84–91.

Malyuk, B. I., 1985, The origin of komatiite magmas: Petrochemical test of models: Geokhimiya, no. 6, pp. 785–795.

Marston, R. J., 1984, Nickel mineralization in Western Australia: Mineral Resources Bulletin 14, Geological Survey of Western Australia, Perth, 271 pp.

Marston, R. J., and Groves, D. I., 1981, The metallogenesis of Archaean base metal deposits in Western Australia; *in* Glover, J. E., and Groves, D. I. (eds.), Archaean Geology, Proceedings of the

Second Symposium on Archaean Rocks, Special Publication 7: Geological Society of Australia, Sydney, pp. 409–420.

Marston, R. J., and Kay, B. D., 1980, The distribution, petrology and genesis of nickel ores at the Juan complex, Kambalda, Western Australia: Economic Geology, v. 75, pp. 546–565.

Marston, R. J., Groves, D. I., Hudson, D. R., and Ross, J. R., 1981, Nickel sulfide deposits in Western Australia: A review: Economic Geology, v. 76, pp. 1330–1363.

Martyn, J. E., 1987, Evidence for structural repetition in the greenstone of the Kalgoorlie district, Western Australia: Precambrian Research, v. 37, pp. 1–18.

McNaughton, N. J., Frost, K. M., and Groves, D. I., 1988, Ground melting and ocellar komatiites: a lead isotopic study at Kambalda, Western Australia: Geology Magazine, v. 125, pp. 285–295.

McQueen, K. G., 1981a, Volcanic-associated nickel deposits from around the Widgiemooltha dome, Western Australia: Economic Geology, v. 76, pp. 1417–1443.

McQueen, K. G., 1981b, The nature and metamorphic history of the Wannaway nickel deposit, Western Australia. Economic Geology, v. 76, pp. 1444–1468.

McQueen, K. J., 1979, Experimental heating and diffusion effects in Fe–Ni sulfide ore from Redross, Western Australia: Economic Geology, v. 74, pp. 140–148.

Muir, J. E., and Comba, C. D. A., 1979, The Dundonald deposit: an example of volcanic-type nickel-sulfide ore in ultramafic lavas: The Canadian Mineralogist, v. 17, pp. 351–360.

Naldrett, A. J., 1966, The role of sulphurization in the genesis of iron-nickel sulphide deposits of the Porcupine District, Ontario: Canadian Institute of Mining and Metallurgy, Transactions, v. 69, pp. 147–155.

Naldrett, A. J., 1973, Nickel sulphide deposits—their classification and genesis with special emphasis on deposits of volcanic association: Canadian Institute of Mining and Metallurgy, Bulletin, v. 66, pp. 45–63.

Naldrett, A. J., 1979, Partitioning of Fe, Co, Ni and Cu between sulfide liquid and basaltic melts and the composition of Ni–Cu sulfide deposits: a reply and further discussion: Economic Geology, v. 74, pp. 1520–1528.

Naldrett, A. J., 1981, Nickel sulfide deposits: Classification, composition and genesis; *in* Skinner, B. J. (ed.), Seventy-Fifth Anniversary Volume: Economic Geology Publishing Company, El Paso, pp. 628–685.

Naldrett, A. J., 1984, Duke, J. M., Lightfoot, P. C., and Thompson, J. F. H., 1984, Quantitative modelling of the segregation of magmatic sulphides: An exploration guide: Canadian Institute of Mining and Metallurgy, Bulletin, v. 77, pp. 46–56.

Naldrett, A. J., and Cabri L. J., 1976, Ultramafic and related rocks: Their classification and genesis with special reference to the concentration of nickel sulfides and platinum-group elements: Economic Geology, v. 71, pp. 1131–1158.

Naldrett, A. J., and Campbell, I. H., 1982, Physical and chemical constraints on genetic models for komatiite-related Ni-sulphide deposits; *in* Arndt, N. T., and Nisbet, E. G. (eds.), Komatiites: George Allen and Unwin, London, pp. 423–434.

Naldrett, A. J., and Gasparrini, E. L., 1971, Archaean nickel sulphide deposits in Canada: Their classification, geological setting and genesis with some suggestions as to exploration; *in* Glover, J. E. (ed.), The Archaean Rocks, Proceedings of the First Symposium on Archaean Rocks, Special Publication 3: Geological Society of Australia, Adelaide, pp. 331–347.

Naldrett, A. J., and Turner A.R., 1977, The geology and petrogenesis of a greenstone belt and related nickel sulfide mineralization at Yakabindie, Western Australia: Precambrian Research, v. 5, pp. 43–103.

Naldrett, A. J., Hoffman, E. L., Green, A. H., Chou, C.–L., Naldrett, S. R., and Alcock, R. A., 1979, The composition of Ni-sulfide ores, with particular reference to their content of PGE and Au: Canadian Mineralogist, v. 17, pp. 403–416.

Nesbitt, R. W., 1971, The case for liquid immiscibility as a mechanism for nickel sulphide mineralization in the Eastern Goldfields, Western Australia (abstract); *in* Glover, J. E. (ed.), The Archaean Rocks, Proceedings of the First Symposium on Archaean Rocks, Special Publication 3: Geological Society of Australia, Adelaide, p. 253.

Nesbitt, R. W., 1986, Are komatiitic lavas voracious?: Nature, v. 319, pp. 97–98.

Nesbitt, R. W., Sun, S.–S., and Purvis, A. C., 1979, Komatiites: Geochemistry and genesis: Canadian Mineralogist, v. 17, pp. 165–186.

Nisbet, E. G., 1982, The tectonic setting and petrogenesis of komatiites; *in* Arndt, N. T., and Nisbet, E. G. (eds.), Komatiites: George Allen and Unwin, London, pp. 501–520.

Nisbet, E. G., and Chinner, G. A., 1981, Controls on the eruption of mafic and ultramafic lavas, Ruth Well Ni–Cu prospect, West Pilbara: Economic Geology, v. 76, pp. 1719–1735.

Nisbet, E. G., and Walker, D., 1982, Komatiites and the structure of the Archaean mantle: Earth and Planetary Science Letters, v. 60, pp. 105–113.

Nisbet, E. G., Bickle, M. J., and Martin, A., 1977, The mafic and ultramafic lavas of the Belingwe greenstone belt, Rhodesia: Journal of Petrology, v. 18, pp. 521–566.

Nisbet, E. G., Bickle, M. J., Martin, A., Orpen, J. L., and Wilson, J. F., 1982, Komatiites in Zimbabwe; *in* Arndt, N. T., and Nisbet, E. G. (eds.), Komatiites: George Allen and Unwin, London, pp. 97–103.

Ohmoto, H., 1986, Stable isotope geochemistry of ore deposits; *in* Valley, J. W., Taylor, H. P., Jr., and O'Neil, J. R. (eds.), Stable Isotopes in High Temperature Geological Processes, Reviews in Mineralogy, Volume 16: Mineralogical Society of America, Washington, D.C., pp. 491–559.

Page, M. L., and Schmulian, M. L., 1981, The proximal volcanic environment of the Scotia nickel deposits: Economic Geology, v. 76, pp. 1469–1479.

Parker, P., 1984, The role contamination in the formation of the nickel sulphide ores at Wannaway and Mt. Edwards, Western Australia: Unpublished B.Sc. Honours thesis, University of Western Australia, Perth, 144 p.

Paterson, H. L., Donaldson, M. J., Smith, R. N., Lenard, M. F., Gresham, J. J., Boyack D. J., and Keays, R. R., 1984, Nickeliferous sediments and sediment-associated nickel ores at Kambalda, Western Australia; *in* Buchanan, D. L., and Jones, M. J. (eds.), Sulphide deposits in mafic and ultramafic rocks, Proceedings of International Geological Correlation Program Projects 161 and 91, Third Nickel Sulphide Field Conference, Perth, Western Australia: Institution of Mining and Metallurgy, London, pp. 81–94.

Peredery, W.V., 1979, Relationship of ultramafic amphibolites to metavolcanic rocks and serpentinites in the Thompson belt, Manitoba: Canadian Mineralogist, v.17, pp. 187–200.

Peredery, W.V., and Geological Staff,1982, Geology and nickel sulfide deposits of the Thompson belt, Manitoba; *in* Hutchinson, R. W., Spence, C. D., and Franklin, J. M. (eds.), Precambrian Sulfide Deposits, Special Paper 25: Geological Association of Canada, Waterloo, pp.165–210.

Philpotts, A.R., 1961, Textures of the Ungava nickel ores: Canadian Mineralogist, v.6, pp. 681–688.

Porter, D. J., and McKay, K. G., 1981, The nickel sulfide mineralization and metamorphic setting of the Forrestania area, Western Australia: Economic Geology, v. 76, pp. 1524–1549.

Prider, R. T., 1970, Nickel in Western Australia: Nature, v. 226, pp. 691–693.

Pridmore, D.F., Coggon, J.H., Esdale, D.J., and Lindeman, F.W., 1984, Geophysical exploration for nickel sulphide deposits in the Yilgarn block, Western Australia; *in* Buchanan, D. L., and Jones, M. J. (eds.), Sulphide deposits in mafic and ultramafic rocks, Proceedings of International Geological Correlation Program Projects 161 and 91, Third Nickel Sulphide Field Conference,

Perth, Western Australia: Institution of Mining and Metallurgy, London, pp. 22–34.

Robinson, D. J., and Hutchinson, R. W., 1982, Evidence for a volcanogenic-exhalative origin of a massive nickel sulfide deposit at Redstone, Timmins, Ontario; *in* Hutchinson, R. W., Spence, C. D., and Franklin, J. M. (eds.), Precambrian Sulfide Deposits, Special Paper 25: Geological Association of Canada, Waterloo, pp. 211–254.

Ross, J. R., and Hopkins, G. M. F., 1975, Kambalda nickel sulfide deposits; *in* Knight, C. L. (ed.), Economic Geology of Australia and Papua–New Guinea, I. Metals, Monograph 5: Australasian Institute of Mining and Metallurgy, Parkville, pp. 100–121.

Ross, J. R., and Keays, R. R., 1979, Precious metals in volcanic-type nickel sulfide deposits in Western Australia. I. Relationships with the composition of the ores and host rocks: Canadian Mineralogist, v. 17, pp. 417–436.

Ross, J. R., and Travis, G. A., 1981, The nickel sulfide deposits of Western Australia in global perspective: Economic Geology, v. 76, pp. 1291–1329.

Sanders, T. S., 1982, Nepean nickel deposits; *in* Groves, D. I., and Lesher, C. M. (eds.), Regional geology and nickel deposits of the Norseman–Wiluna Greenstone Belt, Western Australia, Publication 7: University of Western Australia Department of Geology and Extension Service, Perth, part C, pp. 35–44.

Schmulian, M. L., 1982, Windarra Nickel Deposits; *in* Groves, D. I., and Lesher, C. M. (eds.), Regional geology and nickel deposits of the Norseman–Wiluna Greenstone Belt, Western Australia, Publication 7: University of Western Australia Department of Geology and Extension Service, Perth, part C, pp. 67–76.

Schmulian, M. L., 1984, Windarra nickel deposits, Western Australia; *in* Buchanan, D. L., and Jones, M. J. (eds.), Sulphide deposits in mafic and ultramafic rocks, Proceedings of International Geological Correlation Program Projects 161 and 91, Third Nickel Sulphide Field Conference, Perth, Western Australia: Institution of Mining and Metallurgy, London, pp. 95–102.

Seccombe, P. K., Groves, D. I., Binns, R. A., and Smith, J. W., 1978, A sulfur isotopic study to test a genetic model for Fe–Ni sulphide mineralization at Mt Windarra, Western Australia; *in* Robinson, B. W. (ed.), Stable Isotopes in the Earth Sciences, Bulletin 220: New Zealand Department of Scientific and Industrial Research, Wellington, pp. 187–200.

Seccombe, P. K., Groves, D. I., Marston, R. J., and Barrett, F. M., 1981, Sulfide paragenesis and sulfur isotopic evidence: Economic Geology, v. 76, pp. 1675–1685.

Shima, H., and Naldrett, A. J., 1975, Solubility of sulfur in an ultramafic melt and the relevance of the system Fe–S–O: Economic Geology, v. 70, pp. 960–967.

Smith, D.H., 1984, Geochemical exploration for nickel sulphides in lateritic terrain in Western Australia; *in* Buchanan, D. L., and Jones, M. J. (eds.), Sulphide deposits in mafic and ultramafic rocks, Proceedings of International Geological Correlation Program Projects 161 and 91, Third Nickel Sulphide Field Conference, Perth, Western Australia: Institution of Mining and Metallurgy, London, pp. 35–42.

Smith, H.S., and Erlank, A.J., 1982, Geochemistry and petrogenesis of komatiites from the Barberton greenstone belt, South Africa; *in* Arndt, N. T., and Nisbet, E. G. (eds.), Komatiites: George Allen and Unwin, London, pp. 348–393.

Stolz, G. W., 1981, A petrographic and geochemical investigation of the Archaean volcanic succession in the vicinity of the Scotia nickel deposit: Unpublished Ph.D. thesis, University of Adelaide, 234 pp.

Stolz, G. W., and Nesbitt, R. W., 1981, The komatiite nickel sulfide association at Scotia: a petrochemical investigation of the ore environment: Economic Geology, v. 76, pp. 1480–1502.

Sun, S. S., 1982, Chemical composition and origin of the earth's primitive mantle: Geochimica et Cosmochimica Acta, v. 46, pp. 179–192.

Takahashi, E., and Scarfe, C. M., 1985, Melting of peridotite to 14 GPa and the genesis of komatiite: Nature, v. 315, pp. 566–568.

Tomich, B. V. N., 1974, The geology and nickel mineralization of the Ruth Well area, Western Australia: Unpublished B.Sc. Honours thesis, University of Western Australia, Perth, 118 pp.

Turner, J. S., Huppert, H. H., and Sparks, R. S. J., 1986, Komatiites II: Experimental and theoretical investigations of post-emplacement cooling and crystallization: Journal of Petrology, v. 27, pp. 397–437.

Ussleman, T. M., Hodge, D. S., Naldrett, A. J., and Campbell, I. H., 1979, Physical constraints on the characteristics of nickel-sulfide ore in ultramafic lavas: Canadian Mineralogist, v. 17, pp. 361–372.

Viljoen, M. J., and Bernasconi, A., 1979, The geochemistry, regional setting and genesis of the Shangani–Damba nickel deposits, Rhodesia; *in* Anhaeusser, C. R., Foster, R. P., and Stratten, T. (editors), A Symposium on Mineral Deposits and the Transportation and Deposition of Metals, Proceedings of Metallogenesis '76, University of Rhodesia, Salisbury, Special Publication 5: Geological Society of South Africa, pp. 67–98.

Viljoen, M. J., Bernasconi, A., Van Coller, N., Kinloch, E., and Viljoen, R. P., 1976, The geology of the Shangani nickel deposit, Rhodesia: Economic Geology, v. 71, pp. 76–95.

Viljoen, M.J., and Viljoen, R.P., 1969, The geology and geochemistry of the lower ultramafic unit of the Onverwacht group and a proposed class of igneous rocks, Special Publication 2: Geological Society of South Africa, pp. 55–85.

Viljoen, M.J., Viljoen, R.P., and Pearton, T.N., 1982, The nature and distribution of Archaean komatiite volcanics in South Africa; *in* Arndt, N. T., and Nisbet, E. G. (eds.), Komatiites: George Allen and Unwin, London, pp. 53–78.

Wentlandt, R. F., 1982, Sulfide saturation of basalt and andesite melts at high pressures and temperatures: American Mineralogist, v. 67, pp. 877–885.

Williams, D. A. C., 1979, The association of some nickel sulfide deposits with komatiitic volcanism in Rhodesia: Canadian Mineralogist, v. 17, pp. 337–349.

Williams, R. K., Veeraburns, M., and Philbrook, W. O., 1972, Thermal conductivities of some Ni–S, Ag–S, and Ag–Te melts: American Institute of Mining Engineers, Metallurgical Transactions, v. 3, pp. 255–260.

Wilson, J. F., 1979, A preliminary appraisal of the Rhodesian Basement Complex: Geological Society of South Africa, Transactions, v. 5, pp. 1–23.

Woodall. R., and Travis, G. A., 1969, The Kambalda nickel deposits, Western Australia: Ninth Commonwealth Mining and Metallurgy Congress, Publication 2, pp. 517–533.

Woolrich, P., Cowden, A., and Giorgetta, N. E., 1981, The chemical and mineralogical variations in the nickel mineralization associated with the Kambalda dome: Economic Geology, v. 76, pp. 1629–1644.

# Chapter 6

# Ores Associated with Flood Basalts

A. J. Naldrett

## INTRODUCTION

Periodically during the earth's history, large-scale rifting of the crust has developed across and within many continents. In some cases this has reached the stage of complete rupture, to produce two separate plates; in others the rifting has been less extreme and the magmas have intruded into and extruded on continental crust. Interaction between the magmas and the crust has given rise to Ni–Cu sulfide ores which form the subject of this chapter. Two very important nickel camps have formed in this way: the Lower Triassic ores of the Noril'sk–Talnakh camp, Siberia; and the 1.1 Ga ores of the Duluth Complex, Minnesota.

## NORIL'SK–TALNAKH

### Tectonic and Geologic Setting

The Noril'sk–Talnakh Cu–Ni deposits are located on the extreme northwestern margin of the Siberian Platform, which has been a stable craton since the end of the Paleozoic (Fig. 6.1). To the north, the platform is separated by the Khatanga trough from a second platform, the Taimyr Peninsula, which has been stable since the Paleozoic. To the west, a third craton, the East European–Urals Block, is separated from the Siberian Platform by the Yenisei trough; this is bordered farther to the west by a large low lying area, the West Siberian Lowlands. The East European–Urals Block has behaved as a craton since middle Permian time.

Within Western Siberia, Lower Paleozoic dolomites, argillites and sandstones of marine origin are overlain by extensive Devonian calcareous and dolomitic marls and sulfate-rich evaporites, and Lower Carboniferous shallow-water limestones (Genkin et al., 1981). These are in turn unconformably overlain by Middle Carboniferous lagoonal and continental sediments, including gravels, conglomerates and coal measures (Smirnov, 1966; Glazkovsky et al., 1977). The emergent sedimentary sequence is covered by an extremely large volume (approximately one million km$^3$) of late Permian and Triassic flood basalt and tuff (Bazunov, 1976; Glazkovsky et al., 1977) known as the Siberian Traps. In the Noril'sk region, these were extruded in four phases, consisting of late Permian quartz and hypersthene normative tholeiites, early Triassic andesite-basalts and olivine-basalts, followed by an early Triassic basaltic and then an early mid-Triassic basaltic phase. Although the magmatic activity is very widespread, the olivine-rich basalts tend to be concentrated in the Noril'sk region, and individual sheets thin towards the flanks of the region. There is an overall thickening of the flood basalts within the troughs imposed on the underlying sediments. Sill-like tholeiitic intrusions, varying in composition from subalkaline dolerite to gabbro-dolerite (Glaskovsky et al., 1977), were emplaced contemporaneously with, and were, in part, feeders to, the extrusive activity.

The structure of the district is dominated by late Permian to Triassic block faulting which was coeval with the igneous activity. Individual faults may be over 500 km in length and have throws of up to 1,000 m (Smirnov, 1966). The principal faults trend north–northeast to northeast (Fig. 6.2).

Following extrusion of flood basalt, considerable subsi-

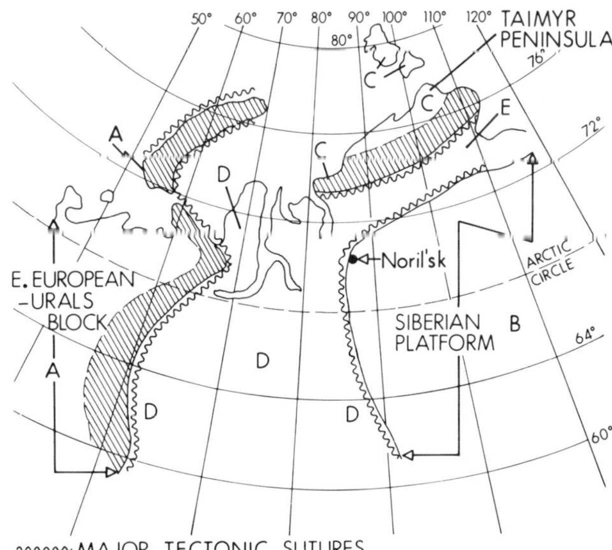


FIGURE 6.1—The main tectonic features of northwestern Siberia (after Maksimov and Rudkevich, 1974).

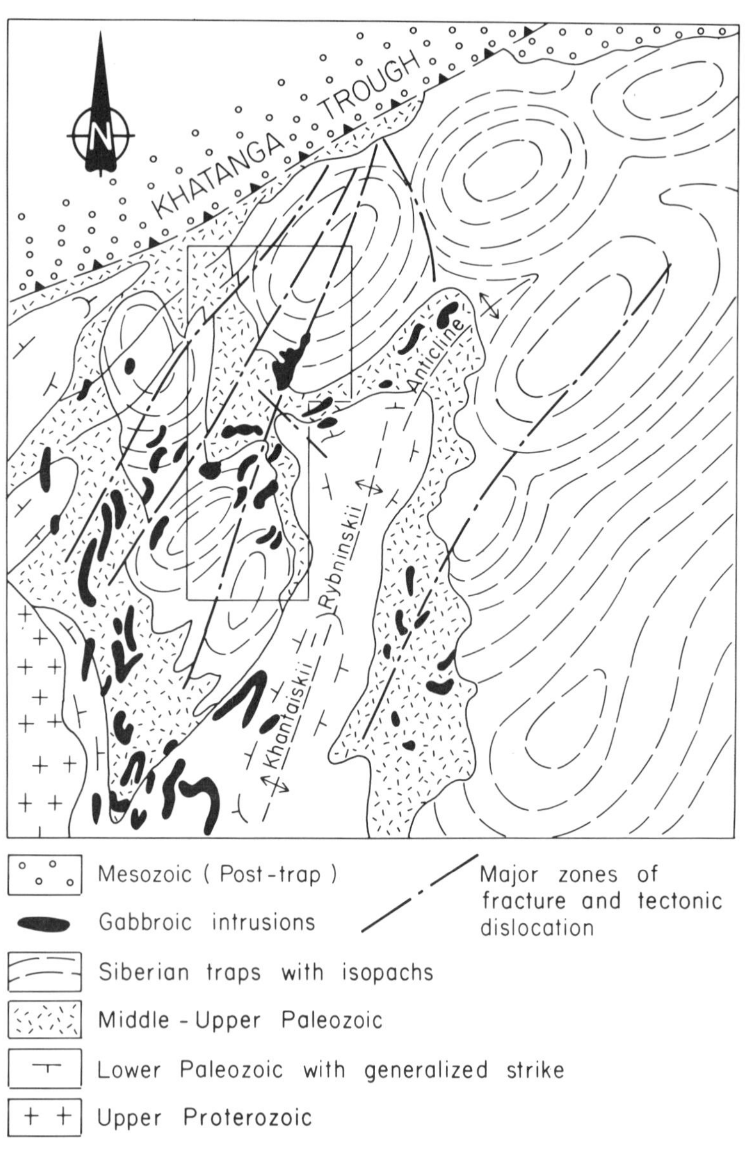


FIGURE 6.2—Geological map of the Noril'sk–Talnakh area showing the main structural elements. After Genkin et al. (1981).

dence took place in the West Siberian Lowlands and Yenisei and Khatanga troughs from the Early Triassic to the mid-Tertiary; trough-filling sediments attain a thickness in excess of 10 km (Tamrazyan, 1971). The zone of subsidence coincides with the Y-shaped structure (see Fig. 6.3), bounded by the block faulting illustrated in Fig. 6.1 (Maksimov and Rudkevich, 1971).

Geophysical measurements of depth to the Mohorovicic discontinuity (Tamrazyan, 1971) indicate that, despite the considerable thickness of Mesozoic and early Tertiary sediments in the area, the discontinuity is significantly shallower beneath the Yenisei and Khatanga troughs than beneath the adjacent platforms (Fig. 6.4). The coincidence of considerable subsidence with crustal thinning beneath the troughs is indicative of a rifting environment. In its early stages, the rifting was accompanied by the widespread intrusions and extrusions already described as occurring on the Siberian platform, and which aeromagmatic surveys (Fig. 6.5) indicate underlie broad areas of the lowlands themselves. Where they can be mapped on the platform, most of the intrusions are found to be unmineralized (Grinenko, 1985 states that only 1–2% are mineralized), and are tabular to subcordant bodies ranging from gabbro-dolerite to trachy-andesite porphyry.

The emergence during the Carboniferous was accompanied in the Noril'sk region by the development of an anticline (the Khantaiiski–Rybrinnskii anticline) and two flanking synclinal troughs (Fig. 6.2). These flexures appear to have reflected large scale fractures that appeared in the basement during the middle and upper Paleozoic, and along which movement persisted until after the Permo–Triassic volcanism. Many of the mineralised intrusive centers of the Noril'sk–Talnakh camp are associated with one of these fractures, the Noril'sk–Kharaelakh fault (Fig. 6.6). It is regarded

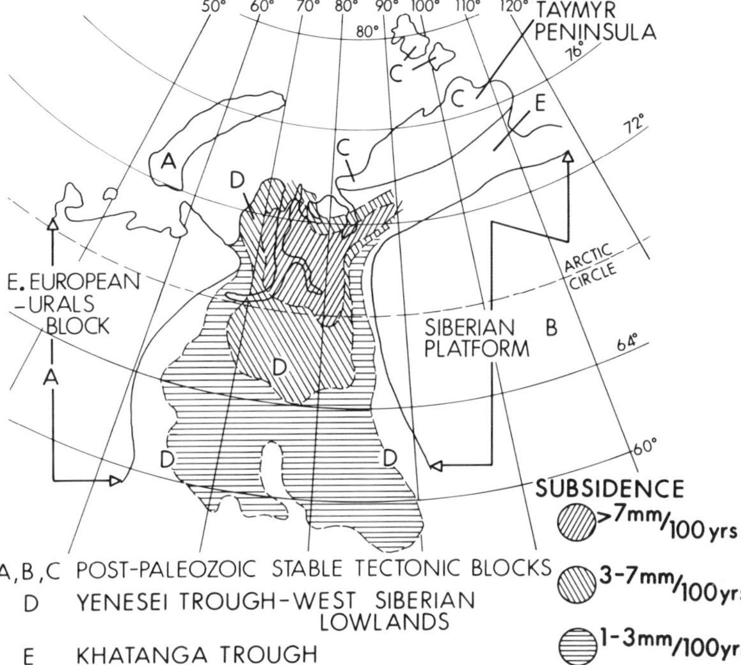


FIGURE 6.3—Subsidence during the early and middle Jurassic in northwestern Siberia. After Tamrazyan (1971).

as a deep-seated fault that acted as the route for upwelling magma (Malich and Tuganova, 1976). Both the Mt. Chernaya deposit to the south and the Talnakh–Oktyabr'sk deposits to the north are located on or adjacent to this fault. The controlling feature for the ore-bearing intrusions in the Noril'sk–Talnakh region appears to be the zones in which this or related faults impinge on the borders of the troughs in the pre-Permian basement which are associated with the thickening of the flood basalts (Genkin et al., 1981).

At Noril'sk, the mineralized Triassic intrusions represent volcanic conduits radiating outward and upward from intrusive centers, cutting the sedimentary sequence and, in places, reaching and apparently acting as feeders to the lavas. Farther to the north, at Talnakh, the intrusions do not reach lavas and terminate within underlying sediments.

A further group of gabbro-dolerite-hosted Cu-Ni deposits, the North Kharaelakh ore field has been described by Dyuzhikov et al., 1976. These deposits, located approximately 375 km north of Noril'sk have an aerial extent of 1,500 km$^2$ and lie close to the northerly extension of the Noril'sk–Kharaelakh fault.

### Host Intrusions

The ore-bearing intrusions are differentiated; some have all of the phases to be expected as a result of differentiation of basic magma, in roughly the expected proportions, with picrite and picritic dolerite overlain by more felsic differentiates; other intrusions are particularly rich in melanocratic components and yet others are leucocratic. Cryptic variation is well developed in most intrusions with olivine normally ranging from $Fo_{82}$ to $Fo_{65}$ (up to $Fo_{87}$ in some of the melanocratic intrusions) and the cores of plagioclase grains from $An_{93}$ to $An_{60}$ (Genkin et al., 1981; see Fig. 6.7).

Genkin et al. (1981) place particular stress on the presence of chrome-bearing spinels, which, at Noril'sk, serve to distiguish mineralized from unmineralized intrusions. Spinels occur primarily in the olivine-bearing rocks, both as sparsely disseminated cumulus grains, and as equant agglomerations. The latter range from a few cm up to 1 m in diameter, contain 25–30 vol% chromite, and are restricted mainly to the lower part of the picrite horizons.

The disseminated chromite is of two types; an early, dark-coloured type containing 15–18 wt% $Al_2O_3$, 32–38 wt% $Cr_2O_3$, 6–20 wt% MgO, 8–20 wt% FeO (as $Fe^{2+}$) and less than 4 wt% $TiO_2$, and a later, lighter coloured type (often forming overgrowths on the earlier type) containing 5–6 wt% $Al_2O_3$, 18–25 wt% $Cr_2O_3$, 2–4 wt% MgO, about 38 wt% FeO (as $Fe^{2+}$) and 10–15 wt% $TiO_2$. Ratios of $Fe^{3+}/Fe^{2+}$ range between 0.5 and 0.8. Chromite of the agglomerations tends to be richer in Cr (up to 44 wt% $Cr_2O_3$) and have a lower $Fe^{3+}/Fe^{2+}$ ratio. The $TiO_2$-rich variants are absent in agglomerations. The space interstitial to the chromite in the agglomerations is occupied by pyroxene, plagioclase and (more rarely) olivine. The interstitial crystals of these minerals are commonly optically continuous with grains forming the fabric of the rock enclosing the agglomerations. One particuarly interesting feature of the agglomerations is the presence of clearly defined globules, with a spherulitic internal structure, rich in prehnite, biotite, anhydrite, carbonate, zeolites, chlorite, talc, occasional sulfides and plagioclase and pyroxene.

Individual sills may attain lengths of up to 12 km and widths of 2km, while thicknesses are typically 200–250 m or less. In cross section the intrusions are lenticular, trough-shaped, or, in some cases, tabular. Genkin et al. (1981) note that while the intrusions are generally sub-concordant, they may cut down steeply to lower levels in the stratigraphy

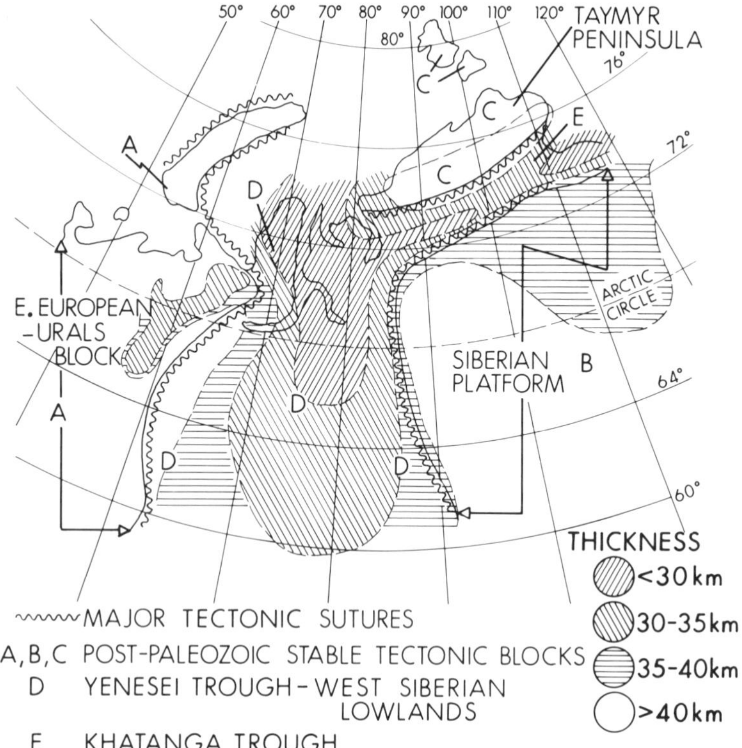


FIGURE 6.4—Thickness of the crystalline crust between the Mesozoic to Holocene sediments and the Moho.

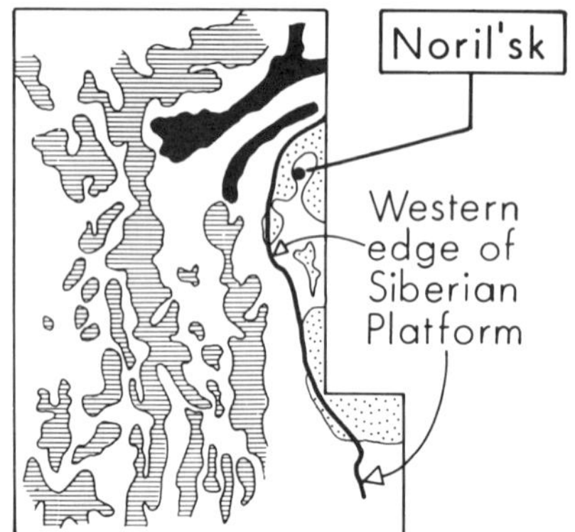


FIGURE 6.5—Aeromagnetic map reflecting buried mafic igneous rocks underlying the area outlined in Fig. 6.1.

where they again become sub-concordant. They suggest that leucocratic and some normal sills are connected to melanocratic sills that are buried at deeper levels in the stratigraphy, and that together they represent ensembles in which the higher sills represent injections of more fractionated magma which, in fractionating, gave rise to the refractory deposits which remain as the melanocratic sills. They report that magma conduits have been established as connecting the normal Talnakh and the melanocratic Nizhne–Talnakh intrusions. They note that the cryptic variation present in the normal and leucocratic sills indicates continuous differentiation in situ, while the uniform composition of olivine in the melanocratic bodies indicates that the magma responsible for these was intruded carrying crystals of uniform composition which were presumably derived from elsewhere.

Relationships between an ore-bearing intrusion and both the surrounding rocks and the ore itself have been well described for the Talnakh mineralized body (Fig. 6.8) by Genkin et al. (1981). Much of the following description is taken from their work.

This intrusion can be considered as consisting of three parts, one lying east of the Noril'sk–Kharaelakh fault, one within the fault zone and one lying to the west of it. That in the east is the simplest, being an elongate tabular body with few apophyses. It has intruded Permo–Carboniferous rocks, except for a central keel which penetrates Devonian limestones. Its roof comes close to the base of the overlying flood basalts. Within the fault zone, the intrusion has been disturbed by later movements along the fault, but it appears also to have been an elongate, tabular body. That portion of the intrusion occurring west of the fault is quite different and has intruded a highly tectonically disturbed sedimentary sequence. It plunges with the general dip of the country rocks; the stratigraphically lowest part is at the centre (where

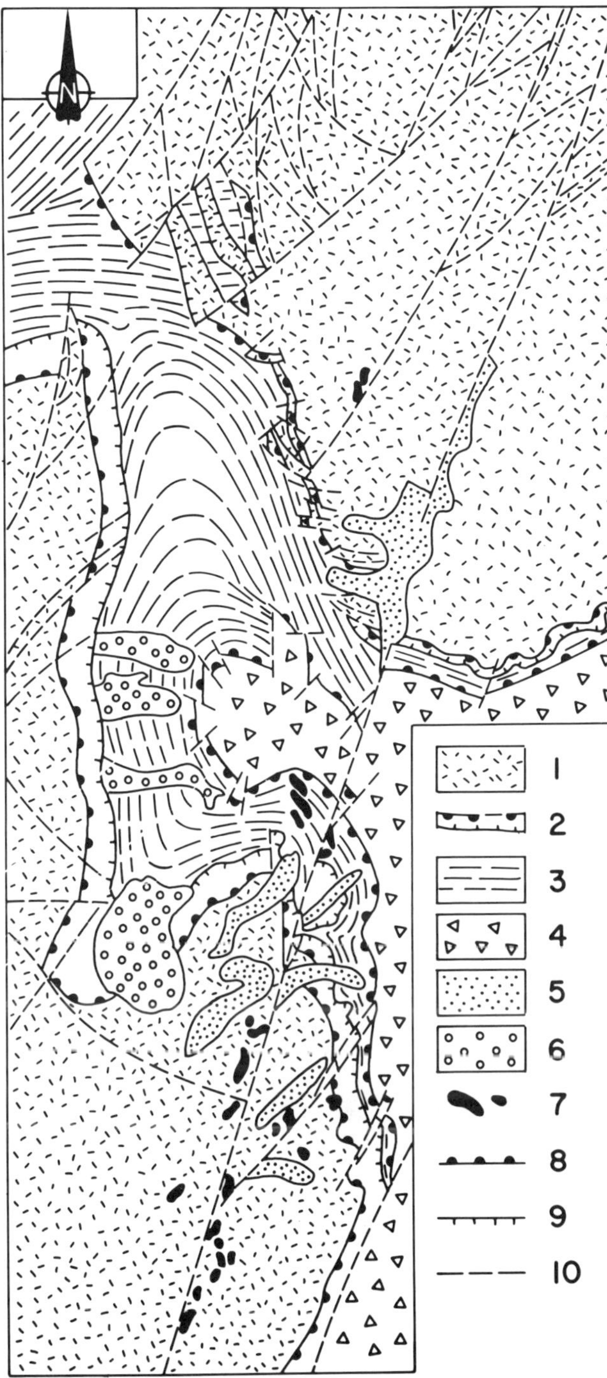


FIGURE 6.6—Distribution of different types of differentiated intrusions in the Noril'sk–Talnakh region. 1 = Upper Samoedskii and Nadeshdinskii tuffs and lavas, 2 = Lower Gudzhikhinskii and Ivakinskii tuffs and lavas, 3 and 4 = sedimentary rocks, 5 = Fully differentiated intrusions, 6 = Differentiated leucocratic intrusions, 7 = Differentiated melanocratic intrusions, 8 = This symbol defines the stratigraphic interval within which differentiated intrusions occur, 9 = Sediment − tuff + lava contact, 10 = Faults. From Genkin et al. (1981).

dykes are observed connecting it with the stratigraphically lower, melanocratic Nizhne–Talnakh body) and it rises away from this keel. It is localized within lower and middle Devonian rocks. Where the flanks of the western portion come into contact with sulfate-evaporite and carbonate rocks, they have an intricate structure, breaking up into a series of apophyses, the larger of which are themselves stratified. In general, the flanking, and stratigraphically higher parts of the western portion are marked by a higher proportion of leucocratic differentiates than the central part. A breccia, which the intrusive rocks have penetrated, characterizes the western and southwestern margins of the western body.

The country rocks are extensively metamorphosed adjacent to the body. Typically, zones of metamorphism, which includes both hornfelsic and metasomatic alteration, have thicknesses of several hundreds of m, which is 2–4 times the thickness of the intrusion and its apophyses that they surround.

The products of both the metamorphism and metasomatism are closely related, even on a very fine scale, to the compositional heterogeneities of the country rocks. The hornfelsing has been largely isochemical, ranging from the melilite–monticellite sub-facies to pyroxene, then biotite–epidote–amphibolite and finally to the muscovite chlorite facies.

The metasomatism includes the development of alkali-feldspar and feldspar-quartz rocks in sandstones, siltstones, argillites, extrusive basalts and olivine-free intrusive roks. Magnesium and calcium-rich skarns characterize the carbonate-rich rocks, anhydrite evaporites and olivine gabbros. The magnesian skarns, some of which are post-intrusive, include forsterite, fassaite, spinel–forsterite and spinel–forsterite–pyroxene rocks; anhydrite-rich and also monticellite-bearing skarns are also present. Calcium-rich skarns are more widespread than the magnsium-rich type, and include the asemblages pyroxene–plagioclase, garnet–pyroxene and wollastonite–vesuvianite.

## Mineralization

### Ore types

The mineralization includes the following principal ore types: Disseminated ores within the intrusion; massive ores localized at the basal contact, both within the country rocks and, to a lesser extent, within the intrusion; veinlets and associated dissemination related to the massive ores; similar mineralization at the top of the intrusion; and breccia and disseminated ores on the western flanks of the intrusion where it is pinching out.

Within the intrusion, the disseminated ores are restricted to the picrite and the contact phase of the gabbro underlying this. Mineralization in the picrite dies out on the flanks as the intrusion breaks up into a series of apophyses before itself pinching out.

Massive ores are present almost everywhere at the lower contact, although in some places they consist only of thin veinlets. The thickest concentrations of massive ore occur along the axes of the intrusion and are elongated in the down-dip direction. Where the basal contact curves sharply, the massive ore may cut from the intrusion into the country

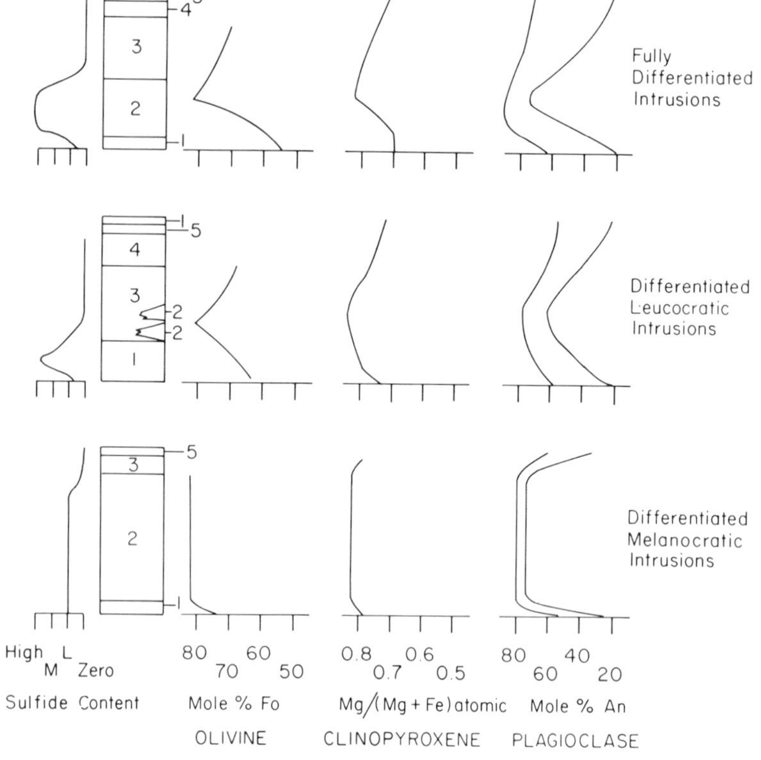


FIGURE 6.7—Cryptic variation in the composition of olivine, clinopyroxene and plagioclase in different types of intrusions of the Noril'sk–Talnakh region. 1 = contact gabbro-dolerite; 2 = picrite and troctolite; 3 = olivine gabbro; 4 = gabbro; 5 = diorite. After Genkin et al. (1981).

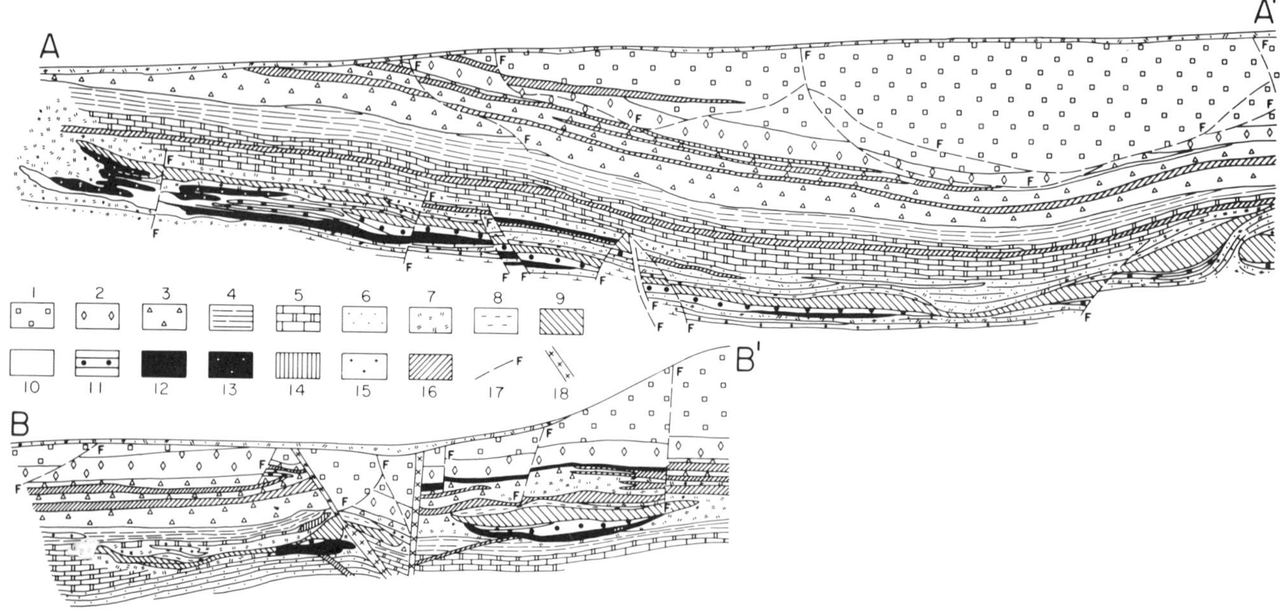


FIGURE 6.8—Geological section through the northwestern part of the Talnakh ore field. 1 = tholeiitic basalts of the Syverminskii suite; 2 = two-feldspar titanaugite basalts of the Ivakinskii suite; 3 = sandstones, siltstones, argillites, carbonaceous argillites and coals of the Tunguska series; 4 = dolomites and limestones of the Kalargonskii suite; 5 = anhydrites and marls of the Manturovskii suite; 6 = argillites of the Razvedochninskii suite; 7 = contact metamorphic and metasomatic rock; 8 = diorite, gabbro-diorite and gabbro; 9 olivine gabbro; 10 = picrite and taxitic gabbro; 11–14 = ores: 11 = disseminated ore in intrusive rocks, 12 = massive ore, 13 = breccia ore, 14 = veinlet ore; 15 = undifferentiated gabbro-dolerite sills; 16 = titanaugite dolerites; 17 = faults. From Genkin et al. (1981).

rocks. In some cases massive ore outside the intrusion is continuous with ore localized within stockworks of vertical and horizontal fractures within the intrusion. The veinlets and their associated dissemination at the base of the intrusion are found primarily in country rocks at places where the massive ores pinch out laterally. Where this type of mineralization occurs in the roof, it is restricted to that part of the intrusion within the Noril'sk–Kharaelakh fault, and is controlled by fissuring related to the fault. Genkin et al. (1981) note that this ore zone plunges northward, in the direction of basal concentrations of ore of similar type, and suggest that it joins up with and is therefore related to these concentrations.

The breccia ores occur only on the western flank of the intrusion, where the country rocks are very brecciated, and where the sulfides have occupied voids in pre-existing breccia. The intrusion is itself very complex in this area, consisting of numerous apophyses and occupying the breccia zone itself, and the ore forms an almost continuous fringe around the flank of the intrusion. Genkin et al. emphasize that the intrusion contains little disseminated sulfide itself, although massive sulfides may occur within and adjacent to breccia ore bodies and, in particular, form veins within the intrusion.

**Mineralogy and mineral zoning**

The main primary minerals are pyrrhotite [both monoclinic ($po_m$) and hexagonal ($po_h$)], troilite (tr), chalcopyrite (cp), pentlandite (pn), cubanite (cb), talnakhite (tk), mooihoekite (mk), bornite (bo), millerite (ml) and magnetite (mt). The disseminated mineralization within the intrusion is zoned. Horizontal zonation is best demonstrated within the picrite horizon where a (tr + tk + mk) assemblage characterizes the centres of ore bodies, and gives way laterally in an outwards direction (Fig. 6.9) to many or all of the following assemblages: (tr + $po_h$ + cp + cb + pn), ($po_h$ + cp + pn + cb), ($po_h$ + cp + pn) and ($po_m$ + cp + pn). A similar zonation is also present in both directions vertically, with the ($po_m$ + cp + pn) assemblage characterizing the contact gabbro beneath the picrite layer and the olivine gabbro above it.

Massive ores consist of two types in so far as their mineralogy and zoning is concerned. In the simplest type, a ($po_{m\ and\ h}$ + pn + cp) characterizes the centre of the ore zone and cp increases in amount to dominate at the margins. In the complex type, a (cb + tk + mk) assemblage occurs at the centre, giving way progressively to (cb + cp + pn + $po_h$ + tr), (cp + cb + pn + $po_h$), (cp + cpn + $po_{h\ and\ m}$) and ($po_m$ + pn + cp). Cubanite-rich assemblages are always found separating

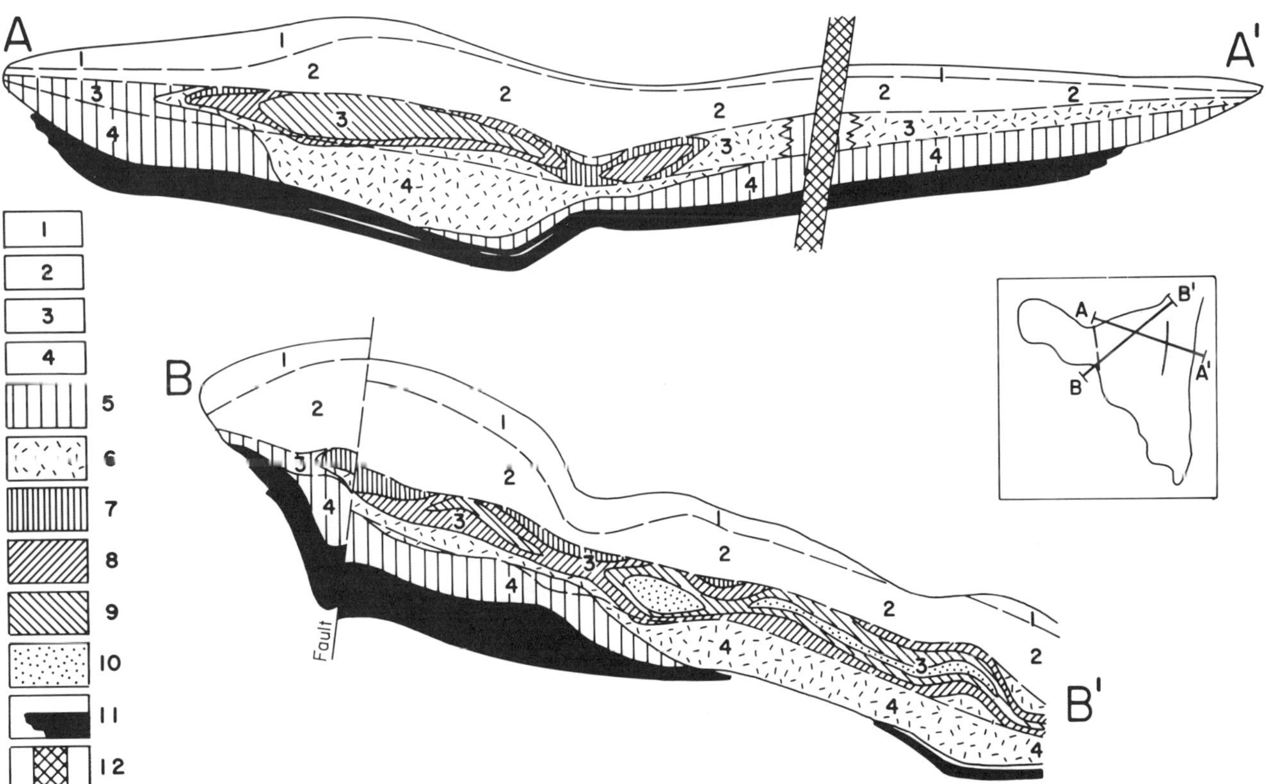


FIGURE 6.9—Distribution of sulfide mineral assemblages in two vertical sections through the Talnakh intrusion. 1 = diorite, gabbro; 2 = olivine gabbro; 3 = picrite; 4 = taxitic gabbro; 5–10 = sulfide mineral assemblages: 5 = pn + cp + minor po, 6 = pn + cp + po, 7 = pn + cub + cp + po, 8 = pn + cub + cp + tr + $po_h$, 9 = pn + cub + minor cp + tr, 10 = pn + cub + minor tn (mh,cp) + tr, 11 = massive sulfide ore, 12 = ruptured zones. Note that pn = pentlandite; cp = chalcopyrite; $po_m$ and $po_h$ = monoclinic and hexagonal pyrrhotite respectively; cub = cubanite; tr = troilite, tn = taenite, mh = mooihoekite. From Genkin et al. (1981).

(tk + mk) assemblages from pyrrohotite-rich assemblages. The sulfur-poor (tk + mk-rich) assemblages in massive ores differ from similar assemblages in disseminated ores in that the latter contain troilite and the former do not.

Veinlet ores are characterized primarily by [$po_m$(60–80%) + cp(5–20%) + pn(5–7%)] assemblages, but a zoning is sometimes present with cp increasing and the total Fe content decreasing in amount away from the intrusion so that mineral assemblages change to (cp + $po_m$ + pn), (cp + pn) and then (cp + ml). Commonly a secondary alteration to valleriite and/or pyrite has been superimposed on the primary veinlet mineralization.

The breccia ores are characterized by assemblages of ($po_m$ + cp + pn) with accessory py, ml, mt and valleriite. As in the veins, cp generally increases in proportion away from the intrusion or its apophyses, laterally and in both directions vertically, with assemblages changing to (cp + $po_m$ + pn), (cp + pn) and then (cp + ml). The pyrrhotite of the veinlet and breccia ores is exclusively monoclinic.

The massive sulfides have their own associated metamorphism, which they appear to have imposed on previously metamorphosed and metasomatized country rocks. These effects include the recrystallization of these rocks to a coarser grain size, and the melting of sandstones displaying potash felspar metasomatism to produce pegamatoidal potash feldspar-quartz aggregates.

### Sulfur isotopes

Grinenko (1967, 1985), Godlevski and Grinenko (1963) and Gorbachev and Grinenko (1973) have studied the isotopic composition of sulfides in un-mineralized and mineralized intrusions of the northwest Siberian platform. The majority of the intrusions are not mineralized or carry only a sparse dissemination of sulfide. Amongst the mineralized intrusions, Grinenko (1985) distinguishes between those that contain economic mineralization and those that do not. Her data are illustrated in Fig. 6.10.

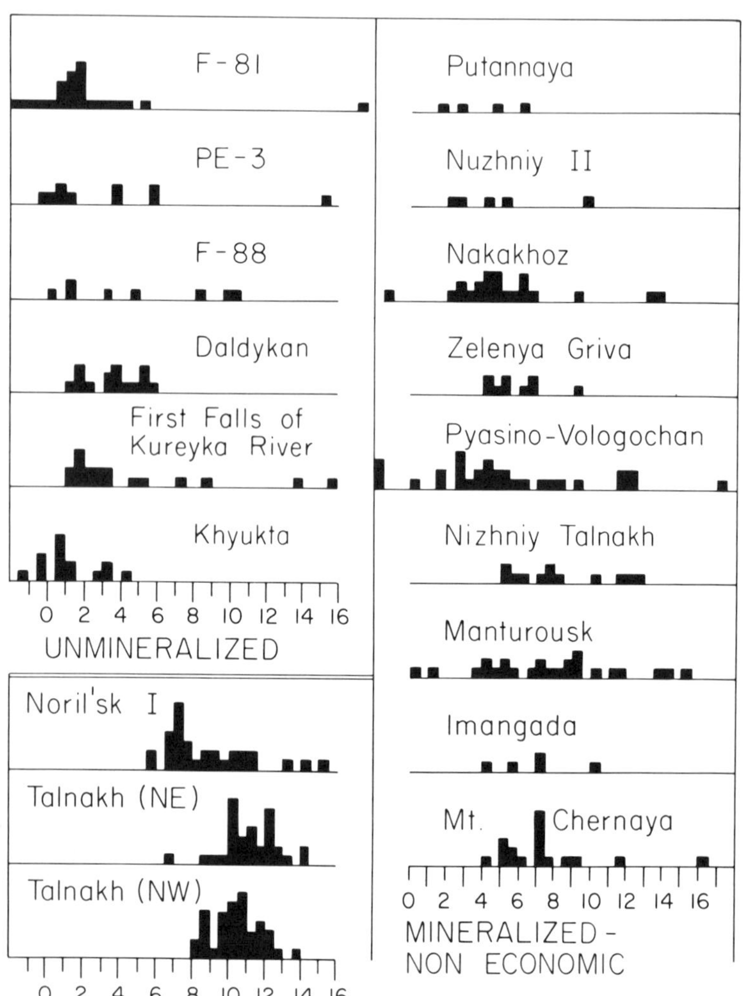


FIGURE 6.10—Isotopic composition of sulfur in unmineralized, mineralized economic and mineralized sub-economic intrusions of the Noril'sk–Talnakh region. After Grinenko (1985).

Representative samples were selected from drill core and subjected to whole-rock sulfur and sulfur isotope analysis. Since, in most cases, less than 20 samples were taken, it is unlikely that the average sulfur values are representative of the whole intrusion, but merely of the samples analysed. Most unmineralized intrusions contain an average of less than 0.1 wt % total sulfur (equal to less than 0.3 percent sulfide) with average $\delta^{34}S$ values ranging from +0.1 to +4.6. Average sulfur contents for the mineralized non-economic intrusions range from 0.17 to 0.34 wt%, with average $\delta^{34}S$ values from +5.5 to +8.4. The samples from the economic mineralized intrusions contain on average 0.95 to 2.2 wt% sulfides, with average $\delta^{34}S$ values from +8.9 to +11.4.

The sulfides from individual intrusions with economic mineralization show much less spread in $\delta^{34}S$ than those of barren or non-economic intrusions, suggesting that those of the first group have been much better homogenized than those of the two latter groups.

**Genesis of mineralization**

Grinenko (1985) noted that there is a direct correlation between the sulfur content and the average $\delta^{34}S$ value of the sulfides of the barren and non-economic mineralized intrusions. She also noted that for these two groups there is a relationship between the nature of the wall rocks and the amount and isotopic composition of the sulfur. Barren intrusions tend to occur in sulfur-free clastic sediments of the Tunguska super-group and in the overlying flood basalts. Amongst the non-economic mineralized group, most of those intruded into sulfate-bearing sediments have higher sulfur contents than those in sulfur-free sediments. She suggested that much of the sulfur has come from the wall rocks, and that the lack of isotopic homogenization shown by the sulfides in these bodies is due to the sulfur contamination having occurred in situ.

Assuming that any original magmatic sulfur had an isotopic composition of zero, adopting the measured value for the $\delta^{34}S$ of the evaporite sulfur of +20, assuming that any contamination was not accompanied by isotopic fractionation, and assuming that the measured average sulfur content of her samples represented that of the magmas responsible for the intrusions, Grinenko (1985) calculated from mass-balance considerations that 20–36 percent sulfur has been added to the non-economic mineralized bodies, and that their initial (before contamination) sulfur content was between 0.12 and 0.23 wt%. This, she argued, is consistent with the solubility of sulfur in mafic magma (this author would regard these amounts, equivalent to 0.5 to 1.0 wt% sulfide, as distinctly high for dry basaltic magma, see section on solubility of sulfide).

Grinenko also noted that if a similar calculation is made for the economic intrusions, also assuming the evaporites to be the source of sulfur, the initial sulfur content is much higher than could have been dissolved in the initial magma. Making an assumption as to the amount that could of been transported in solution in the magma, she calculates that the $\delta^{34}S$ value of the contaminant was +10.7 to +12. She interpreted the homogeneous isotopic composition of these sulfides to be due to the contamination having occurred prior to final emplacement of the igneous bodies, and concluded that the contaminant was $H_2S$ (sour gas), which is known to be common in the region and has an isotopic composition of about +10.

This author is in agreement with the interpretation that the economic intrusions were contaminated prior to their final emplacement, and that the sulfides have become homogenized during this emplacement, but believes that the conclusion as to contamination by sour gas is not proven by her evidence. Amongst others, her conclusion is based on the assumption that the sulfur added as a contaminant has to account for all sulfide in excess of that which could have been dissolved by the basic magma responsible for the intrusions. This assumption is open to doubt, indeed is probably not true. The very high PGE (platinum group element) content of the sulfide ores indicates a high 'R' factor for the sulfides (see discussion in section on PGE deposits), which is to say that they have equilibrated with and concentrated PGE from many times their own mass of magma. Values of $10^{3.5}$ to $10^4$ are indicated for the mass ratio of magma to sulfide for the Noril'sk–Talnakh ores assuming that the magmas were similar in their PGE contents to normal flood basalts, in comparison with the value of 16 or so that arises from the premise behind Grinenko's argument.

Grinenko places the Nizhniy Talnakh (Lower Talnakh) non-economic mineralized intrusion in a separate category to the Talnakh body, implying a different source for the sulfur. As seen above, Genkin et al. (1981) have argued on geologic and petrologic grounds that the two are related and connected, with Nizhniy Talnakh representing a lower chamber in which the magma differentiated, and thus gave rise to ultramafic cumulates, before rising farther in the stratigraphy to form the Talnakh body. Unless the sulfur was added to the magma between the two horizons (which from the descriptions of Genkin et al. seems very unlikely) the source of the sulfur must have been the same for the two bodies.

This author favours the view of Genkin et al. (1981) that Nizhniy Talnakh was an intermediate holding and differentiation chamber, and that the contamination occurred either in this, or prior to the magma intruding to form this. It is suggested in theses notes that a limited proportion of the magma that gave rise to the lower body (Nizhniy Talnakh), enriched in sulfide that had been present within the body and that had therefore had the opportunity to scavenge ore metals from a much larger volume of magma than itself, then rose farther in the stratigraphy to form the Talnakh intrusion.

The way in which such a sulfide-enriched magma was derived from the Nizhniy Talnakh body is not entirely clear at present, although it seems likely that not all of the sulfide was necessarily introduced at the same time as the magma. Distler et al (1986) have argued that the introduction of much of the massive and veinlet sulfide at Talnakh post-dates injection of the main intrusion. They base their arguments on the observations (i) that the intrusion at Talnakh had reacted in response to tectonic stress as a solid mass, and that it therefore gave rise to dilatent zones into which massive sulfides were then introduced (some of these are even present in the roof of the intrusion as outlined above) and (ii) that massive sulfides were accompanied by their own metamorphic effects which were superimposed on the metamorphism and metasomatism due to the intrusion.

Distler et al. (1986) use their conclusion of the later introduction of massive sulfide to support their hypothesis that the separation of sulfide liquid from silicate magma occurred deep in the crust and not within the supercrustal sulfate-rich rocks. Godlevsky and Likachev (1986) believe that these magmas have carried 3–10% sulfide all of the way from the mantle. Thus they do not accept contamination by country-rock sulfur as a significant ore-forming process, ascribing the heavy sulfur of the ores to the presence of heavy sulfur in some parts of the mantle (A. D. Genkin, personal communication, 1982). This author is impressed by Grinenko's (1985) sulfur isotope evidence as to the involvement of crustal sulfur. He suggests that, because of their lower melting temperature, sulfides in and beneath the Nizhniy Talnakh intrusion remained liquid after the ultramafic cumulates of this and parts of the Talnakh body had solidified enough to fracture, and that the liquid sulfides were then squeezed up higher in the stratigraphy into these fractures in the upper body in response to the on-going tectonic movements.

This author suggests that the prehnite + biotite + anhydrite + carbonate + zeolite + chlorite ± sulfide globules that occur within the chromite agglomarations in the picrite of the intrusions could represent melted remnants of partially assimilated country rock. The assimilation of anhydrite-rich rocks, coupled with the reduction of sulfate to sulfide would have introduced considerable oxygen into the silicate melt, which could have been the cause of the precipitation of chrome spinel. Inclusions of anhydrite-rich material, floating in the magma, could have served as locii for chromite crystallization, thus giving rise to the association between the agglomerations and globules.

Genkin et al. (1981) noted that the normal progression of zoning from pyrrhotite-rich to chalcopyrite-rich assemblages is readily explicable in terms of the fractional crystallization of sulfide magma, as, for example, Naldrett et al. (1982) have suggested for the Sudbury ores. Genkin et al. remarked that the existence of large masses of sulfide with talnakite and mooihoeckite assemblages is unique to the Noril'sk–Talnakh camp, and attributed the zoning from pyrrhotite + chalcopyrite to mooihoeckite + talnakhite to the separation of the sulfide melt into two liquids, one iron and sulfur rich and the ther copper-rich and sulfur-poor. They suggested that the cubanite ores are the result of reaction between the two adjacent liquids as they crystallized. Although there is no experimental evidence for liquid immiscibility in the relevant parts of the Cu–Fe–S and Cu–Fe–Ni–S systems, Distler et al. (1986) have described globular masses of sulfide from oceanic basalts in which the sulfides have segregated into two halves, one with the composition of nickeliferous Mss and the other close in composition to cubanite. Similar globules have been described from the Insizwa deposit by Lightfoot et al. (1984) where the copper-rich part is observed to be at the top and to consist largely of chalcopyrite. An alternative explanation is that the compositional separation is due to the crystallization and settling of a refractory, dense Mss phase from a sulfide melt, leaving a copper-rich residual liquid. The separation between the two domains in the globules is very sharp for the latter explanation to be entirely satisfactory. On the other hand, the progressive zoning towards the centre of the disseminated ore zone in the picrite, or the centre of zones of massive ore is also difficult to explain on the basis of either theory.

Perhaps one of the most perplexing questions remaining at Talnakh is the size of the metamorphic and metasomatic aureoles, several times larger than the intrusive bodies themselves. Although most of the intrusive bodies have an aureole, Genkin et al (1981) stressed that the aureoles are much larger about well-differentiated bodies than about others, concluding that the former must have been much richer in volatiles. The aureoles would be easier to explain if the intrusive bodies were all conduits for the overlying volcanism, as seems to be the case at Noril'sk itself, since in this case one could call upon a much larger mass of magma than is present in the intrusions. However the Talnakh intrusions appear to be blind and not linked physically to the overlying volcanics. The resolution to this problem may come from a quantitative evaluation of the heat and mass transfer that has occurred between the Noril'sk–Talnakh intrusions and their surroundings.

## MINERALIZATION OF THE DULUTH COMPLEX

### Geological Setting

The Duluth Complex crops out as an arcuate mass extending about 150 miles northeast from Duluth, Minnesota, to the Canadian border. It is the host to more than $4 \times 10^9$ tonnes of mineralization averaging 0.66 wt% Cu and 0.2 wt% Ni (Listerud and Meineke, 1977).

The Complex is an intrusive part of the Keweenawan flood basalt province of Lake Superior. A midcontinental gravity high extends this province southwards from the western end of Lake Superior through Minnesota to Kansas, and also southeastward from the eastern end of the lake into central Michigan (Fig. 6.11). The zone of anomalous gravity is also marked by numerous strong positive and negative magnetic anomalies (Fig. 6.12). Pre-Paleozoic basement rocks intersected in boreholes within the zone include a high proportion of Keweenawan basalts and mafic intrusions (Fig. 6.13). These and the related gravity and magnetic anomalies are expressions of magmatic activity associated with intracontinental rifting, which had its greatest development in the Lake Superior area with the extrusion of the Keweenawan flood basalts and the intrusion of the Duluth Complex.

Sims and Morey (1972) presented a generalized discussion of the stratigraphic setting of the Complex. It was intruded 1.12 Ga ago (Faure et al., 1969) close to the contact of Keweenawan basalts with older Precambrian basement rocks, which are now exposed to the northwest (Fig. 6.14). These basement rocks consist, in the north, of Archean felsic intrusions and volcanics, primarily the Giants Range Granite, which are overlain unconformably to the south by the south-dipping Biwabik Iron Formation. The iron formation is itself overlain by black argillites, greywackes, siltstones, graphitic slates, and sulfide facies iron-formation comprising the Virginia Formation.

The Duluth Complex is a composite intrusion which the studies of Grout (1918), Green et al. (1966), Bonnichsen (1970, 1972), Phinney (1970, 1972) and Weiblen and Morey (1975, 1980) have shown to consist of two series. An older series of anorthositic rocks is cut towards its western margin by a series of later intrusions consisting predominantly of troctolite. In the central part of the Complex, Foose and Weiblen

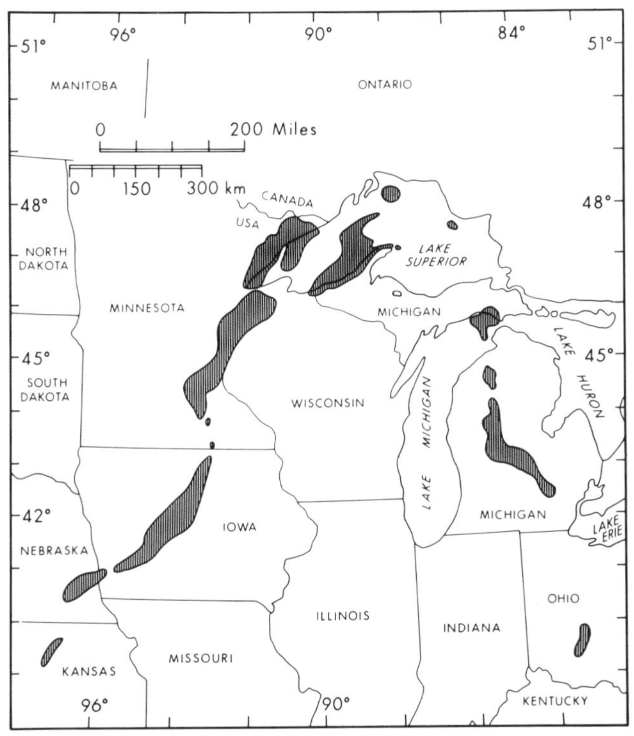


FIGURE 6.11—The midcontinent rift structure showing the associated gravity anomaly. Shading indicates areas of positive Bouguer anomaly (>0 mGal). After Halls (1978).

(1986) describe the troctolitic rocks as comprising three separate intrusions, Bald Eagle, South Kawishiwi and Partridge River, although the distinction between the latter two is made on the basis of subtle variations in texture.

Foose and Weiblen (1986) report that the South Kawishiwi intrusion consists of an upper portion, 1000–1200 m thick, in which troctolite and minor anorthosite are interlayered. Pegmatoidal layers of plagioclase are commonly in sharp contact with underlying troctolite, and form the bottoms of packages of rocks which grade upwards into medium-grained plagioclase cumulates and then plagioclase-olivine cumulates. Constituent minerals show normal and reverse cryptic variation within individual rock packages, but no systematic overall compositional variations exist, leading Foose and Weiblen to suggest that the chamber was repetitively recharged with new magma, possibly some of it from adjacent intrusions.

The layered troctolites are underlain by a heterogenous zone, 70–300 m thick, in which medium to fine grained troctolites, together with picrites, norites, anorthosites and oxide cumulates are the host to disseminated sulfides and numerous hornfels inclusions.

### Mineralization

The mineralization occurs along the western margins (stratigraphic bases) of these troctolitic intrusions. Wager et al. (1969) described the geology of the Spruce deposit of INCO Metals Ltd; Matlock and Watowich (1980), the geology of AMAX Exploration Ltd's Minnamax deposit; Rao and Ripley (1983) and Ripley (1981) the petrology and geochem-

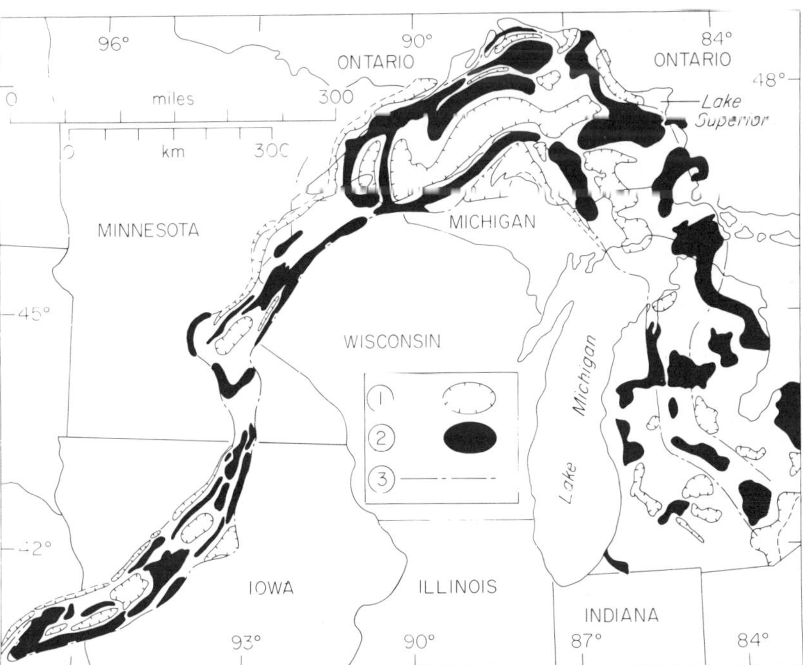


FIGURE 6.12—The midcontinent rift structure showing areas of positive (2) and negative (1) aeromagnetic anomalies as compared to the outline of the associated gravity anomaly (3). After Halls (1978).

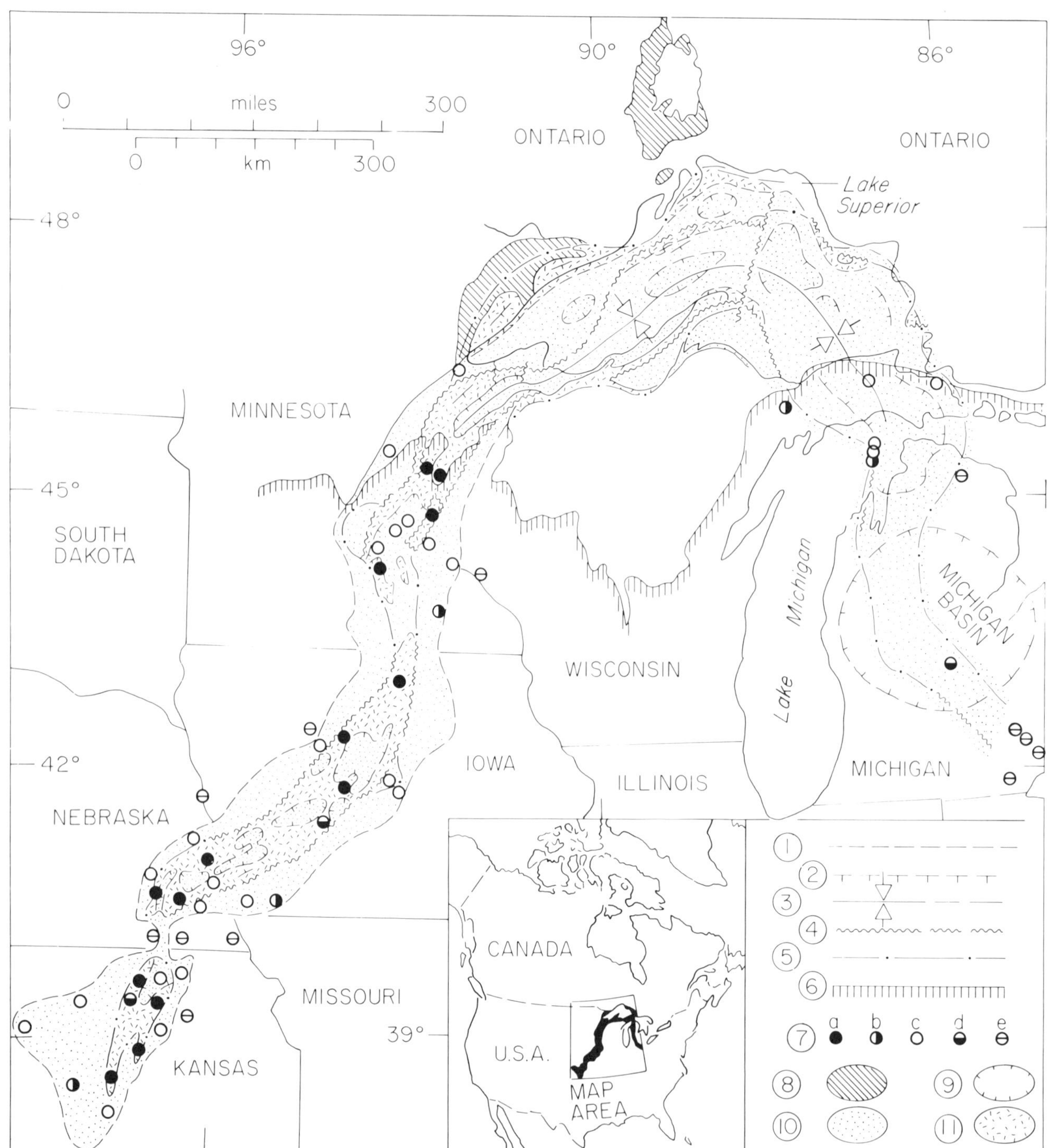


FIGURE 6.13—Distribution of rock types within the midcontinent rift structure. (1) = geological contact; (2) = structural form lines; (3) = synclinal axes; (4) = faults; (5) = boundary of Keweenawan volcanic rocks; (6) = northern limit of Paleozoic cover; (7) rocks intersected by bore holes penetrating to Keweenawan rocks, (a) = Keweenawan mafic igneous rocks, (b) = Keweenawan red clastic and underlying mafic volcanic rocks, (c) = Keweenawan red clastic rocks, (d) = Keweenawan red clastics lying directly on pre-Keweenawan basement, (e) = pre-Keweenawan basement only; (8) = Keweenawan mafic intrusive rocks; (10) area overlain by Keweenawan red clastic rocks; (11) area overlain by Keweenawan volcanic rocks. After Halls (1978).

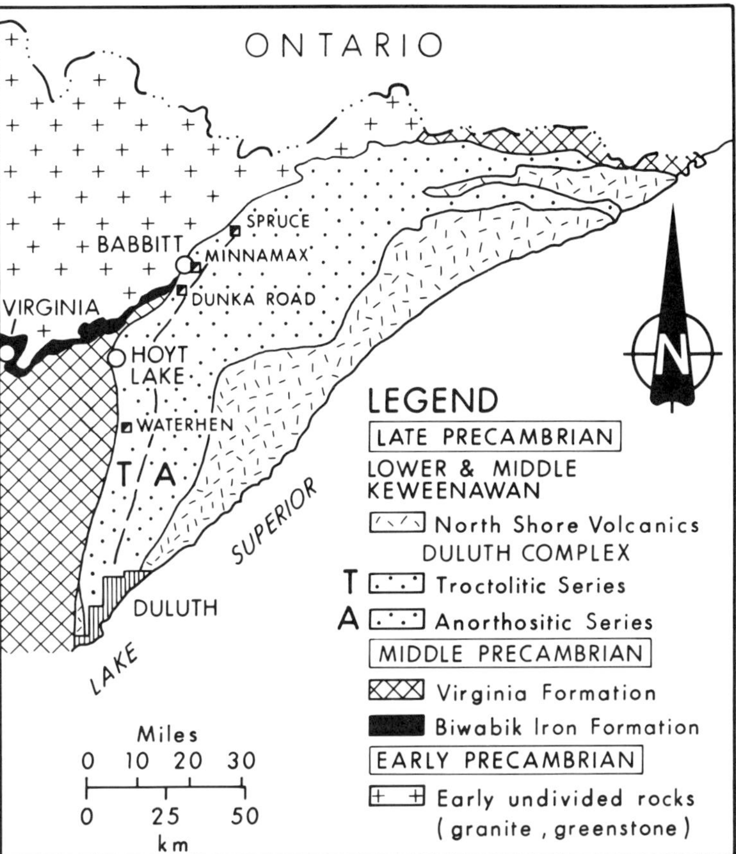


FIGURE 6.14—Map of the western end of Lake Superior showing the location of the Duluth Complex and its relationship to the country rocks.

istry of U.S. Steel Corporation's Dunka Road deposit; and Mainwaring and Naldrett (1977), the mineralization of the Waterhen Complex. The locations of these deposits are shown in Fig. 6.15. In the first three deposits, the mineralization consists of pyrrhotite, chalcopyrite, pentlandite, and cubanite weakly disseminated in troctolite and norite within 300 m of the base of the complex. The mineralized zones are characterized by numerous inclusions of country rocks consisting of hornfels of the Virginia and Biwabik Formations, together with barren gabbro and peridotite. Inclusions of Virginia and Biwabik rocks occur even at the Spruce deposit, where granite forms the footwall and the Virginia and Biwabik do not appear to be present in situ at surface within several kilometers of the deposit (Fig. 6.16). At both the Minnamax and Dunka Road deposits, norite is more common and troctolite less common in the vicinity of the sulfide horizons, attesting to reactions between the troctolite and country-rock inclusions.

The Waterhen Complex is somewhat different in that the sulfides are associated with a zone of dunite and also with peridotite layers within the overlying troctolite (Fig. 6.17). Although the mineralization consists of a normal assemblage of pyrrhotite, pentlandite, and chalcopyrite, occurring interstitial to silicate grains in a characteristic magmatic "net texture", an uncommon feature is the occurrence of as much as 10% graphite in certain zones. Other unusual features of the sulfide environment include the presence of green (Mg–Al) spinel in some of the troctolites, and the local occurrence of cordierite. As in other deposits, numerous partially resorbed remnants of what appear to be hornfelsic Virginia slate occur in the troctolite. These observations led Mainwaring and Naldrett (1977) to question whether the graphite and sulfide might not have been derived from the Virginia Formation. The results of their sulfur isotope study are included in Fig. 6.18. Barren sulfides from the Virginia Formation average about 18 $\delta^{34}S$‰. whereas the Waterhen sulfides range from 11 to 16 $\delta^{34}S$‰. They noted that these results are consistent with a model in which as much as 75 percent of the sulfur was derived from the country rocks.

Ripley's (1981) sulfur isotope data for the ore of the Dunka Road deposit (Fig. 6.18) ranged from 0.2–15.3‰ with a mean of 7.5‰. Pyrrhotite in the underlying Virginia Formation shows a similar range in isotopic composition, also supporting a country rock source for much of the sulfur. The heterogenous distribution of the sulfur isotope values throughout the deposit, coupled with a wide range in the nickel contents of olivine in the mineralized zone lead Ripley (1986) to suggest that sulfur introduction had occurred essentially in situ, and that widespread equilibration between the sulfides and their enclosing silicates had not occurred.

Ripley (1986) found that the $\delta^{18}O$ values at Dunka Road ranged from 5.8–9.6, but that anomalous values (>8‰) only

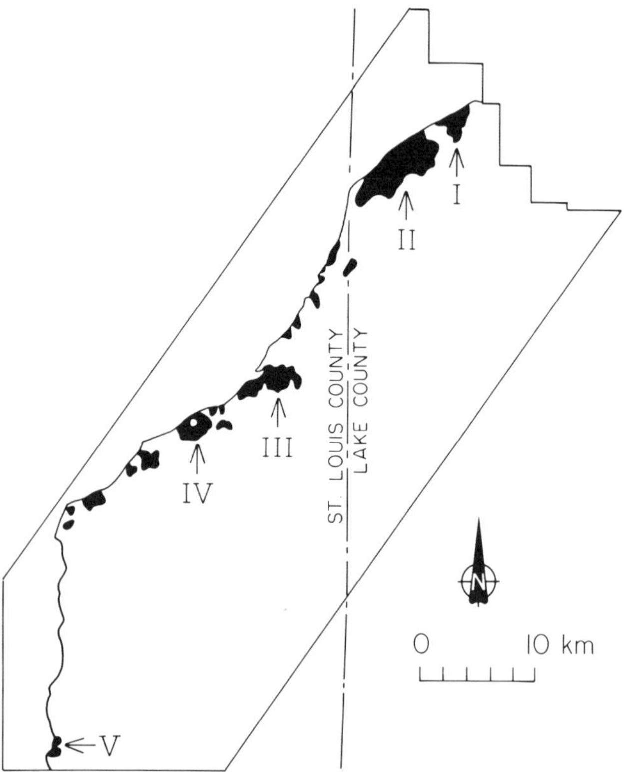


FIGURE 6.15—Map of parts of St Louis and Lake counties, Minnesota, showing the main areas of mineralization outlined to date. I = Spruce Pit Area, 635 × $10^6$ tonnes at >0.5 wt% Cu; II = 2100 × $10^6$ tonnes at >0.5 wt% Cu; III Minnamax area, 725 × $10^6$ tonnes at >0.5 wt% Cu; IV = Dunka Road area; V = Waterhen area. Tonnage figures from Listerud and Meineke (1977).

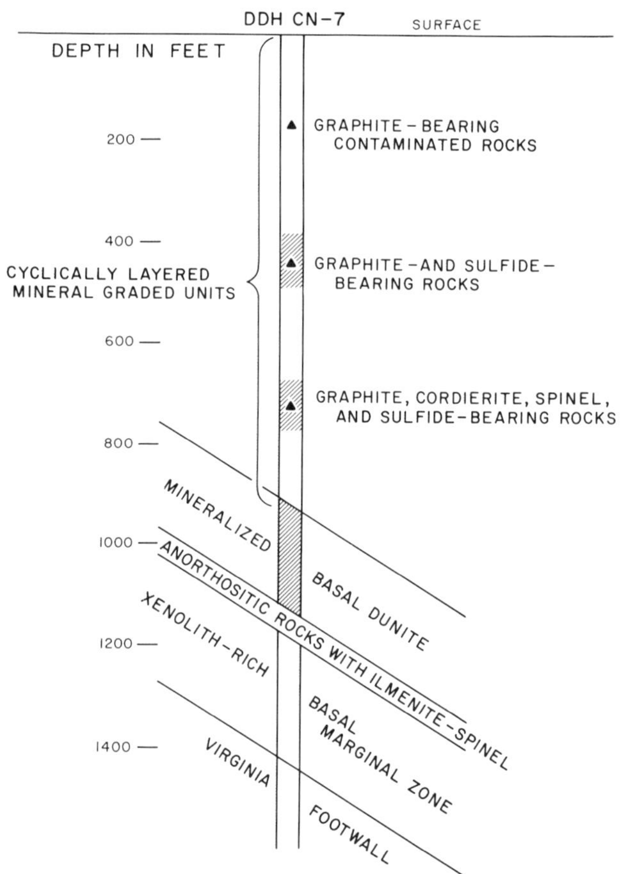


FIGURE 6.17—Simplified drill log through a typical hole (CN–7) in the Waterhen intrusion of the Duluth Complex. From Mainwaring and Naldrett (1977).

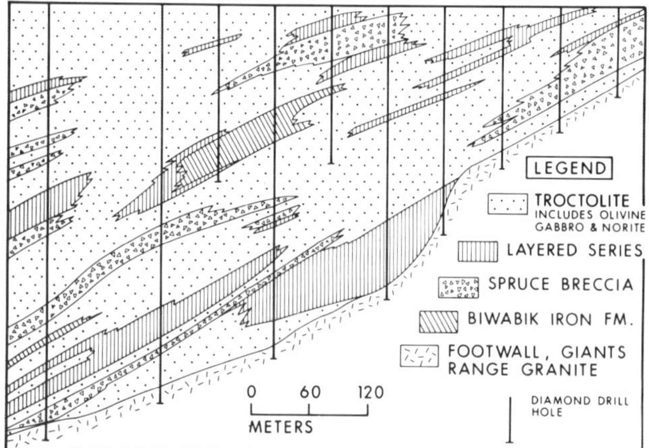


FIGURE 6.16—Vertical section through the Spruce deposit, Duluth Complex, Minnesota. The layered series refers to zones of well-banded picrite, mafic troctolite, and troctolite. The Spruce breccia refers to troctolite with more than 30 percent of inclusions of clinopyroxene–plagioclase–olivine bearing rock. Biwabik Iron Formation, and recrystallized mafic rocks. From Wager et al. (1969).

occurred near (within 3 m) the margins of country rock xenoliths (which themsleves have a range of from 8.6–11.1‰). The olivine of the troctolite gives way to orthopyroxene in these zones, suggesting that $SiO_2$ contamination has accompanied the change in $\delta^{18}O$. The localized nature of the oxygen isotope contamination lead Ripley to suggest that whole-rock assimilation was not a major cause of sulfur contamination. He proposed that the sulfur had been introduced in a volatile phase that was released from the footwall rocks as they were metamorphosed by the intrusion. This is supported by the common association of hydrous minerals such as biotite and amphibole and patches of pegmatite with the mineralization.

At Babbit, a much higher proportion of the igneous rock is isotopically anomalous with respect to oxygen, the sulfides are more abundant, and their PGE contents are higher (indicative of higher 'R' factors). The sulfides are more homogenous isotopically, particularly those of the better grade ores, and significantly higher than those of the adjacent country rocks (Fig. 6.18). These factors lead Ripley (1986) to suggest that contamination had occurred at an earlier stage here than at Dunka Road, possibly during intrusion.

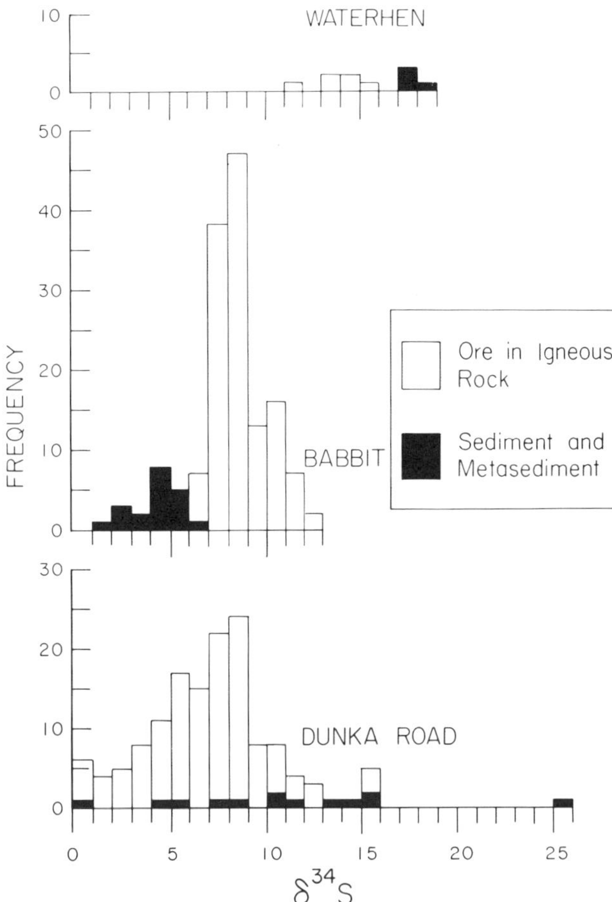


FIGURE 6.18—Isotopic composition of sulfur in Ni–Cu and sedimentary sulfides of deposits in the Duluth complex. Data for Dunka Road (Ripley, 1981), for Babbitt (Ripley, 1986), and Waterhen (Mainwaring and Naldrett, 1977).

## REFERENCES


Bazunov, E. A., 1976, Development of the main structures of the Siberian platform: history and dynamics: Tectonophys., v. 36, pp. 289–300.

Bonnichsen, B., 1970, Geologic map of the Allen, Babbitt, Babbitt NE, Babbitt SE, and Babbitt SW $7^1/_2$ minute quadrangles: Minnesota Geological Survey, Open File Maps.

Bonnichsen, W., 1972, Southern part of the Duluth complex; *in* Sims, P. K., and Morey, G. B. (eds.), Geology of Minnesota: A centennial volume: St. Paul, Minnesota Geological Survey, pp. 361–387.

Distler, V. V., Genkin, A. D., and Dyuzhikov, O. A., 1986, Sulfide petrology and genesis of copper–nickel ore deposits; *in* Freidrich, G. H., Genkin, A. D., Naldrett, A. J., Ridge, J. D., Sillitoe, R. H., and Vokes, F. M. (eds.), Geology and metallogeny of copper deposits: Springer–Verlag, Heidelberg, Berlin, pp. 111–123.

Dyuzhikov, O. A., Fedorenko, V. A., Nestorovoskiy, V. A., and Dermisdovich, V. I., 1976, The new North Kharayelakh ore find and its nickel potential: Doklady, Earth Science Sections 229, pp. 123–125.

Faure, G., Chaudhuri, S., and Fenton, M., 1969. Ages of the Duluth gabbro complex and the Endion sill, Duluth, Minnesota: Jour. Geophys. Res., v. 74, pp. 720–725.

Foose, M., and Wieblen, P., 1986, The physical and petrologic setting and textural and compositional characteristics of sulfides from the south Kawishiwi intrusion, Duluth Complex, Minnesota, USA; *in* Freidrich, G. H., Genkin, A. D., Naldrett, A. J., Ridge, J. D., Sillitoe, R. H., and Vokes, F. M. (eds.), Geology and metallogeny of copper deposits: Springer–Verlag, Heidelberg, Berlin, pp. 8–24.

Genkin, A. D., Distler, V. V., Gladyshev, G. D., Filiminova, A. A., Evstigneeva, T. L., Kovalenker, V. A., Laputina, I. P., Smirnov, A. V., and Grokhovskaya, T. L., 1981, Sulfide copper–nickel ores of the Noril'sk deposits: Moscow, Nauka, 234 pp. (in Russian).

Glazkovsky, A. A., Gorbunov, G. I., and Sysoev, F. A., 1977, Deposits of nickel; *in* Smirnov, V. I. (ed.), Ore deposits of the USSR, Vol. II (Translated into English by D. A. Brown): pp. 3–79.

Godlevsky, M. N., and Grinenko, L. N., 1963, Some data on the isotopic composition of sulfur in the sulfides of the Noril'sk deposit: Geochem., v. 1, pp. 335–41.

Godlevsky, M. N., and Likhachev, A. P., 1986, Types and distinctive features of ore-bearing formations of copper–nickel deposits; *in* Freidrich, G. H., Genkin, A. D., Naldrett, A. J., Ridge, J. D., Sillitoe, R. H., and Vokes, F. M. (eds.), Geology and metallogeny of copper deposits: Springer–Verlag, Berlin, Heidelberg.

Gorbachev, N. S., and Grinenko, L. N., 1973, The sulfur isotope ratios of the sulfides and sulfates of the Oktyabr'sk sulfide deposit, Noril'sk region, and the problem of its origin: Geokhimiya 8, pp. 1127–1136.

Green, J. C., Phinney, W. C., and Weiblen, P. W., 1966, Gabbro Lake quadrangle, Lake County, Minnesota: Minnesota Geological Survey, Misc. Map M2.

Grinenko, L. N., 1967, The sulfur-isotope ratios of sulfides of some copper–nickel deposits and prospects of the Siberian platform; *in* The petrology of the trap-rocks of the Siberian Platform: Nedra Press, Leningrad, pp.221–229.

Grinenko, L. N., 1985, Sources of sulfur of the nickeliferous and barren gabbro-dolerite intrusions of the northwest Siberian platform: International Geology Review, pp. 695–708.

Lightfoot, P. C., Naldrett, A. J., and Hawkesworth, C. J., 1984, The geology and geochemistry of the Waterfall Gorge section of the Insizwa complex with particular reference to the origin of nickel sulfide deposits: Econ. Geol., pp. 1857–1879.

Listerud, W. H., and Meineke, D. G., 1977: Minnesota Dept. of Natural Resources, Div. of Minerals, Minerals Exploration Section, Report 93.

Mainwaring, P. R., and Naldrett, A. J., 1977, Country rock assim ilation and the genesis of Cu–Ni sulfides in the Waterhen intrusion, Duluth Complex, Minnesota: Econ. Geol., v. 72, pp. 1269–1284.

Malich, N. S., and Tuganova, Ye. V., 1976, Distribution patterns of mineral resources in the Siberian plaform cover: Internat. Geol. Rev., v. 18, pp. 417–424.

Matlock, W. F., and Watowich, S. N., 1980, Geology and sulfide mineralization of the Duluth Compex–Virginia formation contact, Minnamax deposit, Minnesota (abs.): Geol. Soc. America, Abstracts with Programs, pp. 477–478.

Phinney, W. C., 1970. Chemical relations between Keweenawan lavas and the Duluth complex, Minnesota: Geol. Soc. Amer., Bull., v. 81, pp. 2487–2496.

Phinney, W. C., 1972, Northwestern part of the Duluth complex; *in* Sims, P. D., and Morey, G. B. (eds.), Geology of Minnesota: A centennial volume: Minnesota Geological Survey, pp. 335–345.

Ripley, E. M., 1981, Sulfur isotopic abundances of the Dunka Road Cu–Ni deposit, Duluth Complex, Minnesota: Econ. Geol., v. 76, pp. 619–620.

Ripley, E. M., 1986, Applications of stable isotope studies to problems of magmatic sulfide ore genesis with special reference to the Duluth Complex, Minnesota; *in* Freidrich, G. H., Genkin, A. D., Naldrett, A. J., Ridge, J. D., Sillitoe, R. H., and Vokes, F. M. (eds.), Geology and metallogeny of copper deposits: Springer-Verlag, Berlin, Heidelberg, pp. 25–42.

Sims, P. K., and Morey, G. B., 1972, Geology of Minnesota: A centennial volume: Minnesota Geological Survey, p. 632.

Smirnov, M. F., 1966, The Noril'sk nickeliferous intrusions and their sulfide ores: Nedra Press, Moscow.

Tamrazyan, G. P., 1971, Siberian continental drift: Tectonophysics, v. 11, pp. 433–460.

Wager, R. E., Podolsky, T., Alckock, R. A., Weiblen, P. W., and Phinney, W. C., 1969, A comparison of the Cu–Ni deposits of the Sudbury and Duluth basins: Proc. 30th. Ann. Min. Symp., 42nd Ann. Mtng, Minnesota Sect., Am. Inst. Min. Eng., pp. 95–96.

Weiblen, P. W., and Morey, G. B., 1980, A summary of the stratigraphy, petrology and structure of the Duluth complex: Am. Jour. Sci., v. 280–A, pp. 88–133.

## Chapter 7

# CONTAMINATION AND THE ORIGIN OF THE SUDBURY STRUCTURE AND ITS ORES

A. J. Naldrett

### INTRODUCTION

The Ni–Cu ores of the Sudbury district are associated with the Sudbury Igneous Complex (SIC), a layered intrusion ranging from quartz norite at the base, through gabbro to a granophyric cap. The purpose of this section is to describe the geological setting at Sudbury, show the extent to which contamination by crustal rocks has modified the composition of the Complex, and how this can account for the profusion of concentrations of magmatic sulfide ore deposits. The chapter draws very heavily on articles by Naldrett et al. (1986a, b).

### GEOLOGICAL SETTING

The Complex is located at the contact between tonalitic gneisses and intrusive quartz monzonites all of Archean age to the north, and rocks of the Proterozoic Southern Province, which overlie the Archean basement unconformably and thicken to the south. The gneisses are granulites and extend around much of the northern and western margins of the Complex (Fig. 7.1). The Proterozoic rocks belong to the Huronian supergroup; in the Sudbury area, they consist of local accumulations of mafic and felsic volcanics, overlain by greywackes and siltstones, which are overlain in turn by arenites. Where clastic units occur at the base of the Huronian, they may contain high concentrations of detrital U- and Th-rich minerals. These reach their maximum development 100 km to the west of Sudbury in the uranium ores of the Eliot Lake area.

Card et al. (1984) have drawn attention to a dominant linear gravity anomaly extending 350 km from Eliot Lake eastward to Engelhardt. The Sudbury Complex straddles this feature and coincides with one of three highpoints along it (Fig. 7.2). Gupta et al. (1984) have analysed the combined residual gravity and magnetic anomaly that marks the Sudbury region itself. After subtracting large-scale 'super regional' gravity trends, they conclude that the broad +20 to +30 mGal anomaly at Sudbury cannot be explained by the rocks of the Complex itself. Their modelling indicates that a large (60 × 40 km) mass of mafic rock, with a density similar to gabbro or gabbro-anorthsite (3.02 ± 0.03 g/cc) underlies the Complex at a depth of at least 5 km, extending beyond it, as is indicated in Fig. 7.3. Their analysis of the regional magnetic data indicates that this can be explained if some of the rocks of this body are partially serpentinized.

The SIC has intruded beneath the Onaping formation (Fig. 7.3), a breccia composed of fragments of country rocks and recrystallized glassy material set in a matrix of glassy shards, and variably interpreted as an ignimbrite or the 'fall-back' breccia resulting from the impact of a meteorite. The SIC and strata overlying it are exposed as a series of concentric, crudely elliptical rings and dip towards the centre, giving rise to the interpretation of the structure as a basin. Many aspects of the local setting suggest that an explosion of unusually large intensity gave rise to a crater at Sudbury. These include:

(i) The basinal shape of the structure as interpreted from surface and underground mapping and drilling.

(ii) The presence of an upturned collar around the basin, as seen particularly in the Huronian rocks along the southern margin (Dressler, 1984).

(iii) Evidence of shock metamorphism in the country rocks around the structure (Dressler, 1984).

(iv) The presence of Sudbury breccia (comparable with the pseudotachylite of the Vredefort and Ries structures) in the country rocks around the structure and Footwall breccia beneath the Complex. This occurs as irregular masses and also as dykes that are radial and concentric to the structure. It occurs intermittently as much as 50 km away from the structure; a zone in which the breccia is common extends from 20 to 25 km north of the northern perimeter (Dressler, 1984).

(v) Evidence of shock metamorphism in country rock inclusions in the Onaping formation (Muir and Peredery, 1984).

(vi) The 1800 m of Onaping formation itself, the lower part of which is variably interpreted as a meteorite fall-back breccia or a pyroclastic flow (Peredery and Morrison, 1984; Muir, 1984).

Opinions are divided between an extra-terrestrial and endogenic origin for the structure. Naldrett (1984) concluded that meteorite impact is the more likely origin, primarily because so many of the features observed at Sudbury are also found at known impact sites. There are, however, many difficulties with such an origin and these are summarized by Muir (1984).

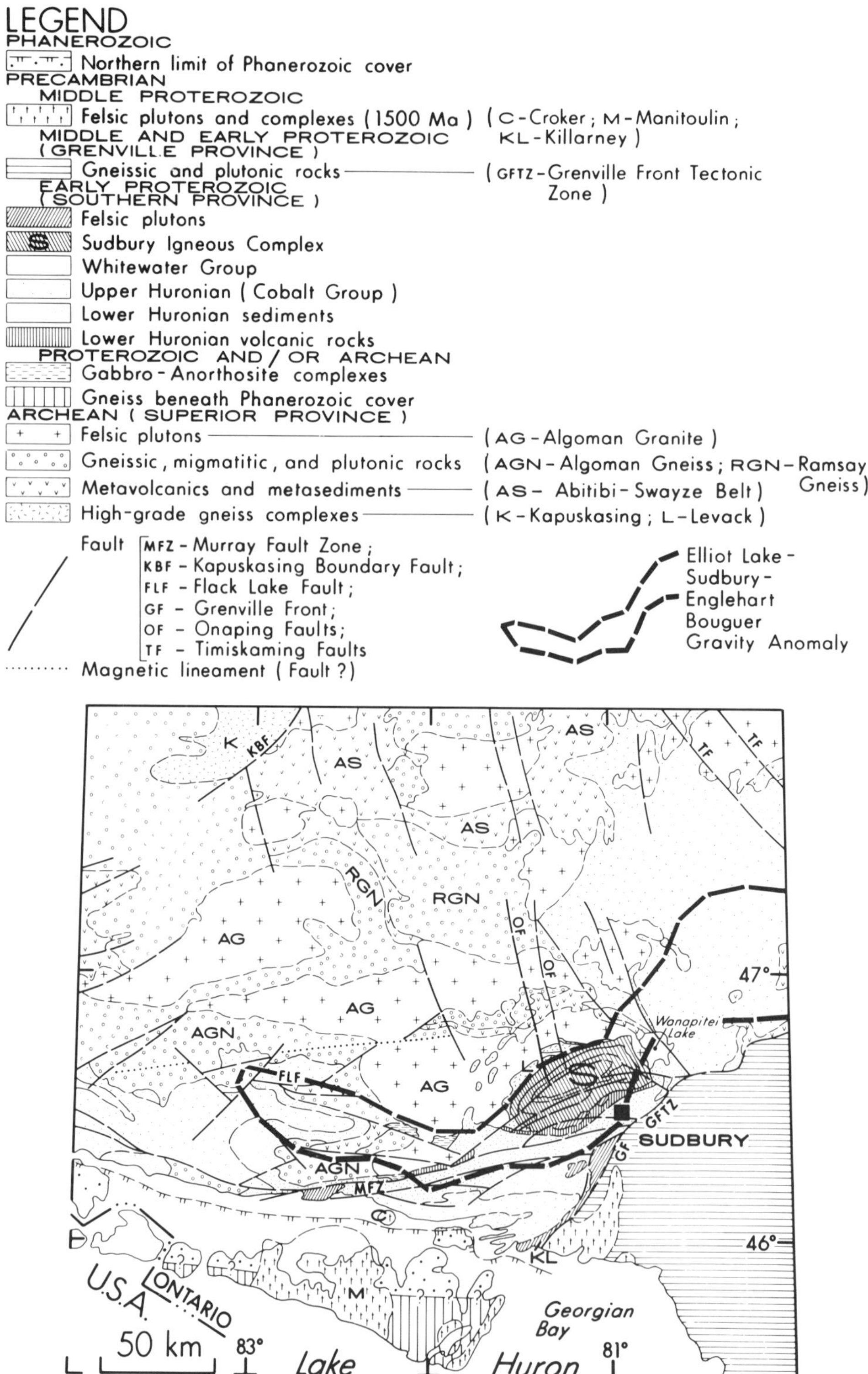


FIGURE 7.1—Regional geological map in the vicinity of Sudbury. The Eliot Lake–Engelhardt regional gravity anomaly is shown by the stippling. (after Card et al., 1984)

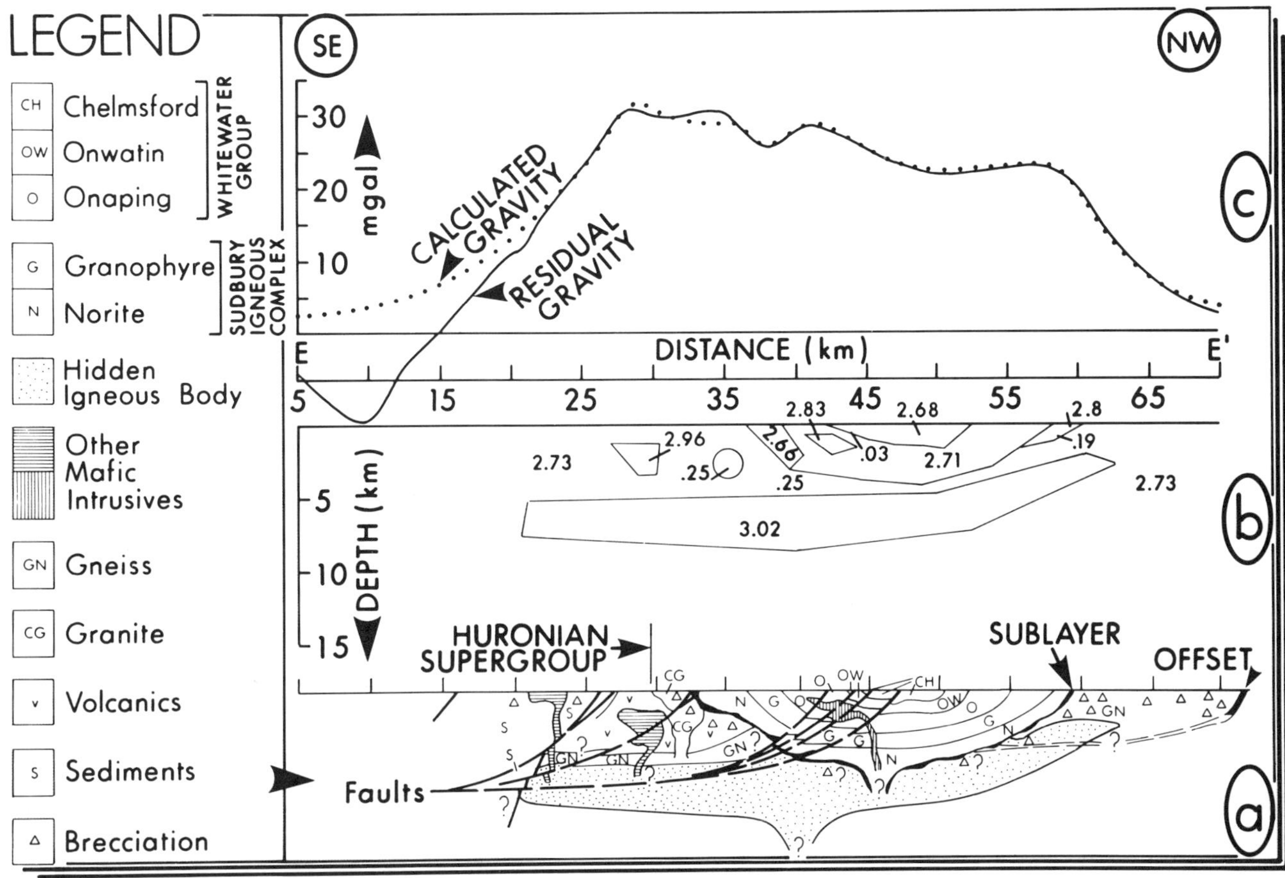


FIGURE 7.2—A northwest–southeast section across the Sudbury district showing the residual and calculated gravity profiles, a 2-dimensional reconstruction of the model used to calculate the gravity profile, and a geological interpretation consistent with the reconstruction. (after Gupta et al., 1984).

## PETROLOGY OF THE SUDBURY IGNEOUS COMPLEX

The main units of the Complex include (i) the Sublayer, (ii) the marginal Quartz-rich Norite of the South Range and Mafic Norite of the North Range, (iii) the South Range Norite and Felsic Norite, (iv) the Quartz Gabbro, and (v) the Granophyre and Plagioclase-rich Granophyre. All except the Sublayer are included within the Main Mass of the Complex.

### Main mass

The marginal unit of the Main Mass on the South Range is the Quartz-rich Norite. In this, the quartz content increases progressively towards the contact over the outer 300m (Naldrett et al., 1970). This is unlikely to be due to local contamination since the increase in quartz occurs as much where the footwall is composed of $SiO_2$-deficient greenstone as it does where it is composed of granite. Thus, if contamination is involved, it is not in situ. The increase in quartz is accompanied by an equally progressive decrease in the average Mg/(Mg + Fe) ratio of the pyroxenes (Naldrett et al., 1970). This decrease is due to the pyroxenes becoming progressively more strongly zoned, with Mg/(Mg + Fe) decreasing towards the edges of grains while the cores retain a constant composition. These observations, coupled with a decrease in grain size towards the margins (Naldrett et al., 1970), indicate that the outer part of the Quartz-rich Norite is a non-cumulate rock that crystallized essentially in situ.

### Sublayer

The Sublayer occurs discontinuously around the Complex. It can be divided logically into those variants which occur close to the outer contact and those which occupy dykes (known locally as offsets) that either radiate outwards from or are concentric to the Complex.

The Contact Sublayer consists of a suite of fine to medium grained norites and gabbros that can be distinguished from the Main Mass Felsic Norite and Quartz Gabbro by their lower quartz content in relation to pyroxene (Naldrett et al., 1972). Sublayer rocks occurring within the offsets generally have the composition of quartz diorite and are referred to as such. Some of the Sublayer has a high (>5 modal percent) sulfide content and, in many cases, has deposits of Ni–Cu sulfide associated with it; this is referred to here as Miner-

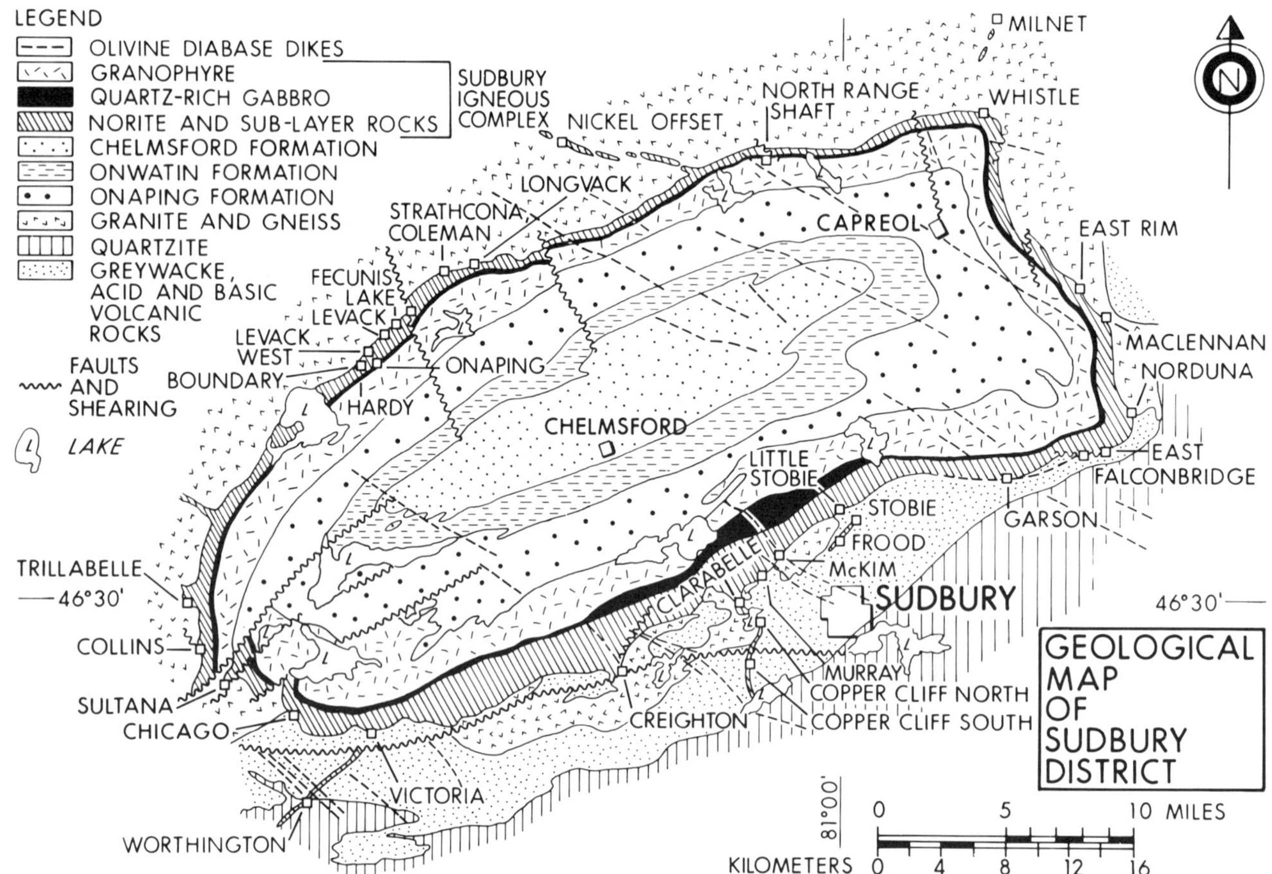


FIGURE 7.3—Geological map of the Sudbury district.

alized Sublayer. Other Sublayer has a lower sulfide content and no associated ore bodies. The sulfide content of the Mineralized Sublayer is much greater than that which can be dissolved in basaltic magma (see Chapter 2). This Sublayer has therefore been enriched in sulfide at some stage before it reached its present position.

Geological relationships between the Sublayer and the Main Mass that bear on the relative ages of the two are conflicting. Inclusions of marginal, Quartz-rich Norite have been observed in the Sublayer and inclusions of Sublayer have been observed in norite of the Main Mass of the Complex. Contacts are never marked by fine grained chill zones, suggesting that whichever was the older at any particular location, it was still warm at the time of intrusion of the younger. On the North Range, the distinction between Main Mass and Sublayer is always clear with the Sublayer having the finer grain size and lower quartz content. This is not always the case on the South Range, where a number of researchers have commented on gradations that are visible between the two (Slaught, 1951; Cochrane, 1984). It would seem, therefore, that the introduction of the Sublayer and Main Mass was a complicated process with one preceding the other and vice versa in different localities.

In some areas the Sublayer is characterized by inclusions. These can be divided into two groups, those of obviously local derivation, and those composed of mafic and ultramafic rocks, few of which outcrop in the Sudbury area. Scribbins et al. (1984) have described the latter group as ranging from peridotite, through clino- and orthopyroxenites, to olivine gabbro and norite. Most of this group of inclusions display either cataclastic or cumulate textures. Olivine within them ranges in composition from $Fo_{86}$ to $Fo_{72}$, with the Fo content decreasing with plagioclase content. Scribbins et al. conclude that the inclusions are derived from layered intrusions that have fractionated at moderate depths in the crust.

## MAJOR AND TRACE ELEMENT GEOCHEMISTRY

The average compositions of the Quartz-rich Norite, Mineralized Sublayer from Levack West, Strathcona and Little Stobie, and North and South Range Unmineralized Sublayer are given in Table 7.1.

### Major elements

Since the SIC is a tholeiitic body emplaced into a cratonic environment, it is reasonable to compare it chemically with continental flood basalts (CFB). Naldrett (1984) pointed out that, judging from field and petrographic criteria, the Quartz-rich Norite is likely to be the rock type closest in composition to that of the magma giving rise to the Complex as it was emplaced along the South Range. He compared it on the basis of MgNo [MgO/(MgO + FeO) atomic ratio] to a series of Keweenawan and Columbia River flood basalts and showed it to be significantly richer in $SiO_2$ (SIC = 57; CFB = 49 wt% $SiO_2$) and $K_2O$ (SIC = 1.5; CFB = 0.249 wt% $K_2O$) and poorer in CaO (SIC = 7; CFB = 11 wt% CaO) and $Na_2O/K_2O$ (SIC = 1.8; CFB = 6 to 12). In Table 7.2 it is shown that the result of mixing relatively unfractionated CFB (the average of 5 Keweenawan olivine tholeiitic basalts KEW, reported by Basaltic Volcanism Project, 1981) with 45% of a 1:2 mixture respectively of quartz monzonite and tonalitic gneiss (QMT), which are the two principal rock types in the basement beneath the Complex and the Huronian, gives rise to a composition that is very similar to that of the Quartz-rich Norite (model QRN).

The $SiO_2$, $K_2O$ and CaO contents and $Na_2O/K_2O$ of Mineralized and Unmineralized Sublayer are compared on the basis of MgNo in Fig. 7.4 with the same CFB analyses as used by Naldrett (1984). Considering Unmineralized Sublayer, this is also distinctly richer in $SiO_2$ and $K_2O$ and lower in CaO and $Na_2O/K_2O$ than CFB. There is relatively little difference between the North and South Range samples. With the exception of one sample with very high MgNo, the Mineralized Sublayer spans a similar range in MgNo and has a similar CaO content to the Unmineralized. However, the $SiO_2$ and $K_2O$ contents of the Mineralized are distinctly lower and the $Na_2O/K_2O$ distinctly higher than those of the Unmineralized Sublayer. There is no systematic variation within either group between these elements and MgNo. On the other hand, it can be seen from Fig. 7.5 that $SiO_2$ is correlated positively with $K_2O$ and negatively with CaO.

### Trace elements

A spectrum of incompatible trace elements representing different units of the SIC are illustrated on a "spidergram" (constructed after Thompson, 1982) in Fig. 7.6. A number of distinctive characteristics are apparent:

1) All rocks are characterized by enrichment at the Ba end of the spectrum, and by Ta and Ti depletion.
2) The North Range Mineralized Sublayer is distinctly lower in Ba, Rb and Th than other rock types. It shows a weak to distinct negative Th anomaly.
3) The South Range Mineralized Sublayer is similar to that of the North Range, except that it is higher in Ba, Rb and Th and shows a positive Th anomaly. It also shows a weak negative Sr anomaly.
4) The unmineralized Sublayer is distinctly richer in trace elements ranging in the spectrum from La to Yb than the other rock units and shows pronounced negative Th and Sr anomalies.

Representative REE diagrams for different units of the Complex are illustrated in Fig. 7.7. Slopes for all units are similar despite variation in the absolute levels of the REE. The Sublayer and Quartz-rich Norite show no Eu anomalies while the Felsic Norite is marked by a positive and the Granophyre by a negative Eu anomaly. Patterns for ultramafic inclusions from the Sublayer show the same steep slopes as other units of the Complex.

### Discussion of major and trace element data

The lack of correlation shown in Fig. 7.4 between MgNo and $SiO_2$, $K_2O$ and CaO indicates that fractionation is not directly responsible for the wide variation in the concentration of these oxides. The unusual major element composition of the SIC has long been explained as the consequence of contamination (Irvine, 1975; Naldrett & Macdonald, 1980). Kuo & Crocket (1979) interpreted the enrichment in LREE exhibited by all rocks at Sudbury to the same cause.

The correlation shown between $SiO_2$ and each of $K_2O$ and CaO in Fig. 7.5 is consistent with the premise that the var-

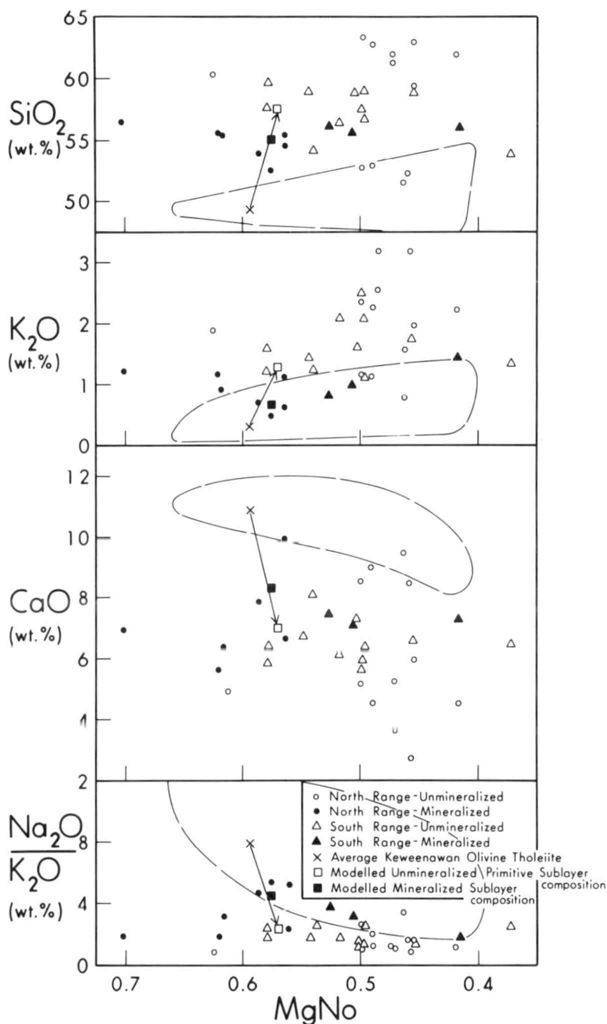


FIGURE 7.4—Comparison of some major element concentrations in the Sublayer with those in typical continental flood basalts. Composition of KEW is shown by the start of the arrow; the model composition of BVR 12 by the cross at the end of the arrow. The model composition of the averge North Range Mineralized Sublayer is shown by the solid square.

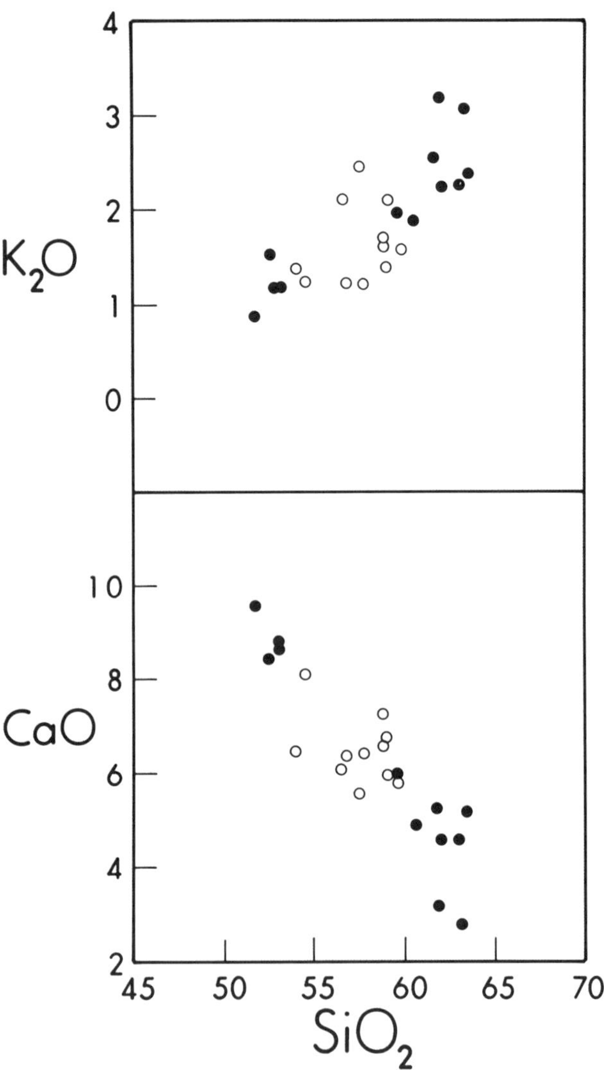


FIGURE 7.5—Variation of $K_2O$ and CaO with $SiO_2$ for samples of Mineralized and Unmineralized Sublayer.

iation shown by these oxides is the result of a variable degree of mixing of basaltic magma with a contaminant rich in $SiO_2$ and $K_2O$ and poor in CaO. The range in MgNo shown by the rocks indicates that the magma has undergone fractional crystallization. Much of the contamination may have accompanied this fractionation, although, as pointed out above, the evidence of the major elements is against the degree of assimilation having been directly proportional to the degree of fractionation.

Fig. 7.8 illustrates the trace element contents of some of the possible contaminants. An average upper crustal contaminant in the vicinity of Sudbury is taken to be QMT. In this, while the major element composition is based on rocks collected at Sudbury, the trace element concentrations are those of Arth & Hanson (1975) for rocks in Minnesota with the same major element compositions as those at Sudbury, coupled, for elements for which Arth & Hanson had no data, with data from Wood (1980) on the upper crustal Laxfordian amphibolite. The similarity between the REE profile of QMT and Shaw et al.'s (1976) profile for the average Canadian Shield is illustrated in Fig. 7.9; this is supportive of the use of QMT is a facsimile for the upper crust. As outlined above, granulites (Levack Gneiss) form the immediate footwall to the Complex along the North Range and are a possible contaminant for the rocks emplaced there. McKim pelites form the footwall along much of the South Range and are a possible contaminant there. Data for the average of 5 Keweenawan olivine tholeiites are again used as a starting point from which to model the effect of contamination in reproducing the composition of the SIC. Points worthy of note in Fig. 7.8 are the low Rb and Th of the Levack Gneisses, which is a characteristic of most lower crustal granulites; the high Ba, Rb, Th and K and negative Ta and Ti anomalies of both QMT and the McKim pelite; and the marked negative Sr anomaly of the McKim. This last characteristic is also present in the Matinenda arenite which underlies the McKim in the Eliot Lake area (Froelich, personal communication, 1984). The negative Ta and Ti anomalies are a well established feature for continental shields and also for many volcanic rocks developed within island arcs. The reason for them is not understood.

The relatively high concentration of Rb and Th in all rocks of the SIC indicate that the Levack Gneiss alone cannot account for the contamination of even the immediatly adjacent North Range Mineralized Sublayer. However, contaminating the Keweenawan basalt (KEW) with 50% of a 2:1 mixture of Levack Gneiss and QMT produces a reasonable match for the average trace element profile of the North Range Mineralized Sublayer, including the negative Th anomaly (Fig. 7.10) The Zr and Hf contents of the model composition appear to be somewhat high, perhaps due to their being compatible in some minerals common in felsic rocks and thus difficult to judge when making estimates of the composition of mega sections of the crust. The major element composition of this mixture is also represented in Fig. 7.4, in which it is seen to fall near the centre of the points representing the North Range mineralized Sublayer.

The higher Rb and Th and slight negative Sr anomaly of the South Range Mineralized Sublayer can be explained as the consequence of contaminating KEW with 40% of a 1:1 mixture of McKim pelite and QMT (Fig. 7.11). The positive Th anomaly present in the South Range Mineralized Sublayer is not reproduced by this model composition, but, as discussed above, the basal Huronian is characterized by localized concentrations of U and Th-bearing minerals, and assimilation of a small amount of this material would cause a very marked increase in the Th content of a magma.

The Quartz-rich Norite, representing the Main Mass of the SIC, has high Ba, Rb, Th, and K and can be modelled well (Fig. 7.12 as the consequence of contaminating KEW with the same proportion of QMT as used to model the major elements (45%), except for the ubiquitous high Zr and Hf of the model curve.

In Fig. 7.13, it is seen that the Unmineralized Sublayer, with its overall high concentrations of incompatible trace elements yet relatively (in comparison with other SIC units)

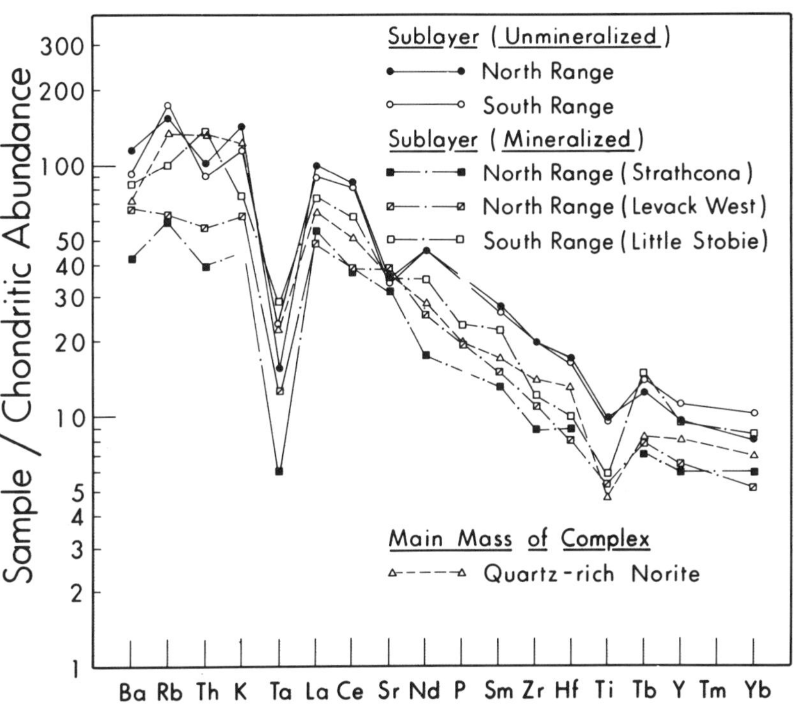


FIGURE 7.6—"Spidergram" showing trace element concentrations in selected units of the Sudbury Igneous Complex. The elements are normalized by their chondritic abundances, except for K and Ti. Abundances used for the normalization in this and succeeding similar diagrams are (in ppm): Ba, 6.9; Rb, 0.35; Th, 0.042; K, 120; Ta, 0.02; La, 0.328; Ce, 0.815; Sr, 11.8; Nd, 0.63; P, 46; Sm, 0.203; Zr, 6.84; Hf, 0.2; Ti, 620; Tb, 0.052; Y, 2, Tm, 0.034; Yb, 0.22 (values from Thompson, 1982).

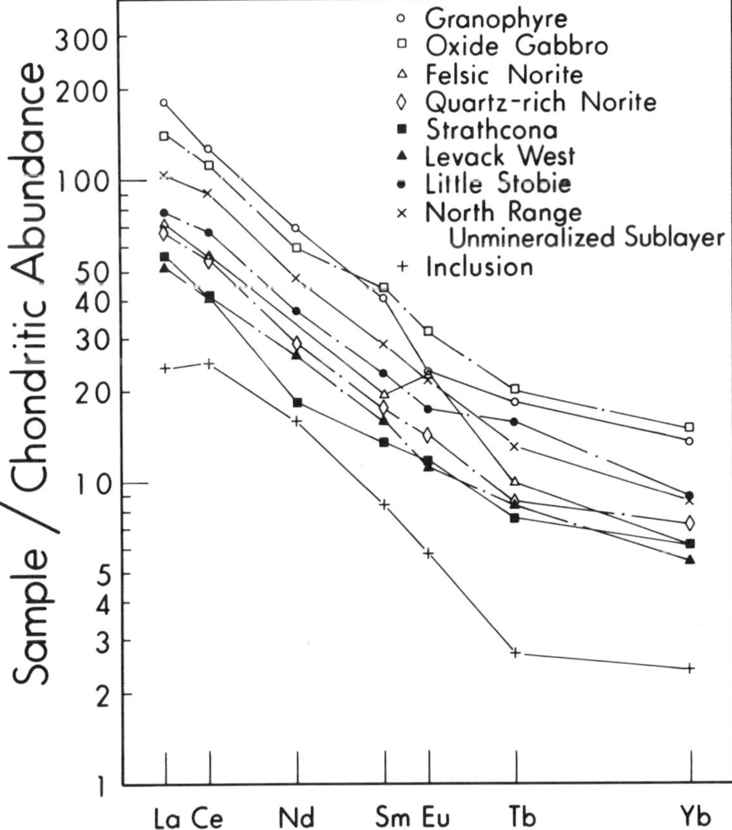


FIGURE 7.7—REE chondrite normalized diagram for selected units of the Sudbury Igneous Complex, including one sample of ultramafic inclusions from the Strathcona Deposit.

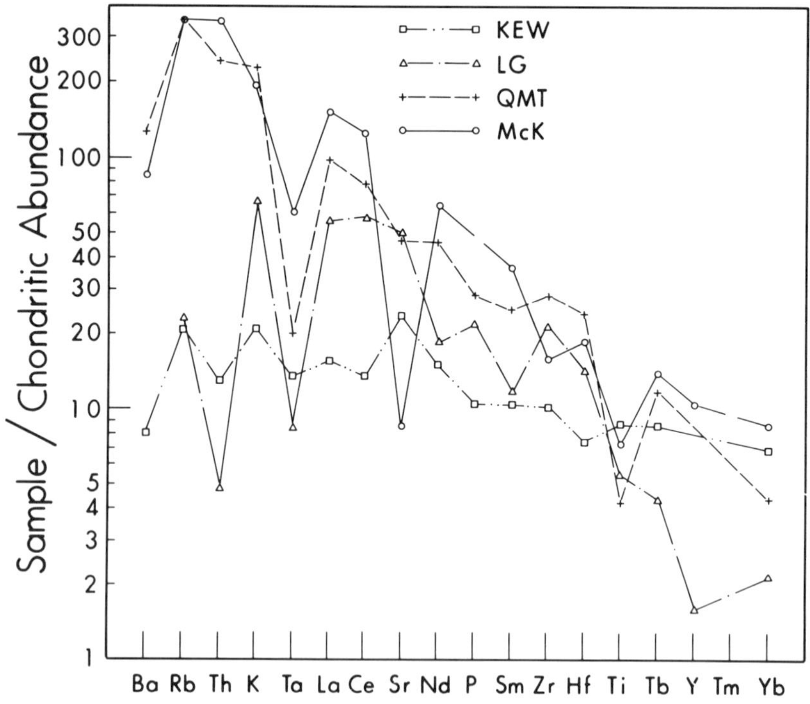


FIGURE 7.8—Trace element variation in some possible contaminants for the Sudbury Igneous Complex. See caption to Fig. 7.6 for further explanation.

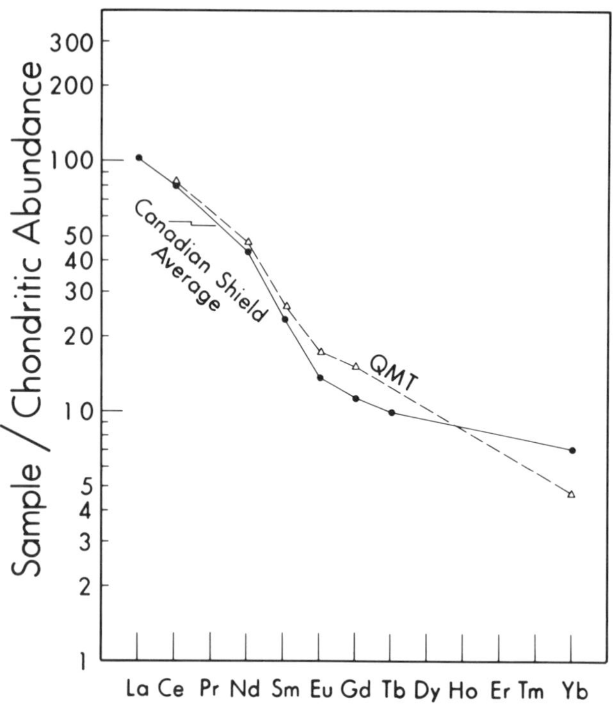


FIGURE 7.9—REE data for the Quartz Monzonite–Tonalite mixture (QMT) used as a contaminant compared with Shaw et al.'s (1976) data for the "average" Canadian Shield.

low Rb/La ratio, and marked negative Sr anomaly, is best modelled as the consequence of mixing equal proportions of KEW, QMT, Levack Gneiss and McKim pelite (the modelling attempted here for the trace elements differs from that for the major elements, in that, for the latter, one of the more primitive samples was modelled, while, for the former, modelling is attempted for the averages of the North and South Ranges. The similarity in the major and trace element contents of both the North and South Range representatives of this unit lends no support to models involving localized contamination, but is suggestive that this magma type had access to both sides of the Complex.

### Summary of conclusions from major and trace element data

1) Rocks of the SIC are highly contaminated, major and trace element modelling indicating that 33 to 75% contamination is involved.

2) The Main Mass of the Complex appears to have been contaminated largely by typical upper crustal rocks.

3) The Mineralized Sublayer has also been affected by typical upper crustal contamination, but a component of more localized contamination appears to be present as well. The North Range Mineralized Sublayer bears evidence of contamination by the lower crustal granulites which form the footwall rocks immediatly along the North Range. Contamination with a component from the McKim formation can account for some of the characteristics of the South Range Mineralized Sublayer.

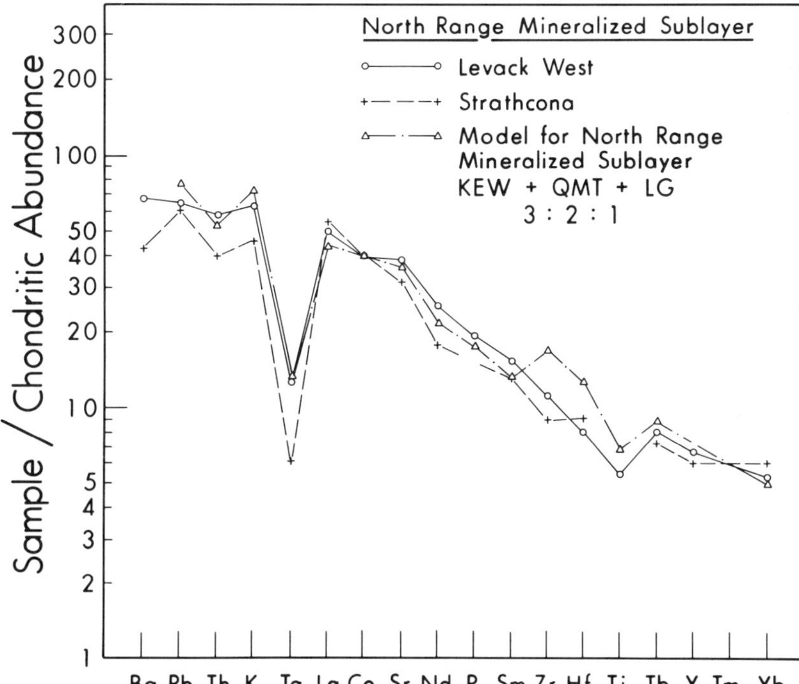


FIGURE 7.10—Model trace element pattern for the North Range Mineralized Sublayer compared with observed patterns. See caption to Fig. 7.6 for further explanation.

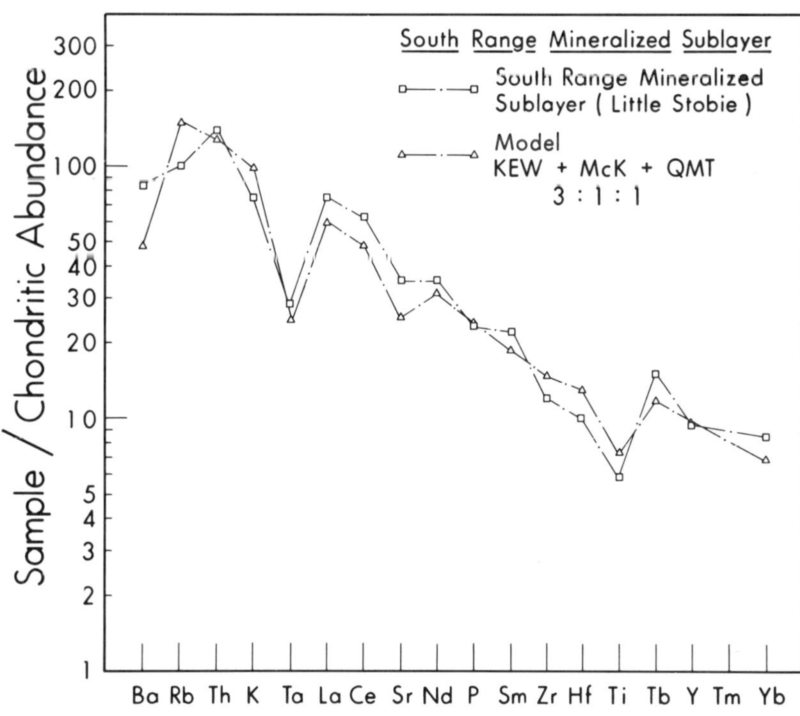


FIGURE 7.11—Model trace element pattern for the South Range Mineralized Sublayer compared with the observed pattern. See caption to Fig. 7.6 for further explanation.

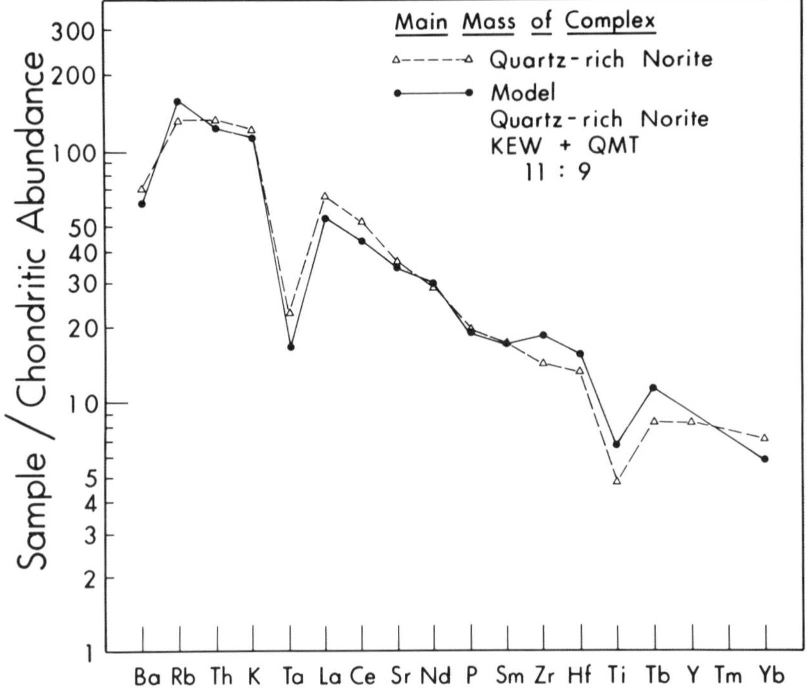


FIGURE 7.12—Model trace element pattern for the Quartz-rich Norite compared to the observed pattern. See caption to Fig. 7.6 for further explanation.

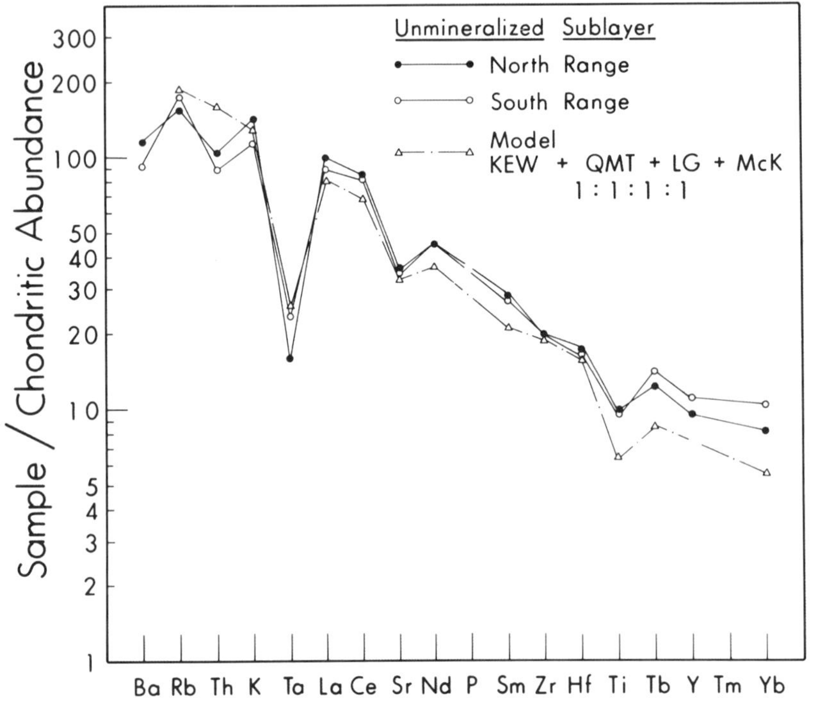


FIGURE 7.13—Model trace element pattern for the Unmineralized Sublayer compared to observed patterns. See caption to Fig. 7.6 for further explanation.

4) The Unmineralized Sublayer has an almost identical profile on both the North and South Ranges, which can be modelled as the consequence of contamination by a mixture of many of the country rocks at Sudbury.

## ISOTOPE GEOCHEMISTRY

Isotope data for Sr and Nd (Rao et al., 1984; Rao et al., in prep.) are presented in Fig. 7.14a and b. All rocks are characterized by negative values of $\epsilon_{Nd}$. Most samples from the Main Mass of the Complex (Fig. 7.14a) have $\epsilon_{Sr}$ values between +40 and +70, although a few samples, largely Granophyre, extend to +100 and beyond. The Main Mass samples lie close to a model line representing the mixing of a magma with an isotopic composition between bulk earth (1.849Ga ago) and somewhat depleted mantle, with material believed to be typical of upper crust (see caption to Fig. 7.14). The estimated degree of contamination varies with the degree of depletion of the primary magma, but is of the order of 40 to 60%. Samples of Mineralized Sublayer from Little Stobie on the South Range lie close to the mixing line with upper crust (Fig. 7.14b). Those for the North Range Mineralized Sublayer (Levack West and Strathcona) are displaced from this line, having much lower $\epsilon_{Sr}$ values. As might be expected from their mineralogical and trace element composition, the Levack Gneisses resemble other lower crustal granulites isotopically (Carter et al., 1978) in having both negative $\epsilon_{Sr}$ and $\epsilon_{Nd}$. The isotopic composition of the North Range Mineralized Sublayer is consistent with the trace element model suggesting contamination by both upper and lower crustal rocks. The present proximity of the North Range Sublayer to the Levack granulites, and the lack of evidence indicating that the Main Mass or South Range Sublayer magmas have interacted with granulites, is strongly indicative that granulite contamination of the North Range Mineralized Sublayer took place locally, that is in the upper rather than in the lower crust. This implies that it post-dated the component of contamination from typical upper crustal rocks.

Turning to the isotopic composition of the ultramafic inclusions, it is argued in the following section that these are related genetically to Sublayer magmas. However, the isotopes indicate that they have crystallized from magmas that had not experienced the same degree of upper crustal contamination as the Sublayer in which they occur.

## RELATIONSHIP OF ORE DEPOSITS TO THE ROCKS OF THE COMPLEX

Although there is considerable variation in the characteristics of different ore deposits (Pye et al., 1984), there are a number of features common to all of them. These include:

(i) Embayments or other irregularities at the base of the SIC. An increase in sulfide content is usually observed at the lower contact throughout the Complex, but it is where irregularities exist that the zone of sulfide thickens and increases in intensity sufficiently to form ore.

(ii) The presence of Sublayer. The spatial relationship of ore to Sublayer is such that the sulfides constituting the ore bodies appear to have settled out of bodies of Sublayer. Minor amounts (up to 5 modal percent) of sulfides also occur within border norites to the SIC, but, except where faulting obscures the original relationship of ore bodies to units of the SIC, the ore deposits are invariably associated with Sublayer.

(iii) Ultramafic inclusions within Sublayer. Both in the contact and the offset environments, the mineralization is

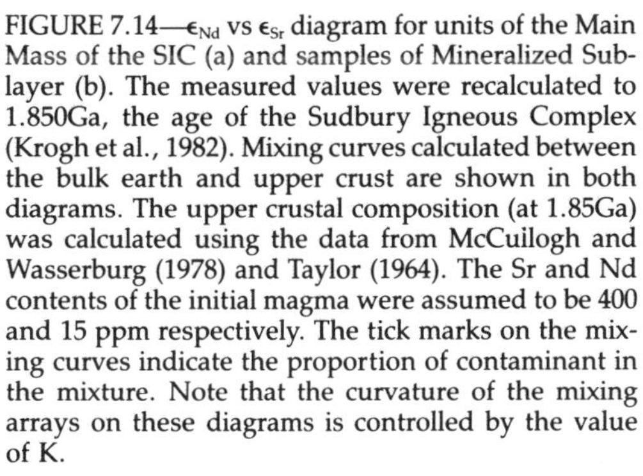

FIGURE 7.14—$\epsilon_{Nd}$ vs $\epsilon_{Sr}$ diagram for units of the Main Mass of the SIC (a) and samples of Mineralized Sublayer (b). The measured values were recalculated to 1.850Ga, the age of the Sudbury Igneous Complex (Krogh et al., 1982). Mixing curves calculated between the bulk earth and upper crust are shown in both diagrams. The upper crustal composition (at 1.85Ga) was calculated using the data from McCuilogh and Wasserburg (1978) and Taylor (1964). The Sr and Nd contents of the initial magma were assumed to be 400 and 15 ppm respectively. The tick marks on the mixing curves indicate the proportion of contaminant in the mixture. Note that the curvature of the mixing arrays on these diagrams is controlled by the value of K.

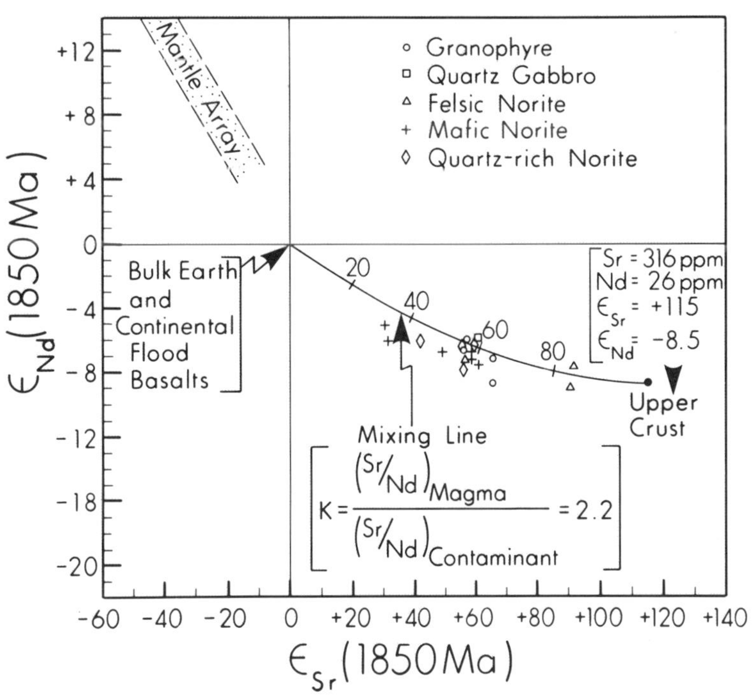

associated only with Sublayer that carries members of the suite of mafic and ultramafic inclusions described above. Scribbins et al. (1984) conclude these inclusions are derived from one or more hidden layered intrusion.

### Relationships among inclusions, sublayer, and main mass of the SIC

As discussed above, the trace element profiles of the Sudbury rocks differ from those of most mafic magmas and indicate that contamination by country rocks has occurred on a large scale. The similarity in the profiles of the Sublayer and the Main Mass rocks is suggestive that the magmas responsible for them have a common heritage. REE profiles for inclusions have the same steep slopes (Fig. 7.7), although the abundances are lower. Thus there is some evidence that the inclusions were derived from a magma that had suffered similar contamination to the remainder of the SIC rocks.

The MgNo of samples of Sublayer are illustrated in Fig. 7.15 (contrary to the previous representation of MgNo, in this figure only 90% of the total Fe has been included in calculating them, to allow for Fe in the magma in the ferric state). Using Roeder and Emslie's (1970) relationship between the Mg/Fe ratio of olivine and of the basaltic liquid that is in equilibrium with it, MgNo of liquids in equilibrium with the olivine of the inclusions have been calculated. It can be seen that these hypothetical liquids span essentially the same range of MgNo as that exhibited by the Sublayer samples. This indicates that the Mg/Fe ratio of the Sublayer samples is consistent with their being liquids derived from a chamber(s) in which the material from which the inclusions were derived was crystallizing as a cumulate(s). Many of the Sublayer samples are rich in $SiO_2$, and olivine would not be stable in equlibrium with a liquid this siliceous. However, Pattison (1979) has reported olivine present in Sublayer at Whistle (northeast corner of the structure) and G. A. Morrison (personal communication, 1984) has observed it at a number of locations. The Sublayer in these localities is not obviously different in composition to that elsewhere, which suggests that olivine may have played a role in the fractionation of much of the Sublayer, and only have disappeared from the liquidus shortly before it was emplaced.

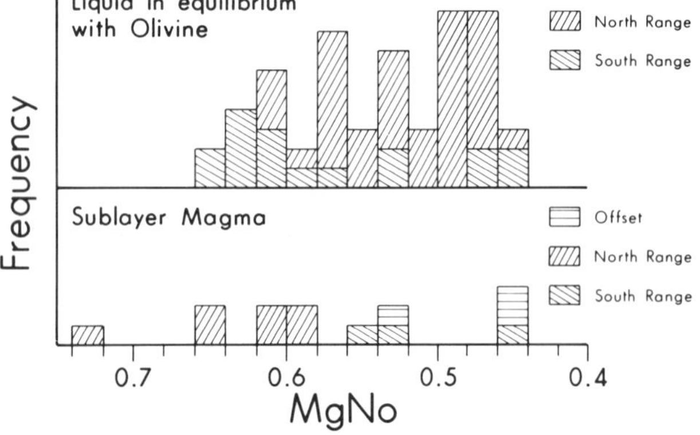


FIGURE 7.15—MgNo of Sublayer samples that are believed to approximate the compositions of liquids, compared with the calculated MgNo of liquids in equilibrium with olivines of the inclusions of the Sublayer.

It would be highly coincidental for the inclusions to be derived from a magma that had suffered the same extreme contamination as the SIC, in the same place, and yet be unrelated to it. The association between the inclusions and their host Sublayer is further evidence of a linkage between them. In the model presented at the end of this paper, it is proposed that the cumulates giving rise to the inclusions formed from Sublayer magma as it fractionated. As pointed out above, olivine compositions are consistent with this hypothesis.

### Segregation of sufides

The SIC differs from other layered complexes in a number of significant ways, including:

(i) Evidence that the area into which it was intruded had been involved in a catastrophic explosion, possibly the consequence of the impact of a meteorite.
(ii) Rocks of the Complex are very siliceous when compared with other intrusions on the basis of the Mg/(Mg + Fe) ratios of their pyroxenes, implying that the magma responsible was unusually siliceous for its state of fractionation.
(iii) Major and trace elements and isotopic analyses that indicate that the high $SiO_2$ content is the consequence of country rock assimilation.
(iv) An unusually large number of occurrences of very concentrated sulfide.

Irvine (1975) pointed out that assimilation of $SiO_2$-rich material by a mafic magma could lower the solubility of sulfur within it, and suggested that this had occurred at Sudbury. Naldrett and Macdonald (1980) pointed out that the $SiO_2$-rich nature of the SIC lent support to this hypothesis. The evidence presented in this paper that extensive country rock assimilation has occurred points to how the $SiO_2$-rich composition has been achieved.

It could be argued that assimilation of an equal mass or more of country rock by a mafic magma is impossible unless the magma was intruded with an unrealistically high degree of superheat. The problem of the heat budget would be helped if the assimilation accompanied fractional crystallization, so that the latent heat of crystallization were available for melting the country rocks. However, even if no heat whatsoever were lost to the surroundings during the assimilation and crystallization process, 50% contamination could not be achieved, since some heat would be used in heating the country rocks up to the temperature of the mafic magma.

It has been pointed out above that the contamination at Sudbury is rather unusual in that it is not particularly selective, but that major and trace element contents and isotope ratios can be modelled reasonably well by calling upon bulk assimilation of country rocks. A process such as this is much easier to envisage if it involves mainly the mixing of magmas, rather than the melting of and mixing with country rocks by a single magma. While Sudbury is unusual in the degree of bulk assimilation that appears to have occurred, it is also unusual in its geologic setting. It is suggested in this paper that pre-heating and extensive melting of the country rocks accompanied the catastrophic explosion that initiated the Sudbury event, thus providing the ideal environment for extensive contamination.

## A MODEL FOR THE FORMATION OF THE SUDBURY IGNEOUS COMPLEX

Faggert et al. (1985) presented REE abundance and Nd isotope data for the SIC which, in their view, conclusively demonstrated that the Complex originated by fusion of crustal rocks as the result of a major metorite impact, with no component of primary magma involved. Their argument rests on three main points. The first is that their Sm–Nd isochron for the SIC is in general agreement with Krogh et al.'s (1984) precise zircon age, but has an unusually negative $\epsilon_{Nd}(T)$. The second is the general similarity of the REE profiles at Sudbury to crustal profiles. With regard to these two points, it has been shown above that since crustal abundances of the light REE, including Nd, are about an order of magnitude higher than in mantle-derived melts, in a mixture containing 50% crustal component both REE abundances and Nd isotope ratios will be dominated by crustal values. The $\epsilon_{Nd}$–$\epsilon_{Sr}$ data of Figs. 14a and b is clearly consistent with the mixing of crustal and mantle components in about the proportions indicated by the trace elements.

Faggert et al.'s (1985) third line of argument is the similarity of $T_{(CHUR)Nd}$ model ages of the SIC rocks to each other and to the age of the pre-existing crust at Sudbury. However it can be shown that, as a mathematical consequence of the isotope systematics and the small range of Sm/Nd fractionation that occurs in nature, any samples defining an Sm–Nd isochron with initial $\epsilon_{Nd}<0$ will have a narrow spectrum of model ages somewhat greater than the isochron age. In claiming that the mean model age of 2.56 Ga is similar to that of Huronian sediment forming the country rocks at Sudbury, Faggert et al. overlooked the fact that $T_{(CHUR)Nd}$ model ages should represent not the age of the sediments or even metamorphism, but rather that of the original igneous rocks giving rise to the sediments, which is about 2.7Ga. A model age of 2.56 is what one would expect from mixing 2.7Ga crust with a 1.85Ga mantle melt.

In this chapter the primitive high MgNo of the SIC magma of 0.61 has already been pointed out. The sulfides at Sudbury contain, on average, 5 wt% Ni and 5 wt% Cu. They also display chondrite-normalized platinum group element (PGE) profiles (Fig. 7.16) typical of those associated with gabbroic magma and resembling those characteristic of flood basalt-related deposits.

If the SIC is entirely derived from the crust as a result of meteorite impact, the high MgNo of the SIC magma demands that the MgO and FeO contents of the resulting melt were dominated by primitive mafic rocks in the target area. Also the source of the Ni, Cu and PGE must either be the meteorite or mafic rocks already present in the target area. While a meteorite could account for the Ni in the ore. it could not have provided the Cu. Furthermore, all meteorites are characterized by near chondritic proportions of PGE, very different to the proportions observed at Sudbury. Thus, if one is to maintain an "all-impact melt" hypothesis for the SIC, one must conclude that the meteorite struck a target area that was rich in primitive rocks containing adequate concentrations of Cu and PGE to account for those in the ores. This coincidence, coupled with the fact that Faggert et al's

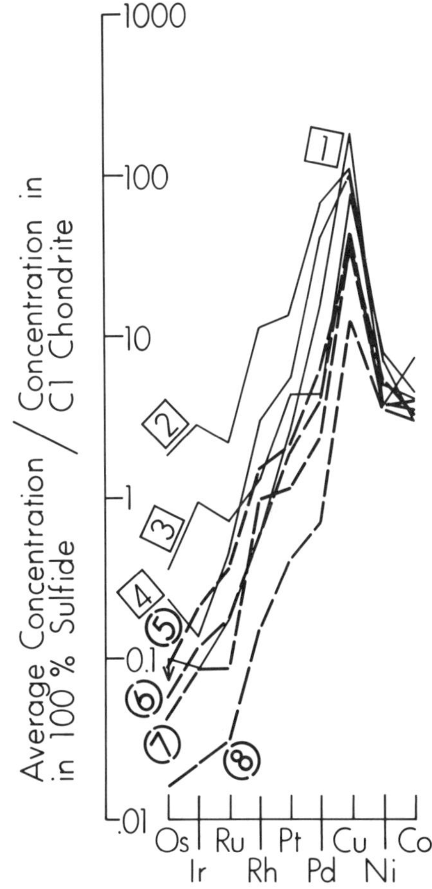


FIGURE 7.16—Chondrite normalized representation of the average concentration of Ni, Cu, Co and PGE in the sulfides of a group of samples collected from 4 Sudbury (heavy dashed lines) and 4 flood basalt-related (fine slid lines) deposits. 1 = Minnamax, Minnesota; 2 = Talnakh, USSR; 3 = Insizwa, Southern Africa; 4 = Great Lakes Nickel, Ontario; 5 = Little Stobie No 2, Sudbury; 6 = Little Stobie No 1, Sudbury; 7 = Levack (McCreedy) West, Sudbury; 8 = Strathcona, Sudbury. Data from Naldrett (1981). Values for normalization are those assembled by Naldrett and Duke (1980).

(1985) data do not demand an all-crustal origin for the SIC, leads this author to favour a model involving mantle-derived magma triggered and highly contaminated as a result of the impact, rather than an all-crustal origin.

It is proposed that the catastrophic explosion that occurred in the Sudbury region 1.85Ga ago gave rise to the Onaping formation, fractured and in part melted the underlying crust, and triggered ascent of a magma similar in its initial composition to continental flood basalt. Magma probably first rose up through the lower crust via a centralized plumbing system. After its initial ascent, the magma halted temporarily within the upper crust, fractionated to some degree, and became contaminated by surrounding rocks and zones of shock-melted material within them. The highly fractured, even pulverized (judging from the Sudbury and Footwall breccias), nature of the country rocks at Sudbury would have greatly facilitated the assimilation, providing a large surface area for contact between the magma and its contaminant. At this stage, some magma spread out lateraly through numerous subsidiary channels within the highly fractured ground beneath the crater (Fig. 7.17). It is possible that these hidden, flanking intrusions account for the buried mass of mafic and ultramafic rock that the analysis of gravity and magnetic data by Gupta et al. (1984) has indicated underlies and extends beyond the present Sudbury structure.

The batches of magma that infiltrated the flanking subsidiary channels cooled, fractionated further, giving rise to olivine and pyroxene-rich cumulates, and became contaminated through reaction with their surroundings. The contamination, coupled with the cooling, caused sulfides to precipitate at an early stage; these collected as massive bodies beneath the cumulates. It is at this stage that the distinctive local trace element and isotopic characteristics that have been documented in this paper started to be impressed on the Sublayer magmas. Because of its greater dimensions, the main body of magma lost heat less quickly than that in the flanking chambers, and thus fractionated less rapidly. Successive injections of this magma occurred along the unconformity between the crater floor and the overlying, relatively unconsolidated and thus low density, Onaping formation to give rise to the Main Mass of the Complex.

The faster cooling, and therefore more fractionated (consistent with the MgNo), magma of the flanking bodies was also squeezed up to intersect the crater floor, where it spread out, exploiting 'troughs' and 'embayments' in the floor and also infiltrating fractures in the basement rocks to form the offset dykes. This magma is responsible for the Mineralized Sublayer. It is likely that further contamination occurred during these events as the magma intruded into fractured, pulverized and probably molten material on the crater floor. In some instances (Fig. 7.17), as it rose up through the basement, it intersected and disrupted overlying bodies, picking up sulfides and ultramafic cumulates in the process, and giving rise to the enrichment in sulfies referred to above and to the very important association between the mineralization and the xenoliths. The offset dykes may represent,

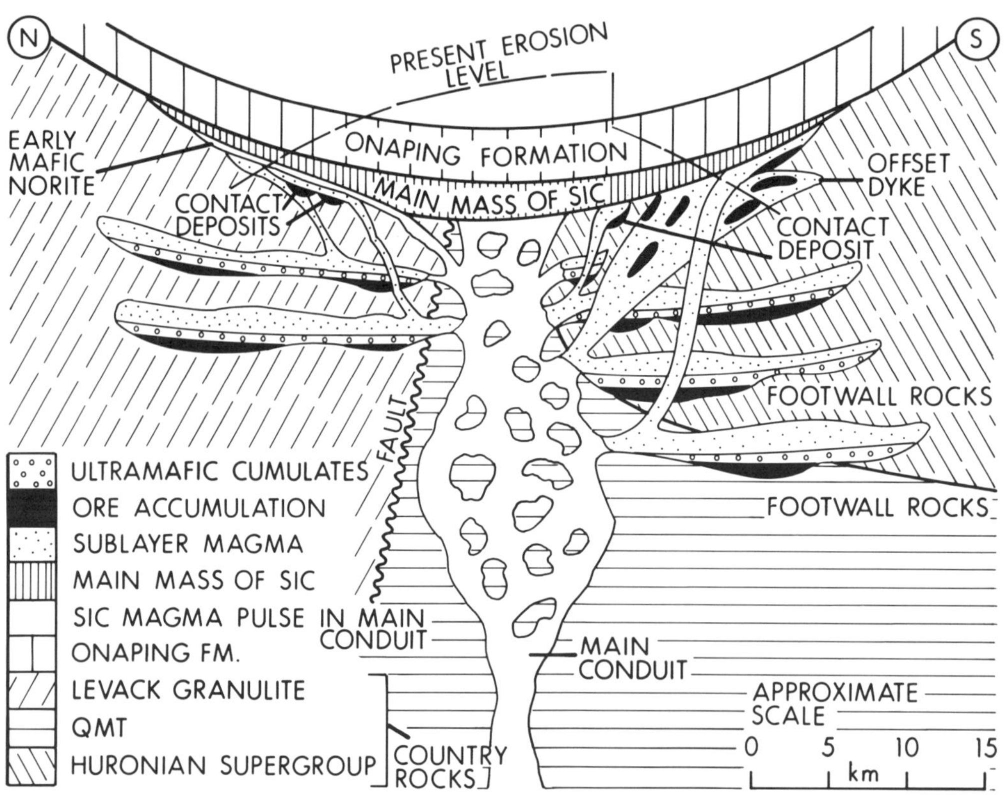


FIGURE 7.17—Schematic cross-section illustrating the emplacement of the SIC. Magma rises up a vertical conduit into a highly fractured zone beneath an explosion (possibly impact) crater. The central body of magma cools somewhat slowly, assimilating much of the fractured country rocks. Subsequently it is emplaced into the crater, beneath the low density Onaping formation, in response to structural readjustments in the area. It is probable that this is a multi-stage process. Magma has also spread out as a series of sills peripheral to the central conduit. Assimilation also occurs within these, giving rise to an imprint of localized contamination. The magma of the sills deposits sulfide and fractionates, giving rise to olivine and pyroxene-rich cumulates. Fractionated magma from the flanking sills also works its way up to the crater floor, intruding to form the Sublayer of the contact and offsets. Where magma from a lower sill has cut and disrupted an overlying one, it has picked up sulfides and inclusions, intruding with these in suspension within itself.

in part, channel-ways up which magma flowed to reach the overlying crater floor; erosion has stripped away the floor, leaving only the underlying feeder dykes. The fact that the Sublayer intrusions were derived from a number of bodies following parallel, but distinct, evolutionary trends accounts for the greater variability of the Sublayer in MgNo and Sr initial ratio in comparison with the Main Mass.

The Unmineralized Sublayer is more problematical. Naldrett (1984) interpreted it as batches of magma from the flanking intrusions that reached the crater floor without intersecting overlying bodies. They were emplaced, therefore, without additional sulfides and carrying only country rock inclusions. The uniformity in trace element composition of this type of Sublayer all around the Complex, in contrast to the differences observed between the Mineralized Sublayer of the North and South Ranges, and the marked compositional differences between the unmineralized and mineralized variants of Sublayer, were unknown to Naldrett at the time he made his suggestion and cast doubt on the hypothesis. It is possible that the Unmineralized Sublayer represents hybrids produced as Main Mass magma rose up and mixed with debris and molten material on the crater floor.

## REFERENCES


Arth, J. G., and Hanson, G. N., 1975, Geochemistry and origin of the early Precambrian crust of northeastern Minnesota: Geochimica et Cosmochimica Acta, v. 39, pp. 325–362.

Card, K. D., Gupta, V. K., McGrath, P. H., Grant, F. S., 1984, The Sudbury structure: Its regional geological and geophysical setting; *in* Pye, E. G., Naldrett, A. J., and Giblin, P. E. (eds.), The geology and ore deposits of the Sudbury structure: Ont. Geol. Surv., Spec. Vol. No. 1, pp. 25–44.

Carter, S. R., Evensen, N. M., Hamilton, P. J., and O'Nions, R. K., 1978, Neodymium and strontium isotopic evidence for crustal contamination of continental volcanics: Science, v. 202, pp. 743–747.

Cochrane, L. B., 1984, Ore deposits of the Copper Cliff Offset; *in* Pye, E. G., Naldrett, A. J., and Giblin, P. E. (eds.), The geology and ore deposits of the Sudbury structure: Ont. Geol. Surv., Spec. Vol. No. 1, pp. 347–360.

Dressler, B. O., 1984, The effects of the Sudbury event and the intrusion of the Sudbury igneous complex on the footwall of the Sudbury structure; *in* Pye, E. G., Naldrett, A. J., and Giblin, P. E. (eds.), The geology and ore deposits of the Sudbury structure: Ont. Geol. Surv., Spec. Vol. No. 1, pp. 97–138.

Fairbairn, H. W., Faure, G., Pinson, W. H., and Hurley, P. M., 1968, Rb–Sr whole-rock age of the Sudbury lopolith and basin sediments: Canad. J. of Earth Sci., v. 5, pp. 707–714.

Green, J. C., 1981, Basaltic Volcanism Study Project, pre-Tertiary continental flood basalts, basaltic volcanism on the terrestrial planets: Pergamon Press, Inc., New York, pp. 30–77.

Gupta, V. K., Grant, F. S., and Card, K. D., 1984, Gravity and magnetic characteristics of the Sudbury structure; *in* Pye, E. G., Naldrett, A. J., and Giblin, P. E. (eds.), The geology and ore deposits of the Sudbury structure: Ont. Geol. Surv., Spec. Vol. No. 1, pp. 381–410.

Irvine, T. N., 1975, Crystallization sequences of the Muskox intrusion and other layered intrusions—II. Origin of chromitite layers and similar deposits of other magmatic ores: Geochim. et Cosmochim. Acta. 39, pp. 991–1020.

Krogh, T. E., Davis, D. W., and Corfu, F., 1984, Precise U–Pb zircon and baddeleyite ages for the Sudbury area; *in* Pye, E. G., Naldrett, A. J., and Giblin, P. E. (eds.), The geology and ore deposits of the Sudbury structure: Ont. Geol. Surv., Spec. Vol. No. 1, pp. 431–447.

Kuo, H. Y., and Crocket, J. H., 1979, Rare earth elements in the Sudbury nickel irruptive: comparison with layered gabbros and nickel irruptive petrogenesis: Econ. Geol., v. 79, pp. 590–605.

McKulloch, M. R., and Wasserburg, G. J., 1978, Sn–Nd and Rb–Sr chronology of continental crust formation: Science, v. 200, no. 4345, pp. 1003–1011.

Muir, T. L., 1985, The Sudbury structure: considerations and models for an endogenic origin; *in* Pye, E. G., Naldrett, A. J., and Giblin, P. E. (eds.), The geology and ore deposits of the Sudbury structure: Ont. Geol. Surv., Spec. Vol. No. 1, pp. 449–490.

Muir, T. L., and Peredery, W. V., 1984, The Onaping Formation; *in* Pye, E. G., Naldrett, A. J., and Giblin, P. E. (eds.), The geology and ore deposits of the Sudbury structure: Ont. Geol. Surv., Spec. Vol. No. 1, pp. 139–210.

Nakamura, E., Campbell, I. H. and Sun, Shen-Su, 1985, The influence of subduction processes on the geochemistry of Japanese alkaline basalts: Nature, v. 316, pp. 55–58.

Naldrett, A. J., Rao, B. V., and Evensen, N. M., 1986, Contamination at Sudbury and its role in ore formation; *in* Gallagher, M. J., Ixer, R. A., Neary, C. R., and Pritchard, H. M. (eds.), Metallogeny of basic and ultrabasic rocks: Spec. Pub. of Inst. of Mining & Metallurgy, London, pp. 75–92.

Naldrett, A. J., Rao, B. V., Evensen, N. M., and Dressler, B. O., 1985, Grant 146: Major and trace element and isotopic studies at Sudbury—a model for the structure and its ores; *in* Milne, V. G. (ed.), Geoscience research programme, summary of research 1984–85: Ont. Geol. Surv., Misc. Publ. 127, pp. 30–44.

Naldrett, A. J., 1984, Summary, discussion and synthesis; *in* Pye, E. G., Naldrett, A. J., and Giblin, P. E. (eds.), The geology and ore deposits of the Sudbury structure: Ont. Geol. Surv., Spec. Vol. No. 1, pp. 533–570.

Naldrett, A. J., and MacDonald, A. J., 1980, Tectonic settings of some Ni–Cu sulfide ores: Their importance in genesis and exploration: Geol. Assoc. Can,. Spec. Paper 20, pp. 633–657.

Naldrett, A. J., Hewins, R. H., and Greenman, L., 1968, The main Irruptive and the sub-layer at Sudbury, Ontario: Proc. 24th Int. Geol. Congress (Section 4), pp. 206–214.

Naldrett, A. J., Bray, J. G., Gasparrini, E. L., Podolsky, T., and Rucklidge, J. C., 1970, Cryptic variation and the petrology of the Sudbury nickel irruptive: Econ. Geol., v. 65, pp. 122–155.

Pattison, E. F., 1979, The Sudbury sublayer: Its characteristics and relationships with the main mass of the Sudbury irruptive: Can. Mineral., v. 17, pt. 2, p. 257–274.

Peredery, W. V., and Morrison, G. G., 1984, Discussion of the origin of the Sudbury structure; *in* Pye, E. G., Naldrett, A. J., and Giblin, P. E. (eds.), The geology and ore deposits of the Sudbury structure: Ont. Geol. Surv., Spec. Vol. No. 1, pp. 491–512.

Rao, B. V., Naldrett, A. J., and Evensen, N. M., 1984, Crustal signature in the Sublayer, Sudbury Igneous Complex: A strontium and neodymium isotope study. (abs.): Ont. Geol. Survey, Geoscience Research Seminar and Open House '84, Volume of Abstracts, p. 3.

Rao, B. V., Naldrett, A. J., Evensen, N. M., and Dressler, B. O., 1983, Contamination and the genesis of the Sudbury ores; *in* Pye, E. G. (ed.), Geoscience res. grant programme, summary of research, 1982–1983: Ont. Geol. Surv., Misc. Pub. 113, pp. 7–12.

Roeder, P. L., and Emslie, R. F., 1970, Olivine–liquid equilibrium: Contr. Mineral. Petrol., v. 29, pp. 275–289.

Scribbins, B., Rae, D. R., and Naldrett, A. J., 1984, Mafic and ultramafic inclusions in the sublayer of the Sudbury Igneous Complex: Canad. Mineral., v. 22, pt. 1, pp. 67–75.

Shaw, D. M., Dostal, Jaroslov, and Keays, R. R., 1976, Additional estimates of continental surface Precambrian shield composition in Canada: Geochim. et Cosmochim. Acta, v. 40, pp. 73–83.

Slaught, W. H., 1951, A petrographic study of the Copper Cliff offset in the Sudbury District; Unpub. M.Sc. Thesis, McGill University.

Taylor, S. R. 1964, Abundances of chemical elements in the continental crust: a new table: Geochim. et Cosmochim. Acta, v. 28, pp. 1273–1285.

Thompson, R. N., 1982, Magmatism of the British tertiary volcanic province: Scottish J. of Geology, v. 18, pp. 1–108.

Wood, D. A., 1980, The application of a Th–Hf–Ta diagram to problems of tectonomagnetic classification and to establishing the nature of crustal contamination of basaltic lavas of the British tertiary volcanic province. Earth & Planet. Sci. Lett., v. 50, pp. 11–30.

Table 1

Average Compositions of Typical Rocks of the SIC

| | Main Mass QRN | Mineralized Sublayer Lev. W. | Mineralized Sublayer Strath. | Mineralized Sublayer Litt. Stobie | Unmineralized Sublayer North Range | Unmineralized Sublayer South Range |
|---|---|---|---|---|---|---|
| $SiO_2$ | 57.0 | 54.9 | 54.23 | 56.0 | 58.74 | 57.43 |
| $TiO_2$ | 1.34 | 0.55 | 0.44 | 0.60 | 1.01 | 0.99 |
| $Al_2O_3$ | 16.4 | 14.80 | 11.91 | 17.19 | 14.63 | 15.93 |
| $FeO^T$ | 7.33 | 9.15 | 10.71 | 8.96 | 9.51 | 9.13 |
| MnO | 0.13 | 0.15 | 0.18 | 0.12 | 0.15 | 0.13 |
| MgO | 6.40 | 7.97 | 11.81 | 4.72 | 5.08 | 5.28 |
| CaO | 7.28 | 7.93 | 7.92 | 7.26 | 5.96 | 6.48 |
| $Na_2O$ | 2.41 | 2.77 | 2.01 | 2.88 | 2.83 | 2.95 |
| $K_2O$ | 1.55 | 0.90 | 0.72 | 1.08 | 2.03 | 1.67 |
| **Trace Elements in ppm** | | | | | | |
| Ba | 492 | 453 | 287 | 582 | 800 | 638 |
| Pb | 46 | 22 | 21 | 39.5 | 54 | 61 |
| Th | 5.5 | 2.4 | 1.647 | 5.8 | 4.3 | 3.8 |
| K | 14600 | 7500 | | 9000 | 16900 | 14000 |
| Ta | 0.43 | 0.25 | 0.12 | 0.56 | 0.31 | 0.46 |
| Sr | 422 | 447 | 367 | 409 | 419 | 402 |
| P | 870 | 870 | | 1050 | | |
| Zr | 96 | 74 | 60 | 82 | 134 | 134 |
| Hf | 2.6 | 1.6 | 1.77 | 1.99 | 3.4 | 3.32 |
| Ti | 2880 | 3300 | | 3600 | 6050 | 5930 |
| Y | 16 | 13 | 12 | 19 | 19 | 23.4 |
| La | 21.3 | 16.2 | 17.7 | 24.7 | 32.4 | 19.5 |
| Ce | 44.7 | 33.9 | 33.3 | 54.1 | 73.8 | 71.5 |
| Nd | 17.4 | 15.8 | 10.9 | 22.0 | 28.6 | 28.7 |
| Sm | 3.36 | 3.07 | 2.60 | 4.39 | 5.49 | 5.32 |
| Eu | 1.02 | 0.81 | 0.84 | 1.27 | 1.57 | 1.58 |
| Tb | 0.42 | 0.41 | 0.37 | 0.78 | 0.64 | 0.73 |
| Yb | 1.51 | 1.14 | 1.29 | 1.87 | 1.8 | 2.26 |

Table 2

Modelling of Trace Element Data

| | KEW | QMT | McK | LG | Model QRN | Model NR Min Sub | Model SR Min Sub | Model Unmin Sub |
|---|---|---|---|---|---|---|---|---|
| Ba | 55 | 878 | 600 | - | 467 | - | - | - |
| Rb | 7.3 | 121 | 124[1] | 8 | 64 | 26.5 | 53.6 | 64.8 |
| Th | 0.54 | 11[3] | 14.6 | 0.2 | 5.77 | 2.17 | 5.44 | 6.59 |
| K | 2490 | 27100 | 23000 | 8100 | 14800 | 8550 | 11500 | 15300 |
| Ta | 0.267 | 0.4[3] | 1.20 | 0.17 | 0.3334 | 0.257 | 0.48 | 0.509 |
| Sr | 275 | 554 | 100 | 585 | 415 | 425 | 296 | 379 |
| P | 480 | 1300[2] | | 990 | 890 | 790 | - | - |
| Zr | 69 | 193[3] | 109 | 147 | 131 | 116 | 102 | 130 |
| Hf | 1.5 | 4.9[3] | 3.7 | 2.83 | 3.2 | 2.51 | 2.62 | 3.23 |
| Ti | 5280 | 2580 | 4500 | 3420 | 3930 | 4210 | 4580 | 3950 |
| Y | - | - | 21.2 | 3.25 | - | - | - | - |
| La | 5.13 | 32 | 50.6 | 18.3 | 18.6 | 34.0 | 19.6 | 26.5 |
| Ce | 11.7 | 68 | 108 | 50.5 | 39.9 | 42.5 | 42.2 | 59.6 |
| Nd | 9.45 | 28.8 | 40.8 | 11.6 | 19.1 | 13.4 | 19.6 | 22.7 |
| Sm | 2.11 | 5.06 | 7.6 | 2.18 | 3.59 | 2.63 | 3.80 | 4.24 |
| Tb | 0.54 | 0.617 | 0.75 | 0.23 | 0.579 | 0.45 | 0.598 | 0.445 |
| Yb | 1.50 | 0.97 | 1.92 | 0.53 | 1.24 | 1.09 | 1.48 | 1.23 |

Data are as follows:

KEW = Average of 5 Keweenawan olivine basalts (Basaltic Volcanism Project (1981))

QMT = Quartz Monzonite - Tonalite Mixture (Arth & Hanson (1975))

McK = McKim Formation (P. Fralick Personal Communication (1984))

LG = Levack Gneiss - This study - Except as indicated

1. From Sr coupled with the Sr/Rb ratio of Fairbairn et al (1969)
2. Sims et al (1984)
3. Figure for Laxfordian gneiss (Wood, 1980)

QRN = Quartz-rich norite

NR Min Sub = North Range Mineralized Sublayer

SR Min Sub = South Range Mineralized Sublayer

# Chapter 8

# Stratiform PGE Deposits in Layered Intrusions

A. J. Naldrett

## INTRODUCTION

Concentrations of platinum group elements (PGE) are found in a variety of geologic settings, ranging from high-temperature magmatic through hydrothermal (McCallum et al., 1976) to late diagenetic or sedimentary (Kucha, 1982). The only deposits from which substantial quantities of PGE are produced are those associated with mafic or ultramafic rocks. Identified resources within these are summarized in Table 8.1.

The deposits are divisible into those in which the PGE are obtained as a byproduct, as is the case with Noril'sk–Talnakh discussed above, and those in which PGE are the principal product. At the time of writing the Merensky Reef of the Bushveld Complex is the most important producer amongst the second class of deposits. Production is also obtained from the somewhat similar J–M reef of the Stillwater Complex. Both of these are stratiform horizons containing from 0.5% to 3% sulfide. Significant production is now also obtained from the UG–2, which is the most important (from the point of view of grade and total contained PGE) of a number of PGE-enriched chromitites in the layered complexes. The Platreef is a sulfide enriched zone at the base of the Bushveld Complex where it transgresses down through quartzites and carbonate-rich country rocks of the Pretoria Series to come to rest on Archean granitic gneisses. It is not currently in production.

Possible other deposits with a similar setting include the mineralized layers of the Penikat layered intrusion of northern Finland (Alapieti and Lahtinen, 1986). A stratabound, PGE-enriched zone also occurs in the Great Dyke of Zimbabwe (Prendergast, 1988), although this is hosted completely in ultramafic rocks and is therefore different in its setting to those mentioned above. This chapter is concerned with the Merensky and J–M Reefs and the UG–2 chromitite.

## STRATIGRAPHY OF THE BUSHVELD AND STILLWATER COMPLEXES

### Bushveld Complex

The Bushveld Complex is divided into four zones (Fig. 8.1). The Lower zone is composed mainly of bronzitites, harzburgites and dunites (Cameron, 1978). It is overlain by the Critical zone, which hosts the economically interesting LG–6 chromitite layer and the PGE-bearing Merensky Reef and UG–2 chromitite (Cameron, 1980; Cameron, 1982). The Critical zone is overlain, in turn, by the norites, gabbros and anorthosites of the Main zone, which are themselves capped by the ferrogabbros and ferrodiorites of the Upper zone (von Gruenewaldt, 1973; Molyneaux, 1974).

The variation in the MgNo of bronzite and olivine (Fig. 8.1) indicates an overall upward Fe-enrichment in the Complex, consistent with fractional crystallization, although in detail there are numerous reversals and large thicknesses across which no systematic variation is present.

The Critical zone (Fig. 8.2) is divided into two parts (Cameron, 1980, 1982). The Lower Critical zone consists primarily of bronzitites, chromitites and some harzburgites. The Upper Critical zone, marked by the incoming of cumulus plagioclase, can itself be subdivided into two parts; (i) a lower part consisting of anorthosites, norites and minor bronzites, that occur in no systematic order, and define no classical cyclic units; and (ii) an upper part (starting with base of the UG–1 chromitite unit) which, in contrast, comprises units made up of a regular succession of rock types, consisting of some or all of chromitite, harzburgite, bronzitite, norite and anorthosite, in most cases in this order.

The different zones and units vary greatly in thickness throughout the Complex and are absent in some areas. The lower zones vary in thickness due in part to their transgressive nature and they appear to have had the most restricted development. In many places, the Critical zone transgresses over a wider area of the Bushveld floor than the Lower zone, and the Main zone a still wider area, which is attributed (von Gruenewaldt, 1979) to the progressive addition of new magma as the Complex crystallized.

### Stillwater Complex

Recent studies of the igneous stratigraphy of the Stillwater Complex include the detailed documentation of the stratigraphy by McCallum et al. (1980), Todd et al. (1982), and Raedeke and McCallum (1984), and the mapping of the whole of the Complex by Segerstrom and Carlson (1982). These have been correlated by Zientek et al. (1985) who divide the Complex into the Basal, the Ultramafic and the Banded series (Fig. 8.1). The Basal series is in turn subdivided into the Basal Norite zone and the Basal Bronzite Cumulate zone, the Ultramafic series into the Peridotite zone and the Bron-

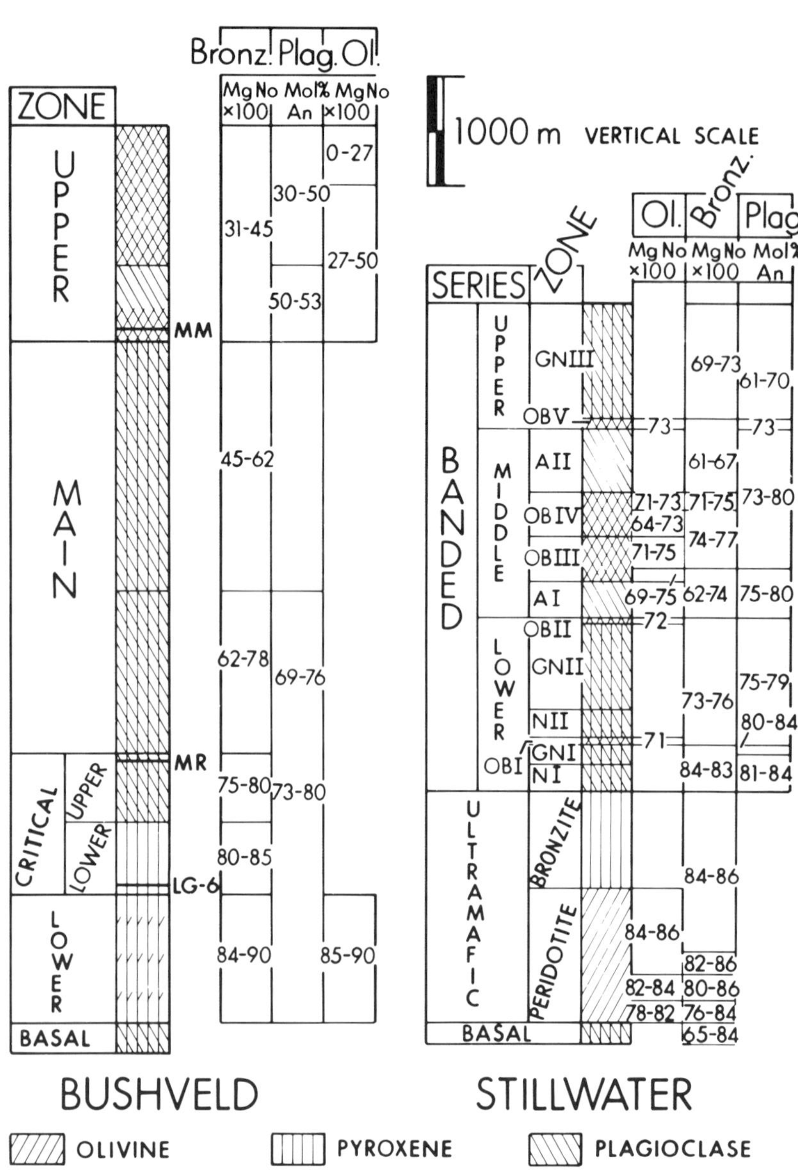


FIGURE 8.1—Generalized stratigraphy of the Bushveld and Stillwater Complexes, showing compositional variation in olivine, bronzite and plagioclase. MM = Main Magnetitite Layer; MR = Merensky Reef; LG-6 = LG-6 Chromitite Layer.Data from Cameron (1978, 1980, 1982), Naldrett et al. (1986), Von Gruenewaldt et al. (1985), Wager and Brown (1968). From Naldrett et al. (1987).

zitite zone, and the Banded series into the Lower, Middle and Upper Banded series. Each of the sub-series in the Banded series is itself divided into a number of zones (Fig. 8.1) (McCallum et al., 1980).

The Basal series is a laterally continuous but petrologically variable unit consisting of bronzite cumulates (bC) or bronzite plus plagioclase cumulates (bpC). The Peridotite zone of the Ultramafic series consists of an alternating sequence of olivine-, chromite-, and bronzite-bearing cumulate layers. Jackson (1970) recognized between 8 and 21 cyclic units in the Peridotite zone, varying in number along strike and changing from ocC (olivine + chromite cumulate) at the base to obC and bc at the top. As a rule, basal ocC contain less than 2 modal% chromite. Thin layered intervals (2 cm to 1 m thick) containing as much as 80 modal% chromite occur in some parts of some cyclic units and yet are missing in other parts of the same units. Two of the chromitite intervals are particularly well developed and have been mined.

The Bronzite zone of the Ultramafic series consists of a succession of laminated and size-graded bronzite cumulates.

The Banded series is marked by the appearance of cumulus plagioclase, and consists of a sequence of norites, gabbronorites and anorthosites. Olivine occurs as a cumulus mineral at 5 places in the stratigraphy; these are referred to as Olivine-bearing zones (OB–I to V). The J–M Reef occurs within the lowest olivine-bearing zone, OB–I.

Todd et al. (1982) and Irvine et al. (1983) have divided the Banded series into 6 mega-cyclic units. The uppermost 4 (above OB–I) display the crystallization order plagioclase, olivine, augite, bronzite in contrast to the sequence olivine,

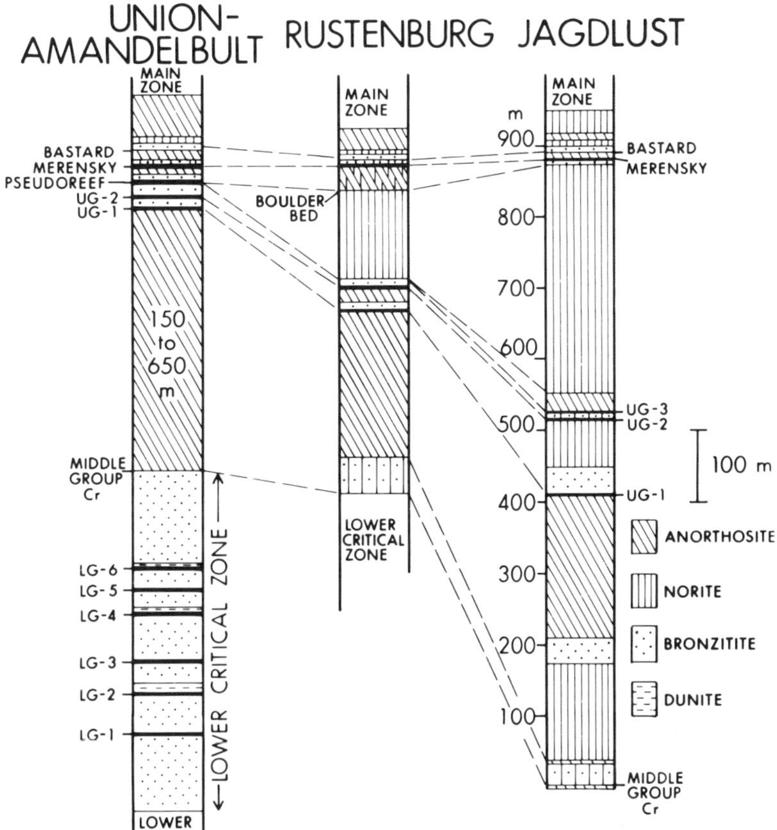


FIGURE 8.2—Stratigraphy of the Critical zone with special refence to the Upper Critical zone in the northwestern (Union–Amandelbult), southwestern (Rustenburg) and eastern (Jagdlust) portions of the Bushveld Complex. Including data from Cameron (1980, 1982), von Gruenewaldt et al. (1985). From Naldrett (1981b).

bronzite, plagioclase, augite shown in the underlying mega-cyclic units. The reality of the mega-cyclic units is supported to some degree by the mineral composition data of Raedeke et al. (1985), particularly the An content of plagioclase (Fig. 8.1), which can be interpreted as decreasing progressively through each unit and increasing abruptly at the base of the overlying unit. Todd et al. (1982) pointed out that the Banded series is particularly plagioclase-rich; they proposed that the Stillwater magma chamber was invaded by pulses of basaltic magma with plagioclase alone on its liquidus at a stage soon after the formation of the Ultramafic series. The onset of this recharge may be documented in the anorthosite layers that occur within the N–I and GN–I zones.

### Comparison of Bushveld and Stillwater

In a general sense, the Lower zone of the Bushveld Complex can be compared with the Peridotite zone of the Stillwater Ultramafic series, and the Bushveld Lower Critical zone with the Bronzitite zone of the Ultramafic series at Stillwater. The comparison is inexact because the Lower zone of the eastern and western Bushveld Complex is much poorer in olivine than the Peridotite zone of the Stillwater. The reason for this almost certainly lies in the nature of the magmas from which the lower portions of the two intrusions crystallized. The Bushveld initial magma is believed to have contained about 56 wt% $SiO_2$ and 13–14 wt% MgO (Sharpe, 1981; Harmer and Sharpe, 1985), while the initial magma at Stillwater contained less than 50 wt% $SiO_2$ and about the same MgO content (Helz, 1985).

At Stillwater, the main chromitite layers occur in the Peridotite zone, only small layers occur near the top of the Bronzitite zone, and chromite is only sparsely distributed at certain horizons above this. In contrast, in the Bushveld, most of the well devloped chromitite layers occur in the Critical zone.

## SETTING OF THE MINERALIZATION

### Bushveld Complex

The geology of the Bushveld Complex is shown in Fig. 8.3. Apart from the Platreef, which lies at the base of the Complex in the northernmost limb, where correlations with the remainder of the Complex are difficult, the principal PGE mineralization in the Bushveld Complex occurs within or very close to the Upper Critical zone. Six to eight chromitite layers comprising the Middle Group occur within 20 m of the contact between the Upper and Lower Critical zone (marked by the incoming of cumulus plagioclase) and contain modest (of the order of 1–3 gm/tonne) concentrations of PGE. The more important PGE concentrations occur in the upper part of the Upper Critical zone. Five or six well-defined cyclic units occur in this part of the stratigraphy

(Fig. 8.4); these are the UG–1, UG–2, UG–3 (in some areas only), Pseudoreef (in some areas only), Merensky Reef, and Bastard Reef units. Concentrations of PGE are associated with the lower parts of most of them. Von Gruenewaldt et al. (1986) report typical values of 3.43, 9.64, and 4.71 ppm (total PGE + Au) respectively for the UG–1, UG–2 and UG–3 chromitites in the Maandagshoek area of the eastern Bushveld. Zones of PGE-rich sulfides occur near the base of the Pseudoreef unit (see below), the Merensky Reef is reported by Buchanan (1979) to grade 8.1 ppm total PGE, and PGE are reported in more modest concentrations at the base of the Bastard unit (see below).

The UG–1 chromitite layer bifurcates and anastomozes at low angles at its lower contact with anorthosite. It is interleaved with and passes up into bronzitite, followed by norite and then anorthosite. In the Union–Amandelbult area, in the northwestern part of the Complex, this unit is beheaded and the plagioclase cumulates are missing.

The UG–2 unit also consists of bronzitite and norite, with several chromitite layers sandwiched within the bronzitite. The norite is succeeded by an anorthosite in the west, but in the central sector of the eastern Bushveld Complex the UG–3 chromitite unit is present (see Fig. 8.2). Three hundred m of norite separate the base of the UG–3 unit from the overlying Merensky Reef at Jagdlust. In the Union–Amandelbult area, this norite is missing, but another cyclic unit, the Pseudoreef unit, intervenes. Viljoen et al. (1986a) emphasize the lateral variation present in the Pseudoreef in the Amandelbult area, where possibly as many as 4 cyclic units are present.

In the Rustenburg area, 130 m of norite are present between the UG–2 unit and a 1 to 2 m zone containing elliptical boulders of bronzitite, known as the Boulder bed. The Boulder bed can be correlated with the Pseudoreef of the Union–Amandelbult area, and evidence is presented below to support this correlation.

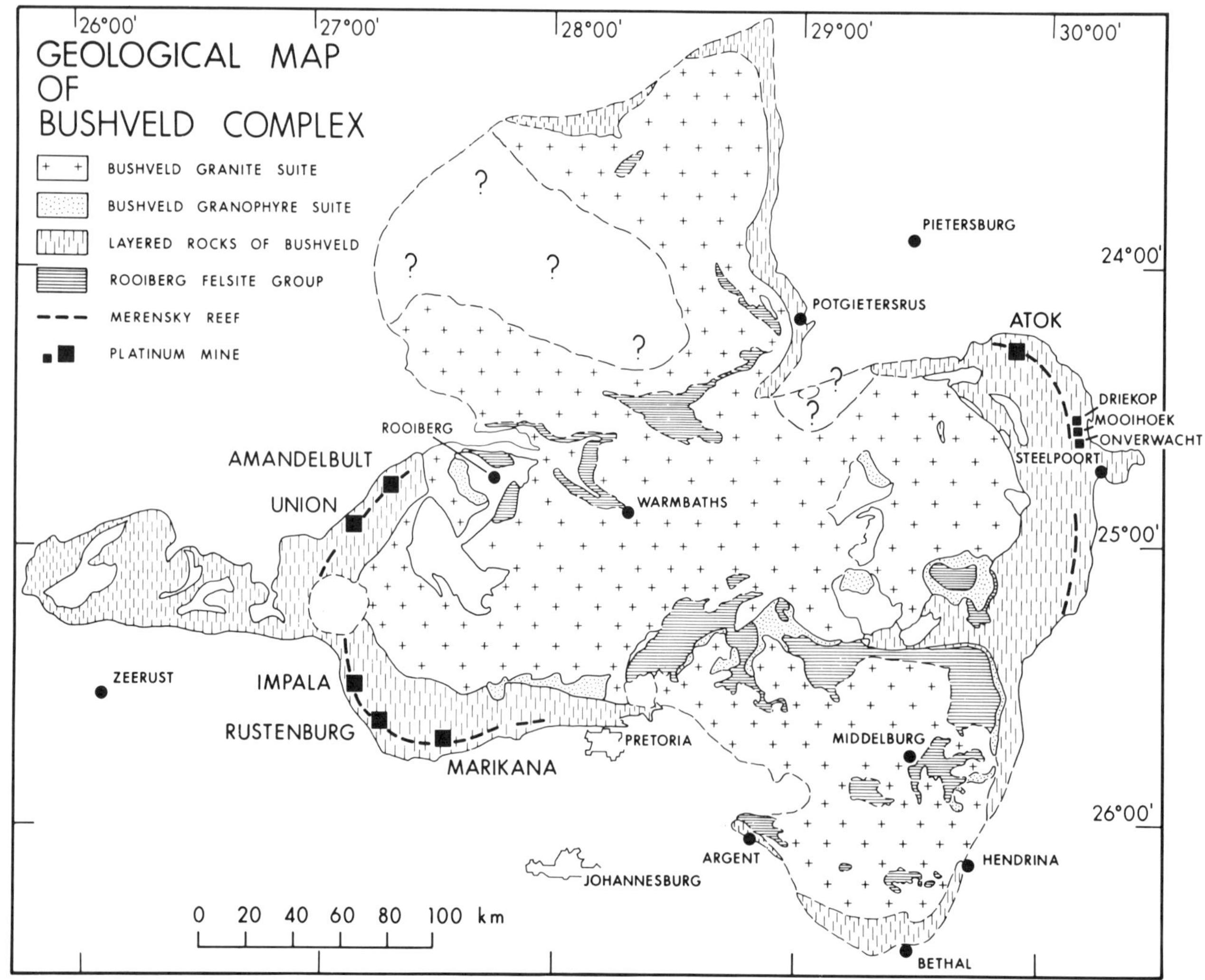


FIGURE 8.3—Generalized sub-outcrop geological map of the Bushveld Complex, showing the main PGE mines in the area. The farm Jagdlust is very close to the Atok mine.

In the western Bushveld, the Merensky cyclic unit consists of bronzitite, norite (which is merely a gradation from bronzitite to anorthosite) and then anorthosite. In the east, the norite is missing and there is an abrupt jump from bronzitite to anorthosite. The uppermost cyclic unit of the series, the Bastard, consists of bronzitite, norite and anorthosite in both the eastern and western Bushveld.

Where the norites are parts of cyclic units, they are mostly transitions from bronzitite to anorthosite (Fig. 8. 4). They do not represent cotectic norites, which should consist of a nearly constant proportion of about 1/3 bronzite to 2/3 plagioclase, and, in this part of the Bushveld stratigraphy, contain about 10 wt% MgO. The principal norites that have cotectic compositions are the large thicknesses that intervene between the Boulder and UG–2 units at Rustenburg and between the Merensky and UG–3 units at Jagdlust.

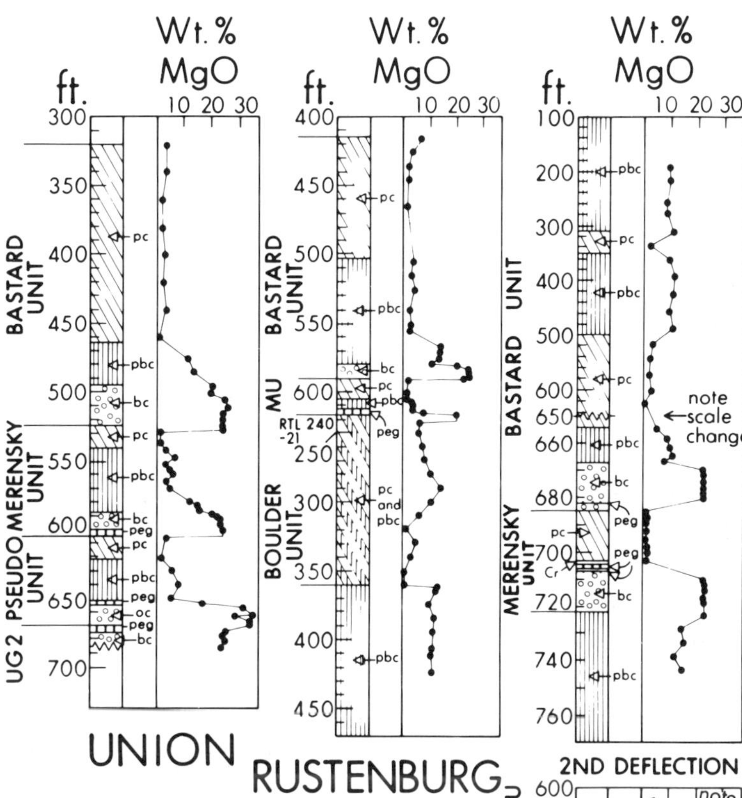


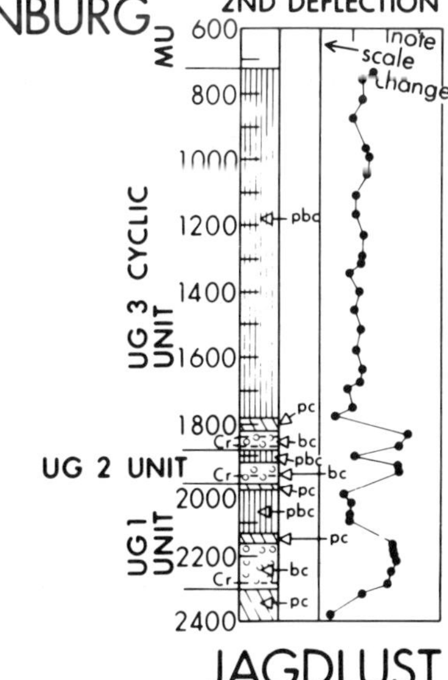


FIGURE 8.4—Plot of MgO vs stratigraphic height. Note the scale changes over the Jagdlust section. Depths indicated are depths in the borehole. Data on Jagdlust from Cameron (1988).

Bronzite in most of the cyclic units shows an overall iron enrichment trend (Fig. 8. 5 and also in Kruger and Marsh, 1985, fig. 3). However, at Union, where individual rock sequences are thicker and variations within them are more easily seen, the upward iron enrichment is much stronger in the norites than it is in the bronzite cumulates.

Both in Fig. 8. 5 and in the diagrams of Kruger and Marsh (1985), plagioclase shows marked zoning in the bronzite cumulates, consistent with an orthocumulate origin for these rocks. In the plagioclase cumulates of the Merensky unit, plagioclase shows relatively little upward enrichment in the albite component, and only slightly more upward enrichment across the anorthosites of the Bastard unit.

There are some systematic changes in the Sr content of plagioclase (Fig. 8.6) in the profiles from Union and Rustenburg. Below the Merensky Reef, there is relatively little change, and plagioclase contains close to 500 ppm Sr. However, even the lowermost norites of the Merensky unit have substantially less Sr than the rocks below them, and there is a decrease in Sr across the norites and the overlying anorthosites of this unit. This decrease continues in the norites of the overlying Bastard unit, although in the anorthosites of the Bastard the Sr contents are relatively constant. The variations illustrated in Fig. 8.7 were first pointed out by Naldrett at al (1984) and are similar to those documented by Kruger and Marsh (1985) in plagioclase separates from the same stratigraphic units at Rustenburg.

In the eastern Bushveld, the major decrease in the Sr

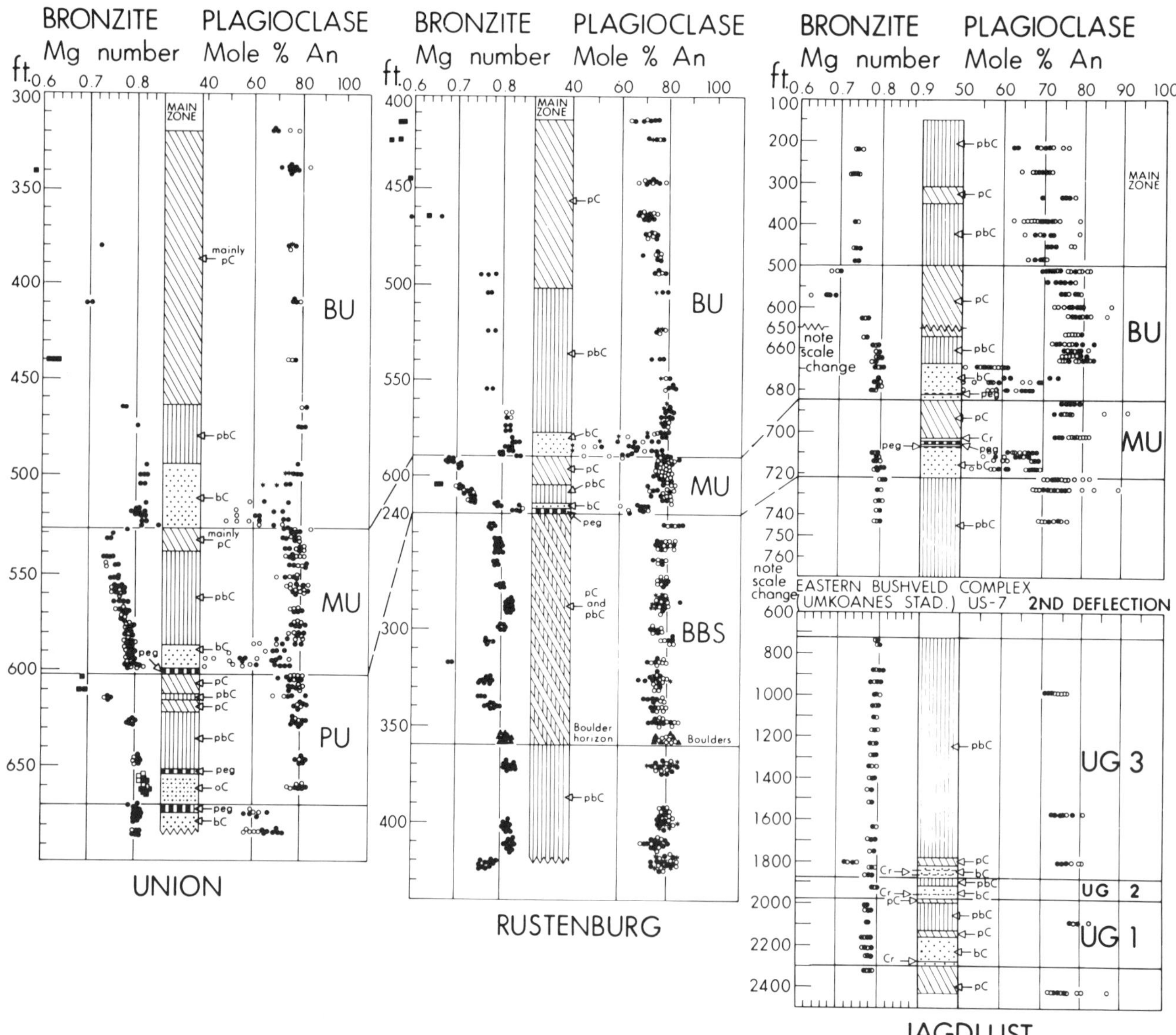


FIGURE 8.5—Variation in the composition of bronzite and plagioclase at three locations in the Bushveld Complex. Note the scale changes over the Jagdlust section. Depths indicated are depths in the borehole. Data on Union and Rustenburg from Naldrett et al. (1986) and on Jagdlust from Cameron (1988).

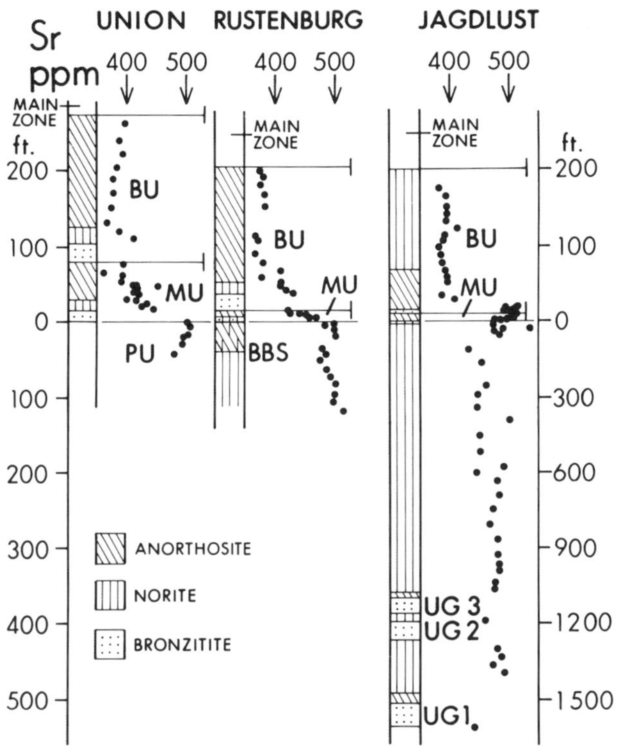


FIGURE 8.6—Change in Sr content of plagioclase over the Union, Rustenburg and Jagdlust sections. PU = Pseudoreef unit; MU = Merensky unit; BBS = Boulder Bed; BU = Bastard unit. Data from Naldrett et al. (1986) and Cameron (1988). From Naldrett et al. (1987).

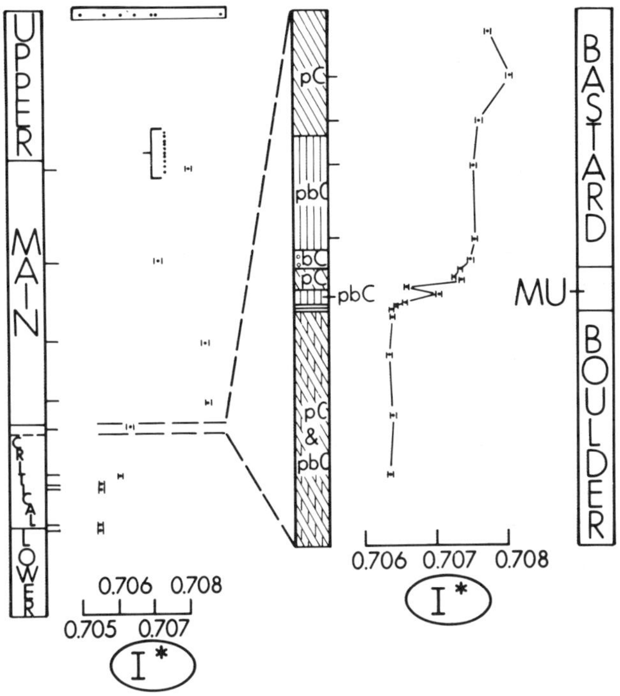

FIGURE 8.7—Variation in initial $^{86}Sr/^{86}Sr$ ratios (I*) in the Bushveld Complex. Data from Hamilton (1977) for the whole of the Complex and from Kruger and Marsh (1982) for the interval around the Merensky Reef in the western Bushveld. The horizontal bar containing dots at the top of the left-hand diagram is the data of Hamilton on the Upper zone which could not be located accurately within the stratigraphic column.

content of plagioclase occurs in the anorthosite of the Bastard unit and not in the Merensky unit.

Naldrett et al. (1987) modelled variations in the MgNo of bronzite and the An and Sr contents of plagioclase in the Merensky and Bastard units, basing their calculations on one of Sharpe's (1981) chilled marginal rocks. They concluded that fractional crystallization of a single mass of magma cannot, by itself, account for the observed compositional variation.

The decrease in Sr in plagioclase, which is proportional to the Sr content of the magma from which the plagioclase is crystallizing, appears to coincide with a major change in Sr isotopic composition of the rocks of the Complex. Fig. 8.7 incorporates data from Hamilton (1977) (from throughout the Bushveld) and from Kruger and Marsh (1982) (from the Western Bushveld). The change in isotopic composition occurs within the norites and anorthosites, precisely where changes occur in the Sr content of the plagioclase and where the bronzite shows upward Fe-enrichment.

Sharpe's (1985) data (Fig. 8.8) for the eastern Bushveld Complex, indicate that there is a similar jump at approximately the same position. However, Sharpe's analyses are not sufficiently closely spaced to indicate whether this occurs in the Merensky unit, or just above in the anorthosite of the Bastard unit, where it has been shown (see above) that the Sr content of the plagioclase changes abruptly. It is likely that, as in the west, the isotopic and bulk Sr changes occur at the same horizon.

A number of authors, including Kruger and Marsh (1982, 1985) and Sharpe (1985) have attributed the isotopic shift to an influx of isotopically distinct magma and have speculated that the presence of the Merensky Reef is related to the influx. If the change in isotopic composition is coincident with the change in Sr content in the eastern Bushveld, it implies that the major influence of the new influx on the cumulate record was felt during formation of the Bastard unit here, not the Merensky. The question arises, therefore, as to whether the onset of crystallization of this isotopically distinct magma is as important to the origin of the Merensky Reef as was previously considered.

## Stillwater Complex

### The J–M Reef

The J–M Reef extends for the whole exposed length of the Complex (Fig. 8.9). It occurs within the Lower Banded series approximately 400 m above its base (see Fig. 8.1). The lowermost rock (the N–I zone) of the series consists of pbC (plagioclase–bronzite cumulate) with 50–30 modal% bronzite. Page et al. (1985) describe these rocks as having modal and locally phase (due to the absence of bronzite) layering.

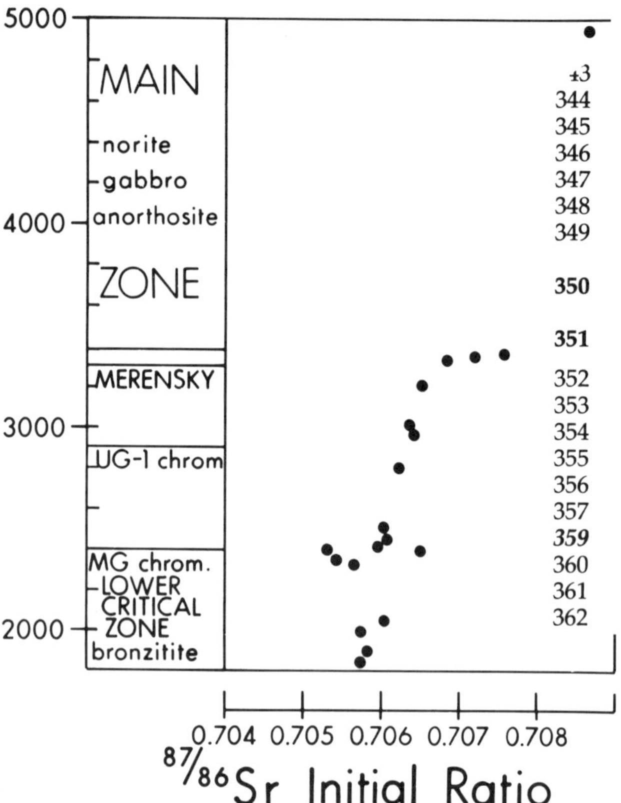


FIGURE 8.8—Initial $^{87}Sr/^{86}Sr$ ratio variations in cumulate rocks from the eastern Bushveld. Data from Sharpe (1985).

They pass up into a sequence of gabbronorites (GN–I) which are interbanded with anorthositic layers. Page et al. (1985a) indicate that the N–I and GN–I zones are divisible into 7 mappable subzones in which the proportion of cumulus pyroxene decreases upwards, although they have not demonstrated any variation in mineral chemistry within individual subzones. Overall the MgNo of bronzite varies between 74 and 83, with a slight upward decrease while the An content of plagioclase is fairly constant at 80 mole% (Fig. 8.10).

The J–M Reef occurs within a sequence of cumulates, which include olivine-bearing layers, and which together constitute Olivine-bearing zone 1 (OB–I). This is the lowermost of the 5 olivine-bearing zones referred to above.

OB–I is overlain by a sequence of norites (N–II). In the Stillwater valley area, these norites differ from the GN–I gabbronorites in that their bronzites are less magnesian (MgNo 73–79 in comparison with 77–84) and less Cr-rich (300–1700 in comparison with 1700–3400 ppm Cr), and their plagioclases are less anorthositic (An 79–85 in comparison with An 82–88) and richer in Sr (150–170 in comparison with 85–110 ppm Sr) than those of the gabbronorites (Barnes and Naldrett, 1986, and see Fig. 8.11). While the less magnesian bronzite and less anorthositic plagioclase of the N–II norites could be attributed to fractional crystallization, the higher Sr contents of the plagioclase could not (Sr partitions into plagioclase and should decrease during fractional crystallization). Barnes and Naldrett (1986) therefore attributed the variations in mineral chemistry to the influx of a second magma. Since there is also a need to explain the formation of rocks in which plagioclase is the sole cumulate phase, or in which it crystallizes first, it is likely, as Todd et al. (1982) and Irvine et al. (1983) have suggested, that the new influx was richer in the plagioclase component than the magma already in the chamber.

## MINERALIZATION

### Bushveld Complex: The Merensky Reef

In the Rustenburg area of the western Bushveld Complex, the ore zone comprising the Merensky Reef is confined to a pegmatoid that occurs at the base of the Merensky cyclic unit (Fig. 8.12). PGE are concentrated in the vicinity of two chromitite layers which occur at the top and bottom of the pegmatoid and are particularly concentrated near the upper chromitite layer (Fig. 13).

In the Union–Amandelbult area, the ore is also associated with chromitite layers within pegmatoid, but here olivine is an important component of the pegmatoid, particularly towards its base (Fig. 8.13).

In the Marikana area, east of Rustenburg Mine, the pegmatoid is less common, occurring as isolated patches within porphyritic bronzitite, and the chromitite layers are much wider apart. The upper chromitite occurs approximately 50 cm below the contact of the bronzitite with the overlying anorthosite, whereas the lower chromitite occurs at the base of the bronzitite. The highest metal concentrations are associated with the upper chromitite layer.

The Merensky Reef is also different in the vicinity of the Atok Mine near Jagdlust in the eastern section of the Complex. Here a pegmatoidal pyroxenite is present within bronzitite, but the chromitite layers and PGE concentrations occur above the pegmatoid, within overlying bronzitite (Fig. 8.13).

The Merensky pegmatoid consists of coarse bronzite crystals (1 to 4 cm in length) enclosed in a matrix of plagioclase. Biotite, apatite and tremolite are common constituents. The lower bounding chromitite layer varies from a few mm to a few cm in thickness. The upper chromitite layer is only a few mm thick and is discontinuously developed in some areas. In many places pegmatoid is developed directly above the upper chromite layer in the overlying bronzite. Elsewhere medium-grained, non-pegmatoidal bronzitite masses occur within the Reef pegmatoid.

Naldrett et al. (1986a) pointed out that the composition of the pegmatoidal reef bronzite is more magnesian than the bronzite immediately overlying the top chromitite layer of the pegmatoid. They used this as supporting evidence for their argument that the Reef was derived from magma more primitive than that which gave rise to the overlying Merensky unit. This argument is discussed more fully below.

Ballhaus and Stumpfl (1985b) have drawn attention to the widespread occurrence of graphite in both the Merensky Reef and in the platiniferous dunite pipes. The graphite has $\delta^{13}C$ in the range of −19 to −21‰, which is permissive but not diagnostic of a crustal source. They found graphite and hydrous silicates to contain significant chlorine. Boudreau et al. (1986) have found that apatites of the PGE-rich zones of the Bushveld and Stillwater Complexes have Cl/

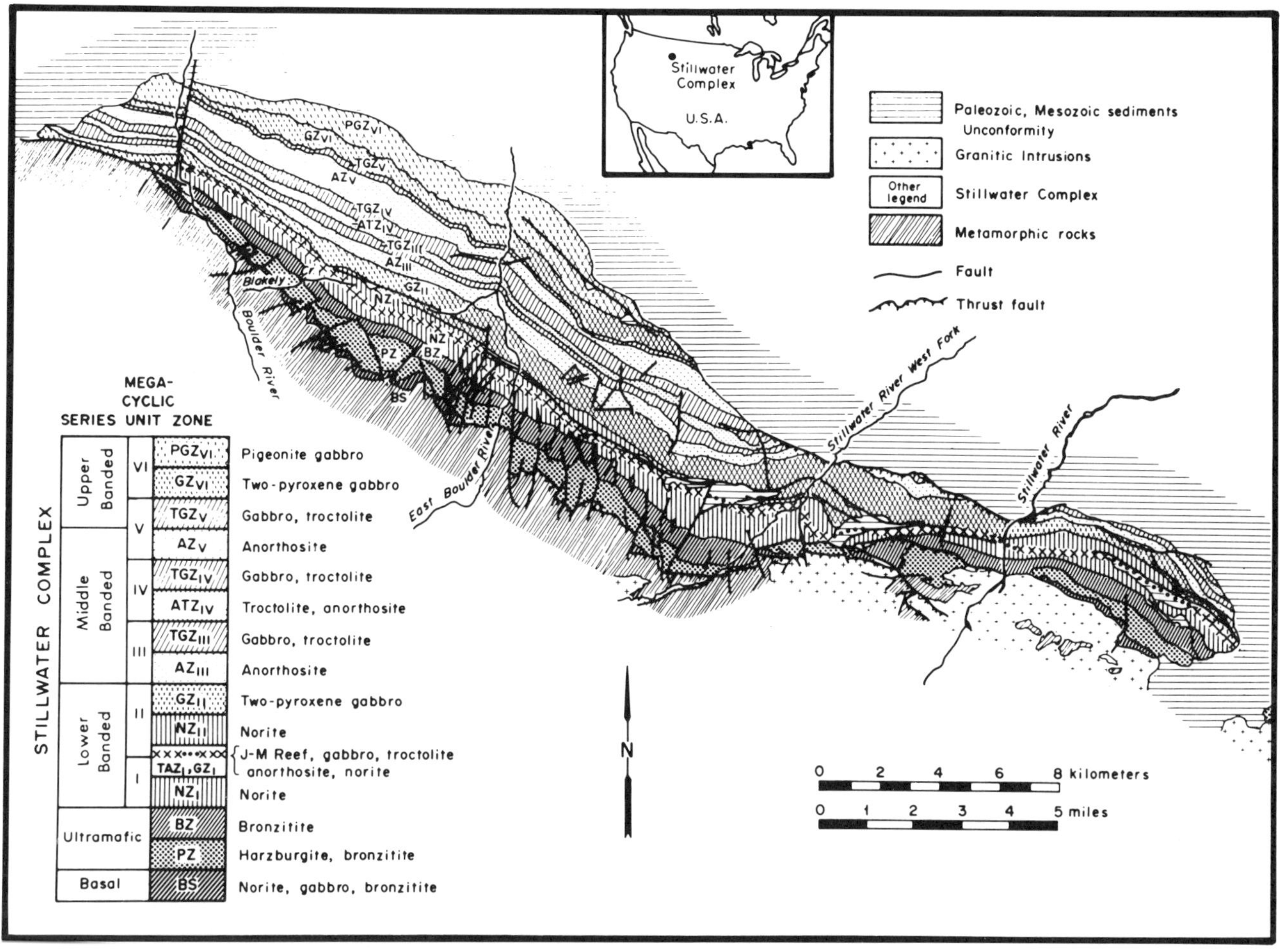


FIGURE 8.9—The geology of the Stillwater complex (from Page et al., 1985a).

(Cl + OH + F) ratios of between 0.45 and 1.0 in comparison with all other non-cumulus and cumulus apatites from the Bushveld, Stillwater, Skaergaard, Kiglapait and Great Dyke intrusions, which have Cl/(Cl + OH + F) ratios <0.2 (Fig. 8.14). Boudreau et al. (1986) also report that the Cl, F, and OH contents of phlogopites and amphiboles from all of these intrusions are more variable, but that Stillwater and Bushveld phlogopites are richer in Cl in comparison with those of the Skaergaard and Kiglapait intrusions. These observations are in agreement with those of Johan and Watkinson (1985) who found up to 1.0 wt% Cl and 1.7 wt% F in mica just above the Merensky Reef.

Boudreau et al. (1986) have argued that, in view of the lower solubility of Cl with respect to F in silicate melts, and the fact that melts with high Cl/F ratios are unknown, their data indicate that the apatites of the PGE-rich zones have equilibrated with Cl-enriched fluids that separated during the solidification of the cumulate sequence.

## Origin of Merensky Pegmatoid

The origin of the pegmatoid that hosts the sulfides of the Merensky Reef in many areas has given rise to considerable debate. Irvine et al. (1983) proposed that it was the "product of postcumulus magmatic replacement of a peridotite layer during upward infiltration of intercumulus liquid relating to the compaction of the cumulate pile". Campbell et al. (1983), pointing to the orthocumulate nature of the base of the bronzitite of the Merensky unit (as is illustrated by the strongly zoned plagioclase that they contain documented in Fig. 8.5), suggested that volatiles, liberated as intercumulus liquid trapped in the underlying crystal pile crystallized, would rise. Because orthocumulate layers were rich in low melting point constituents, they remained partially liquid to much lower temperatures than the remainder of the crystal pile, with the result that the ascending volatiles could not pass across these layers without dissolving in the interstitial liquid. Any halogens carried within the fluids would dissolve within the liquid of these layers, thus enriching them in Cl, as emphasized above. Campbell et al. (1983) suggested that the volatiles promoted recrystallization within the layers. Naldrett et al. (1986a) compared analyses of adjacent pegmatoidal and non-pegmatoidal bronzitite from the Merensky unit at the Western Platinum mine and showed that they had similar compositions. This observation is con-

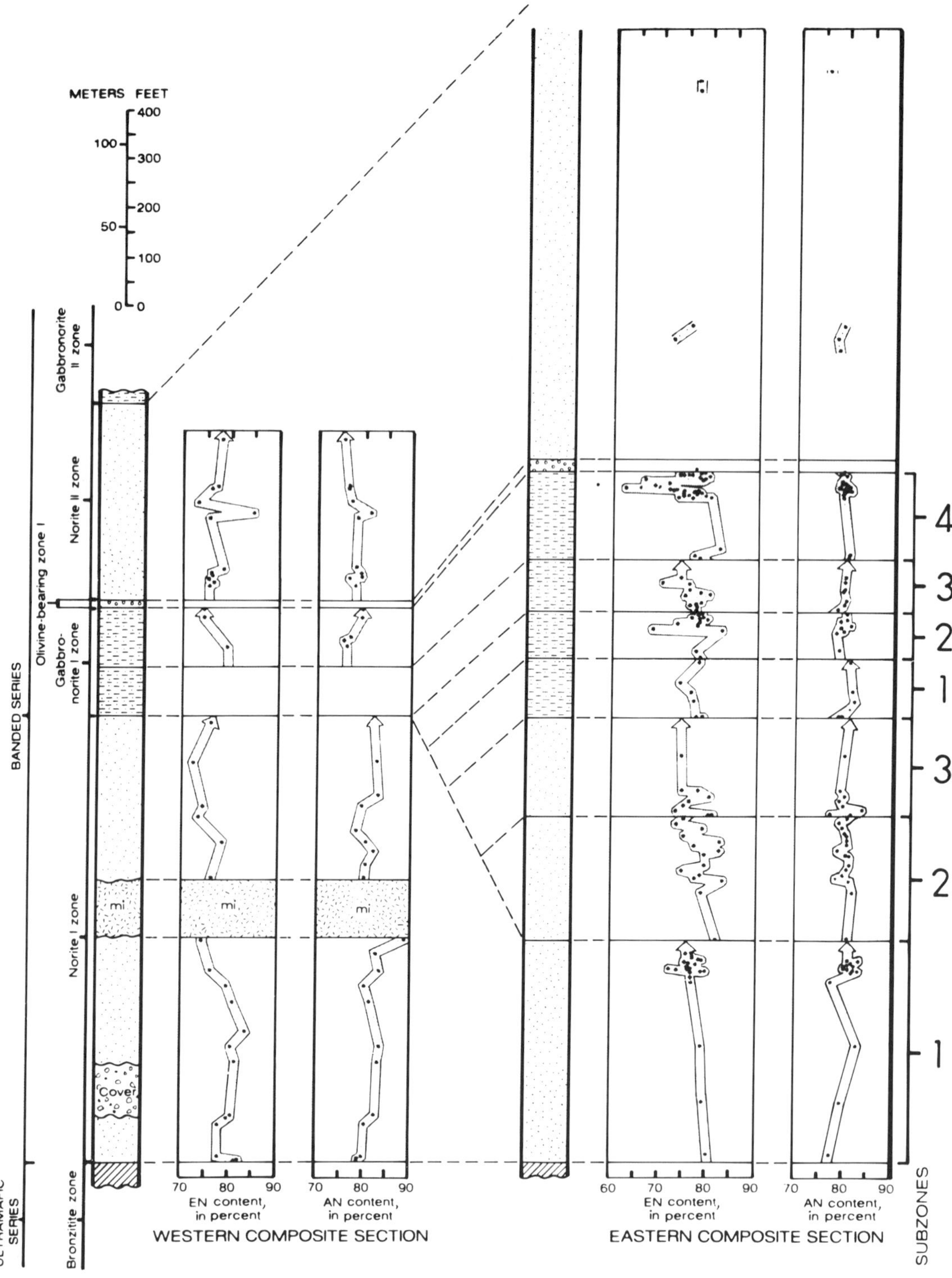


FIGURE 8.10—Variation in the En content of bronzite and An content of plagioclase from the Bronzite zone to the GN–II zone of the Lower Banded series of the Stillwater Complex. mi = intrusive rock. (from Page et al., 1985a).

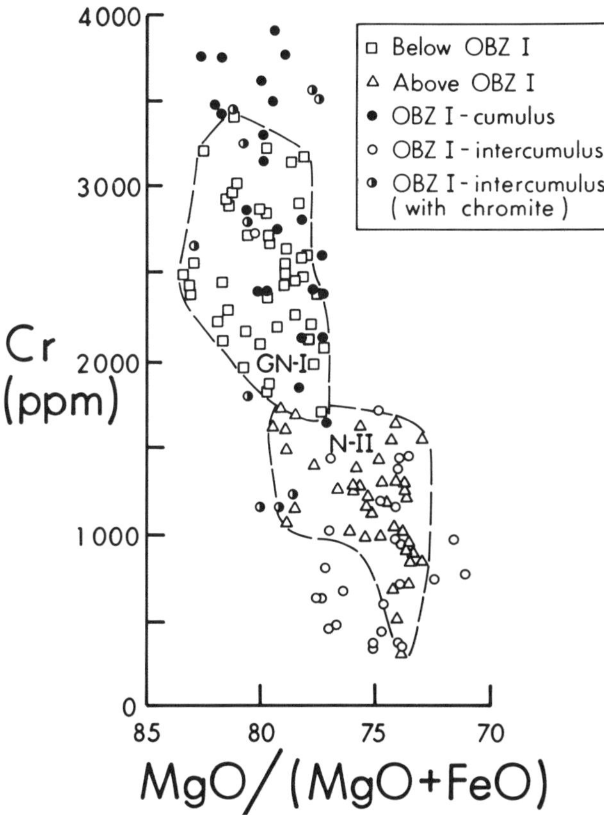


FIGURE 8.11—Cr vs molar MgO/(MgO + FeO) in orthopyroxene in units above and below the J–M Reef of the Stillwater Complex. Data are plotted as analyses of individual grains (single spots for cumulus grains and averages of 2–4 points for large oikocrysts). From Barnes and Naldrett (1986).

sistent with the recrystallization hypothesis, but inconsistent with that calling for the infiltration of expelled intercumulus liquid.

#### Potholes and dimpling

The base of the Merensky Reef is marked by "dimpling" or "scalloping" of the underlying norite or anorthosite, with individual "dimples" having a relief of 5–10 cm and a spacing of about 1 m. The base of the Reef is also the horizon at which a series of local unconformities occur, with the Merensky unit cutting down up to several 10's of m into the underlying strata, in many cases to re-establish itself at certain well-defined horizons within these strata. These unconformable features are known as potholes (Viljoen et al, 1986a and b; Viljoen and Hieber, 1986; Lieb du Toit, 1986).

The walls of potholes may have gentle to steep to overturned dips, with the basal bronzitite of the Merensky unit cutting across layering and lamination in the underlying rocks with little or no disturbance to this layering or lamination. Some potholes are circular, some are elliptical, and many coalesce to produce irregular structures. Diameters vary from a few 10's of m to over 1 km. The Merensky Reef commonly thins abruptly at a pothole margin, is absent on steeply dipping walls, and re-occurs, in some cases in a less-regular form, on the pothole floor. The Merensky bronzitite is continuous and thickens somewhat into a pothole, as do the overlying strata, so that the pothole dies out upwards. In the Union area, much of the Merensky Reef lies on a lower unit within the footwall than is normal, and the Reef is referred to as "pothole reef".

Pegmatitic rocks tend to be particularly common in the vicinity of potholes. Buntin et al. (1985) describe lens-shaped dykes of pegmatitic gabbro, many of which are only a few 10's of m in extent, which are closely associated with potholes. They also report much higher concentrations of native carbon in "pothole" than in normal Reef, and that preliminary measurements of intrinsic oxygen fugacity indicate a low redox "halo" around potholes.

The origin of potholes has been debated for 30 years by geologists at the mines, and is controversial. Most of the hypotheses that have been advanced have called for (i) scouring by magmatic currents, (ii) melting as a result of local increases in the concentration of volatiles in interstitial liquid of the cumulates or overlying magma, or increases in temperature in the overlying magma, or (iii) a combination of both (i) and (ii), and (iv) disruption of the cumulate pile by the upward streaming of intercumulus liquids along circular conduits of increased permeability. The dimpling has been attributed to resorbtion of the cumulate pile by hot magma (Irvine et al., 1983), but Scoon and de Klerk (1987) point out that the upper contact of harzburgite layers are sometimes scalloped and attribute this to a postcumulus process.

#### Composition of the sulfides

Naldrett et al. (1986a) have studied the bulk composition of sulfides in the Reef itself, in the overlying Merensky Unit, and in the flanking Pseudoreef and Bastard Reef and their associated units. The Ni/Cu and Pt/Cu ratios of their samples are illustrated in Fig. 8.15, augmented by data from Lee (1983). Considering Naldrett et al.'s data for the Merensky first, Ni/Cu and Pt/Cu ratios of the sulfides disseminated throughout the rocks of the unit are appreciably less than those of the Reef (in the case of Pt/Cu ratios, a factor of 40 to 13 less). Lee's (1983) data for Pt/Cu ratios show similar differences, although his data for Ni/Cu ratios do not coincide with those of Naldrett et al.

Most of the sulfides in the Pseudoreef and Bastard units have Pt/Cu ratios and, where they have been determined, Ni/Cu ratios very similar to those in the Merensky unit. Both the Bastard and Pseudoreef units have patches of rock enriched in disseminated sulfides close to their basal contacts. Sulfide-enriched zones occur within the Pseudoreef unit in the Pseudoreef itself at the base of the unit and in a harzburgite immediately overlying this that is known as the 'tarentaal'; the zones are characterized by sulfides with very much higher Pt/Cu ratios than the rest of the unit. The same is true but to a much lesser extent of the patches at the base of the Bastard Reef.

The difference in metal ratios is also reflected in the Pt tenor (Pt content in 100% sulfides) of the sulfides. For example sulfides contain about 20 ppm Pt in the Merensky unit, in comparison with 250 to 600 ppm in the sulfides of the

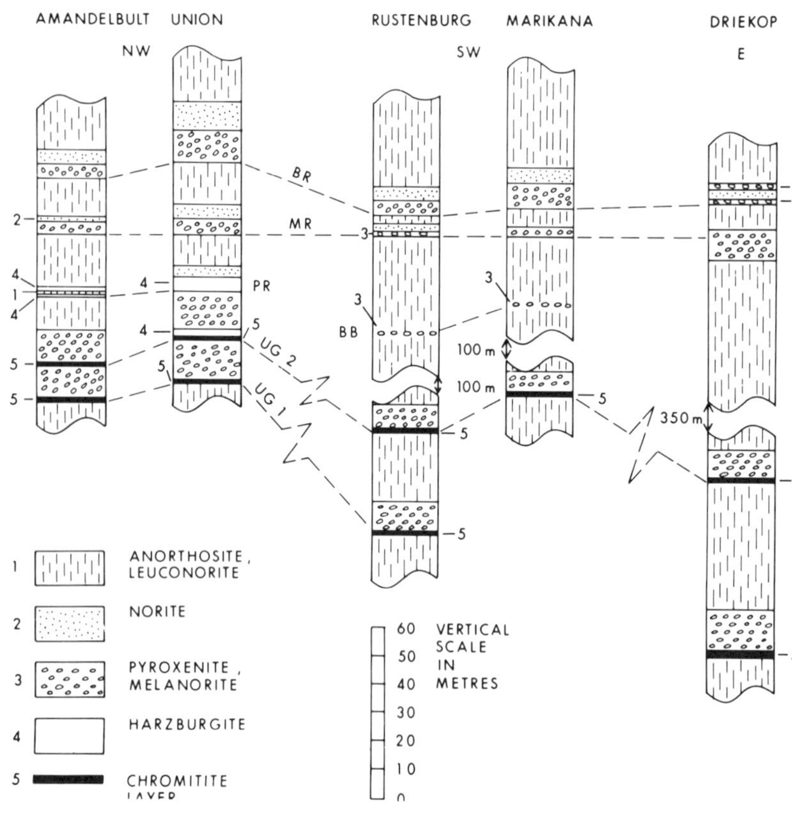


FIGURE 8.12—Cyclic units forming the upper part of the Upper Critical zone of the Bushveld Complex. (from Naldrett, 1981. Data from Vermaak, 1976; Farquhar, 1985; and Gain, 1980) BR, MR, PR and BB refer to Bastard Reef, Merensky Reef, Pseudoreef and Boulder Bed respectively. From Naldrett et al. (1987).

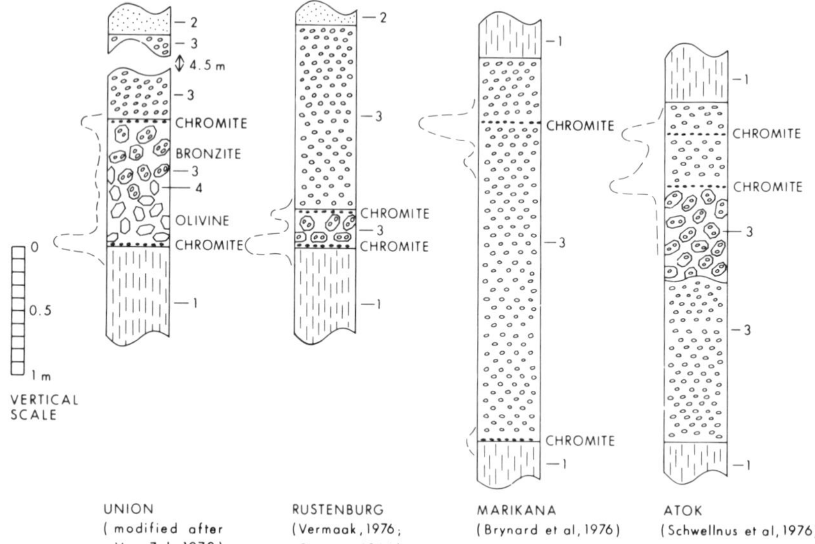


FIGURE 8.13—Variations in the relationship between PGE-rich zones (shown schematically by the dashed lines to the left of each column) and the pegmatoid (olivine shown as 6-sided symbols; bronzite as 8-sided symbols that contain 2 open circles), underlying anorthosite and chromitite, and overlying bronzitite layers in the vicinity of the Merensky Reef in different areas of the Bushveld Complex. PGE values are shown schematically to the left of each section. Vertical shading = anorthosite; open circles = bronzitite, dotted pattern = norite. From Naldrett (1981b).

Merensky Reef itself. Sulfides in the Bastard unit contain 3 to 20 ppm Pt (one highly anomalous value of 50 ppm has been ignored here) in contrast to 15 to 25 ppm in the sulfides of the sulfide-rich patches at the base of the Bastard pyroxenite. Thus the differences in Ni/Cu and Pt/Cu ratios between sulfides in the Reef and the overlying Merensky unit and between the basal patches of sulfide and sulfides in the overlying rocks of the other two units represent differences in the Ni and Pt tenor of the sulfides. In the case of the Merensky Reef, the upper chromitite layer marks a major break in sulfide composition, with sulfides that are extremely enriched in Pt below and sulfides with a high but more

CUMULUS:
△ Skaergaard (Nash, 1976; Brown & Peckett, 1977)
○ Kiglapait (Huntington, 1979)
◊ Bushveld (Grobler & Whitfield, 1970; Boudreau *et al.*, 1986)
NONCUMULUS:
▲ Skaergaard (Nash, 1976; Brown & Peckett, 1977)
◑ Kiglapait (Huntington, 1979)
◘ Stillwater AN II (Boudreau *et al.*, 1986)
▼ Great Dyke Pt Zone (Boudreau *et al.*, 1986)
Cl-RICH:
■ Stillwater OB I (Boudreau *et al.*, 1986)
♦ Bushveld Critical Zone (Boudreau *et al.*, 1986)

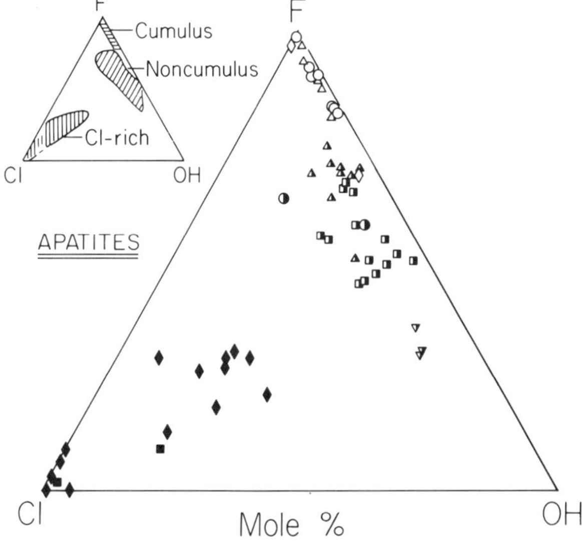


FIGURE 8.14—Data on the OH, Cl and F content of apatite from layered intrusions. From Boudreau et al. (1986).

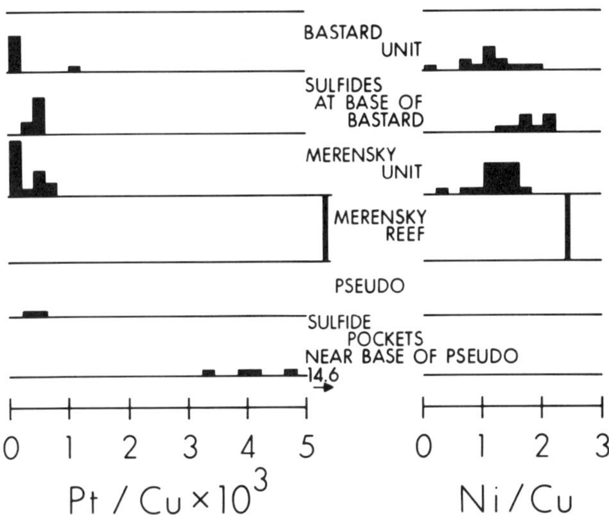


FIGURE 8.15—Variations in Ni/Cu and Pt/Cu ratios in sulfides from the Pseudoreef, Merensky and Bastard reefs and respective cyclic units (the data shown as solid symbols are from Naldrett et al., 1986; those shown with shading, and also the Pt/Cu ratios lying to the right of the bar for the Merensky Reef are from Lee, 1983). Modified after Naldrett et al. (1986).

normal tenor above. Naldrett et al. (1986a) stress that the change occurs over a very few cm; samples 1–2 cm below the chromitite layer have high tenor, those 2–10 cm above have much lower tenor. Any genetic model for the reef must account for these features.

### Stillwater Complex: The J–M Reef

As described above, where olivine cumulate members occur within the OB–1 zone, these commonly are marked by a sharp base and then grade upward through poC to pC. The J–M Reef is a horizon within the thickest of the olivine-bearing members, O–5B, that is relatively enriched in sulfide.

Details of the mineralized Reef (the O–5B member) are shown for the Frog Pond, West Fork and Minneapolis areas in Fig. 8.16. The O–5B member at Frog Pond is similar to that at West Fork. The lower portion consists of about 1 m of olivine cumulate (20–40 modal% olivine at Frog Pond, 50–70% at West Fork) enclosed in large bronzite oikocrysts, together with interstitial plagioclase. Minor amounts of Cl-rich phlogopite, apatite and chromite are also present. At Frog Pond this is overlain by poC in which olivine is irregularly distributed in clusters. At West Fork, the basal oC is both over- and underlain by 15–30 cm of poC and this is overlain in turn by pC containing some olivine in places. Because of its alteration, this pC unit consists of 3 differently coloured bands. The pC is coarse grained, with large augite and smaller bronzite oikocrysts, and rare wispy olivine. It is overlain by poC, with small amounts of olivine as discrete grains and as much as 50 modal% postcumulus augite and bronzite.

The commonest sulfides are chalcopyrite, pyrrhotite, pentlandite and pyrite, which generally occur as blebs or as an interstitial network in the lower part of the Reef, and as a fine dissemination towards the top. At both Frog Pond and West Fork, mineralization is concentrated towards the top of the olivine-bearing zone. The highest grades tend to occur towards the top of the poC (at West Fork within 0.6 to 1.0 m of the top). At West Fork, the upper contact of the mineralization coincides with the contact of the poC with the overlying pC, but at Frog Pond it straddles this contact. The overall grade averages about 19 gm/tonne (Pd + Pt) over 1.9 m, with an average Pd/Pt ratio of 3.6. Mann and Lin (1985) report the presence of coarse sulfide segregations in the basal oC layer, but note that many of these contain only traces of Pt and Pd, although discontinuous areas of high grade mineralization have been observed at this horizon in a few places.

The situation is different in the area of the Minneapolis adit (Turner et al., 1985). The ore-bearing member, which is presumed to be O–5B, lies directly on footwall rocks. In an idealized succession, olivine-bearing pegmatoidal rocks are overlain by poC in which olivine decreases upwards as the rock grades into an oikocrystic "mottled anorthosite", and thence into norite of the N–II zone. The idealized succes-

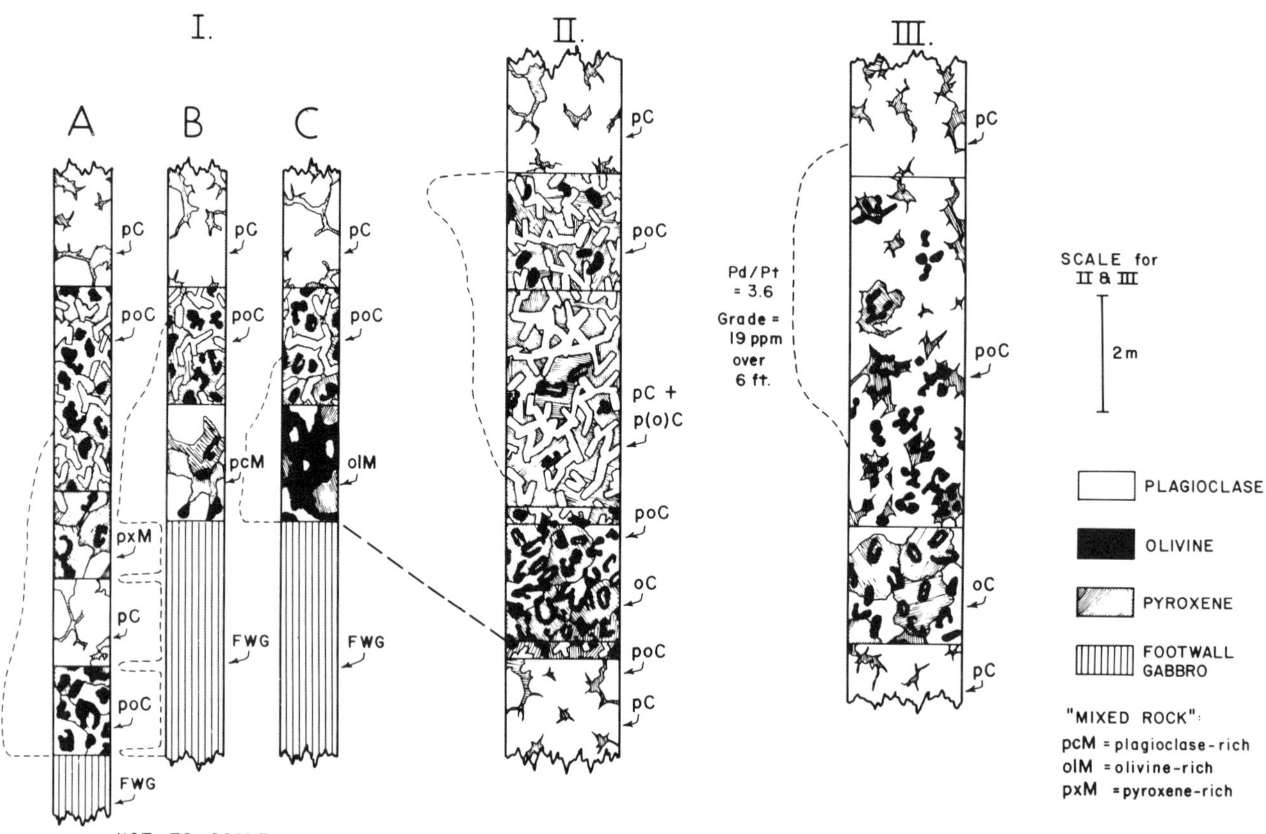


FIGURE 8.16—Variation in the J–M Reef (the O–5B member) from east to west. I = Minneapolis adit – A = within pothole, B = pothole margin, C = away from pothole; II = West Fork adit; III = Frog Pond adit. The dashed lines indicate schematically the distribution of PGE within the succession. Data from Turner et al. (1985); Mann et al. (1985); Leroy (1985). Drawing by B. Murck. From Naldrett et al. (1987).

sion is rarely seen. A common rock-type is that known as "mixed rock". This consists of a matrix of medium to coarse-grained cumulate plagioclase with local intercumulus bronzite and augite forming a net texture, in which there is 20–40 modal% of a coarse ameboidal olivine. Irregular pods and lenses of pegmatitic oC, poC and pC are present. The proportion of olivine in mixed rock is variable, with bronzite oikocrysts increasing at the expense of olivine in the proximity of potholes. In Fig. 8.16 this is illustrated by the olivine mixed rock (olM) giving way to mafic mineral-poor mixed rock (pcM) on the pothole margin to pyroxene mixed rock (pxM) within the pothole.

The poC also contains ameboidal olivine, which commonly shows resorbtion and is jacketed by bronzite.

Raedeke and Vian (1986) describe zones of higher grade mineralisation as having very distinct outlines and as occurring at four different stratigraphic levels (Fig. 8.17). Footwall mineralisation is best developed at the western end of the explored area and occurs as clotty to finely disseminated sulfides in narrow (10–30 cm) anorthosite and norite layers within GN–I. Basal zone mineralisation straddles the GN–I–OB–I contact. Within OB–I, most of the higher grade mineralisation occurs as elongate or crescent-shaped masses, 3–80 m in lateral dimension and spread 60–90 m apart, at 2 or more stratigraphic levels. Only the lower zones are classed as Main zone mineralisation; higher zones are referred to as Upper zone ore.

The base of OB–I is broadly planar, although depressions and elevated areas relative to this plane are present. Although Turner at al (1985) believed that much mineralisation was concentrated around some of the depressions, which they equated with potholes in the Merensky Reef. Raedeke and Vian (1986), with the benefit of a more detailed exploration of the present mine area, believe that no well-defined correlation has been established. It can be seen from Fig. 8.17 that the higher grade areas defined on each of the four different levels of mineralisation tend not to overlie one another, but to be mutually exclusive.

In general, sulfides constitute much less than 2 modal% of the rock, and occur as masses up to 6 mm in diameter. In their work on the Minneapolis mineralization, Barnes and Naldrett (1985) described spherical bodies of sulfide within olivine grains, and cuspate masses of sulfide molded about olivine. They interpreted these textures as indicating that

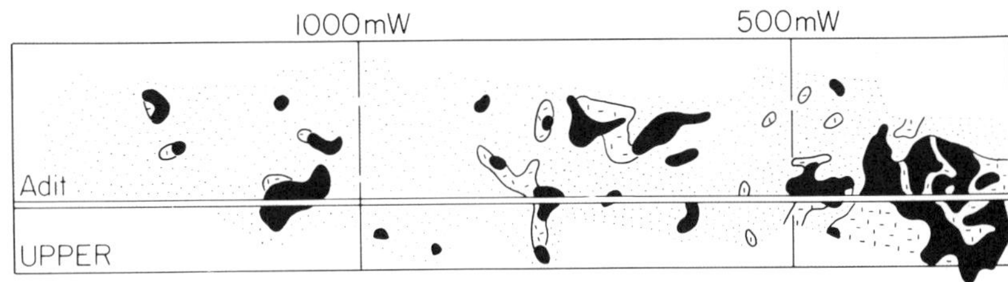

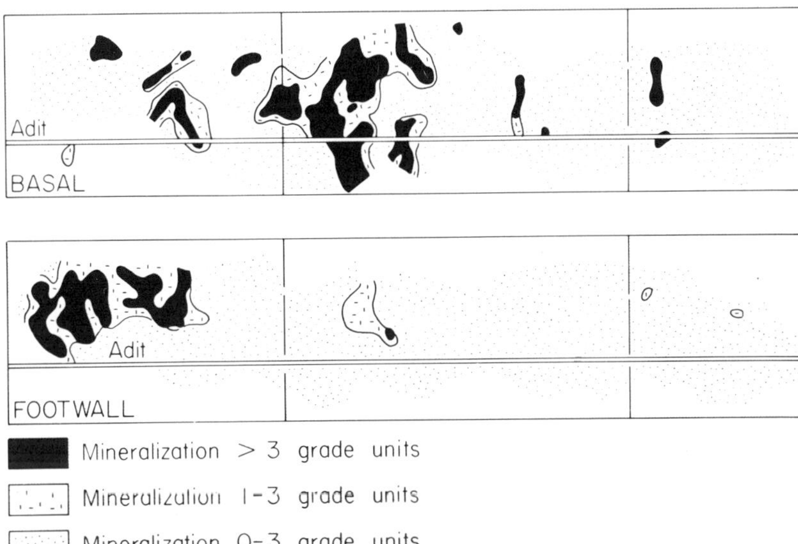


FIGURE 8.17—Distribution of mineralization at various levels within the J–M Reef of the Stillwater Complex. After Raedeke and Vian (1986)

the sulfides were present at the magmatic stage. In some cases, cuspate sulfides are entirely enclosed in oikocrystic bronzite, which also contains remnants of resorbed olivine, indicating that the the resorbtion occurred after the sulfide liquid had solidified (i.e. below 1100°C). In mixed rock, sulfides occur both in olivine-rich "boulders" and within the matrix.

Barnes and Naldrett analysed 40 samples of ore representative of sulfides in oC, mixed rock and pbC, and found a strong correlation between PGE content and percent sulfur (Fig. 8.18), supporting the concept that sulfides have acted as a collector, and emphasizing the homogeneity of the ore with respect to PGE tenor.

Boudreau et al. (1986) have found that apatite from the OB–1 zone, as in the Merensky Reef, has molar Cl/(Cl+F+OH) ratios ranging from 1 to 0.1, greater than for the Upper Zone of the Bushveld, Skaergaard and Kiglapait Complexes.

## DISCUSSION

### Constraints on genetic models

#### Merensky Reef

Any model of origin for the Merensky Reef must account for the following features:

1) The presence of the Reef within a sequence of clearly defined cyclic units.
2) Cryptic variation within these cyclic units that cannot be explained by simple fractional crystallization.
3) The progression in the cyclic units from bC to pbC to pC, which involves crossing the bronzite–plagioclase cotectic, and the fact that norites of cotectic composition are missing from many units.
4) The stratabound nature of the Reef and constant content and proportions of PGE over distances of 100s of km.

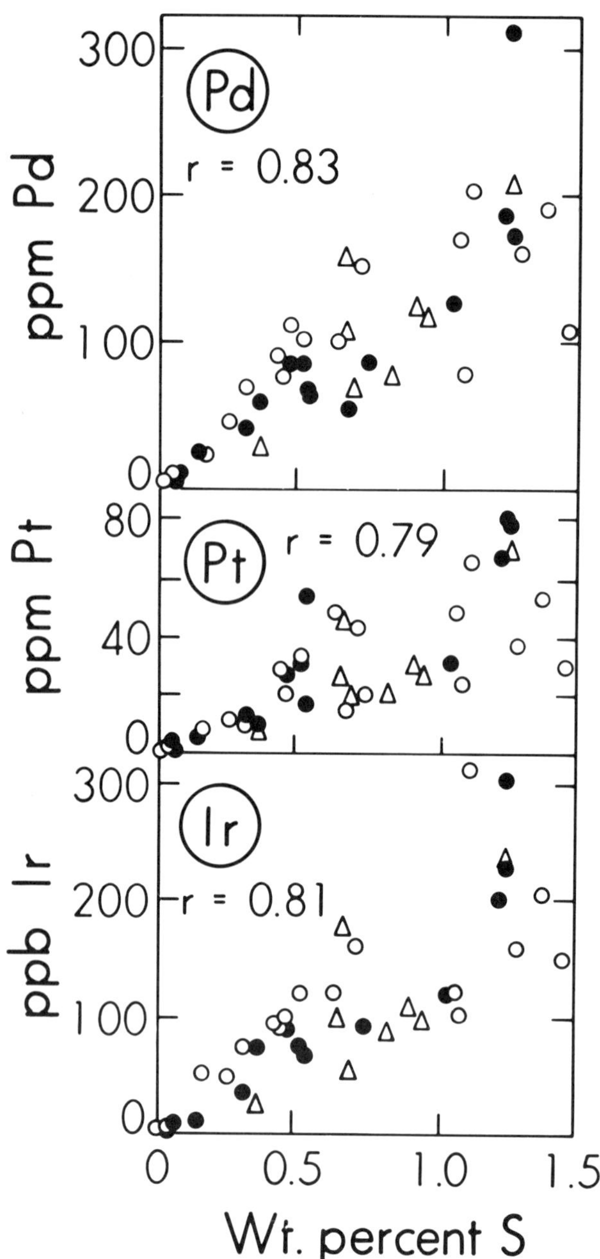


FIGURE 8.18—Pd, Pt and Ir vs S for the J–M Reef samples. Solid dots = sulfides in oC; open circles = sulfide in mixed rock; triangles = sulfides in pbC. From Barnes and Naldrett (1985).

5) The presence of potholes which constitute an exception to the generalization of (4) above.
6) The observation that bronzites in the Merensky pegmatoid are more magnesian than those immediately overlying in the Merensky unit.
7) The presence of the Reef at the base of the Merensky unit.
8) The high Ni/Cu, PGE/Cu and PGE tenor of the sulfides of the Reef in comparison with those in the unit, only a few cm above the top chromitite defining the Reef.
9) The presence of fine chromitite layers at the base and top of the Reef.
10) The presence of graphite within the Reef, particularly in the vicinity of potholes.
11) The high proportion of biotite, apatite and other hydrous minerals that characterize the Reef in comparison with rocks bordering the Reef; and the presence of high Cl concentrations within some of these minerals.
12) The fluid inclusion data discussed by Mathez elswhere in this volume.

**J–M Reef**

The J–M Reef resembles the Merensky Reef in that:
1) It occurs a few 100 m above the level at which plagioclase first appears as a cumulate phase in the intrusion.
2) It is associated with one of a number of cyclic units comprising the OB–1 zone.
3) The Reef is characterized by a higher proportion of hydrous minerals than the surrounding rocks.
4) The presence of graphite in the vicinity of the ore.
5) The presence of high concentrations of halogens in biotite and apatite.
6) Sulfides are the principal collector of the PGE.

In many ways, however, it is very different:
1) The cyclic units are smaller and show no clear trends of cryptic variation.
2) There is no continuous development of pegmatoid.
3) The PGE tenor of the sulfides is 10 times that of the Merensky Reef and of the order of the present tenor of the UG–2.
4) The PGE-rich zone is much less well-constrained by fine details in the stratigraphy of the host rocks than the Merensk Reef, occurring in poC, oC and in some cases in the overlying pC.

**Layered magma chambers**

Following the experiments and/or discussions of Turner and Gustafson (1978), Irvine (1980), Turner (1980), Irvine et al. (1983) and Campbell et al. (1983), it is widely accepted that magma chambers are density stratified into discrete layers.

Exchange of both heat and chemical constituents will take place by diffusion across boundaries between layers, and layers will convect in response to differential buoyancy effects induced by this diffusion. Tank experiments show that if new magma is less dense than resident magma at the base of a chamber, and if it is injected into the chamber through a feeder at its base, it will rise through the denser layers to find its own density level and spread out at this level to form a new convecting layer.

A number of lines of evidence show that both the Bushveld and Stillwater Complexes have received magma inputs of distinctly different composition. The time-transgressive change in composition in the marginal rocks and marginal sills described by Sharpe (1981, 1982) and Harmer and Sharpe (1985) is evidence of a change in the composition of input to the Bushveld. The upward change in Sr isotopic composition (Hamilton, 1977; Kruger and Marsh, 1982; Sharpe,

1985) indicates that the inputs were isotopically distinct as well as distinct in major and trace element composition. Todd et al. (1982) argue that the change in the order of crystallization in the Stillwater Complex, from olivine, bronzite, plagioclase, augite to plagioclase, olivine, augite, bronzite requires a change in the dominant magma type controlling the crystallization, and thus, in all probability, the input of new magma. These inputs will immediatley form discrete, density- and composition-distinct layers, which, if they are compositionally zoned within themselves, will soon break further up into a series of thinner, homogeneous, density stratified layers. Clearly, successively higher layers must be less dense than those beneath them. Furthermore, in a large layered complex in which hundreds or thousands of meters of cumulates underlie tha main body of magma the bulk of the heat is lost through the roof. For this to occur, successive layers must also decrease in temperature upwards. This then is the likely setting in which the Merensky and other neighbouring cyclic units of the Bushveld Complex and the cyclic units of the OB–I zone of the Stillwater Complex, together with their associated ores, must have formed.

### Models for the origin of PGE-rich reefs

One of the key questions with regard to deposits of this type is why the sulfides within them are so highly enriched in PGE. Answers that have been suggested in the literature for the high PGE tenor of the sulfides are summarized in Fig. 8.19. The stratabound nature of the ore has led most people to accept that the sulfides themselves were deposited at the same time as the rocks now enclosing them. There is a divergence of views when it comes to the timing of the deposition of the PGE.

Looking at cartoon (A), Vermaak (1976) and von Gruenewaldt (1979) suggested that highly fractionated, and therefore PGE-enriched, magmatic liquids were filter-pressed out of the cumulate pile and forced upwards as this compacted. On moving upwards, they reacted with and enriched the PGE content of sulfides that were already present at the level of the Merensky Reef.

Cartoons (B) and (C) are based on the observations of fluid inclusions, chloride-rich minerals and hydrated minerals in the Merensky and J–M Reefs referred to above, and illustrate hypotheses calling for the upward transport of PGE by chloride-rich hydrothermal fluids, as suggested by Ballhaus and Stumfl (1985), Johan and Watkinson (1985), Boudreau (1988) and others.

Following the concept of density stratification within magma chambers that has been discussed above, Campbell et al. (1983) suggested the model illustrated in cartoon (D), which is the one favoured by this author and which is discussed in detail below. First, however, some of the other hypotheses will be discussed.

The models described in Fig. 8.19 can be divided into those in which the bulk of the PGE are derived from below

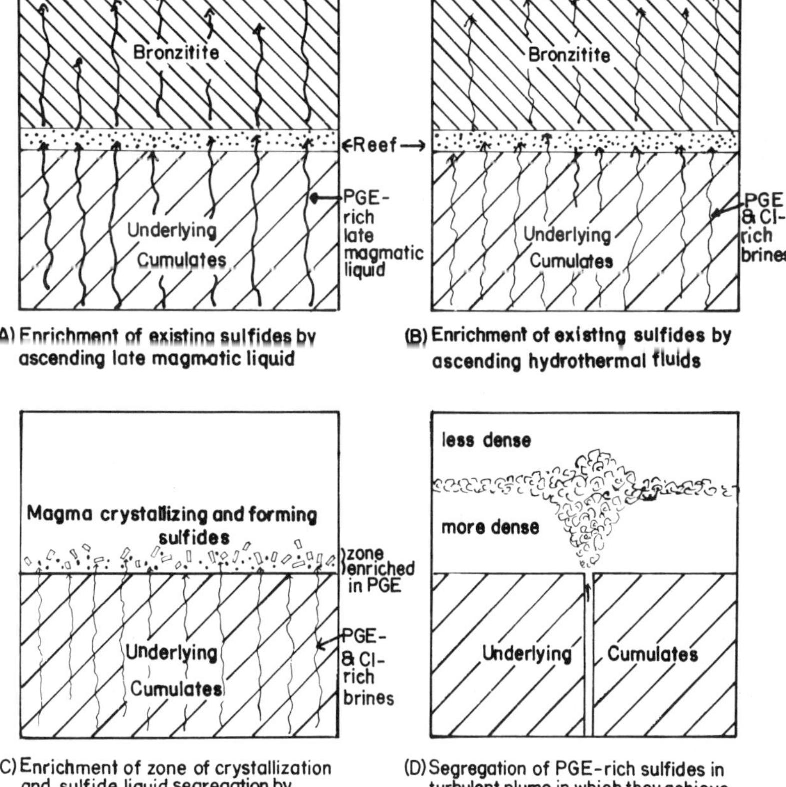


FIGURE 8.19—Cartoon illustrating four suggested mechanisms for the concentration of PGE within the Merensky and J–M Reefs. 1 = suggestion after von Gruenewaldt, 1979; 2 = suggestion after Ballhaus and Stumpfl, 1985; Johan and Watkinson, 1985; 3 = suggestion after Boudreau (1988); 4 = suggestion after Campbell et al., 1983, Naldrett et al., 1986.

(A to C), and those in which they are derived from the overlying magma (D), or, in contempory terminology, into "UPPERS" and "DOWNERS".

In the view of this author, there are a number of observations that are inconsistent with "UPPER" models. Considering first that illustrated in Cartoon A, a rather generous estimate of the maximum amount of pore liquid that could have been trapped in the Bushveld cumulates would be 20% (Most of the cumulates are adcumulates as opposed to orthocumulates). If one assumes that this liquid was trapped containing 15 ppb Pt, as indicated by Davies and Tredoux's (1985) data on non-cumulate marginal and sill rocks (15ppb is the value exceeded by 50% of their data on the $B_1$ type marginal rocks which are the most PGE-rich they encountered), the 5 ppm present over the 1 m mining width of the Merensky Reef would require all of the Pt contained in 1667 m of the underlying cumulate to be deposited within the Reef. Thus the ascending intercumulus liquids must have originated below the Pseudoreef and the UG–2. If one takes the view that, in some manner not presently understood, all of the Pt in the magma that originally occupied the space of the underlying cumulates has been retained in the cumulus pile and is then available to enrich the Merensky Reef, one still requires over 300 m of underlying cumulates to supply sufficient Pt for the Merensky Reef, which means that the liquids must still have originated below the UG–1 chromitite. On moving upwards, the liquid would have encountered sulfides of the Pseudoreef and 'tarentaal', followed by those of the remainder of the Pseudoreef unit, then sulfides within the Merensky Reef, followed by those of the Merensky unit, then sulfides at the base of the Bastard unit, and finally those of the remainder of the Bastard. Under these circumstances, the ascending fractionated magmatic liquids would have progressively relinquished their contained PGE, giving rise to sulfides of highest tenor in the Pseudoreef and 'tarentaal' and progressively lower tenor upwards. This is not what is seen in Fig. 8.15. The cyclical variation in Pt/Cu (proportional to Pt tenor), from high values at the base of a unit, to lower values within the unit, and then back to high values at the base of the overlying unit, cannot be accounted for on the basis of ascending magmatic liquids.

Turning to other "Upper" type hypotheses involving hydrothermal transport of PGE, similar arguments apply. It is very difficult to envisage how hydrothermal fluids could have traversed rocks containing sulfides without also enriching those sulfides in PGE. It could be argued that the composition of sulfides in equilibrium with a hydrothermal fluid is a function of the composition of the fluid, which is itself a function of the rock type that it happens to be traversing. Thus the sulfides in a feldspathic bronzitite might be expected to have a different composition to those in a non-feldspathic peridotite, if their compositions were controlled by reaction with a hydrothermal fluid. However the data presented above indicate that sulfide composition is independant of rock type, and that high and low PGE-tenor sulfides occur in feldspathic bronzitites.

A further difficulty with "Upper" type models in the Bushveld Complex is that one cannot consider the Merensky Reef in isolation, but must also consider the PGE content of all other PGE-rich concentrations, including those in the Middle and Upper Group chromitites (which are discussed in more detail in the next section). Taking a conservative view, the UG–2 contains about 3 ppm Pt over 1 m, the UG–1 about 1.3 ppm over 1 m, the UG–3 about 2 ppm over 1 m. The Middle Group taken together contain about 1 ppm over 8 m. With the Merensky Reef, this amounts to 19 ppm meters. Given 20% trapped liquid in cumulates, over 6 km of underlying cumulates would be required to provide this much Pt, which is 3 times the thickness of the cumulates beneath the Merensky Reef throughout most of the Bushveld. Even if, as discussed before, all of the Pt in the magma originally occupying the position of the underlying cumulates had been scavenged and concentrated in the Pt-enriched zones cited above, and none was left in the cumulates, except within these zones, one would require the Pt in 1.3 km of magma. It is known that all of the Pt has not been removed from the cumulates. High concentrations occur in the Pseudoreef, the Lower Group chromitites contain concentrations in the 100's of ppb range, and Lee and Tredoux (1986) have shown that most of the cumulates below the Merensky Reef contain between 10 and 30 ppb. Thus this author is led to conclude that a major mass balance problem exists for "Upper" hypotheses, and that these theories are inconsistent with what we know believe to be the likely Pt content of the initial Bushveld magma.

One of the principal arguments of those who support "UPPER" models is the evidence indicating the former presence of abundant volatile-rich fluid within the Merensky pegmatoid. It has been argued above (see discussion of the origin of the Merensky pegmatoid) that the orthocumulate nature of the bronzitites at the base of this unit will cause them to be an effective trap for ascending volatiles. As the concentration of volatiles built up, they caused recrystallization of early forming minerals, giving rise to pegmatoid and the observed high proportions of hydrous minerals. For the reasons that have already been outlined, it is unlikely that these volatiles introduced significant amounts of PGE.

### A model for the Merensky Reef

The model presented here is based on that proposed by Campbell et al. (1983) and involves the attainment of a high magma/sulfide ratio by those sulfides forming the ore zone.

#### Magma/sulfide ratio

When a very small amount of sulfide melt segregates from a silicate magma, the concentration of any metal $i$ in the sulfide melt ($Y_i$) is related to the initial concentration in the silicate magma ($C_i$) by the partition coefficient $D_i^{Sul/Sil}$ according to the expression:

$$Y_i = D_i^{Sul/Sil} \times C_i \quad [1]$$

which is merely a rearrangement of expression [iv] in Chapter 2 and where $Y_i$ is the wt% of metal $i$ in the sulfide melt and $C_i$ the wt% of $i$ in the silicate melt. Provided that equilibrium exists between sulfide melt and silicate magma, an expression of this kind always holds, provided that $C_i$ is taken to be the concentration in the silicate magma after the attainment of equilibrium.

It is often more useful to be able to model the composition of the sulfide melt in terms of the initial composition of the silicate magma (i.e., before the segregation of, or reaction with, sulfide occurs). Where the ratio of silicate magma to sulfide melt is very large, expression [1] provides a satisfactory answer. However, as this ratio decreases, a stage is reached where the sulfide has concentrated so much of the metal present in the whole system that it causes a significant drop in the concentraion of this metal in the silicate magma with which it is equilibrating. Campbell and Naldrett (1979) have shown that under these circumstances, it is necessary to use the more general expression:

$$Y_i = \frac{D_i \times C_{0i} \times (R + 1)}{(R + d_i)} \quad [2]$$

where $R$ is the ratio of the mass of silicate magma to the mass of sulfide, and $C_{0i}$ refers to the initial concentration of metal $i$ equilibrating with it. Fig. 8.20 illustrates schematically the effect that variations in $R$ and $D$ have on the Ni, Co and Pt contents of a magma such as that responsible for the Bushveld Complex. Where $R$ is low, in the range of 100 to 2000, the Ni and Co contents of the sulfides will be typical of most Ni sulfide ores, and the Pt concentrations will be relatively low, corresponding to those observed in ores such as those at Sudbury. Where $R$ is in the range of 10,000 to 100,000 the Ni and Co contents will not be much higher than at lower $R$ values, but the Pt concentration will be much higher and in the range of those characterising the Merensky Reef. The key question is how are high $R$ values attained.

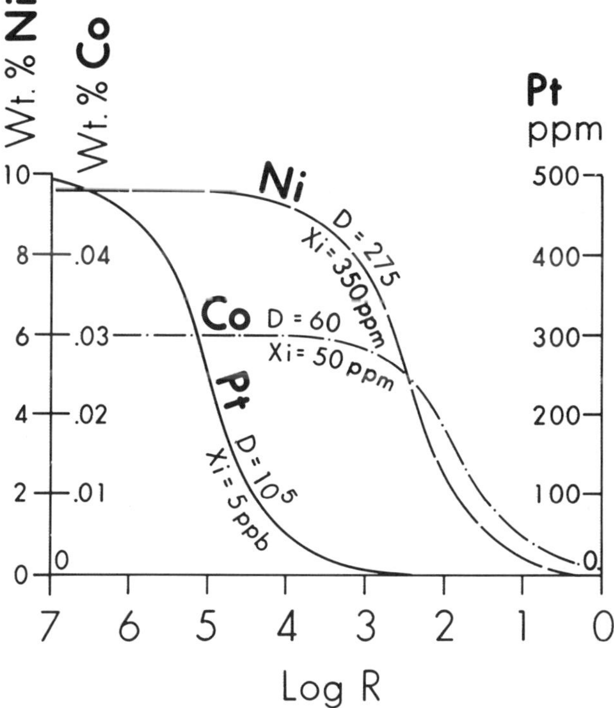


FIGURE 8.20—The effect of variations in silicate magma/sulfide liquid mass ratio on the concentrations of Ni, Co and Pt of sulfides in equilibrium with basaltic magma containing typical concentrations of these elements.

### Details of the model

Cawthorne and Davis (1983) have investigated experimentally the fractional crystallization at 3 kb total pressure of a high-MgO, high-$SiO_2$ liquid that they believe is representative of the parent magma of the lower part of the Bushveld Complex. Barnes and Naldrett (1986) have modelled the compositional changes accompanying this fractionation, and have used the data of Bottinga et al. (1982) to calculate the densities appropriate to the different compositions. Their results are shown in Fig. 8.21. It is seen that even a relatively siliceous magma such as that responsible for the lower part of the Bushveld will achieve a density greater than its initial density once it starts to crystallize plagioclase. A fresh pulse of initial magma ($B_0$ in Fig. 8.22A) will therefore rise through the fractionated old magma (N) to seek its own density level. Sparks et al. (1980) and Campbell et al. (1983) have shown that it will rise as a turbulent plume, entraining surrounding old magma into itself and thus becoming hybridized to a new composition ($B_1$). Drawing on the work of Irvine (1975) and Irvine et al. (1983), Campbell et al. (1983) pointed out that if both magmas were close to saturation with sulfide, the mixture would be likely to exceed saturation, so that immiscible sulfide droplets would form. These droplets would be swirled around in the plume, and subsequently in the convecting layer produced by the hot new magma (Fig. 8.22B).

Naldrett et al. (1986) proposed that the rapid heat loss at the top of the new convecting layer to the overlying, much colder magma would cause the new layer to convect very rapidly and thus turbulently at first. As it lost heat, crystals of bronzite or olivine would form, but so long as convection was turbulent, these, plus any sulfide droplets present, would be kept in suspension. Eventually, as the new layer cooled and the rate of heat loss decreased, convection would become laminar, and the suspended crystals, plus sulfides, plus some entrained liquid would sink through the underlying magma. Drawing on the work of Huppert et al. (1984), they suggested that the mixture would sink as a series of 'downspouts' to spread out over the crystal pile (Fig. 8.22C) as a discrete orthocumulate layer that crystallized as the Merensky Reef.

Since sulfide droplets forming in this environment would have been swirled around within the plume and then in the convecting layer, they would have achieved a high $R$ factor, and thus have developed a high PGE tenor. These sulfides would have reached equilibrium with the relatively primitive hybrid magma produced by the mixing of the new influx and entrained old magma; thus they have a high Ni/Cu ratio. For the same reason the bronzites that have formed at this stage and are part of the Reef have relatively high MgNo.

Once it is capped by hot new magma, the resident magma in contact with the cumulate pile will lose no more heat, and thus all crystallization from it will cease. Irvine (1977) (see detailed discussion below) showed that the mixing of two magmas, one more advanced along its crystallization path than the other, will cause chromite to appear imme-

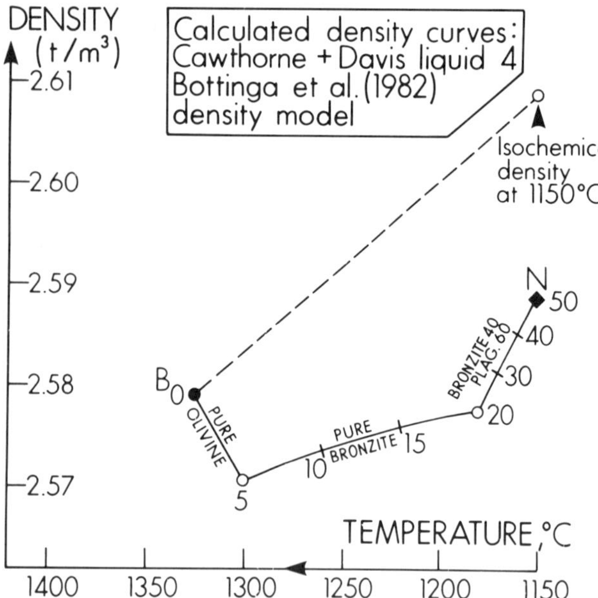


FIGURE 8.21—Density variations in fractionating magma. The variations are based on calculated liquids obtained by the computer-simulated fractional crystallization of Cawthorn and Davies's (1983) best representative (a sample of chilled marginal material) for the magma responsible for the Lower zone of the Bushveld Complex. The simulation was constrained by the sequence and temperatures of phase appearance and disappearance determined by them on this liquid. (modified from Barnes and Naldrett, 1986).

diately on the liquidus while the crystallization of other liquidus phases will not occur until further cooling has taken place. Naldrett et al. (1986) suggested that, prior to 'downspouting', small amounts of chemical diffusion between the overlying layer of new, relatively primitive magma and the underlying more fractionated resident magma caused chromite saturation close to the contact between the layers and gave rise to a gentle rain of crystals which accumulated to form the bottom chromitite layer of the Merensky Reef.

At some stage, perhaps at the time that the downspouting of the crystal and sulfide-rich mixture occurred, perhaps somewhat later, cooling of the new layer raised its density so that it mixed with the underlying layer of resident magma to form a second, less primitive hybrid ($B_2$ in Fig. 8.22D). This then crystallized bronzite, to form the bronzitite of the Merensky unit. The minor amounts of sulfide that separated during this stage of crystallization in response to continued crystallization of silicates would have had a Ni/Cu ratio reflecting the less primitive nature of this hybrid in comparison with that responsible for the Reef itself, and would not have achieved the high magma/sulfide ratio necessary for high PGE tenors. Similarly, the MgNo of the bronzite reflects the less primitive composition of the hybrid.

The second stage of hybridization may also have caused chromite saturation and thus have given rise to the top chromitite layer before the newly formed hybrid as a whole had cooled to its liquidus and had started to crystallize bronzite.

During these events, the whole system had been cooling and the immediately overlying layer ($A_2$ in Fig. 8.22D) could have reached its liquidus. Since it had plagioclase as its first liquidus phase, this started to crystallize. Modelling of the liquid densities indicates (Fig. 8.21) that these are relatively low (<2.60), so that plagioclase (with a density of about 2.68) would have tended to sink into the underlying B-hybrid, so enriching the resulting cumulate in plagioclase. Since the lower layer was likely to have been convecting, the upper part of this would have been somewhat above its liquidus temperature. Some of the settling plagioclase would thus have become resorbed in this part, changing the composition of this layer so that it lay just on the plagioclase side of the plagioclase–bronzite cotectic and thus did not nucleate bronzite. A transition of this kind from a liquid crystallizing bronzite to one crystallizing plagioclase will be marked in the cumulate succession by non-cotectic norites, thus accounting for the observed transitional norites.

Crystallization of plagioclase from the overlying A-type liquid would have raised its density relative to the underlying B-type liquid (Fig. 8.21). A zone of finger-mixing (Irvine et al., 1983) would have developed between the two. If the A-type liquid had a lower MgO/(FeO + MgO) than the B-type, mixing of the two would influence the MgNo of the resulting bronzite in a way unrelated to fractional crystallization, thus accounting for the discrepancies reported by Naldrett et al. (1987) in their attempts to model cryptic variation in bronzite in the bronzitite and the norite using a fractionation model. If the A-type magma had had a lower Sr content than that responsible for plagioclase in underlying cyclic units, a gradual decrease in the Sr content of the resulting cumulate would be expected. This is what is seen in the western Bushveld, but not in the east where the Sr content of plagioclase does not change until the Bastard Reef. This difference is explicable if the A-type liquid overlying the new input that initiated the Merensky cyclic had a different Sr content in the east to that in the west. Perhaps the A-type liquids were layered, as shown in Fig. 8.22, and that in the west it was $A_2$ that had the lower Sr content, and in the east it was $A_1$.

Similarly, if the A-type magma had had a more radiogenic Sr composition, its increasing contribution to the cumulates of a given cyclic unit would have resulted in these cumulates becoming progressively more radiogenic, as observed by Kruger and Marsh (1982).

Once the $B_2$ and A-type liquids had mixed to give rise to a new dominantly A-type hybrid, if crystallization had proceeded uninterupted for some time, the hybrid would have moved from the plagioclase field to the plagioclase–bronzite cotectic, and a cotectic norite would have formed. On the other hand, if an influx of new magma had entered the chamber first, the sequence of events just described above would have been repeated without the intervention of cotectic norite.

This model accounts for many of the features of the Merensky Reef, including: (1) the occurrence of the Reef within a series of cyclic units; (2) the difficulty of accounting for cryptic variation in the Merensky unit with a fractionation model; (3) the apparent crossing of the plagioclase–bronzite cotectic within a given unit; (4) the stratiform nature of the Reef; (6) the compositional difference between the bronzites

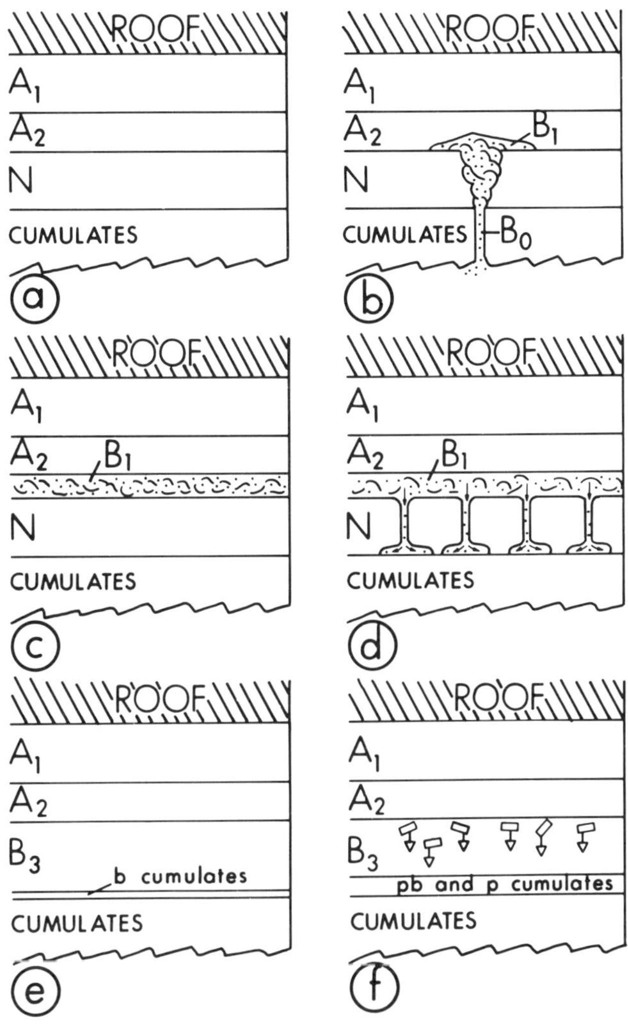


FIGURE 8.22—A proposed model for the formation of the Merensky and other cyclic units, showing the formation of the Merensky Reef. A – liquid N is crystallizing pbC or pC, and is A overlain by two liquid layers ($A_1$ and $A_2$, each with plagioclase alone on the liquidus); B – an injection of B fresh, relatively ultramafic magma ($B_0$) occurs as a turbulent plume, and entrains sufficient N into itself to form liquid $B_1$, which has bronzite on the liquidus, and at the same time segregates a small amount of sulfide; C – $B_1$ C spreads out at its own density level and, because it is hot in comparison with the overlying layers of magma, loses heat to them very rapidly and convects vigorously (turbulently); as $B_1$ cools, it crystallizes bronzite which, together with the sulfide, is kept in suspension by the turbulent convection; D – as $B_1$ loses heat, the convection becomes D less vigorous and eventually laminar rather than turbulent; at this stage bronzite and sulfides settle to break through the underlying layer (N) as a series of downspouts of crystals + sulfides + entrained liquid, to spread out over the cumulate pile as an orthocumulate (which subsequently crystallizes to form the Merensky Reef); E – continued cooling of $B_1$ results in its density exceeding that of N, whereupon the two liquids finger-mix to form hybrid $B_2$; this then gives rise to the bronzitite of the Merensky cyclic unit; F – cooling of the whole system results in liquid $A_2$ arriving at its liquidus, at which stage plagioclase separates and sinks into $B_2$, to join the cumulate pile and give rise to a transitional norite; crystallization of this plagioclase raises the density of $A_2$, which reaches that of $B_2$, and the two then finger-mix to form another hybrid from which the bulk of the anorthosite crystallizes. From Naldrett et al. (1987).

of the Reef and those of the unit immediately above the Reef; (7) the position of the Reef at the base of the cyclic units and (9) the presence of chromitite layers at the top and bottom of the Reef.

Points (10) to (12) relating to volatiles and graphite in the Reef follow from the orthocumulate nature of the bronzite + sulfides + liquid layer that resulted from the "downspouting", and the necessity for ascending volatiles to dissolve in the intercumulus liquid trapped within this, as described above.

Once the resident liquid in contact with the cumulate pile is capped by the layer formed by the hot new influx, it will no longer lose heat, and may even gain heat slightly by conduction across their mutual boundary. All crystallization in this layer will cease and resorbtion may ensue if addition of volatiles from below during this hiatus lowers liquidus temperatures. The resulting breakdown of cement between cumulate grains may result in their easy erosion by and dispersion within overlying magma, thus accounting for potholes. If the ascending volatiles tend to become concentrated in tubular channels of higher permeability, this will also account for the circular shape of many of the potholes.

The Pseudoreef and Bastard units are thought to be due to separate influxes of magma that underwent similar life cycles to that discussed here for the Merensky unit. The Pseudoreef and 'tarentaal' of the Pseudo unit are seen as forming in the same manner as the Merensky Reef, but the proportion of sulfide to cumulate silicate was less, accounting for the lower grade of the mineralization. The Boulder bed at Rustenburg is thought to be a case of 'downspouting' in which the supply of descending material, and the relative viscosities of the descending material and the magma through which it was sinking (Huppert et al., 1984), resulted in the downspouting material breaking up into a series of 'drops', rather like water from a dripping tap.

### Application of the model to the J–M Reef

Barnes and Naldrett (1986) have discussed the formation of the J–M Reef in the light of the model advocated here. They described the gradations in texture (Fig. 8.23) to be expected from the mixing in a plume of olivine + chromite saturated magma with resident A-type, plagioclase-phyric magma. They suggested that liquids dominated by the new input crystallized clumps of olivine and chromite grains, while intermediate mixtures showed partial resorbtion of plagioclase inherited from the resident melt. These plagioclase grains become reversely zoned due to reaction with the hotter liquid, and eventually become trapped within rapidly growing olivines, giving rise to ameboid olivine texture. Further mixing of olivine-bearing liquid with resident

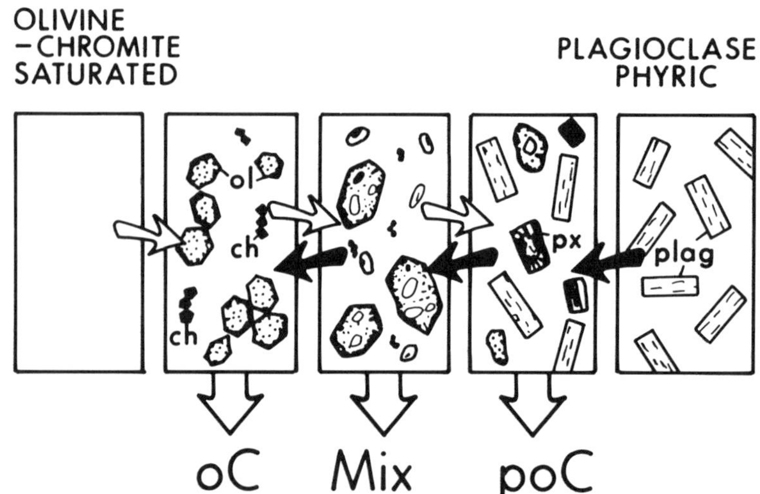


FIGURE 8.23—Cartoon depicting the development of characteristic OB-1 textures during mixing and cooling of an olivine-chromite saturated, plagioclase undersaturated replenishing liquid, and a plagioclase phyric resident melt. Liquids dominated by the replenishing component (left) crystallize clumps of olivine and chromite grains. Intermediate mixtures show partial resorbtion of plagioclase inherited from the resident melt. These plagioclase grains become reversely zoned due to reaction with the hotter liquid, and eventually become trapped within rapidly growing olivines, giving rise to an ameboid olivine texture. Further mixing of an olivine-bearing liquid with resident melt leads to peritectic resorbtion of olivine, and growth of pyroxene as independant grains or jackets on resorbed olivine. This sequence accounts for the observed gradation from olivine cumulates through olivine-rich to pyroxene-rich mixed rock, as observed in the Minneapolis adit. After Barnes and Naldrett (1986).

melt led to peritectic resorbtion of olivine, and growth of pyroxene as independant jackets on resorbed olivine. They pointed out that this hypothetical sequence accounts for the observed gradation from olivine cumulates through olivine-rich to pyroxene-rich mixed rock in the Minneapolis adit.

Barnes and Naldrett (1986) suggested that repeated pulses of fresh primitive basaltic magma were responsible for the different O-members of the OB-I zone. Some of these pulses were relatively small and did not spread throughout the Complex, accounting for the absence of certain members in places. One of the most persistent pulses achieved sulfide saturation, spread over a wide area of the Complex and gave rise to the J-M Reef. They do not invoke "downspouting" but argue that agregates of crystals settled, bringing down with them droplets of sulfide, to mix with the plagioclase cumulates forming from the resident A-type liquid and produce the heterogenous rock types of the J-M Reef at the Minneapolis adit. The greater development of olivine members, and the less heterogenous nature of rock types at West Fork and Frog Pond in comparison with Minneapolis resulted from the closer proximity of a feeder to the first two areas. In an earlier presentation of their ideas, Barnes and Naldrett (1985) stressed the importance of the resident magma being nearly, but not quite, saturated in sulfide in order that the sulfides resulting from the magma mixing might achieve a high PGE tenor. They noted that the cyclic units produced by previous pulses were barren of sulfide, and suggested that any sulfides that might have been associated with these pulses were resorbed into the residual magma, increasing its sulfide content and thus bringing it closer to saturation. The crystallization responsible for these barren cyclic units would have increased the Pt and Pd concentration in the resident magma up to the stage at which sulfides became incorporated in the cumulus pile. The present author notes that the resorbtion of sulfide would also have served to increase the PGE content of the resident magma.

## ASSOCIATION OF PGE WITH CHROMITITE IN LAYERED INTRUSIONS

### Compositional differences between chromitites from ophiolites and those from layered intrusions

The contrast between PGE abundances in chromitites of ophiolites and layered intrusions is illustrated in Fig. 8.24. Ophiolitic chromitite is characterised by Os, Ir and Ru abundances in the range of 0.1 to 1.0 times chondritic and Pt and Pd abundances about 0.01 times chondritic. The Bushveld Middle and Upper Group chromitites are very different with Ir and Ru 0.5 to 1.0 times and Pt and Pd 0.5 to 4 times chondritic. With the exception of the "A" chromitite, the Stillwater chromitites are intermediate between those from ophiolites and those of the Bushveld Upper and Middle Group. The "A" is similar in its concentrations to the Upper and Middle Groups. Typical concentrations in the Merensky and J-M Reefs are also plotted on Fig. 8.24 and they illustrate the very high Pt and Pd and low Ru, Ir and Os that characterise sulfide-dominant ores.

Naldrett et al. (1987) discussed the reasons for the difference between the "A" and other Stillwater chromitites (Fig. 8.25), and showed that if one added 0.06% of the sulfide that characterized the J-M Reef to the "J" chromitite one obtained a reasonable match for the "A" (Fig. 8.26). This led them to suggest that the reason for the "A" being so much richer in Pt and Pd than the others was its original higher sulfide content. It is suggested in this paper that the original abundance of PGE-enriched base metal sulfides is one of the reasons for the difference between chromitites from ophiolites and those from layered intrusions.

It is not suggested that this is the only reason. The origin of podiform chromitites in ophiolites is far from certain, but many authorities believe (Lago et al., 1982; Greenbaum, 1977; Dickey, 1975) that they are the disrupted remnants of masses that have crystallised from mantle magmas. The

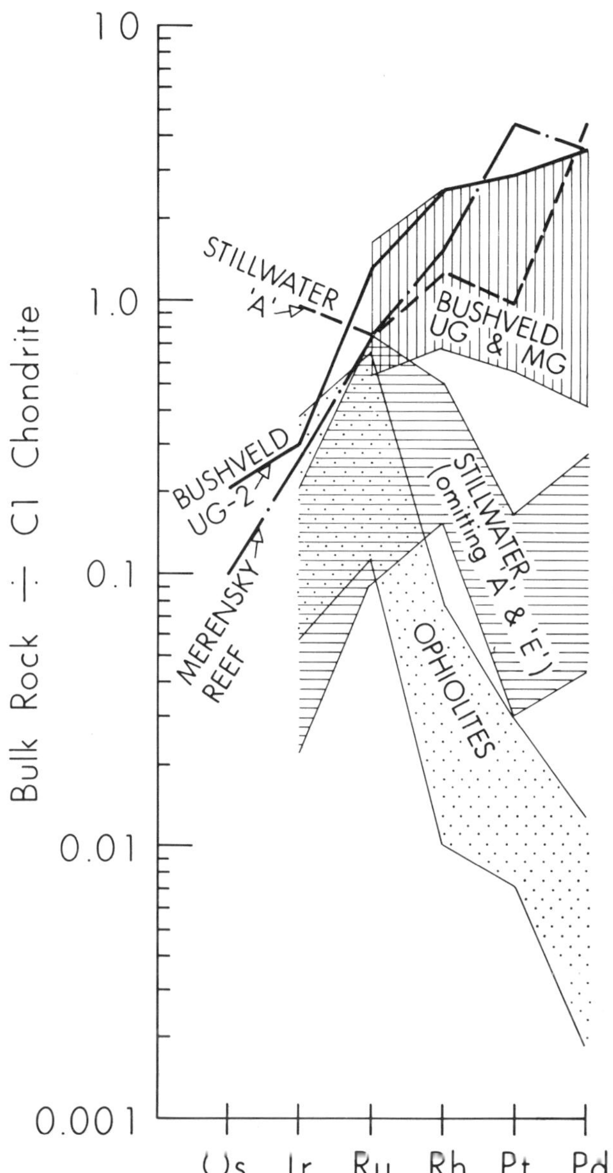


FIGURE 8.24—Chondrite normalized plot of PGE concentrations in bulk rock samples of chromitite zones in the ophiolites of California and Oregon (Page et al., 1986), Newfoundland (Page and Talkington, 1984), the Polar Urals (Page et al., 1983), New Caledonia (Page et al., 1982a), Oman (Page et al., 1982b), the Stillwater Complex (Page, 1985), and the Bushveld Complex (von Gruenewaldt et al., 1986; Gain, 1985). Data for the Merensky Reef (Naldrett and Cabri, 1976) and J–M Reef (Barnes and Naldrett, 1985) are shown for comparison.

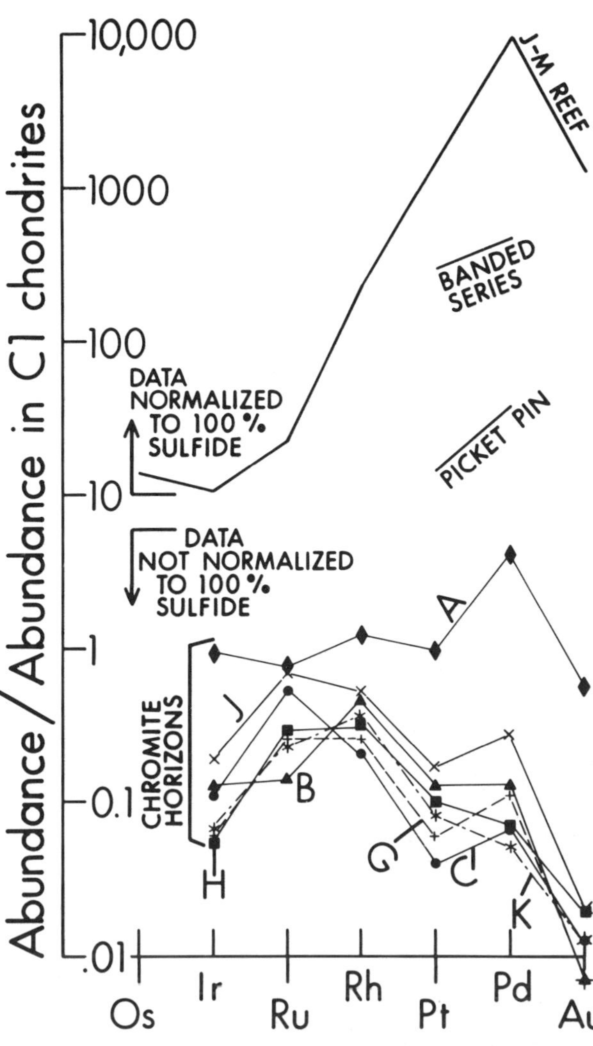


FIGURE 8.25—Chondrite normalized plot of PGE concentrations in bulk rock samples of chromite layers from the Stillwater Complex (data from Page, 1985). Diagram modified after Naldrett et al. (1987).

enclosing rocks are characterised by relatively flat PGE chondrite normalised profiles (Barnes et al., 1985), so that magmas responsible for the chromitites may have had much flatter (possibly nearly chondritic) profiles than those from which the layered intrusions found in the crust have formed. If so, one would expect the chromitites crystallising from these magmas to reflect their PGE profiles to some degree. However, if one takes Davies and Tredoux's (1985) average for the chilled margin of the Upper Critical zone as typical of the Bushveld magma at this stage, one obtains a Pd/Ir ratio of 122, while the average ratio for the UG–2 based on von Gruenewaldt et al.'s (1986) data is 31.9, which means that Pd is 3.8 times less concentrated than Ir in the UG–2 relative to its concentration in the likely source magma. If the source magma for podiform chromitites is assumed to have mantle proportions of PGE, an average of the data compiled by Barnes et al. (1985) for ophiolitic chromitites indicates that in these Pd is 12 times less concentrated than Ir relative to the source. Thus, while a difference in the relative proportions of PGE in the source magma can account for some of the difference between ophiolitic and layered

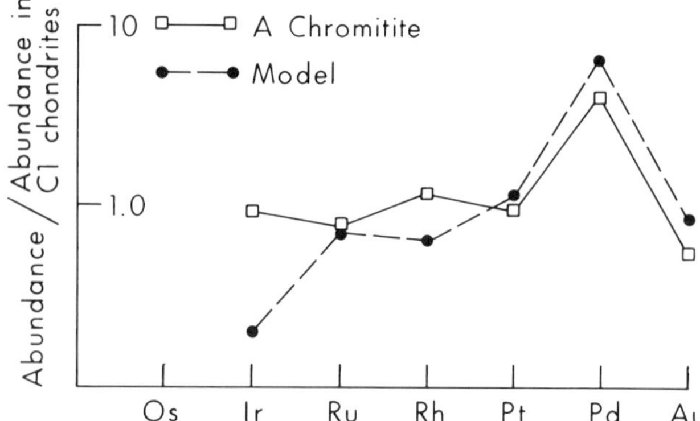


FIGURE 8.26—Chondrite normalized plot of PGE content of the Stillwater "A" chromitite and a model composition obtained by adding 0.06% of J–M Reef sulphide to the Stillwater "J" chromitite. Diagram after Naldrett et al. (1987).

intrusion chromitites, it cannot account for all of the difference, and it is suggested here that the presence of sulfide in the UG–2 (note that the UG–2 is observed to contain about 0.25 wt% sulfide, Gain, 1985) is responsible for the remainder.

### Interaction between sulfide and chromite during cooling

In the above discussion, care has been taken to refer to "original" sulfide concentration, since the composition and abundance of small amounts of sulfides that are trapped within chromitite at high temperature may suffer appreciable modification during cooling. This was first pointed out by von Gruenewaldt et al. (1986), who noted that the Cu content of small amounts of sulfide in the UG–2 was a factor of 2 or 3 higher than was consistent with the direct segregation of sulfide from basaltic magma. Naldrett and Lehmann (1988) investigated the question of non-stoichiometry in chromite; they concluded on the basis of a thermodynamic analysis that the data were consistent with the model that, as chromite cooled, vacancies within its structure took up Fe from any associated base metal sulfide, causing a loss of sulfur to surrounding rocks, and a resultant increase in the Cu, Ni and PGE contents of the remaining base metal sulfide. The limited proportion of vacancies in naturally occurring chromite means that its capacity to take up Fe in this way is severely limited, so that the effect on sulfide composition will only be noticed when the ratio of sulfide to chromite is very low, less than about 200.

Naldrett and Lehmann showed that the removal of half of the mass of the 0.5 wt% sulfide originally contained within the UG–2, with the consequent doubling of the Cu, Ni and PGE concentrations, is feasible thermodynamically. Naldrett et al. (1988) showed that this model can also explain the very high Ni contents of the much smaller proportions of base metal sulfides included within podiform chromitite bodies in ophiolites.

### PGE concentrations in chromites from layered intrusions

The concentration of total PGE in chromitites from the Bushveld and Stillwater Complexes are plotted against stratigraphic height in Fig. 8.27. In some samples, where analyses of Os and/or Ir values are not available, it has been necessary to recalculate the data assuming that Os and/or Ir are present in the same proportion to Ru as they are in the UG–2. Because Ru, Os and Ir are so low in most samples, the uncertainties introduced in this way are very small relative to the total PGE content.

The vertical scale used in Fig. 8.27 is the distance that a given chromitite occurs above or below the level of the first appearance of cumulate plagioclase in the intrusion. Those well above the incoming of cumulate plagioclase have PGE concentrations ranging from 3–7 ppm (gm/tonne). Those well below this level have values generally lower than 1 ppm, except for the Stillwater "A". The Bushveld Middle Group, which occur within 20 m above or below the appearance of cumulate plagioclase contain 2–3 ppm.

It is proposed in this article that the difference in PGE concentrations in the different groups is due to the original concentration of sulfide within them. Those with low total PGE lacked much or any sulfide, those with the highest PGE originally contained the most sulfide.

Because the reaction between chromite and sulfide can lead to a considerable decrease in the actual sulfide content of a sample during cooling, the present content is not a good guide to the original content. A much better guide is the (Pt + Pd)/(Ru + Ir + Os) ratio of the sample. As shown above, sulfides concentrate Pt and Pd effectively and chromitites do not; thus a high (Pt + Pd)/(Ru + Ir + Os) ratio indicates appreciable original sulfide; a low ratio indicates the reverse. Fig. 8.28 shows the ratio plotted against stratigraphic height. Variations of the ratio are very similar to those in the total concentration of PGE, supporting the contention that the present PGE content is a function of the original sulfide content.

### Mixing of a fresh input of magma with that resident in a chamber

As discussed above, Campbell et al. (1983) and Campbell and Turner (1986) have used hydrodynamic theory to model the mixing of an input of fresh primitive silicate magma into a compositional- and density-stratified magma chamber. As already mentioned, the mixing is likely to be turbulent. If it is less dense than the lowest magma layer within the chamber, the fresh input will rise through it as a turbulent plume, entraining and mixing with a high proportion of the resident magma on the way (Fig. 8.29A and B). Another scenario, not yet discussed, is that if it is more dense, the initial momentum of the second magma may carry it up into the overlying magma as a turbulent fountain (Fig. 8.29C), during which stage it can entrain and mix with resident magma, but eventually it will fall back to form a dense layer on top of the cumulate pile. While mixing can occur in either case, momentum is required to produce a fountain, and the amount of mixing will depend on the height of the fountain and thus on the velocity of the input magma. If this is less, less mixing will occur. A plume is therefore the environment in which magma mixing is more certain.

Reference to Fig. 8.21 shows that the density of the resident magma will have increased above that of the original primitive input when the magma in the chamber is well fractionated and plagioclase has been crystallising for some

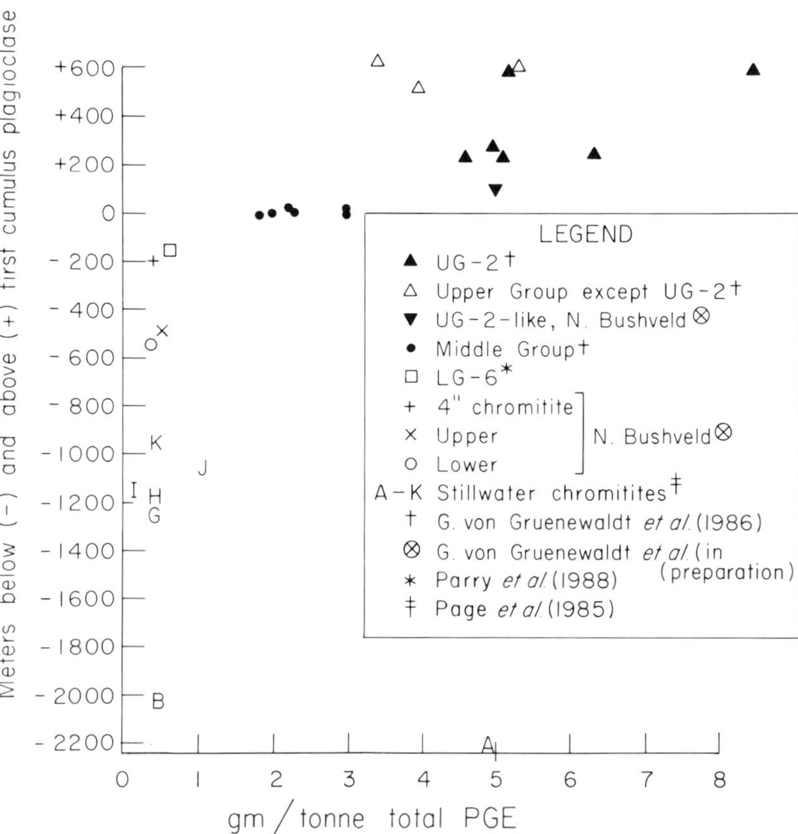


FIGURE 8.27—Plot of gm/tonne (= ppm) total PGE in chromitites from the Bushveld and Stillwater Complexes against height in the stratigraphy. Data are plotted in terms of the vertical height that a given chromitite occurs above and below the first significant appearance of cumulate plagioclase in the intrusion. In some cases where Ir and/or Os values were not available total PGE have been adjusted as described in the text.

time. It is under these circumstances that the plume is most likely to occur. This then will be the situation for the Bushveld Upper Group chromitites, and also for the Merensky and J–M Reefs.

In the case of the Stillwater chromitites and those from the the lower part of the northern limb of the Bushveld Complex (the 4 inch marker, upper chromitite layer and lower chromitite layer), which occur where the magma has been crystallising olivine, or the Bushveld Lower Group chromitites, which occur where it has been crystallising bronzite, the magma in the chamber is likely to have been less dense than the new input, so that a fountain would have developed rather than a plume. The Bushveld Middle Group chromitites occur so close to the level at which plagioclase started to crystallise that fountaining of the new input is likely to have occurred here also.

## Magma mixing and the segregation of sulfide and chromite

As mentioned briefly above, Irvine (1977) showed that the mixing of two magmas, one more fractionated than the other, could inhibit the fractional crystallization of silicate minerals such as olivine and orthopyroxene and permit the crystallization of chromite alone. He proposed that this was the mechanism by which massive chromitite layers could develop, without dilution by cumulate silicates. Fig. 8.30 is based on his diagram; mixing liquid A with liquid D, one on the olivine–chromite cotectic and the other in the orthopyroxene field, gives rise to a hybrid such as AD, which, provided that it lies on or above the liquidus, will crystallize chromite while it moves to point X, at which stage it will continue to crystallize chromite and olivine.

Murck and Campbell (1986) approached chromite crystallisation in a somewhat different way, and showed that the solubility of Cr in basalt magma in equilibrium with chromite decreases more rapidly with falling temperature at high than at low temperatures. Because of the resulting concave upwards curvature of the solubility curve, the mixing of two magmas at different temperatures saturated (or nearly saturated) in chromite places the resultant mixture above the saturation curve.

Campbell et al. (1983) pointed out that if both resident magma and fresh input are saturated or nearly saturated in sulfide, sulfides will form during the mixing. They will be carried in suspension in the magma throughout the tur-

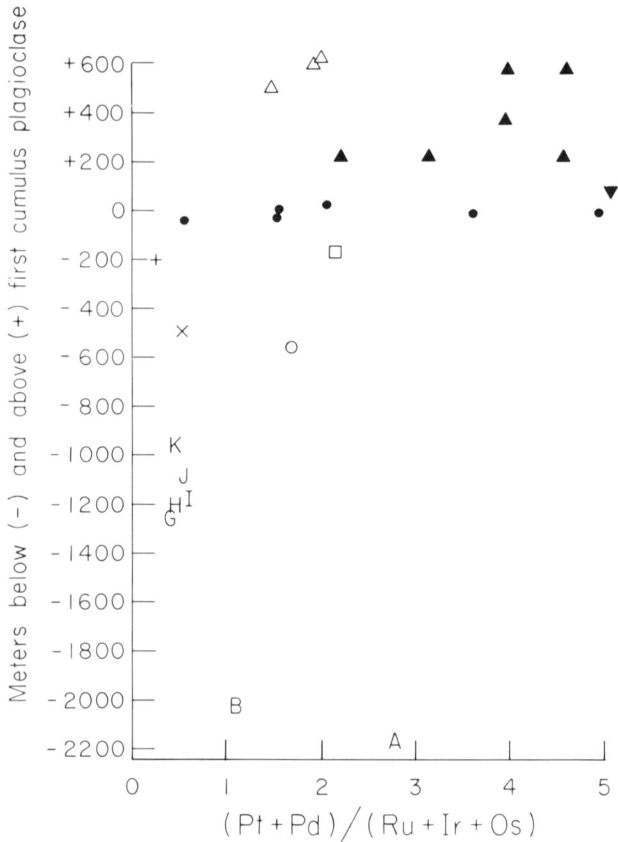


FIGURE 8.28—Plot of (Pt + Pd)/(Ru + Ir + Os) ratio for chromitites against height in the stratigraphy as described for Fig. 8.27. Sources of data are those as given in Fig. 8.27.

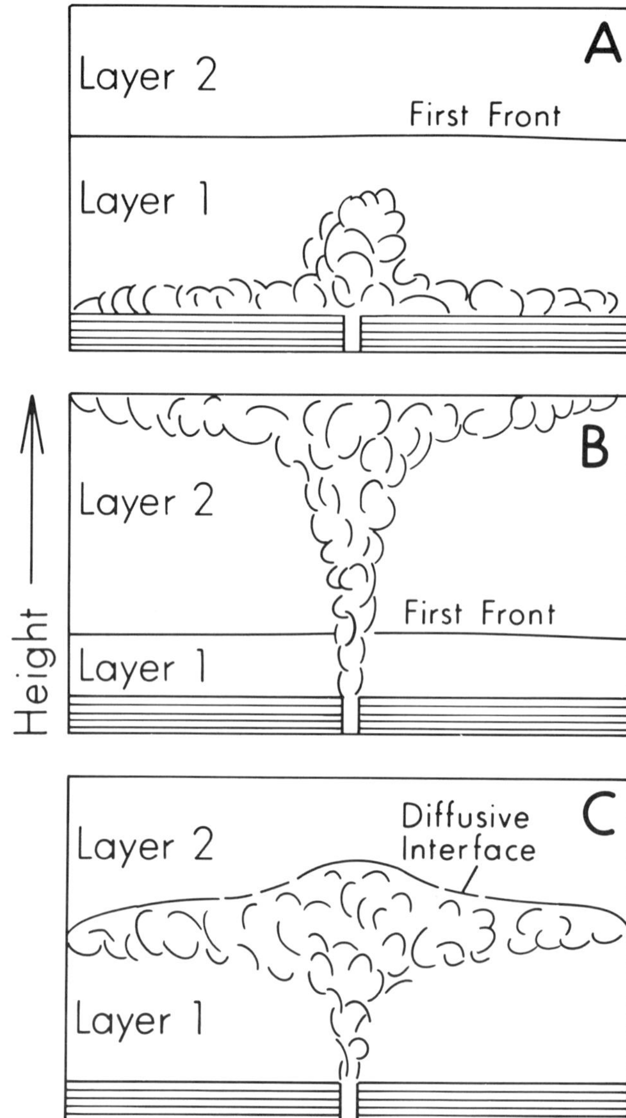


FIGURE 8.29—Behaviour of a new input of magma into a layered intrusion, (A) where input is denser than that overlying the cumulates, (B) where it is less dense than all magma in the chamber and (C) where it is less dense than that immediately overlying the cumulates but more dense than that higher in the chamber. Modified after Campbell and Turner (1986).

bulent stage in the plume or fountain and any subsequent turbulent convection once the new hybrid has spread out as a layer, and thus they will achieve a high *R* factor.

While the style of new inputs may determine whether and to what extent mixing may occur, and thus the *R* factor achieved by any sulfides resulting from this mixing, it is unlikely to be the determining factor as to whether sulfides will segregate or not. The present author suggests that the answer to this question lies in the shape of the solubility curve shown in Fig. 8.31. The derivation of the curve is discussed in Chapter 2.

The figure can be used to propose two slightly different models to explain the precipitaion of PGE-rich sulfides in layered intrusions. If one considers an initial magma input into an intrusion such as the Bushveld Complex corresponding to point A in Fig. 8.31, once this starts to crystallise its sulfide content will rise along the path A to B, reaching saturation after about 8% crystallisation. Further crystallisation will cause the segregation of sulfide and the sulfide still dissolved in the magma will decrease along the solubility curve. The magma will become markedly depleted in PGE during this stage.

Suppose that the magma in the chamber immediately overlying the cumulates has reached point D when a new primitive magma enters the chamber as a turbulent plume. Mixing in the plume will produce a hybrid such as AD. Fig. 8.31 indicates that this will either be saturated in sulfide or, with slight cooling of the hybrid towards its own liquidus (the mixture formed from two silicate liquids at their liquidus temperature can be above its own liquidus temperature), will become saturated without further silicate crystallisation. Because of the turbulent environment in which they segregated, the sulfides will achieve a high *R* value. The PGE

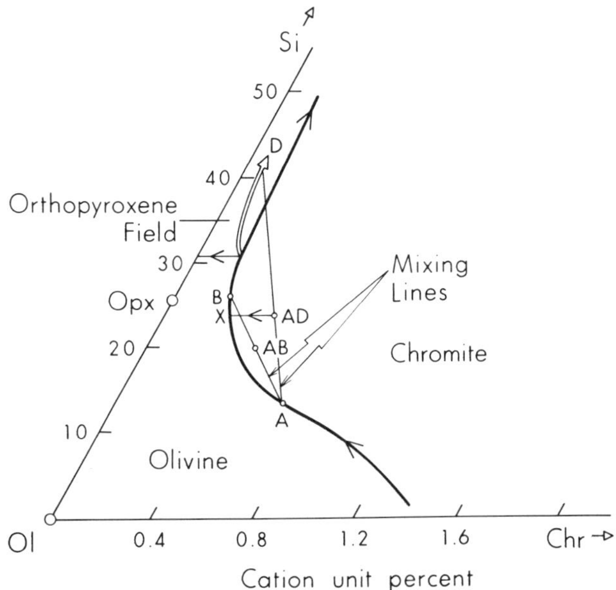


**FIGURE 8.30—Phase relations in the system olivine–silica–chromite as determined by Irvine (1977) and illustrating the consequence of mixing primitive magma (A) with well fractionated (D) and slightly fractionated (B) variants of the same primitive magma.**

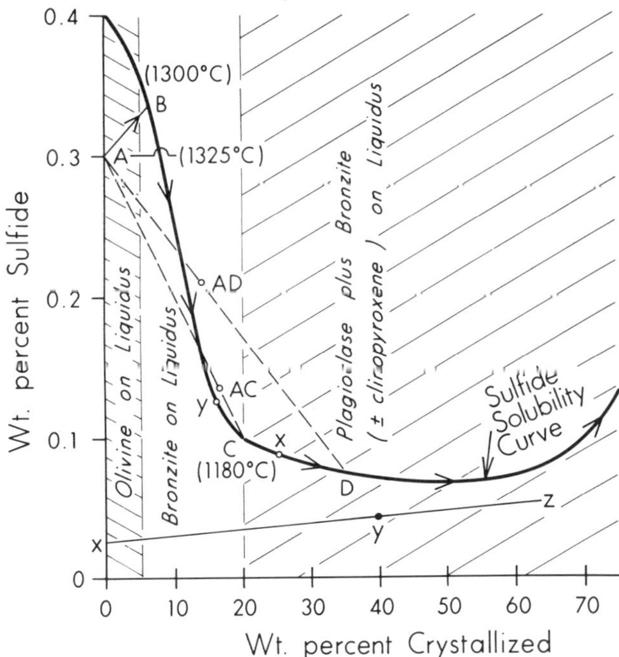


FIGURE 8.31—Schematic diagram illustrating the variation in the solubility of iron sulfide with the fractionation of the Bushveld chilled margin sample referred to under Figure 21. Variations in solubility are based on the data of Haughton et al. (1974), Buchanan and Nolan (1979) and Buchanan et al. (1983) and the compositional calculations of Barnes and Naldrett (1986) as discussed in Chapter 2.

in the hybrid will all have come from the fresh input of A and will therefore depend on the proportion of A in the mixture. In the case of mixing A and D, the hybrid can be relatively rich in A and therefore in PGE.

Provided that the two liquids are saturated or nearly saturated in chromite (as is the case with liquids A and D in Fig. 8.30), they will have the potential to produce a PGE-enriched chromitite. If they are not sufficiently rich in Cr, they may produce a Merensky or J–M Reef.

If the magma overlying the cumulates has not fractionated beyond point Y, mixing with fresh inputs of liquid A will not produce sulfide saturation. They may, however, be saturated in chromite (for example point AB in Fig. 8.30) and give rise to chromitites, in this case not enriched in PGE.

If the magma overlying the cumulates has fractionated nearly to the point of appearance of plagioclase (point C in Fig. 8.31), the mixture can be slightly over-saturated in sulfide, but only at low ratios of A to C; the smaller degree of over-saturation and the low A/C ratio means that the enrichment of the resulting chromitite in PGE may be less than in the case of mixing A and D. Accepting that sulfide solubility varies as predicted above, chromitites that have formed by mixing of primitive magma with that crystallising bronzite or olivine (ie Bushveld Lower Group and most of the Stillwater chromitites) will correspond to the mixing of liquid A with those between B and Y in Fig. 8.31 and no sulfide immiscibility, and thus no Pt and Pd enrichment, will be expected. Those such as the Bushveld Upper Group will correspond to the mixing of A with D, and associated PGE-enriched sulfides are to be expected. The Bushveld Middle Group will correspond to the mixing of liquid A with C, and limited PGE-enriched sulfides could be the result. In this last case, mixing will occur in a fountain rather than a plume, and the amount of mixing, and thus the resultant *R* factor of the sulfides may be lower than would be the case with the plume. There are therefore two possible explanations as to why the Middle Group chromitites are less enriched in PGE than the Upper Group; they may have contained a lower overall original sulfide content, and the sulfides might themselves originally have been less enriched in PGE, due to the lower PGE content of the hybrid from which they segregated and also their possibly lower *R* value.

The Stillwater "A" chromitite appears to be an exception to the model proposed above, until it is appreciated that it lies only a few meters above the first incoming of cumulate olivine, and not far above the norites of the basal zone. The Mg/Fe ratios of the minerals in the vicinity of this chromitite are much lower than higher up (forsterite content of olivine is 78–80 as compared to 84–86 400 m higher, Page et al., 1985). In this respect the Stillwater is no different to most other large layered intrusions, in which the lowermost portions appear to have crystallised from magmas much less primitive than portions farther into the intrusion. In the early stages of the development of most intrusions the initial input is relatively fractionated. This is followed by the addition of progressively more primitive magma which overides the effect of fractional crystallisation and results in the resident magma, and thus the resulting cumulates, becoming progressively more magnesian. Eventually fractional crystallisation becomes the overiding factor, the composition of the magma in the intrusion becomes progressively less primitive with time despite continued inputs of new magma, and plagioclase ultimately joins olivine or pyroxene as a cumulate. An input of primitive magma at an early stage such as that represented by the Stillwater A chromitite thus finds itself in a similar environment to one occurring much later, at the stage represented, for example, by the Bushveld Upper Critical zone (where the Upper Group chromitites occur). Under these conditions the fresh primitive input can give rise to sulfide segregation in addition to chromitite formation.

An alternative to the model proposed above is to suggest that the magma in the chamber had never achieved saturation in sulfide and had evolved along a path such as X to Z in Fig. 8.31. Mixing of A with a composition such as Y would also produce a sulfide-saturated mixture and in this case, since neither magma would have previously been sulfide-saturated, the PGE would be contributed by both. Scenarios analogous to those described for the first model can also be devised involving the mixing of A with different liquids between X and Y to explain the other situations covered by the first model. The present authors regard this second model less favourably than the first since it is distinctly ad hoc in requiring two magmas with widely varying initial sulfide contents.

### Conclusions and applications to exploration

Many chromitites from layered complexes contain much higher proportions of Pt and Pd relative to Ru, Ir and Os than do those from ophiolite complexes; this is attributed in part to the chromitites from layered intrusions originally containing more sulfide than those from ophiolites.

Within layered intrusions, the chromitites richest in PGE are those occurring close to or above the level at which plagioclase first appears as a cumulate phase. Those situated well within the ultramafic cumulates are characterised by much lower PGE concentrations. The relative abundances of different PGE in these chromitites strongly supports the hypothesis that those richest in PGE were originally the most enriched in sulfide.

Magma mixing would appear to be the best mechanism to explain the formation of massive chromitite layers in layered intrusions. Magma mixing can also account for sulfide immiscibility, althought the available data on the controls of sulfide solubility indicate that the mixing of primitive magma with an advanced differentiate that is already crystallising plagioclase is most likely to produce immiscible sulfides, while the mixing of two relatively primitive magmas may not. This can explain the observed association of PGE with particular chromitite layers.

These observations and hypotheses lead to the conclusion that chromites within or close to the level at which norites, gabbros and anorthosites first appear in an intrusions are better prospects for PGE enrichment than those deep within ultramafic cumulates.

The Stillwater "A" chromitite is evidence that chromitites situated close to the base of an intrusion, and forming at a stage in the crystallisation when the resident magma in the intrusion had not reached its maximum MgO content, may also have contained original sulfides and thus be PGE-enriched.

It is stressed that because high temperature reaction between chromite and sulfide during cooling can lead to the removal of FeS from the sulfide, and in this way to a several-fold diminution in its mass, chromitites with little or no visible sulfide may neverthelss contain high PGE concentrations.

## REFERENCES


Alapieti, T. T., and Lahtinen, J. J., 1986, Stratigraphy, petrology and platinum-group element mineralization of the early Proterozoic Penikat layered intrusion, northern Finland: Econ. Geol., v. 81, no. 5, pp. 1126–1136.

Ballhaus, C. G., and Stumpfl, E. F., 1985a, Fluid inclusions in Merensky and Bastard Reefs, western Bushveld Complex (abs.): Canad. Mineral., v. 23, p. 294.

Ballhaus, C. G., and Stumpfl, E. F., 1985b, Graphite, platinum and the C–O–H–S system (abs.): Canad. Mineral., v. 23, pp. 293–294.

Barager, W. R. A., and Scoates, R. F. J., 1981, The Circum–Superior belt: a Proterozoc plate margin? *in* Kroner, A. (ed.), Precambrian plate tectonics: Elsevier, Amsterdam, pp. 293–294.

Barnes, S.-J., Naldrett, A. J., and Gorton, M. P., 1985, The origin of the fractionation of platinum-group elements in terrestrial magmas: Chem. Geol., v. 53, pp. 303–323.

Barnes, S. J., and Naldrett, A. J., 1985, Geochemistry of the J–M (Howland) Reef of the Stillwater Complex, Minneapolis adit area—I. Sulfide chemistry and sulfide–olivine equilibrium: Econ. Geol., v. 80, pp. 627–645.

Barnes, S. J., and Naldrett, A. J., 1986, Geochemistry of the J–M (Howland) Reef of the Stillwater Complex, Minneapolis adit area—II. Silicate mineral chemistry and petrogenesis: Jour. Petrol., v. 27, pp. 791–825.

Bottinga, Y., Weill, D. F., Ricket, P., 1982, Density calculations for silicate liquids—I. Revised method for aluminosilicate compositions: Geochim. et Cosmochim. Acta., v. 46, pp. 909–919.

Boudreau, A. E., Mathez, E. A., McCallum, I., 1986, Halogen geochemistry of the Stillwater and Bushveld Complexes: Evidence for transport of the platinum-group elements by Cl-rich fluids: J. Petrol., v. 27, pp. 967–986.

Boudreau, A. E. 1988, Investigations of the Stillwater Complex: Part IV. The role of volatiles in the petrogenesis of the J–M Reef: Can. Mineral., v. 26, pp. 193–208.

Brown, G. M., and Peckett, A., 1977, Fluorapatites from the Skaergaard Intrusion, East Greenland: Min. Mag., v. 41, pp. 227–232.

Brynard, H. J., de Villers, J. P. R., and Viljoen, E. A., 1976, A mineralogical investigation of the Merensky reef at the Western Platinum mine near Marikana, South Africa: Econ. Geol., v. 71, pp. 1299–1307.

Buchanan, D. L., 1979, Platinum metal production from the Bushveld complex and its relationship to world markets: Bureau for Min. Studies, Univ. of Witwatersrand, Johannesburg, Rep. no. 4, 31 pp.

Buntin, T. J., Grandstaff, D. E., Ulmer, G. C., Gold, D. P., 1985, A pilot study of geochemical and redox relationships between potholes and adjacent normal Merensky Reef of the Bushveld Complex: Econ. Geol., v. 80, pp. 975–987.

Cameron, E. N., 1978, The lower zone of the eastern Bushveld Complex in the Olifants River trough: Jour. Petrol., v. 19, pp. 437–462.

Cameron, E. N., 1980, Evolution of the Lower Critical Zone, central sector, eastern Bushveld Complex and its chromite deposits: Econ. Geol., v. 75, pp. 845–871.

Cameron, E. N., 1982, The Upper Critical Zone of the eastern Bushveld Complex—precursor to the Merensky Reef: Econ. Geol., v. 77, pp. 1307–1327.

Cameron, G. H., 1988, A geochemical investigation into the origin of the Upper Critical Zone of the eastern Bushveld Complex, South Africa: Unpub. M.S. thesis, Univ. of Toronto.

Campbell, I. H., Naldrett, A. J. and Barnes, S. J., 1983, A model for the origin of the platinum-rich sulfide horizons in the Bushveld and Stillwater Complexes: Jour. Petrol., v. 24, pp. 133–165.

Campbell, I. H., and Turner, J. S., 1986, The role of convection in the formation of platinum and chromitite deposits in layered intrustions; *in* Scarfe, C. M. (ed.), SHort course on silicate melts: Mineral. Assoc. Canada, v. 12, pp. 236–278.

Cawthorn, R. G., Davies, G., 1983, Experimental data at 3 K bars pressure on parental magma to the Bushveld Complex: Contrib. Mineral. Petrol., v. 83, pp. 128–135.

Cousins, C. A., 1964, The platinum deposits of the Merensky Reef; *in* Haughton, S. H. (ed.), The geology of some ore deposits in southern Africa, v. II: Geol. Soc. S. Africa, Johannesburg, pp. 225–238.

Davies, G., and Tredoux, M., 1985, The platinum group element and gold contents of the marginal rocks and sills of the Bushveld Complex: Econ. Geol., v. 80, pp. 838–848.

Dickey, J. S., 1975, A hypothesis of origin for podiform chromite deposits: Geochim. Cosmochim. Acta, v. 39, pp. 1061–1074.

du Toit, L., 1986, The Impala platinum mines; *in* Anhaeusser, C. R., and Maske, S. (eds.), Mineral deposits of southern Africa, v. 2: Geol. Soc. S. Africa, Johannesburg, pp. 1091–1106.

Farquhar, J., 1985, The geology of Western Platinum mine; *in* Vermaak, E., and von Gruenewaldt, G. (compilers, Bushveld excursion), Some ore deposits in southern Africa, Excursion Guidebook: Third Int. Pt. Symp., Geol. Soc. of S. Africa, pp. 16–19.

Gain, S. B., 1980, The geology and PGE distribution in the Upper Group chromatite layers at Maandagshoek 254, K.T., eastern Bushveld Complex: Inst. Geol. Res. on the Bushveld, Res. Rept. 22, 24 pp.

Gain, S. B., 1985, The geologic setting of the platiniferous UG–2 chromitite layer on the farm Maandagshoek, eastern Bushveld Complex: Econ. Geol., v. 80, pp. 925–943.

Greenbaum, D., 1977, The chromitiferous rocks of the Troodos ophiolite complex, Cyprus: Econ. Geol., v. 72, pp. 1175–1194.

Halls, H. C., 1978, The late Precambrian central North American rift system—a survey of recent geological and geophysical investigations; *in* Ramberg, I. B., and Newmann, E.–R. (eds.), Tectonics and geophysics of continental rifts: NATO Advanced Study Inst., Series C, v. 37, pp. 111–123.

Hamilton, J., 1977, Sr. isotope and trace element studies of the Great Dyke and Bushveld, mafic phase and their relation to early Proterozoic magma genesis in Southern Africa: Jour. Petrol., v. 18, pp. 24–52.

Harmer, R. E., and Sharpe, M. R., 1985, Field relations and strontium isotope systematics of the marginal rocks of the eastern Bushveld Complex: Econ. Geol., v. 80, pp. 813–837.

Helz, R. T., 1985, Compositions of fine-grained mafic rocks from sills and dikes associated with the Stillwater; *in*Czamanske, G. K., and Zientek, M. L. (eds.), Stillwater Complex: Montana Bureau of Mines and Geology, Spec. Pub. No. 92, pp. 147–209.

Huntington, H. D., 1979, Kiglapait mineralogy I: Apatite, biotite and volatiles: J. Petrol., v. 20, pp. 625–652.

Huppert, H. E., Sparks, R. S. J., and Whitehead, J. A., 1984, Plumes: Magma chamber replenishment by light inputs; *in* Dungan, M. A., Gove, T. L., and Hildreth, W. (eds.), Proc. of the Conf. on Open Magmatic Systems: Inst. for the Study of Earth and Man, and Dept. of Geol. Sci. of Southern Methodist Univ., pp. 83–84.

Irvine, T. N., 1974, Petrology of the Duke Island ultramafic complex, southeastern Alaska: Geol. Soc. America, Mem. 138, 240 pp.

Irvine, T. N., 1975, Crystallization sequences of the Muskox intrusion and other layered intrusions—II. Origin of chromitite layers and similar deposits of other magmatic ores: Geochim. et Cosmochim. Acta, v. 39, pp. 991–1020.

Irvine, T. N., 1977, Origin of chromite layers in the Muskox intrusion and other stratiform intrusions: a new interpretation: Geology, v. 5, pp. 273–277.

Irvine, T. N., 1980, Experimental modelling of convection in layered intrusions: Carnegie Inst. of Washington, Year Book 79, pp. 247–251.

Irvine, T. N., Keith, D. W., and Todd, S. G., 1983, The J–M platinum palladium reef: II, Origin by double-diffusive convective magma mixing and implications for the Bushveld Complex: Econ. Geol., v. 78, pp. 1287–1334.

Jackson, E. D., 1970, The cyclic unit in layered intrusions, a comparison of repetitive stratigraphy in the ultramafic parts of the Stillwater, Muskox, Great Dyke and Bushveld Complex (with discussion); *in* Visser, D. J. L., and von Gruenewaldt, G. (eds), Symposium on the Bushveld Igneous Complex and Other Layered Intrusions: Geol. Soc. of Africa, Spec. Pub. No. 1, pp. 391–424.

Johan, S., Watkinson, D. H., 1985, Significance of a fluid phase in platinum-group element concentration: evidence from the Critical Zone, Bushveld Complex (abs.): Canad. Mineral., v. 23, pp. 305–306.

Kruger, F. J., Marsh, J. S., 1985, The mineralogy, petrology and origin of the Merensky cyclic unit; *in* The mineralogy, petrology and origin of the Merensky cyclic unit in the western Bushveld Complex: Econ. Geol.. v. 80, pp. 958–974.

Kruger, F. J., and Marsh, J. S., 1982, Significance of $Sr^{87}/Sr^{86}$ ratios in the Merensky cyclic unit of the Bushveld complex: Nature, v. 298, pp. 53–55.

Kucha, H., 1982, Platinum-group metals in the Zechstein copper deposits, Poland: Econ. Geol., v. 77, pp. 1578–1591.

Lago, B. L., Rabinowicz, M., and Nicolas, A., 1982, Podiform chromite orde bodies: A genetic model: J. Petrol., v. 23, pp. 103–125.

Lee, C. A., 1983, Trace and platinum-group element geochemistry and the development of the Merensky unit of the western Bushveld Complex: Mineralium Deposita, v. 18, pp. 173–190.

Lee, C. A., and Tredoux, M., 1986, Platinum-group element abundances in the Lower and Lower Critical Zones of the eastern Bushveld Complex: Econ. Geol., v. 81, no. 5, pp. 1087–1095.

Leroy, L. W., 1985, Tractolite–anorthosite Zone I and the J–M Reef: Frog Pond adit to the Graham Creek area; *in* Czamanske, G. K., and Zientek, M. L. (eds), Stillwater Complex: Montana Bureau of Mines and Geology, Spec. Pub. No. 92, pp. 325–333.

Mann, E. L., Lipin, B. R., Page, N. J., Foose, M. P., and Loferski, P. J., 1985, Guide to the Stillwater Complex; *in* Czamanske, G. K., and Zientek, M. L. (eds), Stillwater Complex: Montana Bureau of Mines and Geology, Spec. Pub. No. 92, pp. 231–246.

Mann, E. L., and Lin, C.–P., 1985, Geology of the West Fork adit; *in* Czamanske, G. K., and Zientek, M. L. (eds.), Stillwater Complex: Montana Bureau of Mines and Geology, Spec. Pub. No. 92, pp. 247–252.

McCallum, I. S., Raedeke, L. D., Mathez, E. A., 1980, Investigations in the Stillwater Complex, Part I. Stratigraphy and structure of the landed zone; *in* Irving, A. J., and Dungan, M. A. (eds.), The Jackson Volume: American J. Sci., v. 280–A, pp. 59–87.

Naldrett, A. J., 1981, Pt group element deposits; *in* Cabri, L. C. (ed.), Platinum group elements: mineralogy, geology, geochemistry: Can. Inst. Min. Met., Special Volume 23, pp. 197–232.

Naldrett, A. J., and Cabri, L. J., 1976, Ultramafic and related mafic rocks: their classification and genesis with special reference to the concentration of nickel sulfides and platinum-group elements: Econ. Geol., v. 71, pp. 1131–1158.

Naldrett, A. J., Cameron, G., von Gruenewaldt, G., and Sharpe, M. R., 1987, The formation of stratiform PGE deposits in layered intrusions; *in* Parsons, I. (ed.), Origins of igneous layering: NATO Advanced Sc. Inst., Series C., v. 196, pp. 313–397.

Naldrett, A. J., Gasparrini, C., Barnes, S. J., Sharpe, M. R., and von Gruenewaldt, G., 1984, The petrology of the upper part of the critical zone of the Bushveld Complex and its bearing on the origin of the Merensky Reef: Proc. 27th Int. Geol. Congress, Moscow, v. 9, pp. 373–389.

Naldrett, A. J., Gasparrini, E. C., Barnes, S. J., von Gruenewaldt, G. and Sharpe, M. R., 1986, The upper critical zone of the Bushveld Complex and a model for the origin of Merensky-type ores: Econ. Geol., v. 81, pp. 1105–1118.

Naldrett, A. J., and Lehmann, J., 1988, Spinel non-stoichiometry as the explanation for Ni-, Cu-, and PGE-enriched sulphides in chromitites; *in*Prichard, H., Potts, P., Bowles, J., and Cribb, S. (eds.), Geoplatinum '87: Elsevier, London, (in press).

Naldrett, A. J., Lehmann, J., and Auge, T., 1988, Reactions between non-stoichimetric chromite and sulfide as the explanation for Ni, Cu and PGE-enriched sulfides: examples from layered intrusions and ophiolites: Inst. of Min. & Metall., Proc. 5th Magmatic Sulfide Conf., Harare, Zimbabwe.

Nash, W. P., 1976, Fluorine, chlorine and OH-bearing minerals in the Skaergaard intrusion: Am. J. Sci., v. 276, pp. 546–557.

Page, N. J., Aruscavage, P. J., and Haffty, J., 1983, Platinum-group elements in rocks from the Voikar–Syninsky ophiolite complex, Polar Urals, USSR: Mineral. Deposita, v. 18, pp. 443–445.

Page, N. J., Cassard, D., and Haffty, J., 1982, Palladium, platinum, rhodium, ruthenium and iridium in chromitites from the Massif du Sud and Teibaghi Massif, New Caledonia: Econ. Geol., v. 77, pp. 1571–1577.

Page, N. J., Pallister, J. S., Brown, M. A., Smewing, J. D., and Haffty, J., 1982b, Palladium, platinum, rhodium, iridium and ruthenium in chromite-rich rocks from the Samail ophiolite, Oman: Canad. Mineral., v. 20, pp. 537–548.

Page, N. J., Singer, D. A., Moring, B. C., Carlson, C. A., McDade, J. M., and Wilson, S. A., 1986, Platinum-group element resources in podiform chromitites from California and Oregon: Econ. Geol., v. 81, pp. 1261–1271.

Page, N. J., and Talkington, R. W., 1984, Palladium, platinum, rhodium, ruthenium and iridium in peridotites and chromitites from ophiolite complexes from Newfoundland: Canad. Mineral., v. 22, pp. 137–149.

Page, N. J., Zientek, M. L., Lipin, B. R., Raedeke, L. D., Wooden, J. L., Turner, A. R., Loferski, P. J., Foose, M. P., Moring, B. C., Ryan, M. P., 1985a, Geology of the Stillwater Complex exposed in the Mountain View area and on the west side of the Stillwater Complex; *in* Czamanske, G. K., Zientek, M. L.(eds.), Stillwater Complex: Montana Bureau of Mines and Geology, Spec. Pub. No. 92, pp. 147–210.

Page, N. J., Zientek, M. L., Czamanske, G. K., Foose, M. P., 1985b, Sulfide mineralization in the Stillwater Complex and underlying rocks; *in* Czamanske, G. K., Zientek, M. L. (eds.), Stillwater Complex: Montana Bureau of Mines and Geology, Spec. Pub. No. 92, pp. 93–96.

Parry, S. J., Sinclair, I. W., and Asif, M., 1988, Evaluation of the nickel sulphide bead method of fire assay for the PGE; *in* Prichard, H., Potts, P., Bowles, J., and Cribb, S. (eds.), Geoplatinum '87: Elsevier London. (in press)

Prendergast, M. D., 1988, The geology and economic potential of the PGE-rich Main Sulfide Zone of the Great Dyke, Zimbabwe; *in* Pritchard, H., Potts, P. J., Bowles, J. F. W., and Cribb, S. J. (eds.) Geoplatinum '87: Elsevier, London. (in press)

Raedeke, L. D., McCallum, I. S., 1984, Investigations in the Stillwater Complex, Part II. Petrology and petrogenesis of the Ultramafic series: J. of Petrol., v. 25, pp. 395–420.

Raedeke, L. D., and Vian, R. W., 1986, A three-dimensional view of mineralization in the Stillwater J–M Reef: Econ. Geol., v. 81, pp. 1187–1195.

Schwellnus, J. S. I., Hiemstra, S. A., and Gasparrini, E., 1976, The Merensky reef at the Atok platinum mine and its environs: Econ. Geol., v. 71, pp. 249–260.

Scoon, R. N., and de Klerk, W. J., 1987, The relationship between olivine cumulates, mineralization and cyclic units in part of the critical zone of the western Bushveld Complex: Canad. Mineral., v. 25, pp. 51–77.

Segerstrom, K., and Carlson, R. R., 1982, Geological map of the landed upper zone of the Stillwater Complex and adjacent rocks, Stillwater, Sweet Grass and Park Counties, Montana: US Geol. Survey, Map I–1383, 2 sheets, scale 1:24,000.

Sharpe, M. R., 1981, The chronology of magma in fluxes to the eastern compartment of the Bushveld Complex as exemplified by its marginal border groups: J. Geol. Soc. London, v. 138, pp. 206–307.

Sharpe, M. R., 1982, Noble metals in the marginal rocks of the Bushveld Complex: Econ. Geol., v. 77, pp. 1286–1295.

Sharpe, M. R., 1985, Strontium isotopic evidence for preserved density stratification from the main zone of the Bushveld Complex, South Africa: Nature, v. 316, pp. 119–126.

Sparks, R. S. J., Meyer, P., and Sigurdsson, H., 1980, Density variation amongst mid-ocean ridge basalts: implications for magma mixing and the scarcity of primitive lavas: Earth & Planet Sci. Lett., v. 46, pp. 419–430.

Todd, S. G., Keith, D. W., LeRoy, L. W., Schissel, D. J., Mann, E. L. and Irvine, T. N., 1982, The J–M platinum–palladium reef of the Stillwater Complex, Montana, I. Stratigraphy and Petrology: Econ. Geol., v. 77, pp. 1454–1480.

Turner, A. R., Wolfgram, D., Barnes, S. J., 1985, Geology of the Stillwater County sector of the J–M Reef, including the Minneapolis adit; *in* Czamanske, G. C., and Zientek, M. L. (eds.), Stillwater Complex: Montana Bureau of Mines and Geology, Spec. Pub. No. 92, pp. 210–230.

Turner, J. S., 1980, A fluid dynamical model of differentiation and layering in magma chambers: Nature, v. 285, pp. 213–215.

Turner, J. S., and Gustafson, L. B., 1978, The flow of hot saline solutions from sulfides and other ore deposits: Econ. Geol., v. 73, pp. 1082–1100.

Vermaak, C. F., 1976 The Merensky reef—thoughts on its environment and genesis: Econ. Geol., v. 71, pp. 1270–1298.

Viljoen, M. J., and Hieber, R., 1986, The Rustenburg section of Rustenburg Platinum Mines Limited with reference to the Merensky Reef; *in* Anhaeusser, C. R., and Maske, S. (eds.), Mineral deposits of southern Africa, v. 2: Geol. Soc. South Africa, Johannesburg, pp. 1107–1134.

Viljoen, M. J., Theron, J., Underwood, P., Walters, B. M., Weaver, J., and Peyerl, W., 1986a, The Amandelbult section of Rustenburg Platinum Mines Limited with reference to the Merensky Reef; *in* Anhaeusser, C. R., and Maske, S., (eds.), Mineral deposits of southern Africa, v. 2: Geol. Soc. South Africa, Johannesburg, pp. 1041–1060.

Viljoen, M. J., de Klerk, W. J., Coetzer, M. P., Hatch, N. P., Kinloch, E., and Peyerl, W., 1986b, The Union section of Rustenburg Platium Mines Limited with reference to the Merensky Reef; *in* Anhaeusser, C. R., and Maske, S. (eds.), Mineral deposits of southern Africa, v. 2: Geol. Soc. South Africa, Johannesburg, pp. 1061–1090.

van Zyl, C., 1959, An outline of the geology of the Kapalagubu complex, Kungwe Bay, Tanganyika Territory and aspects of the evolution of layering in basic intrusives: Trans. Geol. Soc. S. Afr., v. 62, pp. 1–31.

von Gruenewaldt, G., 1973, The main and upper zones of the Bushveld complex in the Roossenekal area, Eastern Transvaal: Trans. Geol. Soc. S. Afr., v. 76, pp. 207–227.

von Gruenewaldt, G., 1979, A review of some recent concepts of the Bushveld complex with particular reference to the sulfide mineralization: Canadian Mineral., v. 17, no. 2, pp. 233–256.

von Gruenewaldt, G., Hatton, C. J., Merkle, R. K. W., and Gain, S. B., 1986a, Platinum-group element-chromitite associations in the Bushveld Complex: Econ. Geol., v. 81, pp. 1067–1079.

von Gruenewaldt, G., Sharpe, M. R., and Hatton, C. J., 1985, The Bushveld Complex introduction and review: Econ. Geol., v. 80, pp. 803–812.

Wager, L. R., and Brown, G. M., 1968, Layered igneous roscks: London, Oliver and Boyd.

Weiblen, P. W., and Morey, G. B., 1975, The Duluth complex—a petrographic and tectonic summary: 36th Ann. Minn. Min. Symp., Dept. of Conferences and Continuing Ed., Univ. of Minnesota, pp. 72–95.

Zientek, M. L., Czamanske, G. K., and Irvine, T. N., 1985, Stratigraphy and nomenclature for the Stillwater Complex; *in* Czamanske, G. K., and Zientek, M. L. (eds.), Stillwater Complex: Montana Bureau of Mines and Geology, Spec. Pub. No. 92, pp. 21–32.

Zurbrigg, H. F., 1963, Thompson mine geology: Trans. Can. Inst. Min. Met., v. 66, pp. 227–236.

*Relative Proportions, Grade (in gm/tonne) and Identified Resources (in $10^6$ gm) of PGE in Major Deposits*

| | Bushveld Complex | | | | | | Noril'sk-Talnakh USSR | | Stillwater J-M Reef | | Great Dyke | | Sudbury, Canada | | Columbia | | |
|---|---|---|---|---|---|---|---|---|---|---|---|---|---|---|---|---|---|
| | Merensky Reef | | UG-2 | | Plat. Reef | | | | | | | | | | | | |
| | Prop | Res | Prop | Res | Prop | Res | Prop | Res | Prop | Res | Prop | Res | Prop | Res | Prop | Res | Total Resources |
| Pd | 25 | 4,371 | 35 | 11,315 | 46 | 5,425 | 71 | 4,402 | 66.5 | 713 | 37.1 | 2936 | 40 | 112 | 1 | | 29,274 |
| Pt | 59 | 10,323 | 42 | 13,547 | 42 | 4,960 | 25 | 1,550 | 19 | 217 | 52.8 | 4130 | 38 | 105 | 93 | | 34,925 |
| Rh | 3 | 527 | 8 | 2,573 | 3 | 372 | 3 | 186 | 7.6 | 84 | 4.7 | 367 | 3.3 | <31 | 2 | | 4,109 |
| Ru | 8 | 1,395 | 12 | 3,875 | 4 | 465 | 1 | 62 | 4.0 | 43 | 4.7 | 367 | 2.9 | <31 | — | | 6,207 |
| Ir | 1 | 186 | 2.3 | 744 | 0.8 | 93 | — | — | 2.4 | <31 | 0.6 | 46 | 1.2 | <31 | 3 | | 1,069 |
| Os | 0.8 | 155 | — | — | 0.6 | 62 | — | — | — | — | 0.6 | 46 | 1.2 | <31 | 1 | | 263 |
| Grade in gm/tonne total PGE | 8.1 | | 8.71 | | 7.27 | | 3.8 | | 22.3 | | 4.7 | | 0.9 | | | | |
| Total resources | 16,957 | | 32,054 | | 11,377 | | 6,200 | | 1,057 | | 7,892 | | 217 | | | | |

Prop, proportion; Res, reserves.

Data from Buchanan and Nolan (1979) except for that on the Great Dyke which is from I. G. Anderson (personal communication, 1981).

# Chapter 9

# INTERACTIONS INVOLVING FLUIDS IN THE STILLWATER AND BUSHVELD COMPLEXES: OBSERVATIONS FROM THE ROCKS

E. A. Mathez

> . . . it is of considerable significance with respect to the mineral deposits to compare the Stillwater complex with the extraordinarily similar group of rocks known as the Bushveld complex of Transvaal in the Union of South Africa . . . Anyone interested in the search for ore in the Stillwater complex would do well to read with care the splendid publications of Wagner and Hall. . . It might be noted that the platinum-bearing horizons of the Banded Zone were found as the result of a search directed by the liklihood of sulphides occurring in the same relative "stratigraphic" position in Montana as in South Africa. Professor A.F. Buddington first suggested this analogy, and Howland and Peoples first found the sulphides of the Banded Zone in Montana.
>
> Howland et al., 1936

## INTRODUCTION

Nearly all mafic igneous bodies exhibit evidence for the presence of volatile-rich melts and fluids. This is true for thin sills, in which localized pods of pegmatitic and granophyric rocks are common, as well as for the layered intrusions. In the Bushveld and Stillwater Complexes, volatile elements, either as a separate vapor or simply dissolved in the melt, were clearly involved in the petrogenesis of the PGE-rich horizons because the constituent rocks are pegmatitic and contain magmatic, volatile-bearing minerals. There is also circumstantial evidence that volatiles were involved in processes of PGE enrichment or redistribution. The specific roles of fluid in these processes are enigmatic because there is a lack of knowledge about the composition of fluid under near-solidus conditions, the ability of fluid to transport PGE and the physical and chemical interactions of fluid with partially molten to completely crystalline cumulates.

A conception of how fluids and volatile-rich melts behave in the environment of a crystallizing layered intrusion must originate from observations of the rocks themselves. The point of this chapter is to bring together such observations. To make this task managable, the description is restricted to the Stillwater and Bushveld Complexes. A reference stratigraphic section of the former is presented in Fig. 9.1, and for the Bushveld reference may be made to von Gruenewaldt et al. (1985). Implicit here is the conception, first expressed by Howland et al. (1936) in the above quotation, that the processes which guided the evolution of the two complexes were essentially identical. Therefore, a generalized model applicable to both may be developed from observations made from each individually. It will be left to the reader to develop his or her own suspicions regarding the role of fluids in the genesis of the PGE ores.

## REPLACEMENT BODIES IN THE STILLWATER COMPLEX

### Discordant dunites

The discordant dunite bodies in the Stillwater appear to be most easily understood and therefore are considered first. The dunites are irregularly shaped, localized masses in pyroxene-bearing lithologies. In this respect they appear to be similar to the discordant dunites in Alaskan-type ultramafics (e.g., the Duke Island Complex [Irvine, 1974]) and in ultramafics of ophiolites (e.g., at Canyon Mt., Oregon [Dungan and Ave Lallemant, 1978]). In the Stillwater, discordant dunites are particularly well developed in orthopyroxenite in the Chrome Mountain area, where they have been described by Raedeke and McCallum (1984, p. 401): The dunites are

> "...fine-grained, extensively serpentinized olivine-rich rocks (± chrome spinel). Contacts between the discordant dunite and primary cumulate rocks are sharp, with the secondary olivine replacing the primary cumulus mineral (olivine or orthopyroxene), in wormy, digital embayments. Orthopyroxene oikocrysts occur in some secondary dunites and are thought to be relic from original cumulate dunite. The discordant dunites commonly enclose patches of ultramafic cumulates. In such cases, the primary igneous lamination in the relic patches shows the same orientation as that in the surrounding cumulate, suggesting in situ replacement of the cumulate. The secondary dunites are not restricted to any primary lithology, i.e. they form cross-cutting pipes and pods in cumulate dunite, harzburgite and/or orthopyroxenite. These dunites are not

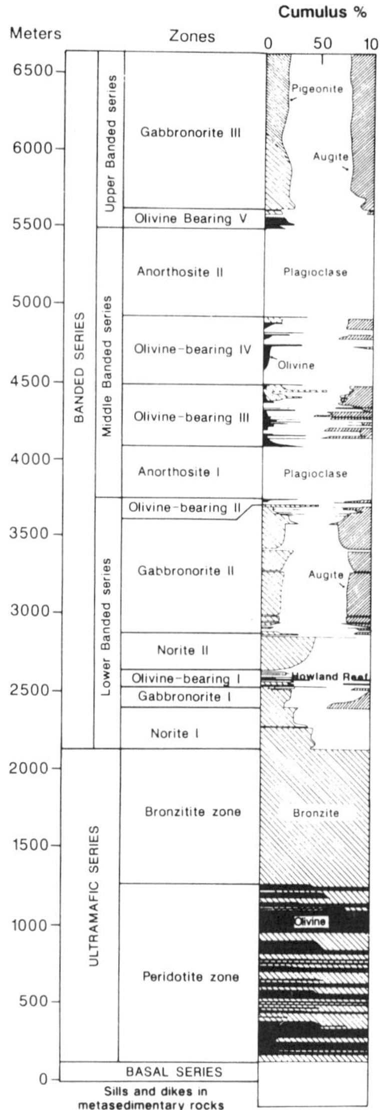


FIGURE 9.1—Composite stratigraphic section of McCallum et al. (1980) and Raedeke and McCallum (1984) of the Stillwater Complex.

> restricted to any single stratigraphic level, but rather occur throughout the Peridotite zone at Chrome Mountain and are particularly abundant near major faults."

Raedeke and McCallum also noted that pegmatoids consisting of orthopyroxene, clinopyroxene and minor olivine are commonly associated with the dunites. Unlike constituent minerals in most discordant bodies in the Bushveld Complex (see below), olivine, bronzite and chromite in the Stillwater dunites are compositionally identical to their igneous counterparts in the surrounding cumulates (Raedeke and McCallum, 1984; Nicholson and Lipin, 1985). The field relations are illustrated by plate 1C of Raedeke and McCallum (1984) and figs. 2 and 4 of Nicholson and Lipin (1985).

It was first proposed by Hess (1960) that the discordant dunites formed by prograde metamorphism of partially serpentinized cumulates. Although there are many documented cases of this, the process results in olivine more magnesian that original igneous olivine (e.g., Trommsdorff and Evans, 1974; Frost, 1975). Page (1977) and Nicholson and Lipin (1985) have taken the view that the dunites are igneous, having been intruded into bronzitite country rock in a partially molten state. The latter workers cited in support of this position the fact that the amounts of modal chromite in the dunites are typically higher than can be accounted for by simple breakdown of orthopyroxene to form olivine. It should be pointed out, however, that most mafic pegmatitic rocks, both those associated with the discordant dunites and present elsewhere, also contain more chromite than surrounding cumulate rocks (Page et al., 1985). This implies that chromium is transported and concentrated in the pegmatitic areas; therefore, the significance of the chromite in the discordant dunites is unclear.

Raedeke and McCallum (1984) argued that the field relations described above imply that the dunites are replacement rather than intrusive features. The replacement of pyroxenite or harzburgite by dunite is readily understood from relevant experimental investigations. Enstatite (En) dissolves incongruently in water to forsterite (Fo) + vapor (v) according to the reaction (Nakamura and Kushiro, 1974)

$$2En + 2H_2O(v) = Fo + Si(OH)_4(v). \quad [1]$$

Above the solidus, enstatite may also melt incongruently to an $H_2O$-unsaturated assemblage of forsterite and hydrous melt (L). This may be represented as

$$En + H_2O(v) = En + Fo + L, \quad [2]$$

where the stoichiometry requires that

$$X^{L}_{SiO2} > X^{En}_{SiO2} \quad [3]$$

(Kushiro et al., 1968). The fundamental relationships are illustrated in Fig. 9.2. It is evident that replacement of a pyroxene-bearing assemblage by one dominated by olivine may occur in the presence or absence of a silicate melt. This possibility was emphasized by Irvine (1980), who envisioned that the intercumulus silicate melt itself was the metasomatizing fluid for similar occurrences in the Muskox Intrusion. Whether the process of dunite formation involved melt, melt + vapor or simply vapor should not obscure the important point, namely that both the field and experimental evidence provide a rational basis for interpreting the process to be mainly one of replacement of the pre-existing pyroxene-dominated assemblage.

### Ameboidal troctolite horizons

There exists in several specific stratigraphic horizons in the Stillwater Complex an unusual rock termed pillow or ameboidal troctolite (Hess, 1960; McCallum et al., 1980; Foose, 1985). The lithology is best developed at the base of AN–I, where it comprises the horizon OB–II (Fig. 9.1). The troctolite forms a sharp, planar contact with overlying anor-

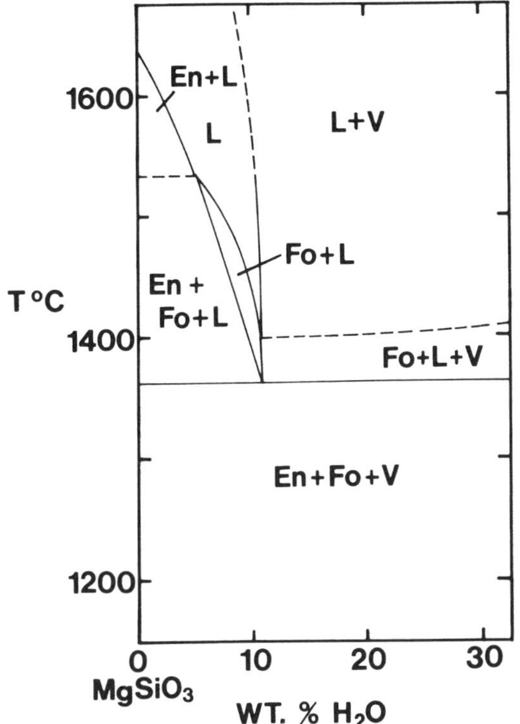


FIGURE 9.2—A portion of the system enstatite–water at 10 kb. En = enstatite, Fo = forsterite, L = silicate melt, V = vapor. Note that enstatite may breakdown to forsterite in the presence of vapor or of vapor + melt. After Kushiro et al. (1968).

thosite and a sinuous, irregular contact with the gabbro below. Irregularly shaped masses of gabbro are enclosed by troctolite and visa versa, and the igneous laminations in the enclosed gabbro masses possess the same orientation as that in the gabbro outside. The troctolite consists of ameboidal masses, typically 20–30 cm across, of polycrystalline coarse-grained (1 cm) olivine in a matrix of fine-grained (1–2 mm) plagioclase. The latter makes up about two thirds of the rock. The megascopic texture is not igneous [See field photographs of Hess (1960, fig. 2 and 3 of plate 5), McCallum et al. (1985, fig. 7C) and Foose (1985, figs. 8 and 9)]. Troctolites at the tops of OB–III and OB–IV, both of which are immediately overlain by massive anorthosites, are lithologically similar. The field and textural relationships clearly indicate that the ameboidal troctolites are also replacement features.

It has been suggested that the ameboidal troctolites formed by reaction of partially molten gabbro with $H_2O$-rich fluid to produce the olivine–plagioclase assemblage in a manner analogous to the formation of the discordant dunites, the fluid being trapped in specific zones where the overlying anorthosite acted as an impermeable cap (McCallum et al.,1977). The problem with this idea is that although the field of olivine stability increases with addition of water, that of plagioclase does not. In fact, in the synthetic diopside–anorthosite system, addition of water decreases the stability field of anorthosite relative to that of diopside (Yoder, 1965). The phase relations led Irvine (1980) to suggest that plagioclase-rich mafic pegmatites in the Skaergaard Intrusion formed by loss of water. Given that the Stillwater rocks are troctolitic, this mechanism does not seem to be appropriate either. A self-consistent hydrothermal mechanism for the formation of the ameboidal troctolite is not at hand. An alternative idea, which has yet to be fully developed, is that the troctolites formed by reaction of partially molten cumulate and anorthositic magma (Czamanske and Scheidle, 1985).

## THE BUSHVELD DISCORDANT PEGMATOIDS AND PIPES

The discordant pegmatitic bodies in the Bushveld Complex come in a bewildering variety, and a clear understanding of their petrogenesis or how they are related to each other has yet to emerge. They include the platiniferous dunite pipes, which are well known from the original description of Wagner (1929) and because they were once important sources of platinum; iron-rich ultramafic pegmatites, which are exposed in most of the mines; and unique features such as the Vlakfontein nickel pipes. There are also non-platiniferous, magnesian, dunite bodies in the Bushveld (Viljoen and Scoon, 1985). These may be similar to the Stillwater replacement dunites but are not described in detail. The following account is gleened from the literature.

### Platiniferous dunite pipes

The platinum-rich Driekop, Onverwacht and Mooihoek pipes are in the eastern Bushveld. The Onverwacht (Fig. 9.3) intersects Lower Critical Zone pyroxenites, and the Mooihoek and Driekop pipes intersect norite higher in the Critical Zone. Wagner described them as carrot-shaped bodies of coarse-grained hortonolitic dunite surrounded by finer grained and larger masses of magnesian dunite and wehrlite. He believed that they represent intrusions of dunitic magma.

The first systematic argument for a replacement origin was based on field and petrographic observations. Cameron and Desborough (1964) interpreted the structure in the Onverwacht dunite to be relic igneous layering and suggested that the pipes formed by reaction [1]. A similar conclusion was reached by Tarkian and Stumpfl (1975) and Stumpfl and Rucklidge (1982) based on more detailed petrographic and mineralogical studies. The Driekop pipe has also been described by Schiffries (1982), who argued that its formation primarily involved transformation of norite cumulates to dunite. Plagioclase dissolved in aqueous fluid, and orthopyroxene reacted with the fluid to form olivine according to reaction [1] at a temperature of 600 ± 150°C and pressure of 3.5 ± 1.0 kb. He computed that the process was accompanied by a reduction in the volume of the original rock by 67%, thereby accounting for the downwarping of the adjacent country rocks. The metasomatizing fluid was chloride-rich, and the assemblage was shifted to a more iron-rich composition as magnesium was preferentially leached from it by the fluid. The process can be represented by the exchange reaction,

$$Mg_2SiO_4(fo) + 2FeCl_2(v) = Fe_2SiO_4(fa) + 2MgCl_2(v). \quad [4]$$

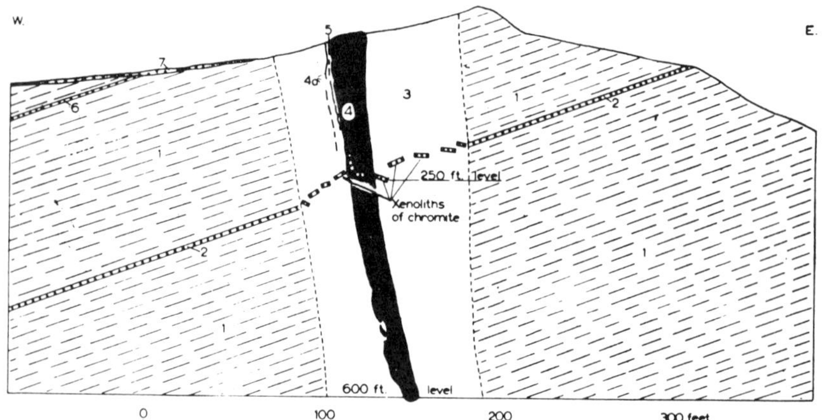


FIGURE 9.3—Schematic cross-section of the Onverwacht Pipe. The numbered lithologies are (1) bronzitite of the lower Critical Zone, (2) Steelpoort chromititie, (3) Mg-rich dunite and wehrlite, (4) hortonolitic dunite and (4a) "veins" of hortonolitic dunite. Originally drawn by P.A. Wagner and modified by Cameron and Desborough (1964). Reproduced with permission from Cameron and Desborough (1964).

The original paper provides additional details, and it remains the principal account of hydrothermal metamorphism in the Bushveld Complex.

### Iron-rich pegmatites

Iron-rich pegmatitic bodies are common in the Critical Zone and throughout most of the overlying stratigraphy. In the western Bushveld they are reported in all the mines (Viljoen et al., 1986a; Viljoen et al., 1986b; Viljoen and Hieber, 1986; Leeb–du Toit, 1986; Farquhar, 1986); in the eastern Bushveld they are particularly common near the Steelpoort and Dwars River faults (Cameron and Desborough, 1964). Most of the following information is summarized from the work of Viljoen and Scoon (1985), in which additional references to earlier descriptions can be found.

The pegmatitic rocks are represented by several lithologies. Most of those in the Critical and Main zones are massive bodies of dunite or wehrlite containing various proportions of hortonolite, hedenbergite and minor amounts of Ti-magnetite and ilmenite. Orthopyroxene, plagioclase, hornblende and biotite are never more than minor constituents, the sulfides tend to be iron-rich and may include troilite, and graphite is common.

One of the largest of these is the Townlands pipe (Viljoen and Hieber, 1986). It is an ovoid body with a dimension in plan view of $\approx 1.0 \times 1.5$ km. The pipe consists of a core of dunite surrounded in turn by wehrlite, clinopyroxenite and feldspathic clinopyroxenite. It contains a diverse collection of ore minerals, such as arsenides and lead-, bismuth- and antimony-bearing phases, that imply fluid metasomatism and reequilibration with cooling.

Some of the characteristics of the pegmatite occurrences are illustrated by Viljoen and Scoon's schematic cross-section (Fig. 9.4). The bodies may vary in form from discordant pipes and veins to quasi-stratiform pods and sheets. The feldspathic lithologies in the cumulate sequence appear to be selectively replaced compared to harzburgite and pyroxenite. Viljoen and Scoon suggested that horizons of the latter act as impermeable barriers to the upward movement of volatile-rich fluid or melt.

Two types of Fe–Ti oxide pegmatites are also reported. The first are sheet-like bodies in the Critical Zone. They are composed of magnetite having compositions intermediate between pure titanomagnetite and Critical Zone chromite. Cameron and Glover (1973) argued that these compositions developed by metasomatism of chromite-rich rocks, which resulted in progressive replacement of Cr, Mg and Al by $Fe^{2+}$, $Fe^{3+}$ and Ti. The second are "normal" Fe–Ti pegmatites, the classic accounts being those of Willemse (1969) and Molyneux (1970). These are discordant, pipe-like bodies composed of Cr-poor, Fe–Ti oxides and are restricted to the Middle and Upper zones (Cawthorn and Molyneux, 1986). Those in stratigraphic positions near the vanadium-rich magnetite cumulates at the base of the Upper Zone are also vanadium rich (e.g., Kennedy's Vale).

Following Wagner, Viljoen and Scoon conceived of platinum-free pegmatitic bodies as distinct from the platiniferous dunite pipes because the latter have as a dominant lithology dunite more magnesian than the surrounding cumulates, whereas all lithologies of the iron-rich pegmatites are much more iron-rich than the surrounding rocks. However, it has already been noted that the Driekop, Mooihoek and Onverwacht pipes have hortonolite cores, and at least at Driekop there exists a border phase of iron-rich wehrlite (Schiffries, 1982), so the distinction does not appear to be justified from a petrogenetic point of view.

### Vlakfontein nickel pipes

The Vlakfontein nickel pipes, once important as a source of nickel, are a group of pipe-like sulfide-rich chimneys in the Vlakfontein area of the western Bushveld. They have been described in detail by Liebenberg (1970) and Vermaak (1976). The pipes are hosted by the orthopyroxenite and harzburgite of the upper Lower Zone. Their map distribution suggests that they formed along or at intersections of faults. The pipes consist of cores of nearly pure, coarse-grained, massive pyrrhotite ($\approx$ 80%) and nickel + copper sulfides ($\approx$20%) surrounded by a poikilitic rock in which individual crystals of olivine and orthopyroxene are enclosed in a sulfide matrix. This in turn grades outward to pegmatitic pyroxenite or harzburgite containing disseminated sulfide. Copper is enriched in the poikilitic and disseminated ore, forming a halo around the nickel-rich core. As for nearly all the other pegmatite bodies, the silicates of the sulfide-rich rocks are iron-rich compared to those of the adjacent cumulate rocks.

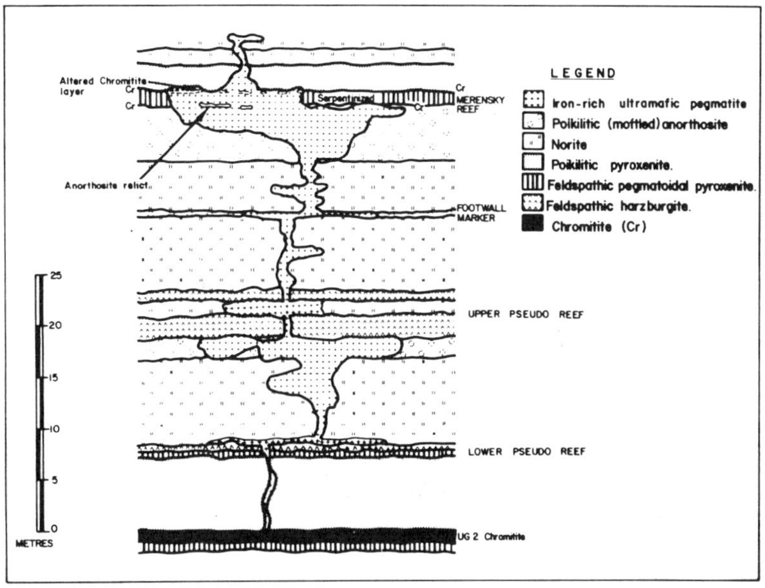

FIGURE 9.4—Schematic cross-section illustrating the distribution of iron-rich ultramafic pegmatite in upper Critical Zone rocks in the Amandelbult mine, northwest Bushveld. Reproduced with permission from Viljoen and Scoon (1985).

### High-temperature veins

Schiffries and Skinner (1987) have classified veins in the Bushveld based on their cross-cutting relationships and vein-filling assemblages. The high-temperature veins are pegmatitic and composed predominantly of calcic amphibole and plagioclase, with lesser amounts of clinopyroxene and Fe–Ti oxides and accessory orthopyroxene, biotite and sulfide. The assemblage corresponds to upper amphibolite facies ($T > 600°C$). Similar veins in the mines in the western Bushveld (Viljoen et al., 1986a) are locally known as "gash veins" or "flame structures." The veins typically exhibit a comb structure in which the long axes of vein minerals are perpendicular to the vein walls, as if the minerals grew inward from them. In the mines, most veins are observed to be approximately vertical and parallel to the regional dike and fault direction. Boudreau et al. (1986, fig. 1D) described an apatite-rich, microscopic vein composed of the high-temperature assemblage, and photographs of megascopic veins in mine wall exposures are presented by Viljoen et al. (1986a, fig. 23). Plagioclase–calcic amphibole–phlogopite veins are also observed in underground exposures in the Stillwater Complex. The assemblage suggests that the veins crystallized from volatile-rich melt.

Schiffries and Skinner (1987) also describe veins composed of greenschist to prehnite–pumpellyite facies minerals. However, they emphasize that specific vein assemblages are primarily controlled by host-rock compositions and that there is no evidence for "temporal discontinuities" in the vein formation processes, which they envision to be continuous during cooling. The Stillwater is cut by similar low-temperature vein systems, but these have not been studied in detail.

### Problems concerning petrogenesis

Three characteristics of the Bushveld pegmatites stand out from the literature descriptions. First, there is a clear relationship between the lithologies and bulk compositions of the pegmatites and those of the surrounding cumulates. Thus, magnesian dunite is described in orthopyroxenite and harzburgite in the Lower and lower Critical Zones; the platinum-rich dunites are restricted to the Critical Zone; the iron-rich dunite–wehrlite pegmatites are found in the upper Critical and Main Zones; and the magnetite bodies are present in the upper Critical, Main and Upper Zones (but generally absent above the magnetite-rich cumulates near the base of the Upper Zone). Second, wherever the local structure is known, the bodies are spatially associated with or aligned in directions parallel to faults and dikes. Third, nearly all the pegmatites are more iron-rich than the cumulates that host them. In this respect they are different from the Stillwater dunites.

These gross characteristics as well as the detailed field relationships indicate that the major process of pegmatite formation was infiltration metasomatism, and nearly all the workers who have in the last two decades described the rocks have adopted that view (e.g., Cameron and Desborough, 1964; Vermaak, 1976; von Gruenewaldt, 1979; Stumpfl and Rucklidge, 1982; Schiffries, 1982; Viljoen and Scoon, 1985; Farquhar, 1986; Leeb–du Toit, 1986). Missing from our understanding of the petrogenesis is the theoretical framework that provides a means of making sense of the profusion of descriptive details. Specific processes and how they operated need elucidation. Did metasomatism involve a compositionally evolved, volatile-rich intercumulus silicate melt, a vapor or a mixture of both; what were the compositions of the phases involved; could the partially molten cumulates have deform brittly and how were local structures involved; what was the source of the volatiles; what were the specific metasomatic reactions; are the pegmatites end products of a continuous chemical evolution which occurred over a wide temperature interval, perhaps spanning the solidus; what were the energetics of these processes? These basic questions remain unanswered.

## CHROMITITES, FLUIDS AND PGE

### PGE in the chromitite horizons

Most of the Stillwater and Bushveld chromitites are enriched in PGE compared to cumulates in which chromite is not a major phase. The A and portions of the J chromitites of the Stillwater (Page et al., 1976) and the UG–2, UG–3 and Merensky Reef of the Bushveld are particularly PGE-rich. (The Merensky usually includes two well-defined chromitite seams and is therefore rightfully included in this discussion). Ruthenium, iridium, and osmium, which exist primarily as the sulfide laurite, are clearly enriched in chromite in the Merensky Reef and Critical Zone chromitite layers (Kinloch, 1982; Lee and Fesq, 1986). PGE and chromite are also associated in ophiolites (e.g., Oshin and Crocket, 1982; Page and Talkington, 1984). Various reasons have been proposed to explain this association, among them that chromite may carry PGE in solution. The PGE–chromitite association is considered by Naldrett (this volume).

### Peculiarities of the UG–2

The high PGE concentrations and certain other geochemical characteristics of the UG–2 cannot be easily rationalized by processes such as accumulation of sulfides or high PGE solubility in igneous chromite. One problem is that the distribution of PGE within the chromitite is peculiar. Throughout the Bushveld, the UG–2 contains two PGE-rich horizons, only the top one of which is sulfide-rich (Fig. 9.5) (McLaren and DeVilliers, 1982; Gain, 1985); sharp internal breaks in Pd/Pt ratios are observed with vertical stratigraphic postion (Hiemstra, 1985); and high concentrations of PGE have been observed in the irregular mafic pegmatite that makes up the UG–2 footwall (e.g., Gain, 1985). Second, the relative concentrations of PGE in the bulk rock appear to be different than those of the chromite fraction alone. This possibility is suggested by the fact that laurite inclusions in UG–2 chromite are richer in Os than laurite occurring as interstitial grains (McLaren and DeVillier, 1982). An analogous relationship has been reported for Stillwater chromitite by Talkington and Lipin (1986), who found that although the PGE inclusions in chromite are mostly Ru–Ir–Os sulfides, the relative bulk-rock concentrations are such that Pd>Pt>Ru>Ir. Third, the dPGE/sulfide and (Ru + Ir + Os)/dPGE ratios of the UG–2 are much higher than those of the Merensky Reef and other sulfide-dominated PGE-rich zones (e.g., Hiemstra, 1979).

Recognizing these problems, recent investigations of the UG–2 have concentrated on unravelling postcumulus igneous and subsolidus recrystallization processes. Eales and Reynolds (1986) reported complex and small-scale heterogeneities in chromite compositions within the UG–2 and argued that these features developed from extensive reaction of chromite with intercumulus melt and with silicates at subsolidus temperatures. Von Gruenewaldt et al. (1986) suggested that the UG–2 experienced sulfur and iron loss during recrystallization. Naldrett and Lehmann (1987) showed that this could be accomplished by a subsolidus reaction involv-

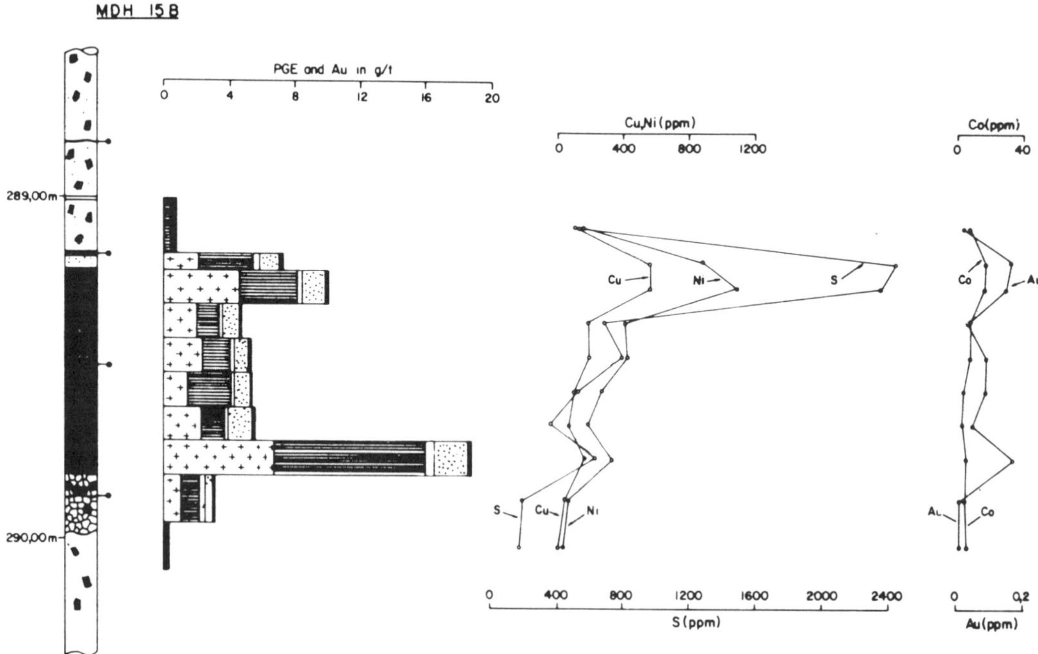


FIGURE 9.5—The distribution of the PGE, chalcophile elements and sulfur from whole-rock analyses through UG–2 chromitite illustrating that there are two distinct horizons enriched in PGE, only the top one of which is associated with sulfides. Patterns: Crosses (Pt), lines (Pd), blank (Rh), dots (Ru) and black (Os, Ir, Au). Reproduced with permission from Gain (1985).

ing pyrrhotite and chromite to produce more Fe-rich chromite, thereby releasing sulfur.

### The chromitite-mafic pegmatite association

One of the extraordinary features of Stillwater and Bushveld chromitites is their association with pegmatoidal rocks. In the Stillwater Complex, this was first noted by Peoples and Howland (1940). Mafic pegmatites are particularly common around the G and H chromitites, which are the two thickest horizons, and the J chromitite is usually encased in pegmatite (Jackson, 1969; Sampson, 1969). In the Bushveld, mafic pegmatite forms the footwall of the UG–2 (e.g., McLaren and DeVilliers, 1982; Gain, 1985; Viljoen et al., 1986a). Pegmatite is also usually observed with the Steelpoort seam (Ireland, 1986), the UG–3A chromitite (Gain, 1985), and the lower and upper "pseudoreef" marker horizons in the western Bushveld mines (Viljoen et al., 1986a), and, of course, the Merensky Reef itself is mostly pegmatitic.

The mafic pegmatites exhibit textural and modal diversity. As described by Howland et al. (1949, p. 70) in the Stillwater Complex:

> "Irregular bodies of a rock of coarse and irregular texture occur at several places in the ultramafic zone. Most of them consist mainly of 1- to 8-inch crystals of bronzite and diopside, accompanied by 10 to 20 percent of interstitial plagioclase. Some, however, consist entirely of bronzitite, or entirely of diopside, or of both pyroxenes with as much as 30 percent of feldspar. In the feldspar-rich varieties the dominant pyroxene is diopside, and from 2 to 10 percent of interstitial chromite is present."

The mineral compositions are identical to those of the finer grained host rocks. The mafic pegmatites associated with the Bushveld chromitites are similar to the Stillwater occurrences.

The reason for the chromitite–pegmatite association is not obvious. Jackson (1969) suggested that the chromitite horizons acted as impermeable layers to upwardly migrating fluids. In the case of the Merensky Reef and the Bastard Reef above it, the hanging wall is composed of massive pyroxenite, which may also have acted as a local cap (Lauder, 1970). This, of course, does not explain why these horizons also contain chromite. Page et al. (1985) noted that mafic pegmatite may occur above, below or within any individual seam, that there is no correlation between thickness of the seam and the amount of pegmatite associated with it and that pegmatites are typically in irregular and patchy occurrences.

Several additional observations make the association even more curious. First, in most chromitites phlogopite is relatively abundant, comprising up to 2% of the modes of some Stillwater rocks (Jackson, 1961). Second, not all chromitites have associated pegmatites. For example, the to the best of this writer's knowledge, mafic pegmatite is not observed with the Bushveld UG–1 or UG–3 chromitites. Third, there are many pods of mafic pegmatite not associated with chromitite seams. However, at least in the Stillwater, as noted above, podiform pegmatite bodies tend to contain more chromite than surrounding, finer grained rocks (Page et al., 1985).

### Evidence for influence of fluid on PGE

#### PGE mineral variations in the UG–2 and Merensky

Of particular interest are the data of Peyerl (1982), who studied PGE mineralogy of the UG–2 in several drill cores through the UG–2 in the vicinity of the Driekop Pipe. The PGE mineralogy of the pipe itself is dominated by platinum–iron metal, sperrylite ($PtAs_2$) and other arsenides (Stumpfl, 1961), and that of UG–2 distant from the pipe is comprised mainly of sulfides. Peyerl found that the PGE mineralogy of the UG–2 adjacent to the Driekop Pipe is similar to that of the pipe itself and that the PGE mineral grain-size progressively decreases away from the pipe. The relationships are summarized in Table 9.1.

The variations exhibited by the UG–2 around the Driekop have been termed by Kinloch (1982, p. 1330) a "microcosm" of those observed in the Merensky and UG–2 around potholes and larger stratigraphic disturbances. The situation around potholes is summarized in Fig. 9.6 and Table 9.2. The normal reef is usually dominated by platinum- and paladium-bearing sulfides. In contrast, the pothole reef contains mostly platinum–iron metal. Laurite is concentrated in the contact reef, which is thus relatively enriched in ruthenium. The base metal sulfide mineralogy also exhibits a systematic variation. The typical magmatic assemblage of pyrrhotite–pentlandite–chalcopyrite is characteristic of normal reef, the contact reef is chalcopyrite-rich, and the poth-

TABLE 9.1—Relative proportions (in percent) of PGE-bearing minerals in the UG2 at various distances from the Driekop Pipe (Peyerl, 1982).

| Core | Driekop* | V4 | DC5 | DC6 | DC7 | DC8 |
|---|---|---|---|---|---|---|
| Distance from Driekop (m) | | 100 | 1300 | 2500 | 4000 | 4500 |
| Pt-Fe alloys, incl. base metal sulfide intergrowths | 50 | 29.8 | 0 | 0 | 3.1 | 0.4 |
| Sperrylite ($PtAs_2$) and other arsenides, Te, Sb, Pb, Hg phases | 45 | 40.0 | 0.9 | 0.1 | 2.7 | 0.4 |
| Braggite (Pt,Pd,Ni)S, cooperite (PtS), other PGE + Ni and Cu sulfides | 5 | 14.5 | 71.2 | 87.1 | 78.9 | 85.4 |
| Laurite $Ru(Os,Ir)S_2$ | | 15.6 | 27.8 | 12.7 | 15.3 | 13.7 |

*Data from Stumpfl (1961), as cited by Peyerl.

TABLE 9.2—Relative abundances of the PGE minerals in pothole, contact and normal reef. (See Fig. 9.6 for terminology.) Data are from Kinloch (1982).

| | Normal | Contact | Pothole Reef |
|---|---|---|---|
| Pt–Fe alloys + base metal sulfide intergrowths | 0.1 | 0 | 92.6 |
| Pt–Pd sulfides | 89.0 | 11.5 | 0 |
| Laurite $Ru(Os,Ir)S_2$ | 0.1 | 87.0 | 0.4 |
| Tellurides and Cu–Au alloy | 10.8 | 1.5 | 7.0 |

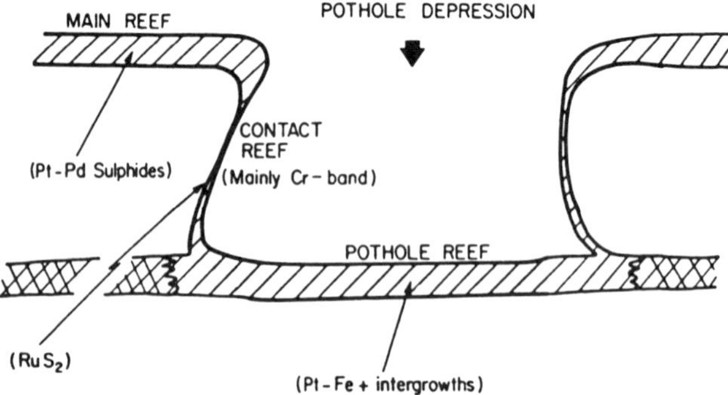


FIGURE 9.6—Schematic cross-section of a pothole in the Merensky Reef illustrating the systematic variation in PGE mineralogy around it. Reproduced with permission from Kinloch (1982).

ole reef contains, in addition to the magmatic sulfide assemblage, sulfur-deficient phases, such as the iron sulfide mackinawite and the iron-rich copper sulfide cubanite. The potholes may also be enriched in graphite, and there may be systematic differences in intrinsic oxygen fugacities of minerals from different parts of the reef (Elliot et al., 1982; Buntin et al., 1985).

Regional variations in PGE mineralogy have also been reported (Kinloch, 1982). Two important features of these variations are first that the mineralogy of the UG–2 and Merensky are approximately similar and vary together along strike and second that the proportions of Pt–Fe metal in both units are much higher than normal in certain localized areas thought to be in the proximity of regional feeders zones, such as at the Union mine (northwest Bushveld).

Kinloch (1982) proposed that both the regional and localized mineralogical variations developed in response to the introduction of relatively oxidized fluid, which resulted in breakdown of sulfide, loss of sulfur, and the reaction of platinum and iron released by the reaction to form Pt–Fe alloy. As emphasized by Kinloch, the alloys are platinum-rich and therefore stable to relatively high fO2s (Taylor and Muan, 1962). The suggestion that the mineralization around potholes is in part related to introduction of fluid is supported by the data of Cawthorn and Poulton (1987), who reported that a specific horizon just below the Merensky is more potassium-rich where it underlies potholes than where it underlies the normal reef. The relations between structural disturbances, petrography, mineral assemblage and geochemistry are not clearly established, and more systematic studies are needed.

It is interesting to note that the idea that sulfur loss was an important mechanism in determining the local character of mineralization has now been developed by several of the Bushveld workers studying different parts of the Complex (e.g., Kinloch, 1982; Gain, 1985; von Gruenewaldt et al., 1986; Naldrett and Lehmann, 1987).

### The Stillwater Picket Pin horizon

The Picket Pin horizon (Boudreau and McCallum, 1986) is a discontinuous zone of disseminated sulfide within and near the top of An–II. Although the horizon is stratabound on a regional scale, the sulfides are concentrated in isolated, podiform and lenticular zones, which are typically meters across. The PGE are concentrated in some of the sulfide-rich zones, and their anorthosite host rock tends to contain less postcumulus pyroxene and more accessory apatite and quartz than the surrounding anorthosite. [See Salpas et al. (1987) for a description and explanation of the relationship between pyroxene mode and geochemistry in the Stillwater anorthosites.] The disseminated sulfide may also be distributed in a pipe-like fashion, extending downward through the stratigraphy 10's to more than 100 meters. The field and petrographic observations are not consistent with accumulation of sulfide-oxide liquid from overlying magma. Boudreau and McCallum proposed that the sulfides and PGE were transported by fluid, which migrated upward through the partially molten rocks and was trapped near the top of An–II because the overlying rocks were completely solidified.

## VOLATILE-RICH PHASES

### Halogen-bearing phases

### Apatite

Apatite occurs as a noncumulus accessory phase in most Stillwater rocks and is probably widespread throughout the Bushveld as well. In the Bushveld Critical Zone and Stillwater OB–I, apatite is observed in intersticies with phlogopite in mafic pegmatite. Apatite is a cumulus phase in the diorite of subzone C of the Upper Zone of the Bushveld (von Gruenewaldt, 1973). An equivalent, highly evolved horizon is missing in the Stillwater Complex, probably because of erosion. Both cumulus and noncumulus apatite have been reported from other intrusions as well (Nash, 1976; Brown and Peckett, 1977; Huntington, 1979).

Boudreau et al. (1986) distinguished three compositional types (Fig. 9.7). Cumulus apatite in all intrusions is essentially fluorapatite with less than 3 mole% chlorapatite. The compositions of most noncumulus apatites are intermediate between the fluorine and OH-endmembers. However, noncumulus apatite in the Stillwater and Bushveld PGE-bearing rocks was found to be extremely chlorine rich. More recently, Boudreau and McCallum (1987) have found that chlorine-rich apatite is not restricted to OB–I of the Stillwater but in fact is present throughout the stratigraphy below it and

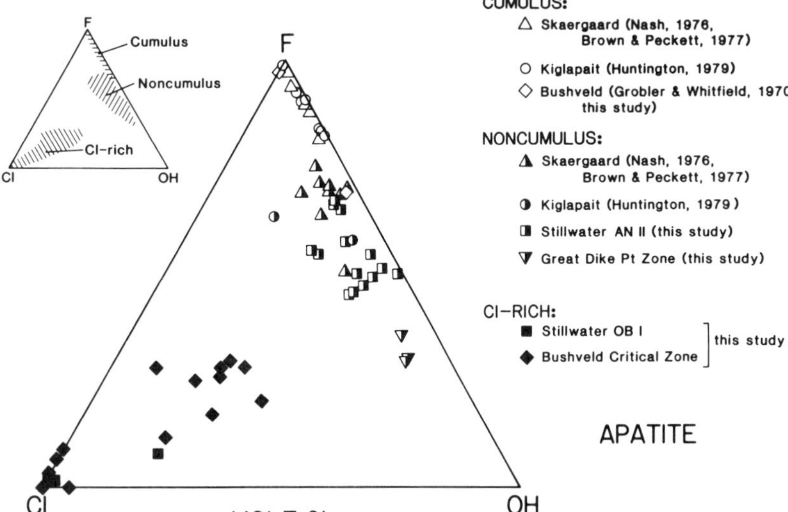


FIGURE 9.7—Compositions of apatite from the Stillwater, Bushveld and other layered intrusions. Reproduced from Boudreau et al. (1986).

absent from overlying rocks. Chlorine-rich apatite is usually enriched in rare earth elements, containing up to 1.3 wt.% $Ce_2O_3$, $La_2O_3$ and $Y_2O_3$. Specific compositional data and photomicrographs are presented by Boudreau et al. (1986).

The chlorapatite was interpreted to indicate the existence of a chlorine-rich vapor at supersolidus conditions. The reasoning is that the partitioning behavior of halogens (Kilinc and Burnham, 1972; Hards, 1976; Korzhinskiy, 1982; Webster and Holloway, 1988) among vapor (v), melt (m) and apatite (ap) is such that

$$(Cl/F)_{ap} < (Cl/F)_{m} < (Cl/F)_{v}. \quad [5]$$

Obviously, chlorine-rich apatite can only crystallize from melts having high Cl/F ratios. However, such compositions are not found in the spectrum of natural melt compositions, as represented by either subaerial lavas or glass inclusions in phenocrysts (e.g., Devine et al., 1984). In addition, continued degassing of a magma chamber will result in residual melts having progressively lower Cl/F ratios. Finally, with cooling below the solidus the fractionation of chlorine from fluorine in a vapor–apatite assemblage becomes even more extreme, and chlorapatites should not form.

### Phlogopite

Phlogopite is a common accessory mineral in Stillwater and Bushveld rocks (Boudreau et al., 1986; Ballhaus and Stumpfl, 1986), especially in mafic pegmatites. In some it is modally important. Parts of the Merensky Reef, for example, consist of more than 5% phlogopite. There appear to be two distinct parageneses. Phlogopite clearly replaces earlier orthopyroxene, probably having formed by the reaction melt + vapor + orthopyroxene = phlogopite + quartz, which may have proceeded at temperatures as low as 750°C (Ballhaus and Stumpfl, 1986). Interstitial phlogopite is also present, usually with chromite and incompatible-element-rich trace phases (e.g., apatite, zircon, sphene, rutile, orthoclase, tourmaline), and presumably crystallized directly from evolved intercumulus melt. Systematic differences exist in the minor element contents of the two petrographic types, with interstitial phlogopite being enriched in titanium, chromium, and barium and depleted in iron relative to replacement phlogopite.

Phlogopite is typically chlorine rich (Ballhaus and Stumpfl, 1985; Johan and Watkinson, 1985; Boudreau et al., 1986) but exhibits considerable compositional heterogeneity within individual thin sections as well as on larger scales (Fig. 9.8). This is unexpected in view of the fact that there are strong crystal-chemical controls on the amount of chlorine that can be incorporated into the mica (or amphibole) structure (e.g., Volfinger et al., 1985). This led Boudreau et al. (1986) to the conclusion that the phlogopite experienced continued reequilibration with pore fluid, which itself was continually evolving as the intrusions cooled. Compositional heterogeneities developed in the pore fluids because they remained isolated and reequilibrated within their microscopic, local environments. This implies that the intergranular fluids were not involved in large-scale circulation. Therefore, fluid permeabilities in the subsolidus environment were not along grain boundaries and microfractures spaced at centimeter or smaller intervals, and these fluids were probably distinct compositionally from those associated with megascopic, through-going fractures.

### Graphite

Disseminated graphite is a common accessory phase in the Bushveld mafic pegmatites (Wagner, 1924), where it is found in intercumulus patches of hornblende, phlogopite and sulfide. The petrographic relations suggest that graphite originally crystallized under supersolidus conditions. This was first suggested by Schneiderhohn (1929), who envisioned graphite to have formed by reaction of "readily fugitive compounds" concentrated in residual interstitial magma.

Also present in the Stillwater (Volborth and Housley, 1984)

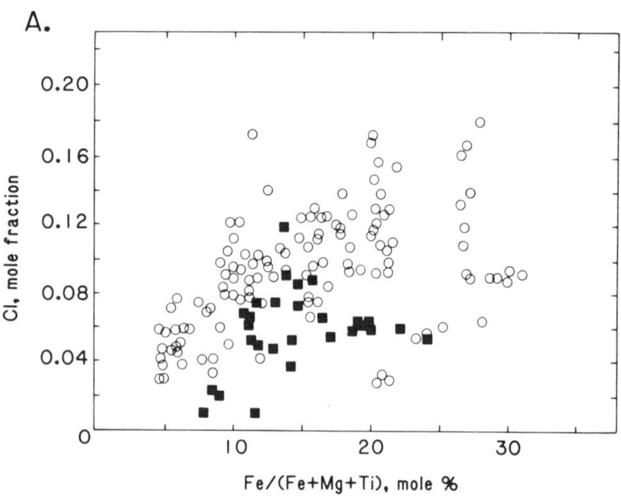


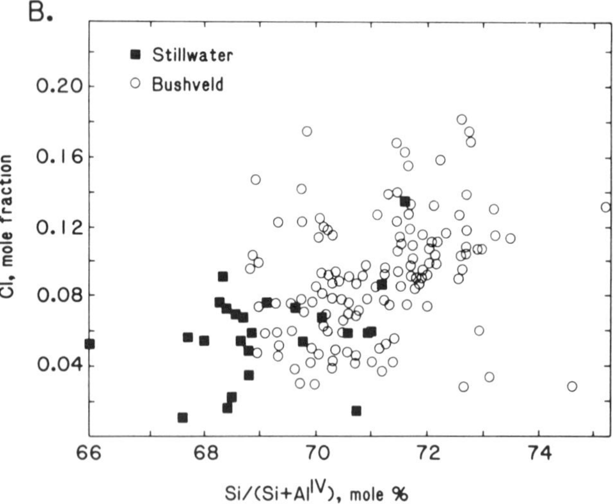


FIGURE 9.8—The relationship for Stillwater and Bushveld phlogopite between chlorine contents and (A) molar Fe/(Fe + Mg + Ti) and (B) Si/(Si + AlIV). The amount of chlorine that can be incorporated into phlogopite is strongly dependent on these compositional variables (e.g., Volfinger et al., 1985). The lack of a well-defined relationship implies that factors such as variability of pore fluid composition also influenced phlogopite composition. Reproduced from Boudreau et al. (1986).

and Bushveld (Ballhaus and Stumpfl, 1985) are massive bodies of graphite and rocks in which graphite is a major constituent. The Stillwater Complex has been subjected to greenschist facies metamorphism, parts of it are moderately serpentinized, and therefore the graphite is difficult to interpret uniquely (Mathez et al., 1988). In contrast, the Bushveld massive graphite must be magmatic because it is distinctly associated with mafic pegmatite. The graphite must have precipitated from magmatic vapor rather than crystallized from melt because vapor is the only phase capable of concentrating large amounts of carbon in one place. The details of how this may have occurred and what were the consequences for fluid evolution are summarized by Mathez (this volume).

### Fluid inclusions

Ballhaus and Stumpfl (1986) describe fluid inclusions in Merensky Reef rocks. The inclusions are 5–50 fm in diameter and disposed in planar arrays in quartz, which is a ubiquitous accessory mineral in the Reef. A variety of inclusion compositions have been identified (Fig. 9.9). The compositions have been interpreted to indicate minimum trapping temperatures of > 750°C. The wide range of compositions also implies that inclusions formed throughout a wide temperature interval. In addition to the compositions shown on Fig. 9.9, $CH_4$-rich inclusions have been identified.

The heterogeneity of fluid compositions is consistent with the notion based on phlogopite compositions that fluids reequilibrated in local, microscopic environments during cooling. The highly saline character of some fluid inclusions is also in agreement with the mineralogical data indicating that high-temperature fluid was chlorine rich.

## CONCLUSION

The original magmatic assemblages and textures of Stillwater and Busvheld rocks have been locally modified and in some cases completely obliterated by processes involving volatiles. Some of these processes are moderately well understood. A replacement origin for the Stillwater dunites, for example, has a foundation in experiment and is consistent with field and petrographic observations. The petrogeneses of other rocks, however, are only vaguely understood, if at all. In particular, there is no suitable hypothesis to explain why mafic pegmatites are associated with chromitites or even why pegmatites are developed at certain stratigraphic levels, and the role of infiltration metasomatism in the formation of the Bushveld iron-rich pegmatites is not defined by specific thermochemical arguments.

There must be a continuum of reactions involving volatiles. Modifications to original cumulate texture must commence with reaction of intercumulus volatile-rich melt and cumulate minerals, and further modifications presumably involve as well an exsolving magmatic vapor. Overprinted on these are the subsolidus reactions between pore fluid and the crystalline assemblage, which appear to reflect equilibria established over microscopic regions and result in compositional heterogeneities of fluids and volatile-bearing phases on that scale. Subsequent retrograde effects may involve fluids flowing through fractures on a regional scale. At least in the Stillwater and Bushveld complexes, these retrograde effects appear to be localized to fractures themselves.

Exactly how the high-temperature fluids and partially molten to completely solidified rocks interact is made even more complicated if, as is indicated by the model of fluid evolution summarized by Mathez (this volume), the fluids themselves undergo dramatic changes in composition with cooling. As a fluid evolves from $CO_2$- to $H_2O$-rich compositions, its ability to dissolve metals and other components of the silicate assemblage probably increases. The transient nature of the fluid is probably one reason it is so difficult to

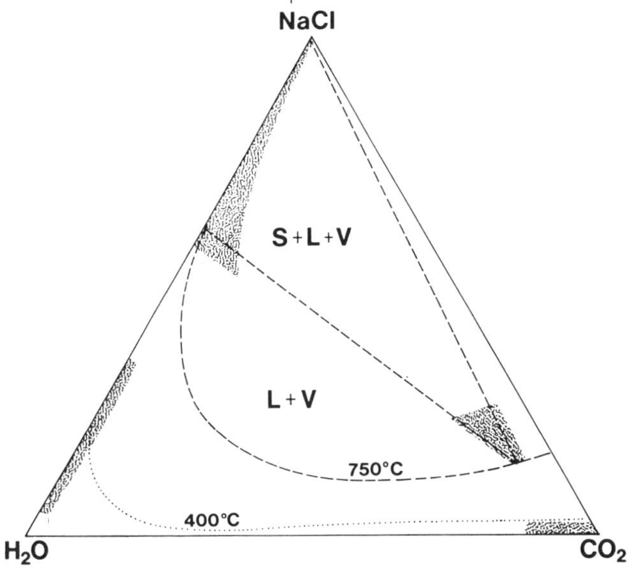


FIGURE 9.9—Range of compositions (stippled) of $NaCl–H_2O–CO_2$ fluid inclusions from the Merensky Reef. Methane-rich inclusions are also present. The isobaric phase relations are shown for 2 kb and 750°C (dashed) and 400°C (dotted). S = solid; L = liquid; V = vapor. After Ballhaus and Stumpfl (1986).

characterize the near-solidus environment and specifically deduce how fluid interacts with the condensed assemblage.

Finally comes the question of how fluids may have redistributed the PGE. The geochemical characteristics of the Bushveld platiniferous pipes and adjacent cumulate host rocks, the distribution and mineralogy of the PGE in the Merensky Reef and the PGE enrichments in the Stillwater Picket Pin horizon provide clear evidence for fluid involvement in redistribution processes. However, there is simply insufficient knowledge of PGE solubilities in complex geologic fluids at near solidus temperatures to develop specific hypotheses, which for the moment must rest only on field, petrographic and geochemical observations. Whether or not fluids were partly responsible for initial concentration of the PGE in the stratiform horizons is a matter for speculation.

Acknowledgements—Reviews of this paper by A. J. Naldrett, A. E. Boudreau and C. L. Peach are gratefully acknowledged. This work was supported by NSF grant EAR8720982.

## REFERENCES

Ballhaus, C. G., and Stumpfl, E. F., 1985, Occurrence and petrologic significance of graphite in the Upper Critical Zone, western Bushveld Complex, South Africa: Earth Planet. Sci. Lett., v. 74, pp. 58–68.

Ballhaus, C. G., and Stumpfl, E. F., 1986, Sulfide and platinum mineralization in the Merensky Reef: evidence from hydrous silicates and fluid inclusions: Contrib. Mineral. Petrol., v. 94, pp. 193–204.

Boudreau, A. E., Mathez, E. A., and McCallum, I. S., 1986, Halogen geochemistry of the Stillwater and Bushveld Complexes: Evidence for transport of the platinum-group elements by Cl-rich fluids: J. Petrol., v. 27, pp. 967–986.

Boudreau, A. E., and McCallum, I. S., 1986, Investigations of the Stillwater Complex: III. The Picket Pin Pt/Pd deposit: Econ. Geol., v. 81, pp. 1953–1975.

Boudreau, A. E., and McCallum, I. S., 1987, Halogens in apatites from the Stillwater Complex: Preliminary results (abs.): American Geophysical Union, EOS, Transactions, v. 68, p. 1518.

Brown, G. M., and Peckett, A., 1977, Fluorapatites from the Skaergaard intrusion, East Greenland: Miner. Mag., v. 41, pp. 227–232.

Buntin, T. J., Grandstaff, D. E., Ulmer, G. C., and Gold, D. P., 1982, A pilot study of geochemical and redox relationships between potholes and adjacent normal Merensky Reef of the Bushveld Complex: Econ. Geol., v. 80, pp. 975–987.

Cameron, E. N., and Desborough, G. A., 1964, Origin of certain magnetite-bearing pegmatites in the eastern part of the Bushveld Complex, South Africa: Econ. Geol., v. 59, pp. 197–225.

Cameron, E. N., and Glover, E. D., 1973, Unusual titanium-chromian spinels from the eastern part of the Bushveld Complex, South Africa: Am. Mineral., v. 58, pp. 172–188.

Cawthorn, R. G., and Molyneux, T. G., 1986, Vanadiferous magnetite deposits of the Bushveld Complex; *in* Anhaeusser, C. R., and Maske, S. (eds.), Mineral deposits of South Africa: Geological Society of South Africa, Johannesburg, pp. 1251–1266.

Cawthorn, R. G., and Poulton, K., 1987, Evidence for fluid in the footwall to potholes below the Merensky Reef, Bushveld Complex (abs.): Geo-Platinum 87 Symposium, Open University, Milton Keynes, Paper F12.

Czamanske, G. K., and Scheidle, D. L., 1985, Characteristics of the Banded-series anorthosites; *in* Czamanske, G. K., and Zientek, M. L. (eds.), The Stillwater Complex, Montana: Geology and Guide: Montana Bureau of Mines and Geology, Special Publication 92, pp. 334–345.

Devine, J. D., Sigurdsson, H., Davis, A. N., and Self, S., 1984, Estimates of sulfur and chlorine yield to the atmosphere from volcanic eruptions and potential climatic effects: J. Geophys. Res., v. 89, pp. 6309–6325.

Dungan, M. A., and Ave Lallemant, H. G., 1978, Formation of small dunite bodies by metasomatic transformation of harzburgite in the Canyon Mountain Ophiolite, northeastern Oregon; *in* Dick, H. J. B. (ed.), Magma genesis: Oregon Department of Geology and Mineral Industries, Bulletin 96, pp. 109–128.

Eales, H. V., and Reynolds, I. M., 1986, Cryptic variations within chromitites of the upper Critical Zone, northwest Bushveld Complex: Econ. Geol., v. 81, pp. 1056–1066.

Elliott, W. C., Grandstaff, D. E., Ulmer, G. C., Buntin, T., and Gold, D. P., 1982, An intrinsic oxygen fugacity study of platinum-carbon associations in layered intrusions: Econ. Geol., v. 77, pp. 1493–1510.

Farquhar, J., 1986, The Western Platinum mine; *in* Anhaeusser, C. R., and Maske, S. (eds.), Mineral deposits of South Africa: Geological Society of South Africa, Johannesburg, pp. 1135–1142.

Foose, M. P., 1985, Primary structural and stratigraphic relations in Banded-series cumulates exposed in the East Boulder Plateau–Contact Mountain area; *in* Czamanske, G. K., and Zientek, M. L. (eds.), The Stillwater Complex, Montana: Geology and Guide: Montana Bureau of Mines and Geology, Special Publication 92, pp. 305–324.

Frost, B. R., 1975, Contact metamorphism of serpentinite, chloritic blackwall and rodingite at Paddy–Go–Easy Pass, Central Cascades, Washington: J. Petrol., v. 16, pp. 272–313.

Gain, S. B., 1985, The geologic setting of the platiniferous UG–2 chromitite layer on the farm Maandagshoek, eastern Bushveld Complex: Econ. Geol., v. 80, pp. 924–943.

Hards, N., 1976, Distribution of elements between the fluid phase and silicate melt phase of granites and nepheline syenites; *in* Biggar, G. M. (ed.), Progress in experimental petrology: Natural Environment Research Council, Publication Series D, No. 6, pp. 88–90.

Hess, H. H., 1960, Stillwater igneous complex, Montana—a quantitative mineralogical study: Geological Society of America, Memoir 80, 230 pp.

Hiemstra, S. A., 1979, The role of collectors in the formation of the platinum deposits of the Bushveld Complex: Can. Mineral., v. 17, pp. 469–482.
Hiemstra, S. A., 1985, The distribution of some platinum-group elements in the UG–2 chromitite layer of the Bushveld Complex: Econ. Geol. v. 80, pp. 944–957.
Howland, A. L., Garrels, E. M., and Jones, W., 1949, Chromite deposits of Boulder River area, Sweetgrass County, Montana: U.S. Geological Survey, Bulletin 948–C, pp. 63–82.
Howland, A. L., Peoples, J. W., and Sampson, E., 1936, The Stillwater igneous complex and associated occurrences of nickel and platinum group metals: Montana Bureau of Mines and Geology, Miscellaneous Contribution 7, 15 pp.
Huntington, H. D., 1979, Kiglapait mineralogy I: Apatite, biotite, and volatiles: J. Petrol., v. 20, pp. 625–652.
Ireland, K. L., 1986, The Winterveld (TCL) Chrome Mine Limited, eastern Bushveld Complex; *in* Anhaeusser, C. R., and Maske, S. (eds.), Mineral deposits of South Africa: Geological Society of South Africa, Johannesburg, pp. 1183–1188.
Irvine, T. N., 1974, Petrology of the Duke Island Ultramafic Complex, southeastern Alaska: Geological Society of America, Memoir 138, 240 pp.
Irvine, T. N., 1980, Magmatic infiltration metasomatism, double-diffusive fractional crystallization, and adcumulus growth in the Muskox Intrusion and other layered intrusions; *in* Hargraves, R. B. (ed.), Physics of magmatic processes: Princeton University Press, Princeton, N.J., pp. 325–383.
Jackson, E. D., 1961, Primary textures and mineral associations in the Ultramafic Zone of the Stillwater Complex Montana: U.S. Geological Survey, Professional Paper 358, 106 pp.
Jackson, E. D., 1969, Chemical variation in coexisting chromite and olivine in chromitite zones of the Stillwater Complex: Society of Economic Geologists, Monograph 4, pp. 41–75.
Johan, Z., and Watkinson, D. H., 1985, Significance of a fluid phase in platinum-group-element concentration: Evidence from the Critical Zone, Bushveld Complex: Can. Miner., v. 23, pp. 305–306.
Kilinc, I. A., and Burnham, C. W., 1972, Partitioning of chloride between a silicate melt and coexisting aqueous phase from 2 to 8 kilobars: Econ. Geol., v. 67, pp. 231–235.
Kinloch, E. D., 1982, Regional trends in the platinum-group mineralogy of the Critical Zone of the Bushveld Complex, South Africa: Econ. Geol., v. 77, pp. 1328–1347.
Korzhinskiy, M. A., 1982, Apatite solid solutions as indicators of the fugacity of HCl° and HF° in hydrothermal fluids: Geochemistry International 1981, pp. 44–60.
Kushiro, I., Yoder, H. S., Jr., and Nishikawa, M., 1968, Effect of water on the melting of enstatite: Geological Society of America, Bulletin, v. 79, pp. 1685–1692.
Lauder, W. R., 1970, Origin of the Merensky Reef: Nature, v. 227, pp. 365–366.
Lee, C. A., and Fesq, H. W., 1986, Au, Ir, Ni and Co in some chromitites of the eastern Bushveld Complex, South Africa: Chem. Geol., v. 62, pp. 227–237.
Leeb–du Toit, A., 1986, The Impala Platinum mines; *in* Anhaeusser, C. R., and Maske, S. (eds.), Mineral deposits of South Africa: Geological Society of South Africa, Johannesburg, pp. 1091–1106.
Liebenberg, L., 1970, The sulphides in the layered sequence of the Bushveld igneous complex: Geological Society of South Africa, Special Publication 1, pp. 108–207.
Mathez, E. A., Dietrich, V. J., Holloway, J. R., and Boudreau, A. E., 1988, Carbon distribution in the Stillwater Complex and evolution of vapor during crystallization of Stillwater and Bushveld magmas: J. Petrol. (in press).
McCallum, I. S., Raedeke, L. D., and Mathez, E. A., 1977, Stratigraphy and petrology of the Banded Zone of the Stillwater Complex, Montana (abs): American Geophysical Union, EOS, Transactions, v. 58, p. 1245.
McCallum, I. S., Raedeke, L. D., and Mathez, E. A., 1980, Investigations of the Stillwater Complex: Part I. Stratigraphy and structure of the Banded Zone: Am. J. Sci., v. 280–A, pp. 59–87.
McCallum, I. S., Raedeke, L. D., Mathez, E. A., and Criscenti, L. J., 1985, A traverse through the banded series in the Contact Mountain Area: Montana Bureau of Mines and Geology, Special Publication 92, pp. 293–304.
McLaren, C. H., and DeVilliers, J. P. R., 1982, The platinum-group chemistry and mineralogy of the UG–2 chromitite layer of the Bushveld Complex: Econ. Geol., v. 77, pp. 1348–1366.
Molyneux, T. G., 1970, The geology of the area in the vicinity of Magnet Heights, eastern Transvaal, with special reference to magnetite iron ore: Geological Society of South Africa, Special Publication 1, 228–241.
Nakamura, Y., and Kushiro, I., 1974, Composition of the gas phase in $Mg_2SiO_4$—$SiO_2$—$H_2O$ at 15 kbars: Carnegie Institute of Washington, Year Book 73, pp. 255–258.
Naldrett, A. J., and Lehmann, J., 1987, Spinel non-stoichiometry as the explanation for Ni-, Cu-, and PGE-enriched sulphides in chromitites (abs.): Geo-Platinum 87 Symposium, The Open University, Milton Keynes, Paper T10.
Nash, W. P., 1976, Fluorine, chlorine and OH-bearing minerals in the Skaergaard intrusion: Am. J. Sci., v. 276, pp. 546–557.
Nicholson, S. W., and Lipin, B. R., 1985, Guide to the Gish mine area; *in* Czamanske, G. K., and Zientek, M. L. (eds.), The Stillwater Complex, Montana: Geology and Guide: Montana Bureau of Mines and Geology, Special Publication 92, pp. 358–367.
Oshin, I. O., and Crocket, J. H., 1982, Noble metals in Thetford mines ophiolites, Quebec, Canada—Part I: Distribution of gold, iridium, platinum and palladium in the ultramafic and gabbroic rocks: Econ. Geol., v. 77, pp. 1556–1570.
Page, N. J., 1977, Stillwater Complex, Montana: Rock succession, metamorphism and structure of the Complex and adjacent rocks. U.S. Geological Survey, Professional Paper 999, 79 pp.
Page, N. J., Rowe, J. J., and Haffty, J., 1976, Platinum metals in the Stillwater Complex, Montana: Econ. Geol., v. 71, pp. 1352–1363.
Page, N. J., and Talkington, R. W., 1984, Pd, Pt, Rh, Ru, and Ir in peridotites and chromites from ophiolite complexes in New Foundland: Can. Mineral., v. 22, pp. 137–149.
Page, N. J., Zientek, M. L., Lipin, B. R., Raedeke, L. D., Wooden, J. L., Turner, A. R., Loferski, P. J., Foose, M. P., Moring, B. C., and Ryan, M. P., 1985, Geology of the Stillwater Complex exposed in the Mountain View area and on the west side of the Stillwater Canyon: Montana Bureau of Mines and Geology, Special Publication 92, pp. 147–209.
Peoples, J. W., and Howland, A. L., 1940, Chromite deposits of the eastern part of the Stillwater Complex, Stillwater County, Montana: U.S. Geological Survey, Bulletin 922–N, pp. 371–416.
Peyerl, W., 1982, The influence of the Driekop dunite pipe on the platinum-group mineralogy of the UG–2 chromitite in its vicinity: Econ. Geol., v. 77, pp. 1432–1438.
Raedeke, L. D.,and McCallum, I. S., 1984, Investigations in the Stillwater Complex: Part II. Petrology and petrogenesis of the Ultramafic Series: J. Petrol., v. 25, pp. 395–420.
Salpas, P. A., Haskin, L. A., and McCallum, I. S., 1987, Trace-element distributions among subunits of a Stillwater anorthosite boulder (abs.): Lunar Planet. Sci. XVIII, pp. 872–873.
Sampson, E., 1969, Stillwater, Montana chromite deposits—(Discussion of "Chemical variation in coexisting chromite and olivine in chromite zones of the Stillwater Complex," by E.D. Jackson, 1969): Society of Economic Geologists, Monograph 4, pp. 72–75.
Schiffries, C. M., 1982, The petrogenesis of a platiniferous dunite pipe in the Bushveld Complex: Infiltration metasomatism by a chloride solution: Econ. Geol., v. 77, pp. 1439–1453.
Schiffries, C. M., and Skinner, B. J., 1987, The Bushveld hydrothermal system: Field and petrologic evidence: Am. J. Sci., v. 287, pp. 566–595.
Schneiderhohn, H., 1929, The mineragraphy, spectrography, and

genesis of the platinum-bearing nickel–pyrrhotite ores of the Bushveld complex; *in* Wagner, P. A., Platinum deposits and mines of South Africa: Oliver and Boyd, London, pp. 206–246.

Stumpfl, E. F., 1961, Some new platinoid-rich minerals, identified with the electron microanalyzer: Mineral. Mag., v. 32, pp. 833–847.

Stumpfl, E. F., and Rucklidge, J. C., 1982, The platiniferous dunite pipes of the eastern Bushveld: Econ. Geol., v. 77, pp. 1419–1431.

Talkington, R. W., and Lipin, B. R., 1986, Platinum-group-minerals in chromite seams of the Stillwater Complex, Montana: Econ. Geol., v. 81, pp. 1179–1186.

Tarkian, M., and Stumpfl, E. F., 1975, Platinum mineralogy of the Driekop mine, South Africa: Mineralium Deposita, v. 10, pp. 71–85.

Taylor, R. W., and Muan, A., 1962, Activities of iron in iron–platinum alloys at 1300°C: Metall. Soc. Am. Inst. Mining Metall. Eng. Trans., v. 224, pp. 500–502.

Trommsdorff, V., and Evans, B. W., 1974, Alpine metamorphism of peridotitic rocks: Schweiz. Mineral. Petrol. Mitt., v. 54, pp. 333–352.

Vermaak, C. F., 1976, The nickel pipes of Vlakfontein and vicinity, western Transvaal: Econ. Geol., v. 71, pp. 261–286.

Viljoen, M. J., and Hieber, R., 1986, The Rustenburg section of Rustenburg Platinum Mines Limited, with reference to the Merensky Reef; *in* Anhaeusser, C. R., and Maske, S. (eds.), Mineral deposits of South Africa: Geological Society of South Africa, Johannesburg, pp. 1107–1134.

Viljoen, M. J., and Scoon, R. N., 1985, The distribution and main geologic features of discordant bodies of iron-rich ultramafic pegmatite in the Bushveld Complex: Econ. Geol., v. 80, pp. 1109–1128.

Viljoen, M. J., De Klerk, W. J., Coetzer, P. M., Nash, N. P., Kinloch, E., and Peyerl, W., 1986a, The Union section of Rustenburg Platinum Mines Limited, with reference to the Merensky Reef; *in* Anhaeusser, C. R., and Maske, S. (eds.), Mineral deposits of South Africa: Geological Society of South Africa, Johannesburg, pp. 1061–1090.

Viljoen, M. J., Theron, J., Underwood, B., Walters, B. M., Weaver, J., and Peyerl, W., 1986b, The Amandelbult section of Rustenburg Platinum Mines Limited, with reference to the Merensky Reef; *in* Anhaeusser, C. R., and Maske, S. (eds.), Mineral deposits of South Africa: Geological Society of South Africa, Johannesburg, pp. 1041–1060.

Volborth, A. A., and Housley, R. M., 1984, A preliminary description of complex graphite, sulphide, arsenide, and platinum group element mineralization in a pegmatoid pyroxenite of the Stillwater Complex, Montana, U.S.A.: Tschermaks Mineral. Petrogr. Mitt., v. 33, pp. 213–230.

Volfinger, M., Robert, J.-L., Vielzeuf, D., and Neiva, A. M. R., 1985, Structural control of the chlorine content of OH-bearing silicates (micas and amphiboles): Geochim. Cosmochim. Acta, v. 49, pp. 37–48.

von Gruenewaldt, G., 1973, The main and upper zones of the Bushveld Complex in the Roossenekal area, eastern Transvaal: Trans. Geol. Soc. S. Afr., v. 71, pp. 1324–1336.

von Gruenewaldt, G., 1979, A review of some recent concepts of the Busvheld Complex, with particular reference to sulfide mineralization: Canadian Mineral., v. 17, pp. 233–256.

von Gruenewaldt, G., Hatton, C. J., Merkle, R. K. W., and Gain, S. B., 1986, Platinum-group element–chromitite associations in the Bushveld Complex: Econ. Geol., v. 81, pp. 1067–1079.

von Gruenewaldt, G., Sharpe, M. R., and Hatton, C. J., 1985, The Bushveld Complex: introduction and review: Econ. Geol., v. 80, pp. 803–812.

Wagner, P. A., 1924, On magmatic nickel deposits of the Bushveld igneous complex in the Rustenburg District of the Transvaal: South African Geological Survey, Memoir 21, 181 pp.

Wagner, P. A., 1929, Platinum deposits and mines of South Africa: Oliver and Boyd, London, pp. 206–246.

Webster, J. D., and Holloway, J. R., 1988, Experimental constraints on the partitioning of Cl between topaz rhyolite melt and $H_2O$ and $H_2O$ + $CO_2$ fluids: New implications for granitic differentiation and ore deposition: Geochim. Cosmochim. Acta (in press).

Willemse, J., 1969, The vanadiferous magnetic iron ore of the Bushveld igneous complex: Econ. Geol., Monograph 4, pp. 187–208.

Yoder, H. S., Jr., 1965, Diopside–anorthosite–water at five and ten kilobars and its bearing on explosive volcanism: Carnegie Institute of Washington, Year Book 64, pp. 82–89.

# Chapter 10

# Introduction: Ore Deposits Associated with Silicic Rocks

J. A. Whitney

Years ago it was assumed that nearly all metallic ore deposits were associated in some way with magmatic systems. Even deposits such as the Mississippi Valley lead/zinc deposits were hypothesized to be related to some granitic source at depth. Then, in the early 1970's, magmatic models began to fall out of favor as many economic geologists recognized that igneous processes were not required for the generation of many ore deposits. Today, the pendulum of opinion has returned to a middle position. We recognize that there are a number of metallic ore deposits whose existence does not necessarily require a direct magmatic input. There are, however, a number of classes of deposits whose close temporal and spatial relationships suggest a genetic connection with silicic plutonic or volcanic events.

Porphyry systems are disseminated deposits of copper, molybdenum, lead, and zinc with relatively low gold and silver. The mineral assemblages are mainly oxidized sulfide assemblages. In many cases there are several periods of mineralization involved in the formation of an economic deposits. In most cases, a protore stage in which the ore minerals are deposited in an intricate series of healed veins and veinlets is first developed. Where this stage is preserved, the fluids appear to have been hot and saline. Isotopic studies suggest that these early volatile phases were partly magmatic in origin. Thus, the protore stage of porphyry systems may be thought of as a deuteric magmatic process. Many porphyry systems are thought to be the roots of old volcanic centers, so they may be thought of as the underlying magmatic portion of volcanic epithermal systems.

Skarn systems are formed where the volatile phases from a magma interact with the country rock. This is a dynamic disequilibrium process in which the prior history of the magma, the nature of the country rock, and the chemistry of the fluids involved are all important. The economic mineralogy varies extensively from copper, lead, zinc, iron with varying amounts of gold and silver, to tin, tungsten, and molybdenum. Since the deposition process is dependent on so many factors including magmatic, hydrologic, and lithologic characteristics, skarns are highly variable in mineralogy, structure, and occurrence. Therefore, detailed study of any skarn must include the magmatic history of the associated igneous rock, the nature and reactivity of the country rock, and the hydrologic processes by which fluids are introduced and/or mixed in the depositional environment. Above all it must be remembered that equilibrium cannot be assumed to have prevailed over any significant distance in such a diverse physiochemical environment.

Volcanic epithermal and surficial deposits are formed in the vent and caldera areas of active edifices. In these areas, sulfur- and chlorine-rich magmatic emanations are rising and mixing with local meteoric circulation driven by the volcanic heat source. Such deposits may therefore have varying amounts of magmatic input, yielding a variety of fluids from highly sulfurous and saline magmatic gases to relatively dilute, less acid meteoric systems. The metals deposited also vary from copper, lead, zinc, gold, silver and iron to arsenic, mercury, and uranium.

Exhalative volcanogenic sulfide deposits of the Kuroko type are spatially associated with felsic domes and vents. The solutions, however, are thought to be mainly recirculated sea water which has been heated by the volcanic source and has interacted with the underlying basaltic to dacitic rocks. Most present workers believe there is relatively little direct magmatic input in such systems, although some of the sulfur and acidity could be inherited from a small magmatic gas input. Much of the chemistry of these systems is thought to reflect low temperature alteration of the underlying formations.

To discuss the magmatic contribution to the origin of such deposits it is first necessary to discuss the processes of magma generation and fractionation. These processes control the concentration or depletion of both cations and anions which are of subsequent importance in the formation of a deposit. Second, the process of element fractionation in the magmatic vapor phase and its role in metallogenesis must be discussed. Finally, the interaction of these fluids with the surrounding country rock is important. The following series of papers review and discuss these aspects of magmatic processes. We have not attempted to duplicate previous discussions on hydrothermal systems. It is recommended that the reader consult Volumes 1 and 2 of this series for extensive discussions of the lower temperature hydrothermal environment.

# Chapter 11

# ORIGIN AND EVOLUTION OF SILICIC MAGMAS

J. A. Whitney

## INTRODUCTION

Silicic magmatic processes which contribute to the formation of ore deposits can be subdivided in three stages: 1. the origin of the silicic magma; 2. intrusion and fractionation processes; 3. volatile evolution and separation. This chapter will deal with the physical and chemical processes involved magmatic evolution. Subsequent chapters will deal with chemical fractionation and ore deposition processes.

Evaluating the origin of silicic magmas is an indirect process. Most plutons have undergone extensive post-solidification re-crystallization to some degree. In addition, the extensive modifications caused by upward-migration through the crust and crystallization have altered the evidence of magma origins. Thus, in discussing the physical conditions of magma generation it helps to use volcanic analogs and experimental data.

## VOLCANIC ANALOGS OF GRANITIC BATHOLITHS

Large volume ash-flow tuff eruptions (>500 km³) are the volcanic equivalent of silicic batholiths. Unlike plutons, however, these magmas have been quenched at various points in their development and therefore represent silicic batholiths frozen in time. Studies of these systems have yielded extensive data on magmatic parameters (e.g. Smith, 1960, 1979; Smith and Bailey, 1966; Hildreth, 1977, 1979; Lipman, 1971; Fridrich and Mahood, 1987; Whitney and Stormer, 1983, 1985; Whitney et al., 1988). This effort has been made possible in the last 20 years by the development of various thermodynamic models which allow us to quantify such variables as temperature, pressure, and the activities of gaseous components (e.g. Burnham and Davis, 1974; Burnham et al., 1969; Buddington and Lindsley, 1964; Spencer and Lindsley, 1981; Andersen and Lindsley, 1985; Stormer, 1975; Brown and Parsons, 1981; Stormer and Whitney, 1985; Ghiorso, 1984; Price, 1985; Green and Usdansky, 1986; Wones, 1972; Bohlen et al., 1980; Whitney, 1984). These data, combined with experimental work on both naturally occurring and synthetic granitic materials, allow us to determine the probable volatile content and thermal conditions of silicic magmas responsible for batholith-generating magma systems. Table 11.1 summarizes probable conditions for several such systems as determined by various authors.

The temperatures of such systems can probably best be estimated from the composition of coexisting magnetite and ilmenite (Buddington and Lindsley, 1964; Spencer and Lindsley, 1981; Andersen and Lindsley, 1988; Lindsley, 1976; Haggerty, 1976). Two-feldspar geothermometry is also promising but has a significant pressure dependency (Stormer and Whitney, 1985). Large volume ash-flow tuffs yield temperatures between 750 to 850°C with smaller volume more alkaline systems going up to 900°C. In plutonic systems, the iron–titanium oxides re-equilibrate during cooling, and the alkali feldspars often exsolve or homogenize to potassium-rich compositions. Therefore, direct geothermometry is much more difficult in plutonic systems.

Water content can be estimated two different ways. First, the activity of water can be calculated through the composition and stability data for hydrous phases such as biotite (Wones, 1972; Wones and Eugster, 1966; Bohlen et al., 1980). Such calculations, however, have serious potential errors or uncertainties. It is difficult to determine the ferrous/ferric ratio in biotite at the time of magmatic equilibration due to oxidation and lack of micro-analytical techniques. The fluorine content in the hydroxyl site must also be carefully analyzed, and even with modern microprobes the error is significant. Finally the experimental data upon which the thermodynamic model is based have inherent uncertainties. Even given these uncertainties, several studies have yielded values that are probably good to 0.5 to 1.0 log units. Unfortunately, 1 log unit in the range of 2.0 to 3.0 is the difference between 100 and 1000 bars water fugacity, which corresponds to approximately 0.3 to 3.0 wt% water.

A second approach, which is less numeric but at least as accurate, is the comparison of mineral assemblages and percent crystallization with the results of phase equilibria studies of relative compositions. As discussed above, the temperature range for most large volume silicic magmas is relatively restricted. Since these magmas are 60 to 90% liquid at this stage, analogies with laboratory experiments (e.g. Whitney, 1975; Naney, 1977; Naney and Swanson, 1980; Robertson and Wyllie, 1971) require between 2 and 6 wt% water in the silicic melt phase (Whitney and Stormer, 1985; Whitney et al., 1988). Similar evaluation of less silicic systems such as that erupted at Mt. St. Helens (Melson and Hopson, 1981) at somewhat hotter temperatures requires less water, down to about 2 to 3 wt% in the melt phase (Rutherford et al., 1985).

An important lesson from such volcanic studies is that the parameters of temperature, water content, and degree

TABLE 11.1—Estimated water content and magmatic conditions in some ash-flow tuffs.

| | Temp. (°C) | % melt | log $f_{O_2}$ | log $f_{H_2O}$ | $X_{H_2O}$, melt | $X_{H_2O}$, bulk |
|---|---|---|---|---|---|---|
| Bishop Tuff | 750±30 | 85 | −15±1 | 2.7±0.5 | 2.5 | 2.2 |
| Fish Canyon Tuff | 790±20 | 60 | −12±1 | 3.2±0.5 | 6.5 | 4.0 |
| Carpenter Ridge Tuff | 780±30 | 90 | −13.5±1 | 2.9±0.3 | 3.5 | 3.2 |
| St. Helens Dark Pumice | 950±30 | ≈85 | −9.5±0.5 | ≈2.6 | ≈2.3 | ≈2 |

Note: Values recalculated from the following authors: Bishop, Hildreth (1977); Fish Canyon, Whitney and Stormer (1985); Carpenter Ridge, Whitney et al. (1988); St. Helens, Melson and Hopson (1981); using solubility and thermodynamic data of Day and Fenn (1982); Burnham and Davis (1974); Burnham et al. (1969). Values for water from St. Helens are approximations based on mineral assemblage and degree of crystallization as reported by Lipman and Mullinax (1981).

of crystallization are not independent parameters (Marsh, 1981). If a magma erupts with 10% crystals present and a known temperature, then the water content required can be estimated. Unless very large concentrations of fluorine, boron, or some other component which lowers the melting temperature of granite are present, the required water content is directly related to the temperature. If the magma is undersaturated with respect to water, this relationship is only moderately dependent on pressure.

Another important corollary is the possible contribution of $CO_2$ as a primary magmatic gas in such magma systems. Since $CO_2$ is quite insoluble in calc-alkaline granitic melts, it is strongly fractionated into the volatile phase and lowers the activity of water (Holloway, 1976). If $CO_2$ is present in too great an abundance the melt cannot have the low percentage of crystals present at the temperatures determined (Swanson, 1979). Therefore, although we do not have good thermodynamic measurements of $CO_2$ in such systems, the maximum amount possible can be estimated by analogy with synthetic systems using mixed gas phases.

It is difficult to do the same types of determination of magmatic parameters in plutons because minerals re-equilibrate during cooling from magmatic temperatures. Iron–titanium oxides appear to re-equilibrate at least down to solidus temperatures, and then continue to exsolve at lower temperatures. Some useful results have been obtained (Whitney and Stormer, 1976; Baldasari, 1981). Feldspars also exsolve, but tend to re-equilibrate more slowly. Although calculations on co-existing feldspars have numerous uncertainties, including exsolution of the alkali feldspars, some results have been reported (Whitney and Stormer, 1977a, 1977b, 1976). The stability of the mafic phases may offer better indications of magmatic conditions. Biotite is limited in stability to below about 850°C in the presence of quartz, or slightly higher in the presence of a high silica activity magma (Luth, 1967; Naney 1977). Hornblende has a more complicated stability field highly dependent on bulk composition. Most studies have resulted in magmatic temperatures in the range of 850 to 650°C for calc-akaline granitic rocks. These temperatures again require the same range of water content as determined from volcanic analogs. In certain compositions, highly buffered by co-existing minerals, hornblende may also be an important geobarometer (Hammerstrom and Zen, 1986).

## EXPERIMENTAL STUDIES OF SILICIC SYSTEMS

In the last thirty years many experimental studies have been published on haplogranite to haplogranodiorite systems. Experimentation on the "granite" system, $NaAlSi_3O_8$–$KAlSi_3O_8$–$SiO_2$–$H_2O$, has been carried to higher pressures (Luth et al., 1964). Other components have been added to the system including $Mg_2SiO_4$ (Luth, 1968), $CaAl_2Si_2O_8$ (James and Hamilton, 1969). In addition, a number of studies have been conducted with controlled water contents (Whitney, 1969; Robertson and Wyllie, 1971; Steiner et al., 1975; Whitney, 1975a,b; Naney, 1977) or with the activity of water reduced by an insoluble gas, usually $CO_2$ (e.g. Swanson, 1979; Eggler, 1972; Eggler and Burnham, 1973). Both types of study allow the separation of water content and confining pressure as variables.

Perhaps the effect of water on melting relationships can best be illustrated through the use of phase diagrams which plot the stable phase assemblage with respect to two of the three variables, pressure, temperature, and total water content, with the third held constant. Since it is the phase assemblage that is being plotted, such diagrams may be referred to as Phase Assemblage Diagrams to distinguish them from more conventional phase diagrams (Whitney, 1972). Fig. 11.1A is a temperature–$X_{H_2O}$ diagram at 8 kb confining pressure for a synthetic composition within the system ternary feldspar–$SiO_2$ (R4 of Whitney, 1975a,b). Also shown in Fig. 11.1B are contours of approximate volume percent melt. Similar diagrams at 2 kb pressure are shown in Fig. 11.2A and 11.2B. The chemical compositions of R4 and other materials mentioned in this paper are summarized in Table 11.2. On the *T–X* diagrams the dramatic effect of water content on degree of crystallization in the vapor-absent region is apparent.

Consider the temperature paths A–A′ and B–B′ in Fig. 11.1. A–A′ (12 wt% water) is oversaturated with respect to a free volatile phase (V). Under these conditions, this granitic composition does not begin to crystallize until temperatures below about 750°C. Crystallization is slow until below 700°C, with the majority of crystallization occurring over a very short temperature range between 630 and 650°C. In contrast, a composition with 2% water (B–B′) has plagioclase stable to temperatures in excess of 1100°C. Crystallization is slow, but continuous between 1000 to 630°C. The rapid crystalli-

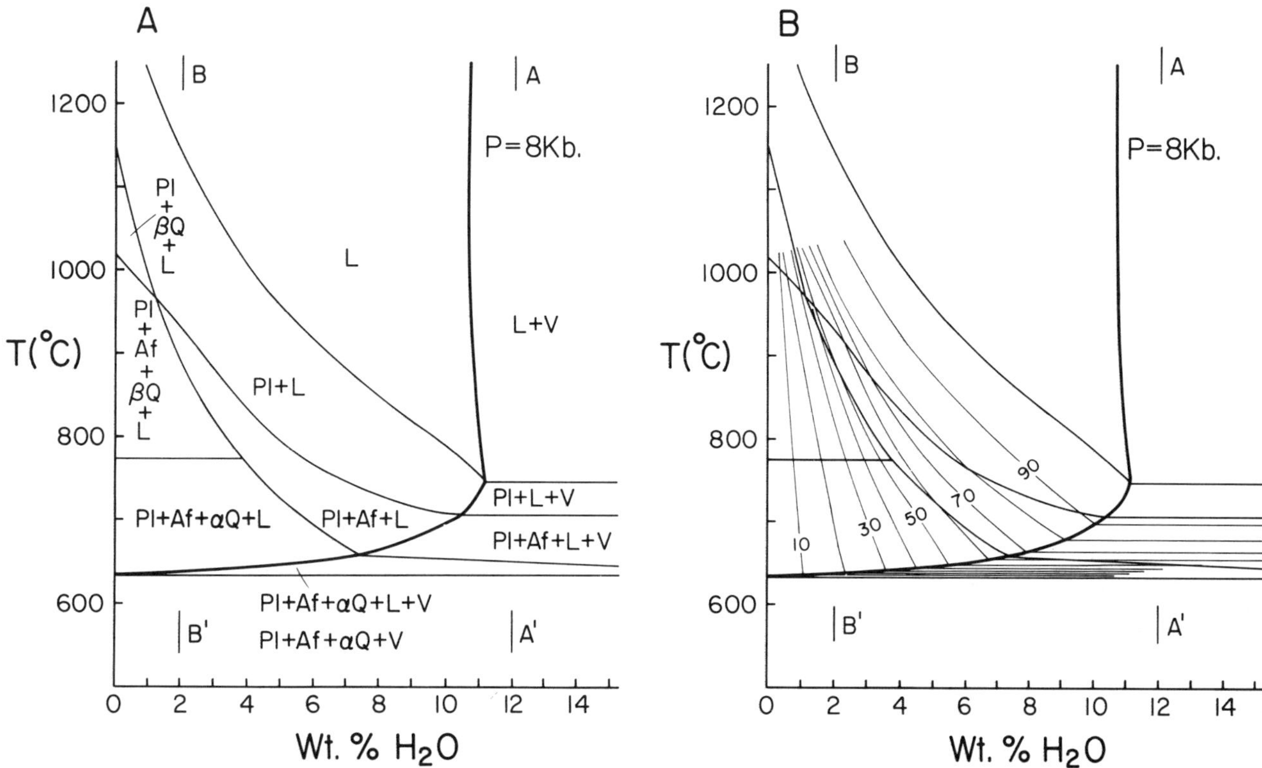


FIGURE 11.1—Temperature–$X_{H_2O}$ diagrams for synthetic granite (compositions R4 of Whitney, 1972, 1975a, 1975b). See Whitney (1972, 1975a) for experimental procedures and results. Abbreviations are the same as in Table 11.3. **A.** Stable phase assemblages at 8 kb confining pressure. **B.** Phase assemblage diagram contoured for volume percent melt. Phase fields are the same as in A. Contours represent approximate percent of melt present. Position of contours are semi-quantitative.

zation of 50% of the magma over a small, 10–20°C, temperature range only occurs under water vapor saturated conditions. The same phenomena can be observed at 2 kb (Fig. 11.2, A–A', B–B'), but now vapor saturation only requires 6 to 7 wt% water. Therefore, in general with other factors such as nucleation rate being equal, and assuming that heat flux out of the system is sufficient that latent heat will not arrest the cooling process, crystallization in the water vapor undersaturated region will lead to longer temperature ranges of crystallization and coarser grain size. Crystallization under vapor saturated conditions will cause a shorter temperature range and finer grain size. The crossing of the vapor saturation surface during cooling will cause an increase in the rate of crystallization, and in general smaller grain size. An example of such a phenomena may be the porphyritic, subporphyritic, and non-porphyritic phases of the Pinos Altos pluton, Silver City, New Mexico (Owen, 1983; Jones et al., 1967; El-Hindi, 1977). Other cases may be the fine-grained versus coarse grained phases of the Siloam Granite, Georgia (Speer, 1977) and the Liberty Hill pluton, North Carolina (Speer et al., 1980).

Another way to suddenly reduce the confining pressure and volatile content is through venting the the surface. In this case, the pressure in the magma chamber is reduced to the pressure exerted by the magma column. In the case of a vesiculated magma, this pressure can be substantially lower that the lithostatic pressure (Whitney and Stormer, 1986). Such a phenomena is probably responsible for the hiatal textures in many shallow porphyries.

Another diagram useful for understanding phase relations during upward migration of a magma is the isothermal Phase Assemblage Diagram which plots pressure and water content as independent variables. Fig. 11.3A and 3B are such diagrams for the same synthetic R4 composition. Again A–A' represents vapor saturated conditions, while B–B' is vapor undersaturated throughout most of its path.

As pressure decreases along A–A' the magma crystallizes. This process is slow throughout the plagioclase field, with the majority of crystallization occurring within 1 kb of the solidus in a manner similar to classic eutectic crystallization. At pressures below the solidus the melt is metastable. This phenomena is the familiar pressure quench effect discussed by Tuttle and Bowen (1958). Such rapid crystallization due to decreasing confining pressure is probably the cause of many fine-grained aplites, aphanitic groundmass in some porphyries, and fine-grained homogeneous plutons such as the Elberton Batholith, Georgia (Stormer and Whitney, 1980). Because of this phenomena, low-temperature granitic mag-

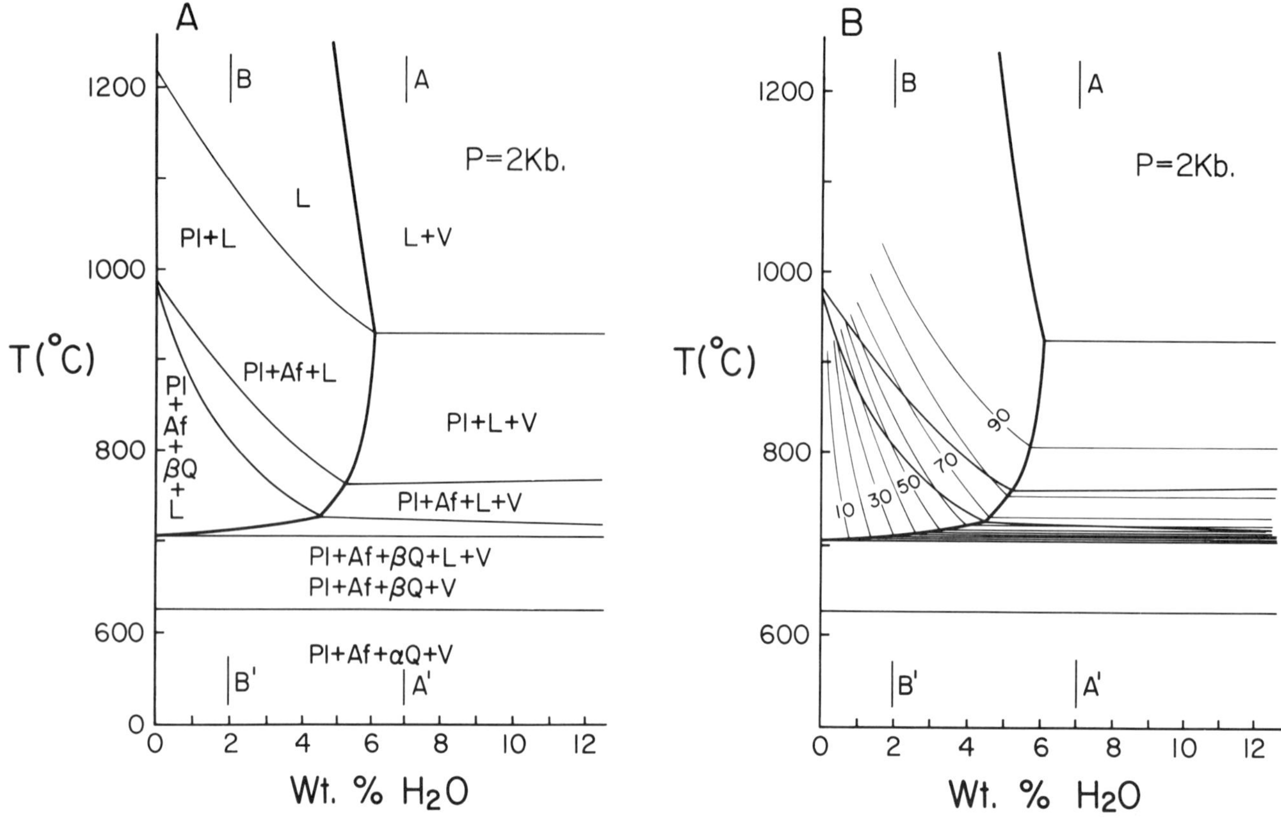


FIGURE 11.2—Temperature–$X_{H_2O}$ diagrams for synthetic granite (R4) at 2 kb confining pressure. **A.** Stable phase assemblage. **B.** Diagram contoured for volume percent melt. Positions of contours are semi-quantitative.

TABLE 11.2—Chemical composition of experimental materials.

| Oxide | R4 | Nockolds' Average Hb-Bi Qtz. Monzonite | Cape Ann Granite | Westerly Granite | Mount Airy Leucogranodiorite |
|---|---|---|---|---|---|
| $SiO_2$ | 70.99 | 65.88 | 77.61 | 72.34 | 71.03 |
| $TiO_2$ | – | 0.81 | 0.25 | 0.26 | N.D. |
| $Al_2O_3$ | 17.21 | 15.07 | 11.94 | 14.34 | 16.52 |
| $Fe_2O_3$ | – | 1.74 | 0.55 | 0.68 | – |
| FeO | – | 2.73 | 0.87 | 1.13 | 1.75* |
| MnO | – | 0.08 | – | 0.02 | N.D. |
| MgO | – | 1.38 | Tr | 0.37 | Tr |
| CaO | 2.83 | 3.36 | 0.31 | 1.52 | 1.93 |
| $Na_2O$ | 4.00 | 3.53 | 3.80 | 3.37 | 5.53 |
| $K_2O$ | 5.07 | 4.64 | 4.98 | 5.97 | 3.71 |
| $P_2O_5$ | – | 0.26 | N.D. | 0.11 | N.D. |
| $H_2O^+$ | – | 0.52 | 0.23 | 0.30 | 0.19 |
| Total | 100.0 | 100.0 | 100.54 | 100.30 | 100.66 |

*Total iron reported as FeO.

Note: R4 and Nockolds' average, of which R4 represents the normative feldspar and quartz components, are taken from Whitney (1975). The other analyses are as listed in Whitney (1969).

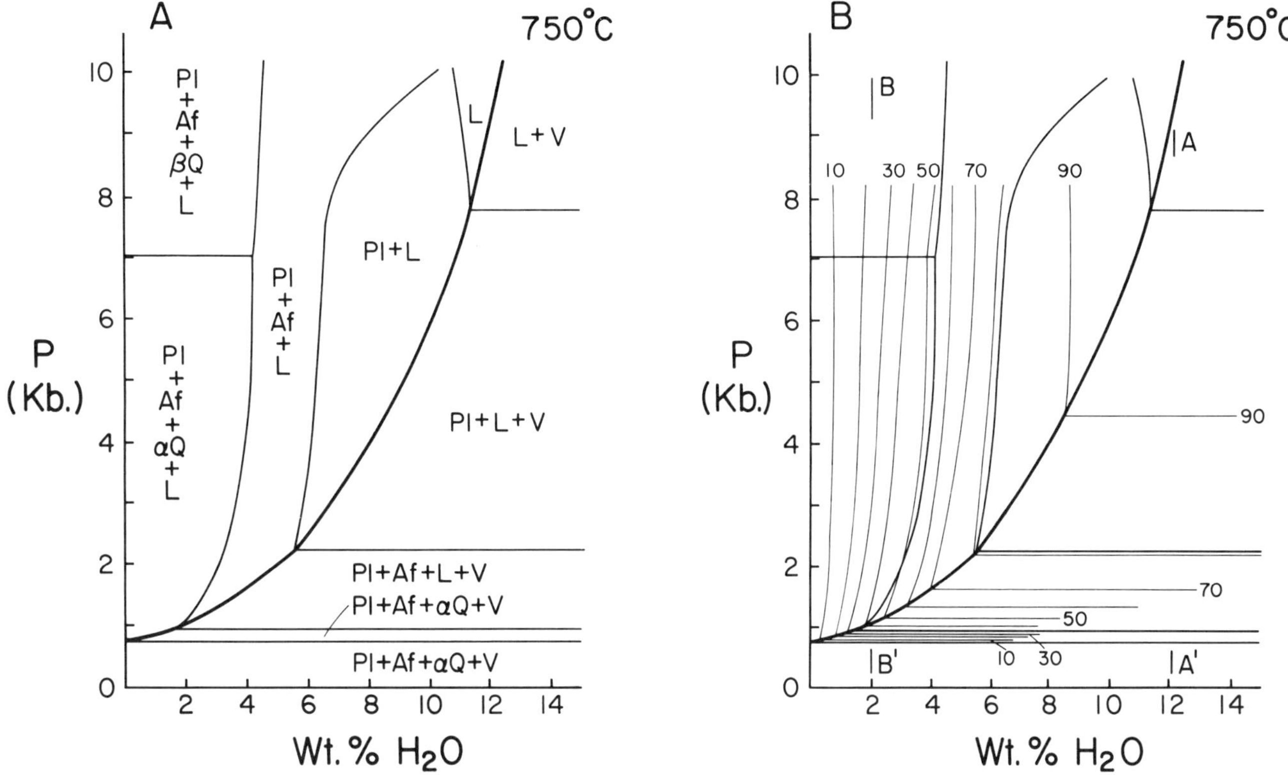


FIGURE 11.3—Pressure–$X_{H_2O}$ diagrams for synthetic granite (R4) at 750°C. **A.** Phase assemblage diagram showing experimental results and phase assemblages. **B.** Same diagram as A contoured for volume percent melt. Positions of contours are semi-quantitative.

mas are not stable under near surface conditions. Granitic magmas with temperatures below about 650°C cannot rise above the upper mid-crust before they tend to crystallize. Thus, it is rare to find extensive ash-flow sheets with magmatic temperatures below about 750°C. Such magmas would tend to freeze before they reached shallow crustal depths.

At lower water contents (e.g. 2 wt%, B–B' on Fig. 11.3B) the crystallization history is quite different. As long as the magma remains isothermal and undersaturated with respect to water, the percentage of melt either increases or remains nearly the same. In the case of quartz, the amount degree of melting increases causing resorption. This occurs because the quartz saturation surface within the granite system moves to higher silica content with decreasing pressure as originally shown by Tuttle and Bowen (1958; see also Luth et al., 1964; Whitney, 1975a, 1975b; Nekvasil and Burnham, 1987). Alkali feldspar also appears to decrease in abundance slightly for a number of granitic compositions (Whitney, 1975a). Plagioclase, however, either remains nearly the same, or crystallizes somewhat, depending on the bulk composition. This occurs due to the shift in the plagioclase–alkali feldspar liquidus boundary within the granite system (Whitney, 1975a; Yoder et al., 1957), and the resulting shift of phase volumes in the water undersaturated region. Eventually, as the pressure drops, the magma will become vapor saturated and pressure quenching will again occur over a very short pressure range. Thus, the melt phase in most large ash-flows and rhyolites is metastable by the time it reaches the surface. With the exception of very dry, high temperature rhyolites, most silicic melts tend to crystallize rapidly to aphanitic products within 1 km of the surface.

The paragenesis of volcanic phenocrysts can also be understood in terms of isothermal decompression. Quartz phenocrysts are ubiquitously rounded or embayed. Such an occurrence is predicted by isothermal decompression in the water vapor undersaturated region. Such a phenomena will not occur under vapor saturated conditions. Similarly, early sanidine is sometimes rounded in silicic volcanics. Plagioclase, on the other hand, is rarely rounded or embayed. When plagioclase is seen to be resorbed, there are usually many other signs of phenocryst disequilibrium involving the mafic phases. In such cases, field and petrographic evidence often indicates disequilibrium processes such as magma mixing or assimilation. Under vapor saturated decompression, resorption should not occur for most granitic compositions. (Note: There are exceptions to this rule in some alkaline granites or syenites due to the position of the "unique fractionation curve" in the granite system, but such a detailed phase equilibrium discussion is beyond the scope of this general paper.)

We can further use experimental phase equilibrium to evaluate the crystallization history of some common granites. Figs. 11.4, 11.5, and 11.6 shows expanded 2 kb isobaric Phase Assemblage Diagrams for some common granite types

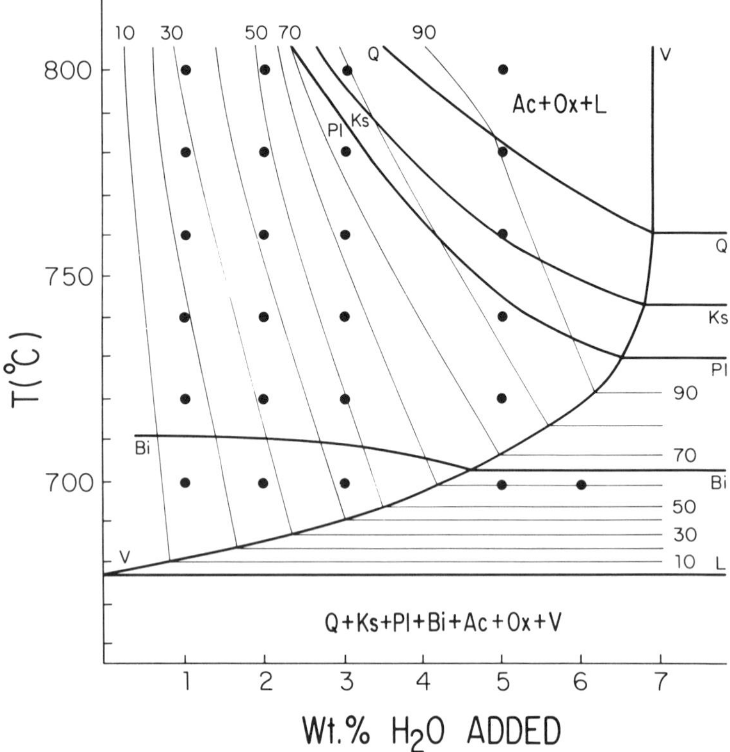


FIGURE 11.4—Temperature–$X_{H_2O}$ phase assemblage diagrams for the Cape Ann Granite (Quincy Type) at 2 kb confining pressure. Abbreviations are given in Table 11.2. Contours represent volume percent of melt present. In each case, the sub-solidus assemblage is listed. The various boundaries are shown with the phase which appears or disappears along that boundary listed on the side of the curve on which it occurs. Percentage melt contours are best fits to the volumetric data in Table 11.3. See Whitney (1969) for experimental methods. All phase assemblages have been re-evaluated from the original work using methods and criteria given in Whitney (1972, 1975a).

taken from Whitney (1988b). These are the Cape Ann Granite (Fig. 11.4, a body of the Quincy-type) from Cape Ann, Massachusetts, the Westerly Granite from Westerly, Rhode Island (Fig. 11.5), and the Mount Airy leucogranodiorite from Mount Airy, North Carolina (Fig. 11.6). Two of these (the Quincy and Westerly) were chosen by Tuttle and Bowen (1958) themselves to represent alkaline (Quincy) and calc-alkaline (Westerly) compositions. The Mount Airy (Deitrich, 1961) is strongly peraluminous. Percentage of glass was determined by modal analyses of powdered samples. Replicate analyses suggests that precision is approximately ± 5% glass.

Again note the contrast between vapor absent and vapor-saturated crystallization. In the case of the Cape Ann Granite, 90% crystallization occurs between 720 and 680°C under vapor-saturated conditions. If only 2% water is present, however, crystallization starts well above 800°C (probably above 900 by comparison with other data) and continues to 680°. At two kilobars, this process is nearly constant until the composition is about 75% crystalline at 685°C at which point the composition becomes vapor-saturated and finishes crystallization within 10°C. The Westerly and Mount Airy crystallize over a slightly greater temperature range under vapor-saturated conditions, but still the majority of the melt crystallizes within 20 to 30°C of the solidus in a manner analogous to classic eutectic crystallization.

Thus, coarse-grained granites that evidence a long crystallization history with initial high temperature paragenesis probably crystallized in the absence of a vapor phase with relatively low total water content. Fine-grained ones which have an abbreviated crystallization history starting at low temperature probably formed under vapor-saturated conditions with relatively higher water contents. The critical factor is the rate of cooling combined with the rate of crystallization. The rate of cooling is controlled by heat flow out of the body versus heat generated by latent heat of crystallization.

## SOURCES OF WATER FOR THE GENERATION OF GRANITIC MELTS

A critical factor in the generation of granitic rocks is the availability of water. Water is required for the formation of granitic rocks under the conditions in which we see them (see Burnham, 1979 for an extended discussion). The granitic solidus is depressed by as much as 400°C by the addition of water, as compared to only about 200° for basalt. Without water, granites do not begin to melt until 1000°C at crustal pressures. To duplicate the paragenesis of mineral crystallization we see in natural rocks requires between about 2 wt% water for dioritic rocks, up to 4% or more for granitic rocks. Therefore, essential factors in the formation of granitic rocks are the sources and availability of water.

There are a number of sources of water for melting in the

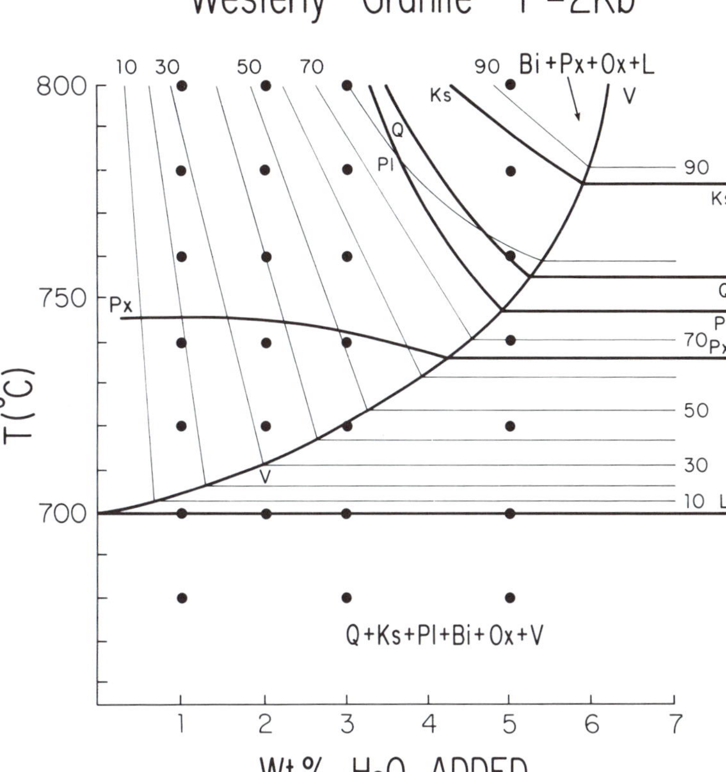


FIGURE 11.5—Temperature–$X_{H_2O}$ phase assemblage diagrams for the Westerly Granite, Westerly, Rhode Island, at 2kb confining pressure. Contours are volume percent melt. See Figure 4 for symbols and explanation.

lower crust. First is the dehydration of hydrated phases which are present in high grade metamorphic rocks (Hyndman, 1981). We also have very small amounts of intergranular volatile phases which may be called upon to help in that melting process. This intergranular water, however, is very low in abundance and yields only 0.1 to 0.3 wt% of the 2 to 4 wt% water we need to form most common granitic rocks. So, in anatexis of lower crustal rocks, the dehydration of hydrated phases is highly important and can be used to describe the nature of the melting process.

A second source of volatiles is from subducted crust and upper mantle which has been hydrated during hydrothermal alteration at the ridge crest. These volatiles must be transported from the mantle into the crust, possibly by the formation of hydrated andesites and basalts from the subducted plate which eventually crystallize in the lower crust. Since many large granitic batholiths are associated with subduction zone magmatism, such a mechanism could be highly important in those areas.

### Dehydration reactions

First, consider melting associated with the dehydration of hydrous crystalline phases, starting with muscovite. The dehydration melting of muscovite occurs along a series of incongruent melting reactions in which muscovite reacts with albite or plagioclase, K-feldspar, and quartz to form a liquid. These reactions originate at a series of invariant points generated by the intersection of the muscovite plus quartz reaction with the beginning of melting for the assemblage involved (see Thompson, 1982; Thompson and Algor, 1977 for a more thorough discussion of Schreinemakers' constructions). For anatexis, the so-called "vapor-absent" reactions which do not require the presence of a hydrous volatile phase are particularly important. These reactions represent the melting of the hydrous assemblage without the input of additional water. Such reactions are the best experimental analogs to anatexis under crustal conditions.

For example, within the Inner Piedmont province of eastern Georgia, the reaction

Muscovite + Quartz + Na–Plag =
Sillimanite + K-spar + Liquid [1]

appears to be responsible for the majority of migmatization. Layers which lack significant Na-rich plagioclase have not melted and remain competent even though they contain muscovite and quartz. Therefore, in this case, it appears as if the reaction

Muscovite + Quartz = Sillimanite + K-spar + Liquid [2]

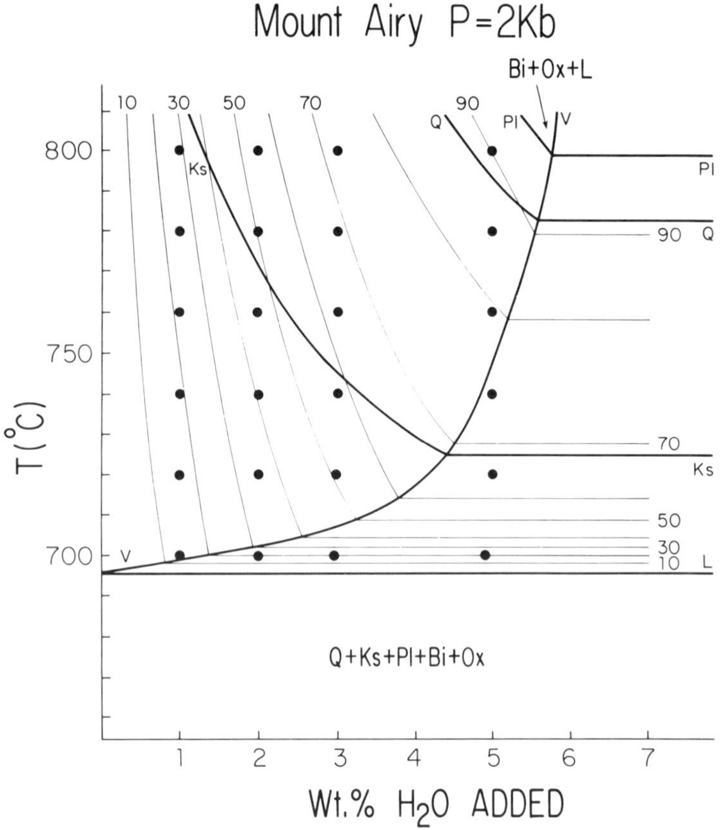


FIGURE 11.6—Temperature–$X_{H_2O}$ phase assemblage diagrams for the Mount Airy Leucogranodiorite, Mount Airy, Virginia, at 2 kb confining pressure. Contours are volume percent melt. See Fig. 11.4 for symbols and explanation.

(Day, 1973) has not been exceeded. Thus, with rising temperatures, muscovite-generated migmatization will first begin in rocks containing quartz, albitic plagioclase and muscovite, and proceed at higher temperatures into rocks lacking plagioclase. The range of conditions for such muscovite-driven melting is shown on Fig. 11.7.

The melts generated by muscovite anatexis are expected to have certain chemical characteristics. First, they tend to be relatively high in potassium with K/Na atomic ratios of 1 or more. Second, they are granitic in composition. Third, they are strongly peraluminous. Finally, they have relatively low initial temperatures of below about 750°C. Since the most common muscovite-rich lithologies are derived from continental pelitic sediments, most muscovite-generated granitoids will have high initial strontium ratios (0.705 or higher) and high delta $^{18}O$ values (usually around 10 or more). These granites correspond to the ilmenite-series plutons described by Ishihara (1977). Since such sediments often contain organic material which become graphite under metamorphic conditions, the initial oxygen activity may be low (Ishihara, 1977). However, since the volatile system is within the ternary system C–O–H, and not simply C–O, a variety of oxygen fugacities are possible in the presence of graphite, and once the magma separates from the source area these melts contain very little iron due to their low temperature, with biotite being the most common mafic

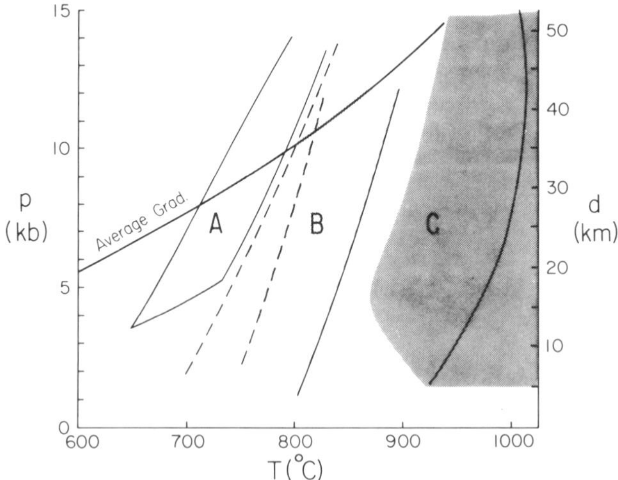


FIGURE 11.7—Pressure–Temperature diagram showing the approximate conditions of dehydration melting of muscovite, biotite, and hornblende. See text for references. **A.** Conditions of direct dehydration melting of muscovite-bearing assemblages. **B.** Conditions of direct dehydration melting of biotite-bearing assemblages. **C.** Conditions of direct dehydration melting of hornblende-bearing assemblages.

phase. Therefore during cooling there is little control of oxygen activity in these silicic, peraluminous melts and such granitoids may attain higher oxygen activities late in their crystallization history. The low iron content and initial low oxygen activity means that magnetite is rare since most iron goes into silicate phases. Accessory minerals such as tourmaline and garnet are common. Schlieren or mafic enclaves carried within the magma may be formed from more refractory sedimentary layers. Common metamorphic minerals inherited from the source area or assimilated crust would include biotite, garnet, and sillimanite. The Stone Mountain Granite (Whitney et al., 1975) may be an example of such a granite.

Melting reactions involving biotite occur along a similar set of reactions at higher temperatures (Fig. 11.7). The reactions involving the highest temperature stability of the magnesium end-member phlogopite have been studied by Luth (1967).

Phlogopite + Quartz = K-feldspar + Enstatite + Liquid. [3]

Similar reactions occur for iron-rich biotites (Wones and Eugster, 1965), but at somewhat lower temperatures. Such reactions are also dependent on oxygen activity.

The melts generated by biotite dehydration will vary in geochemical characteristics depending on the composition of the protolith. Melting will again begin in lithologies containing Na-rich plagioclase through reactions of the form

Biotite + Na-plag + Quartz =
K-feldspar + Pyroxene + Liquid. [4]

The K/Na ratio of the melt will vary, but in general will be near 1 or less. Depending on the abundance of sodium and calcium in the source region, the resulting plutons may be granitic to granodioritic in composition. The melt may be meta-aluminous to peraluminous, but would rarely be as rich in alumina as muscovite-generated melts unless muscovite had played an early roll in beginning anatexis. The initial temperature for biotite-generated melts would be about 750 to 850°C. Since many protoliths contain biotite, other characteristics will be quite variable. If the source is metasedimentary, initial strontium and oxygen ratios may be high. If, on the other hand, the protolith were relatively young metavolcanic and volcaniclastic rocks these ratios may be lower. Similarly, parameters such as the oxygen activity will depend on the protolith. Graphitic metasediments may yield low initial oxygen activity, while more oxidized metavolcanic units would yield higher values. The activity of oxygen is important in determining the abundance of magnetite as an accessory mineral. Under low oxygen activity the iron remains in silicate phases, while under higher activity it forms magnetite. Therefore, magnetite may or may not be important in biotite-generated granitoids. Mafic schleiren or enclaves may have a variety of mineralogies, but are not extremely common. The Elberton Granite (Stormer and Whitney, 1980) and the so-called "New Hampshire series" plutons of northeastern Vermont may be of this type.

The most refractory of the common hydrous phases is hornblende. It is rare to see anatexis involving hornblende without the introduction of water. Migmatite within the aureole of the Bear Mountain Pluton on the Smith River of the Klamath Mountains (Snoke et al., 1981) may be an example. Such reactions may be more important in the deep crust where temperatures are higher. The approximate conditions of hornblende melting in gabbroic compositions determined by Wyllie (1971) are also shown on Fig. 11.7.

The composition of hornblende-generated melts will depend on protolith and temperature (Helz, 1979). In general, the K/Na ratio is low, with the resulting magma being granodioritic to dioritic in composition. The magmas generated will be meta-aluminous to peralkaline in composition. The initial temperatures may be quite high, and may range up to 1,000°C. Since melting is occurring at such a high temperature, the initial magmas are quite dry. Crystallization therefore occurs over a long temperature range. Early accessory minerals such as pyroxene or even fayalitic olivine may be present, but the stability of magnetite versus fayalite will depend on oxygen activity. The most likely protoliths which would be high in amphibole content would be amphibolites and hornblende gabbros. Amphibole anatexis could be important in some alkaline granitic provinces such as the White Mountain Magma Series of New Hampshire, or the Quincy-type granites in Massachusetts. It is also possible to generate both quartz-normative and nepheline-normative silicic magmas by step-wise fractional melting (Presnall and Bateman, 1973) of amphibole. At first melting, all available quartz and some feldspar would react with the amphibole to form a high-temperature, granitic melt. If this melt were removed, the residual amphibole and pyroxene would be peralkaline. Subsequent higher temperature melting of the residue could yield nepheline-normative melts.

Thus, the origin of anatectic granites can be discussed in terms of the hydrous phases involved (Hyndman, 1981). Table 11.3 summarize some of the characteristics expected from various melting reactions. Most of the granites classified as "S type" in the original classification of Chappel and White (1974) would be generated by dehydration melting of muscovite and biotite. At low to moderate pressures, cordierite would be a product of the melting reaction at high temperatures. However, at lower temperatures and moderately higher pressures cordierite would not form (Zen, 1987). Some of Chappel and White's "I-types" could be generated by biotite or biotite plus hornblende melting reactions. The type of protolith which could be involved is limited, however, to non-sedimentary materials by the low strontium and oxygen isotope ratios in such granites. Some "I-types" and some more alkaline granites could be generated by hornblende melting reactions. However, for the large-volume silicic batholiths so commonly associated with subduction zones, such as found in the Sierra Nevada range or the Andes of South America, it is very probable that other sources of volatiles exist that are related to the large volumes of andesites and basalts being generated from below the crust. In these terrains there are very small percentages of true "granites" in the sense used by Tuttle and Bowen (1958; rocks in which 2/3 or more of the feldspar is alkali feldspar). The most common lithologies are intermediate granites (for-

TABLE 11.3—Characteristics of various anatectic granites.

| Hydrated phase involved in melting | Muscovite | Biotite | Hornblende |
|---|---|---|---|
| Alkali/Al characteristics | Peraluminous | Peraluminous meta-aluminous | Meta-aluminous peralkaline |
| K/Na ratio | High, >1 | Intermediate, ≈1 or so | Low, <1 |
| Initial temperature | 650–750°C | 750–850°C | >900°C |
| $\delta^{18}O$ | High, ≈10 | Variable, up to 10 or so | Lower, 7–8 |
| $(^{87}Sr/^{86}Sr)_O$ | High, >0.710 | Variable, often >0.710 | Lower, 0.703–0.710 |
| Accessory minerals | Epidote, allanite, tourmaline, topaz, garnet, sillimanite, andalusite, ilmenite | Garnet, allanite, sillimanite, cordierite, ilmenite, pyroxene | Pyroxene, magnetite, sphene |

merly termed adamellite or quartz monzonite), granodiorites, and tonalites.

### Volatiles from subducted oceanic crust and mantle

One of the great modern advances has been our understanding of plate tectonics. We now know that many of the great granitic batholiths are formed on compressive plate margins where they underlie basalt–andesite–rhyolite volcanic belts. Further, we now know from the study of hydrothermal processes at mid-ocean ridges and obducted ophiolites that the subducted oceanic crust and upper mantle contains significant quantities of volatiles in the form of hydrated alteration minerals (perhaps as much as 2% of the crust). Therefore, these volatiles are a potential source of water for the formation of granitic melts in the crust. The method of transport and exchange, however, is potentially complex (see Marsh, 1984 for a quantitative evaluation of problem of magma ascent). Recently, numerous authors have recognized that mafic enclaves found within calc-alkaline batholiths represent magma mixing during which more mafic magmas are quenched against the silicic host (Eichelberger, 1980; Gerlack and Grove, 1982; Van Bergen et al., 1983; Bacon and Metz, 1984; Bacon, 1986; Cantagrel et al., 1984; Dorais, 1987). Such a process could be important in the transport of volatiles.

Fig. 11.8 illustrates the hydrothermal processes at mid-ocean ridges and surrounding sea floor crust. During this process, water is fixed in the crust and upper mantle as hydrous minerals including chlorite, amphibole, and serpentine. Studies of alteration mineral chemistry as well as mass balance of hydrothermal fluids suggest that chlorine as well is present in hydrous minerals. Sulfur is also deposited in the form of sulfides in vein fillings and exhalative deposits. We know that this alteration system extends to rather great depths in the oceanic crust and upper mantle. Taylor (1983) and Gregory and Taylor (1981) have shown that pervasive hydrothermal exchange with seawater has occurred throughout the upper 6 km of oceanic crust, and locally has penetrated into the tectonized peridotites of the upper mantle. Therefore, by the time the oceanic crust and upper mantle have cooled and moved away from the ridge, they are highly hydrated.

When oceanic crust and upper mantle are subducted, most of the sediments are left behind in the trench area. These materials are low in density and seismic studies suggest they are accumulated as a deformed sedimentary wedge. As the altered crust is subducted it slowly heats up as it interacts

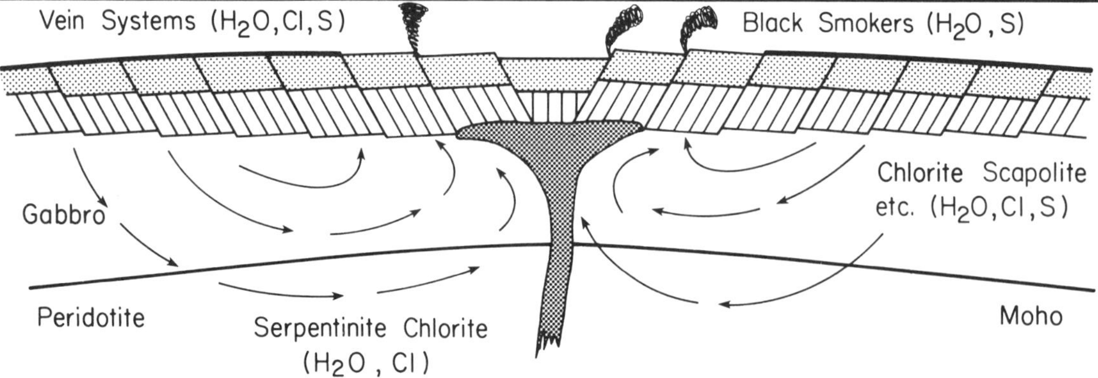


FIGURE 11.8—Hydrothermal circulation around a mid-ocean ridge. Alteration of the upper crust deposits water and chlorine in the alteration assemblages. Deeper circulation deposits water and chlorine through the formation of chlorite and serpentine. The crust is also relatively oxidized by the process. Hydrothermal vents on the ocean floor and vein fillings in the upper crust also fix sulfur. Circulation model after Gregory and Taylor, 1981.

with the mantle. As it does, the hydrous minerals will begin to break down, first through dehydration reactions and then through melting. The products will be either tholeiitic basalts or andesites depending on the availability of water (Green and Ringwood, 1968; Green, 1976; Mysen and Boettcher, 1975; Wyllie, 1979). However, since the amount of water available is only 1 to 2%, most of the primary melts are probably basalt, with andesites being generated at shallower depths (e.g., Maaloe and Petersen, 1981). These pass through the overlying continental lithosphere and perhaps part of the asthenosphere. The products are the copious basalts and andesites associated with island arcs and compressive plate margins. Melting of peridotite, however, does not form true granite. Unless the continental lithosphere has a much higher silica activity (i.e., is made of quartz eclogite rather than peridotite) the generation of most granites must involve the lower crust. Thus we find that true granites are absent in island arcs which do not involve continental crust.

When magmatism first begins in a compressive plate margin, the mafic magmas encounter lower crust at temperatures of around 500 to 700°C or so. This process is schematically shown in Fig. 11.9A. Under these conditions, much of the

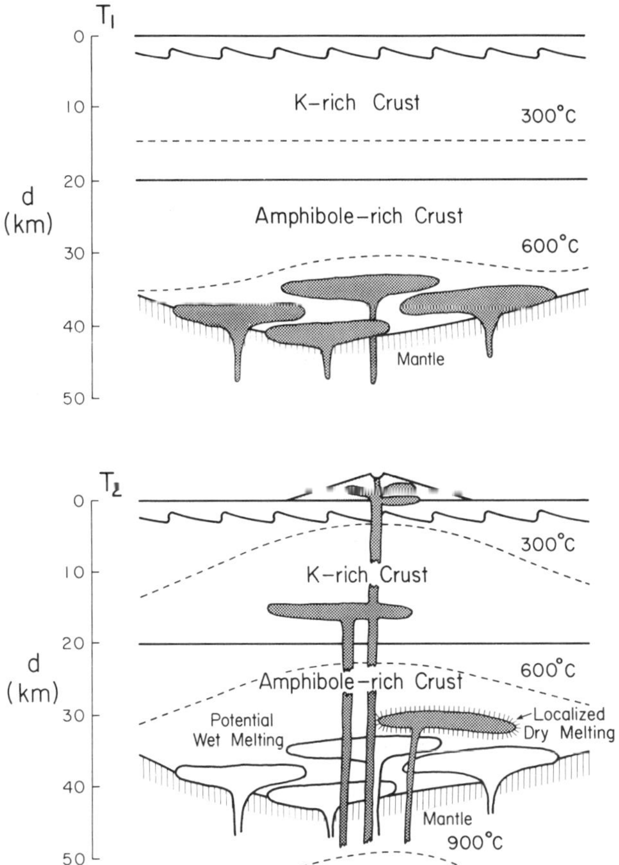


FIGURE 11.9—Thermal and magmatic development in the crust above an evolving subduction zone. $T_1$, early in the volcanic history. $T_2$, after substantial period of calc-alkaline magmatism.

early mafic magmas will tend to crystallize in the lower crust (Marsh, 1978). During this process, the crust may be thickened and partially underplated with hydrous gabbroic material. Although some of the water will be fixed as hornblende and/or biotite, part of the dissolved volatiles may be released to interact with the lower crust. As the lower crust heats up, anatexis can begin where water is available. The melting relations in the Smith River area described by Snoke et al. (1981) may be a small-scale analog of what might happen around a crystallizing pluton in the lower crust. As the lower crust continues to heat up, anatexis can begin over larger regions as the dehydration reactions of various phases are surpassed (Fig. 11.9B). Thus, the passage of subduction related magma through the crust may transfer heat and perhaps volatiles to the crust to cause melting is sufficient basalt traverses the crust (Marsh, 1984).

It is possible, however, that the mafic melts and their dissolved volatiles may interact more directly with granitic melts once silicic magmas begin to coalesce within the crust. Once a body of granitic magma begins to coalesce into a sizable coherent body it forms a density and thermal barrier for upwelling mafic magmas. If a hydrous andesite or basalt penetrates a granitic magma chamber the denser mafic magma will tend to underplate the less dense silicic magma (Fig. 11.10). Since the mafic magma has a higher crystallization temperature, it will also tend to partially quench against the cooler granitic pluton. Such a process has been hypothesized by a number of authors (e.g. Eichelberger, 1980; Bacon, 1986; Dorais, 1987) to explain the mafic enclaves that are so common in the Sierra Nevadas and other great calc-alkaline batholiths associated with compressive plate margins. Recently, Dorais (1987) has conducted a detailed geochemical study, including ion-probe data, on early phenocrysts of such enclaves in the Dinkey Creek pluton of the Sierra Nevada batholith and a volcanic analog from the San Juan Mountains, Colorado. Fig. 11.11 is Dorais' model for such a system. Under these conditions, both thermal energy and volatiles can be transferred to the granitic system. Convection and mixing in the granitic magma will transport thermal energy throughout the body, aiding in additional assimilation, and at times disrupting the mixed, semi-solid boundary layer strewing fragments of the mixed zone throughout the magma chamber. The chemistry and isotopic systematics of the resulting pluton will depend on the relative contributions of old continental crust, young hydrous gabbroic crust, and subduction-derived more mafic magmas to the system, as well as subsequent fractionation and assimilation processes in the upper crust.

In this model, the volatiles responsible for large-scale melting of the lower crust in compressive plate margins would at least in part be derived from altered, subducted oceanic crust and mantle transported into the lower crust and released during crystallization of the hydrated mafic magmas, which then directly interacted with cooler, less dense silicic magmas. Such a system could explain why chlorine is so prominent in calc-alkaline plutons in compressive plate margins, while being far less common in tensional or anorogenic environments.

Granitic plutons formed by such a mixing process can be quite variable in composition since their characteristics depend on the nature of granitic melts formed in the crusts, the composition of more mafic magmas which interact with them,

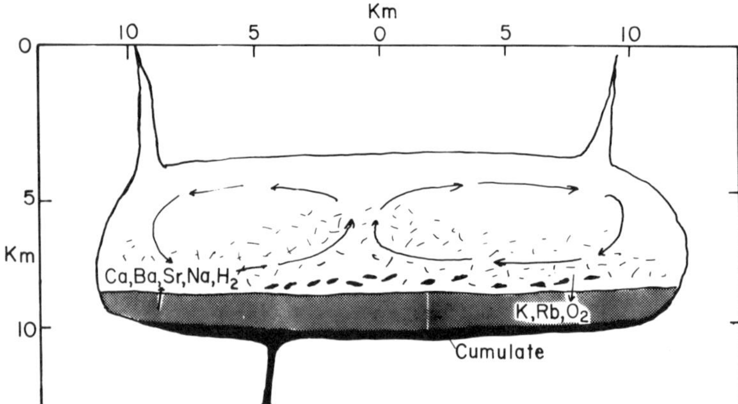


FIGURE 11.10—Underplating of a silicic magma chamber with a more dense mafic magma of higher density. Modeled from the Carpenter Ridge Tuff, Whitney et al., 1988. More mafic magma underplates the chamber, partially quenches against the cooler silicic melt. Temperature and mass transfer are accomplished through mechanical and chemical diffusive methods. In the case of the Carpenter Ridge, the mafic enclaves become enriched in K, Rb, and Oxygen, while the granitic magma becomes enriched in Ca, Ba, Sr, Na, and hydrogen.

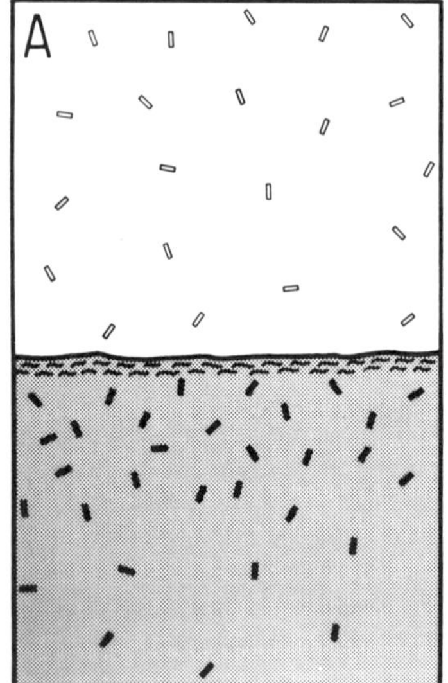


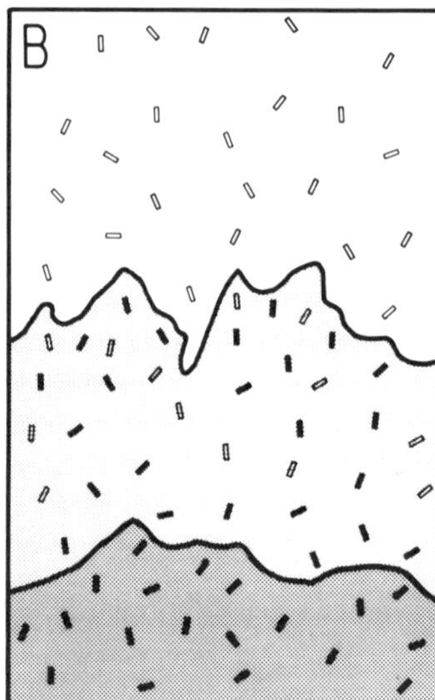

FIGURE 11.11—Detail of the partially quenched boundary layer between silicic and mafic magma. After Dorais, 1987. Boundary layer develops **A.**, interacts through chemical diffusion **B.**, and is partially dismembered by convective movement in the overlying magma **C.**, yielding mafic enclaves which are themselves a chemical/mechanical mixture between the two parent magmas.

the relative quantity of such mafic magmas, and the degree of interaction. First, since basaltic and andesitic systems are rich in sodium, these systems will tend to have a lower K/Na ratio than granitic magmas formed from crustal anatexis. If the degree of interaction is slight, they would still be intermediate granites, but with greater quantities of mafic component they would be granodioritic to tonalitic or trondhjemitic. Second, they would tend to be meta-aluminous (i.e. $K_2O + CaO + Na_2O > Al_2O_3 > Na_2O + K_2O$) in most cases. Their initial temperature would vary depending on the relative magnitude of mafic versus felsic magma, and thus the amount of heat transferred, but would in general be higher than anatectic systems (i.e. 800°C or more) involving muscovite or biotite. The initial isotopic ratios would also be variable, but in general would be more primitive. Common mafic minerals would be hornblende and possibly pyroxene accompanying biotite. The activity of volatiles such as chlorine and sulfur might be expected to be higher than in many anatectic granites. The state of oxidation could also be variable. If the more mafic melts were derived from oxidized oceanic crust and upper mantle without the opportunity for reduction by primitive mantle, then the oxidation state could be high yielding accessory minerals such as magnetite and sphene as is often seen in arc-related rocks. If on the other hand the more mafic magmas had equilibrated with olivine-rich mantle, then they would be reduced resulting in lower oxygen fugacity.

In most cases, such granites should contain mafic enclaves formed from the mafic magma which contain minerals such as hornblende and sphene with signs of interaction with the granitic melt. These enclaves may contain signs of quenching such as acicular apatite (Wyllie et al., 1962; Vernon, 1983). They also contain a poikilitic mesostasis consisting of quartz, potassium feldspar, and sodic plagioclase. This material represents either residual liquids from partial quenching of the basaltic parent (Bacon, 1986) or commingling with the host silicic melt (Sparks and Marshall, 1986).

## FRACTIONAL CRYSTALLIZATION AND ASSIMILATION

Most magmas which reach the surface, and nearly all plutonic rocks, have been effected by crustal interaction during intrusion and crystallization. These include fractional crystallization and assimilation. Fractional crystallization can occur either at the mineral scale through zoning, or at a larger scale through crystal settling, nucleation and growth on the wallrock interface, or separation of the melt through filter pressing. Textural evidence in granitic rocks does not suggest that crystal settling is as important as in more mafic systems.

Recently, there has been a great deal of discussion of the combined effects of bulk assimilation of wall rock and fractional crystallization euphemistically termed AFC processes (DePaulo, 1981). In such models, the latent heat evolved during crystallization is partly responsible for assimilating country rock. The result on quantitative chemical fractionation models is to spread out the fractionation trend, depending on the percent of assimilation, into a region in composition space. However, assimilation may be more complicated that simple bulk assimilation of country rock.

Suppose a magma of intermediate composition containing phenocrysts of hornblende, biotite, and plagioclase intrudes a granitic gneiss composed of biotite, plagioclase, microcline, and quarts. Since the magma is undersaturated with respect to quartz, it will tend to preferentially dissolve quartz, raising its silica content. During this process energy will be consumed to overcome the latent heat of fusion, thus causing a decrease in magma chemistry. However, since the magma is undersaturated with respect to quartz, it changes to a composition with a lower solidus and liquidus temperature, therefore crystallization may or may not accompany this drop in temperature. The melt is also undersatured with respect to potassium feldspar and may tend to dissolve the alkali feldspar. If the magma is simultaneously crystallizing plagioclase the effect may be to dissolve one feldspar and deposit another. In such cases, the net result on trace elements such as Europium which are concentrated in the feldspar phase may be indeterminable. Rubidium, which is preferentially concentrated in the alkali feldspar, would be enriched, while strontium which follows calcium in the plagioclase would be depleted.

Thus, this type of selective assimilation combined with crystallization can yield more complicated chemical fractionation patterns for various elements, depending on the minerals being crystallized and assimilated. These processes may be referred to as Selective Assimilation/Crystallization (SAC) processes. The residual products of such selective assimilation may be mafic shlieren and enclaves which may have been termed restite in some plutons.

Fractional crystallization can also be a more complicated process than pure fractional or equilibrium models suggest. Some minerals re-equilibrate easily with the melt, for example the oxide and sulfide phases. Elements concentrated in these phases may therefore be modeled as equilibrium crystallization as long as the minerals remain in contact with the melt. Only upon physical separation from the melt by settling, wall-rock crystallization, or filter pressing will they be fractionally removed. Other minerals, such as plagioclase, grow as zoned phases with little re-equilibration with the melt. Elements locked in these phases are effectively fractionally crystallized. Thus, in a crystallizing magma, some elements will continue to re-equilibrate with the melt and must be modeled through equilibrium crystallization, while other elements concentrated into different minerals may be effectively fractionally crystallized.

## SEPARATION OF A VOLATILE PHASE

The separation of a volatile phase from a crystallizing magma has been quantitatively discussed by several authors (Whitney, 1975a, 1975b; Burnham, 1979). The following synopsis is based on these sources.

At the time of intrusion an ascending magma will reach some pressure at which a vapor phase will begin to exsolve. The approximate water content for many silicic melts is 2 to 4 wt% (see previous section). Under these conditions, vapor saturation will occur at about 1.3 kb. If $CO_2$ is present in significant amounts, the depth will be increased. High sulfur fugacities may also cause the zone of initial vapor saturation to be deeper.

As the magmatic volatiles rise, they will interact with other waters forming a hydrothermal plume. The movement of fluids in such a system has been modeled by Henley and McNabb (1978) for porphyry copper deposits. Other authors (e.g., Cathles, 1981) have also reviewed the movement of meteoric fluids around such cooling stocks.

The process of vapor separation can be considered to be composed of two parts. The first is spontaneous vesiculation due to oversaturation of the melt phase in response to declining confining pressure. Since vesiculation is fairly fast, such a process is effectively instantaneous, although some amount of oversaturation must occur before nucleation begins. The second portion of volatile separation occurs due to increasing water content in the melt due to crystallization of anhydrous phases. This process is somewhat slower and depends on crystal nucleation and growth.

As an igneous body cools, vapor generation will continue in the upper portion of the body as crystallization occurs. The equilibrium conditions are shown for a 2 km square stock in Figs. 11.12 and 11.13 with an initial temperature of approximately 800°C. If latent heat and convection are ignored, $t_1$ would be about 1,250 years and $t_2$ would be 12,500 years after intrusion. Latent heat would approximately double the time required to reach these states, while convective upwelling could extend the time much longer if it were continuously active (Whitney, 1975b).

Upwelling of new magma into a stock attached to a larger batholith, either through periodic recharging of the magma chamber or convective overturn, may greatly increase the amount of volatile material available and also renew the concentration of elements preferentially partitioned into the gas phase. Fig. 11.14 represents a model for a convecting system. Through time, the overall volatile content of the

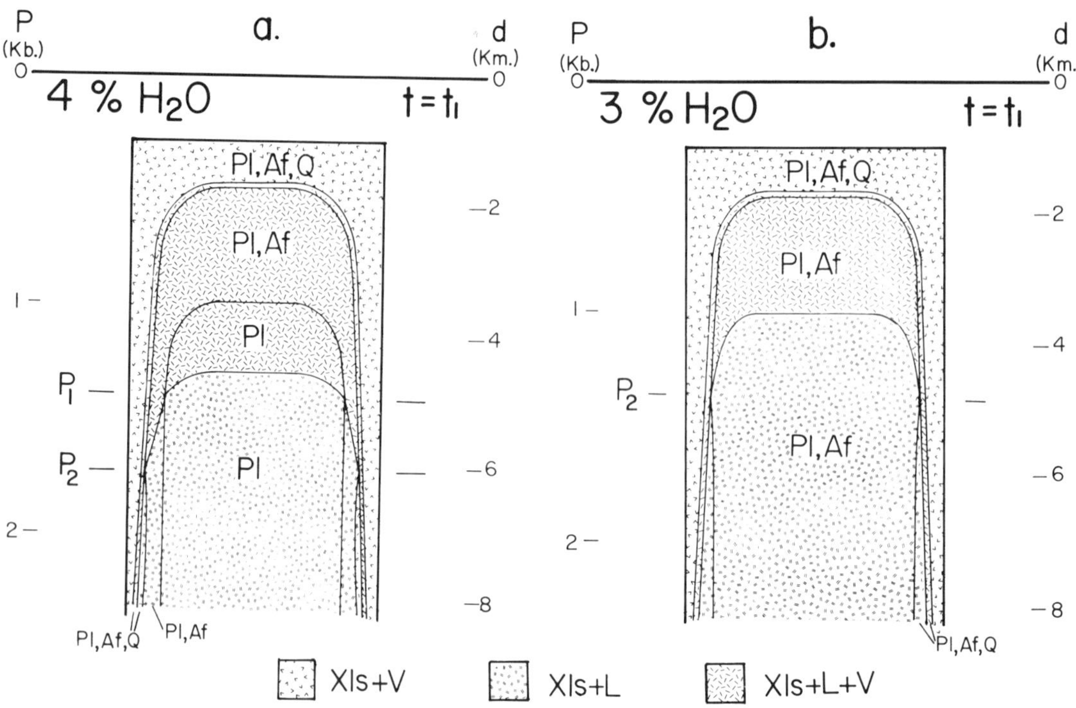


FIGURE 11.12—Equilibrium crystallization model for a 2 km square stock. Based on the model of Whitney, 1975b. Composition is the same as in Figs. 11.1 to 11.3. If latent heat and convection are ignored, the time would be about 1,250 years after intrusion. Latent heat would approximately double the time estimate.

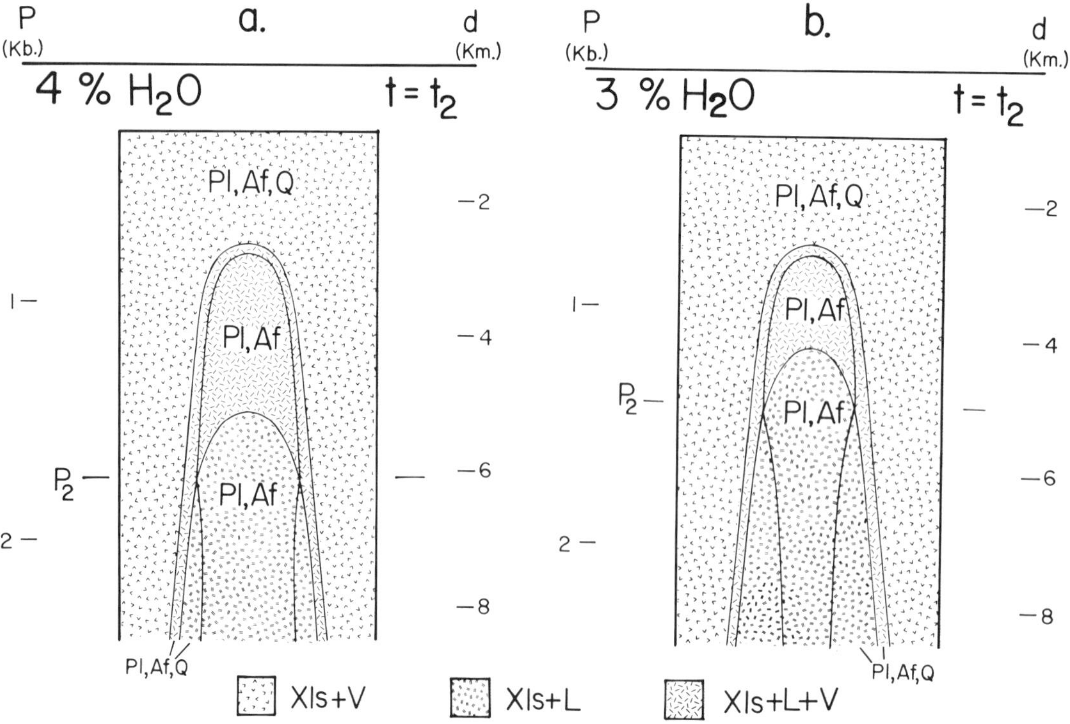


FIGURE 11.13—Equilibrium crystallization model for a 2 km square stock. Time is ten times that in Fig. 11.12.

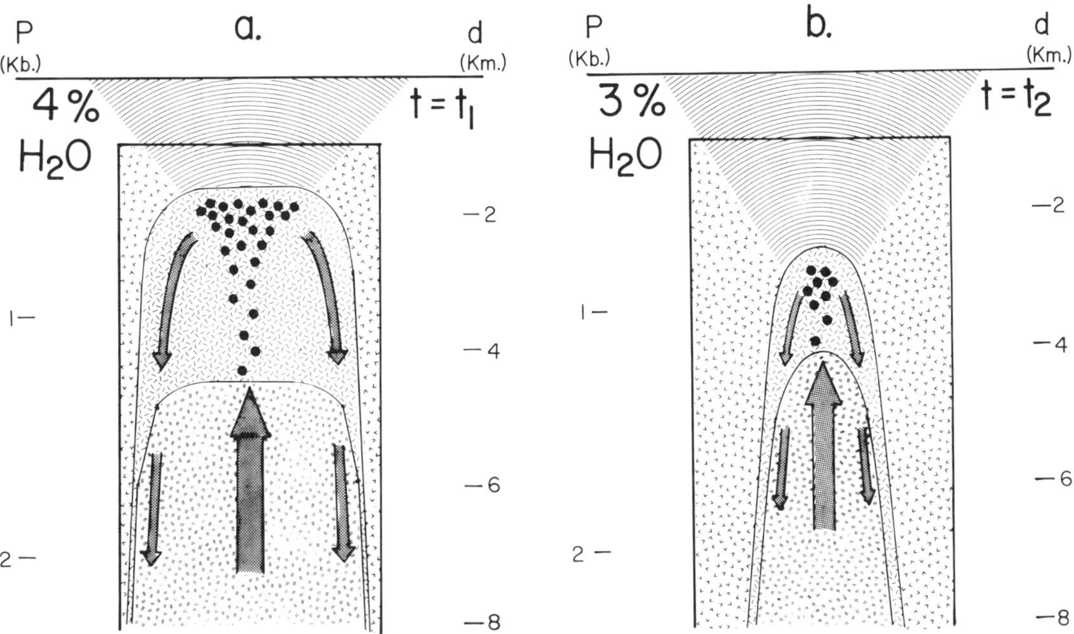


FIGURE 11.14—Crystallization model showing the possible effects of convection or recharge of the reservoir on crystallization model in Figs. 11.12 and 11.13. With time, the stock may lose water from the entire body and conditions move from case a to case b.

entire magma may decrease as part is vented through stocks extending into the vapor-saturated region. Thus Fig. 11.14a may evolve into 11.14b.

## MAGMATIC ORE DEPOSITION

Various types of ore deposits are spatially and temporally related to silicic magmas (see previous chapter). In terms of decreasing magmatic input these include porphyry systems, skarn deposits, volcanic epithermal deposits, and exhalative massive sulfide deposits.

### Porphyry copper systems

Porphyry copper deposits are disseminated deposits of copper, lead, zinc, molybdenum, with minor gold and silver. They are mainly associated with only moderately silicic calc-alkaline intrusions emplaced in the upper few kilometers of the crust. Deposition of the protore stage is as small veins or veinlets in the fractured upper carapace of the intrusion. The early fluids are highly saline chloride solutions which are probably derived from the underlying magma. The plutons do not seem to be highly fractionated and have relatively primitive strontium and oxygen isotopic signatures.

The parent magma is therefore fairly primitive in origin, either formed from melting of an amphibolitic source, or with significant mafic input through the refluxed magma chamber model. The high chlorine content suggest as input from altered oceanic crust. The high chlorine/fluorine abundance suggests that the magma has not undergone previous vapor fractionation since chlorine would be preferentially removed under such processes. The oxygen fugacity is fairly high since many such systems have sphene as an accessory phase. These deposits probably represent deposition from the first evolution of a vapor phase. The magma was probably vapor undersaturated until emplacement, at which time vapor fractionation and separation formed the first stage of ore forming fluids.

It should be emphasized that these magmatic events alone yield a disseminated deposit which may or may not be economic. In many cases, subsequent low temperature hydrothermal activity involving meteoric water driven by the heat of the intrusion is required to generate an economic concentration. The fluids involved in these latter processes are low salinity, low temperature ones which are responsible for much of the alteration halo around such deposits and the secondary concentration which significantly increases the grade.

### Porphyry molybdenum deposits

Porphyry molybdenum systems are quite different from porphyry copper systems. The parent stocks are far more silicic (differentiation indices of around 90) and apparently more differentiated. The magmatic fluids are high in fluorine, but relatively lower in chlorine (although significant amounts of chlorine may still be present). Because of the high fluorine content, the movement of silica in the fluid is greatly enhanced, leading to a great deal of silicification and deposition of quartz veins. The tops of some of these stocks are areas of vapor concentration and deposition where the fluids have accumulated at the top of the system. Primary deposition is again as veins or veinlets within the solidified carapace of the system and surrounding rocks.

The high differentiation index of the parent suggests that

molybdenum porphyry stocks have had extensive differentiation under crustal conditions. The very high silica content of even the most unaltered rocks suggests low pressure processes. The relatively higher fluorine content suggests that these magma may have become vapor saturated prior to final emplacement, and the possibility that there may have been prior loss of magmatic fluids lowering the relative abundance of chlorine. In general, secondary hydrothermal alteration halos and meteoric water circulation are somewhat less important in these systems, but still extensive in their development.

### Skarn systems

The skarn system is characterized by disequilibrium and strong changes in chemical variables over very short distances. In general, skarns are an area where magmatic and high temperature hydrothermal systems are being thrown into a completely different chemical environment generating a heterogeneous system. During this process, the magmatic fluids change dramatically in their chemistry and carrying capacity for a variety of elements. A later chapter in this volume discusses tungsten skarns in detail.

The nature of a skarn deposits depends on three major factors: (1) The origin and evolution of the parent magma; (2) The chemistry of the derived magmatic fluids; (3) The nature and contribution of the country rock. Attempts to classify skarns must consider all three variables. It is no wonder, therefore, that nearly every skarn is in some way unique. It is therefore difficult to make general statements about the magmatic versus country-rock contribution.

### Volcanic epithermal deposits

Volcanic epithermal deposits are formed in calderas and similar shallow vent areas. Deposits may include surficial deposits in caldera lakes and fumaroles, or as vein deposits in the underlying fractured rocks. Volcanic emanations include the sulfurous gases $H_2S$ and $SO_2$ which combine with oxidized waters to form sulfuric acid (Whitney, 1988a). The high temperature equilibration of chloride solutions with feldspar bearing assemblages may also contribute significant HCl activities (Whitney et al., 1985). The interactions of these acid emanations with meteoric water yields a fluid with varying character depending on the relative proportions and individual characteristics. In some cases, the effects of sulfur, chorine, and fluorine in combination can yield a chemically active fluid with significant carrying capacity for elements like silver, gold, and silica.

Magmatic paragenesis controls the availability of cations such as sulfur and chlorine. For maximum concentration of these elements the magma would have input from altered oceanic crust, and not have undergone vapor separation prior to emplacement in the shallow volcanic environment.

### Exhalative massive sulfide deposits

These deposits apparently have little direct contribution from the silicic magma, although they are often associated with silicic volcanic vents. In this case, nearly all of the fluids involved are recycled sea water which as undergone low to high temperature interaction with solidified volcanics. Thus, the magmatic input is not very important, except as a source of heat. Therefore, in classification and study of these deposits the chemical history of the magma is less important than the nature of the country rock through which the solutions circulate, and the temperature and salinities obtained.

## CONCLUSIONS

The origin of granitic melts in the crust requires several weight percent water. The source and amount of water available has a direct effect on the temperatures at which melting occurs, the subsequent chemical composition of the magma, and the subsequent intrusion and crystallization history. Although all groups of rocks form a continuum and therefore can never be classified into neat pigeonholes, an alternate method of classification of associations of granites is one based on the sources of water. Hyndman (1981) suggested such a scheme for granites derived by anatexis of hydrous phases. However, in cases where volatiles from a subducted oceanic plate are involved, it is likely the dehydration of crustal material will also occur, so a mixture of magmatic characteristics may be encountered.

During residence in the crust, fractionation and assimilation processes are important in developing the chemical characteristics we subsequently observe. Besides classical fractional crystallizations, processes such as Assimilation/Fractional Crystallization (AFC) and Selective Assimilation/Crystallization may be important in developing final chemical characteristics.

The evolution of a separate vapor phase is an important phenomena in the upper few kilometers of the Earth's crust. Concentration of volatile elements such as sulfur and chlorine, and the transport of ore-forming elements by these fluids, is an important stage in the development of economic ore deposits. The fractionation of such components is dealt with in subsequent chapters. For a detailed discussion of sulfur, the reader should consult previous publications including Whitney (1984a, 1984b, 1988a).

The magmatic input to various types of ore deposits may be discussed in terms of magma generation, crustal evolution, and vapor separation processes. The ore deposits found as porphyries, skarns, and volcanic epithermal systems may owe their existence to element concentrations derived from these processes.

If the fluxing action of water carried down by subduction is important in the remelting of crustal materials to form granites, then the ultimate development of thick granitic continental crust may be directly related to the fact that the Earth is largely covered by water. Planetologists should investigate whether the differences in crustal development between Venus and the Earth could be affected by surface conditions which do not allow liquid water on our sister planet.

It is interesting to note that the Earth appears to be the only terrestrial planet with a thick continental crust, highly mobile crustal plates, and is also the only one covered with water. As we begin to examine the planet as a complete system we may increasingly discover that surface phenomena and conditions can affect the internal processes of the planet.

ACKNOWLEDGEMENTS—Research supported by NSF grant no. EAR–8719700. Many of the ideas expressed have developed through the interactions with many colleagues and students including J.C. Stormer, Jr., M.J. Dorais, and numerous members of the U.S. Geological Survey. However, errors in fact or concept are soley the responsibility of the author. Portions of this work work were published subsequent to their appearance in the 1988 short course in Whitney (1988b).

## REFERENCES


Andersen, D. J., and Lindsley, D. H., 1985, New (and final!) models for the Ti–magnetite/ilmenite geothermometer and oxygen barometer (abs.): EOS, v. 66, p. 416.

Bacon, C. R., 1986, Magmatic inclusions in silicic and intermediate volcanic rocks: Journal of Geophysical Research, v. 91, no. B6, pp. 6091–6112.

Bacon, C. R., and Metz, J., 1984, Magmatic inclusions in rhyolites, contaminated basalts, and compositional zonation beneath the Coso volcanic field, California: Contributions to Mineralogy and Petrology, v. 85, pp. 346–365.

Baldasari, A., 1981, Iron–titanium oxides in the Elberton Granite: Unpublished M.S. thesis, University of Georgia, 72 pp.

Bohlen, S. R., Pecor, D. R., and Essene, E. J., 1980, Crystal chemistry of metamorphic biotite and its significance in water barometry: American Mineralogist, v. 65, pp. 55–62.

Brown, W. L., and Parsons, I., 1981, Toward a more practical two-feldspar geothermometer: Contributions to Mineralogy and Petrology, v. 76, pp. 369–377.

Buddington, A. F., and Lindsley, D. H., 1964, Iron–titanium oxide minerals and synthetic equivalents: Journal of Petrology, v. 5, pp. 310–377

Burnham, C. W., 1979, Magmas and hydrothermal fluids; *in* Barnes, H. L. (ed.), Geochemistry of hydrothermal ore deposits, 2nd Edition, pp. 71–136.

Burnham, C. W., and Davis, N. F., 1974, The role of $H_2O$ in silicate melts: II. Thermodynamic and phase relations in the system $NaAlSi_3O_8$–$H_2O$ to 10 kilobars, 700° to 1000°C: American Journal of Science, v. 274, pp. 902–940.

Burnham, C. W., Holloway, J. R., and Davis, N. F., 1969, Thermodynamic properties of water to 1000°C and 10,000 bars: Geological Society of America, Special Paper 132, 96 pp.

Cantagrel, J. M., Didier, J., and Gourgaud, A., 1984, Magma mixing: Origin of intermediate rocks and "enclaves" from volcanism to plutonism: Physics of Earth and Planetary Interiors, v. 35, pp. 63–76.

Chappel, B. W., and White, A. J. R., 1974, Two contrasting granite types: Pacific Geology, v. 8, pp. 173–174.

Cathles, L. M., 1981, Fluid flow and genesis of hydrothermal ore deposits: Economic Geology, 75th Anniversary Volume, pp. 424–437.

Day, H. W., 1973, The high temperature stability of muscovite plus quartz: American Mineralologist, v. 58, pp. 255–262.

Day, H. W., and Fenn, P. W., 1982, Estimating the $P$–$T$–$X_{H_2O}$ conditions during crystallization of low calcium granites: Journal of Geology, v. 90, pp. 485–508.

DePaolo, D. J., 1981, Trace element and isotopic effects of combined wallrock assimilation and fractinal crystallization: Earth and Planetary Science Letters, v. 53, pp. 189–202.

Dietrich, R. V., 1961, Petrology of the Mount Airy "granite": Bulletin of the Virginia Polytechnic Institute, Engineering Station Series 144, v. 54, no. 6, 63 pp.

Dorais, M. J., 1987, The mafic enclaves of the Dinkey Creek granodiorite and the Carpenter Ridge Tuff: A mineralogical, textural and geochemical study of their origins with implications for the generation of silicic batholiths: Unpublished Ph.D. dissertation, University of Georgia, Athens, 185 pp.

Eichelberger, J. C., 1980, Vesiculation of mafic magma during replenishment of silicic magma reservoirs: Nature, v. 288, pp. 446–450.

Eggler, D. H., 1972, Water-saturated and undersaturated melting relations in a Paricutin andesite and an estimate of water content in the natural magma: Contributions to Mineralology and Petrology, v. 34, pp. 261–271.

Eggler, E. H., and Burnham, C. W., 1973, Crystallization and fractionation trends in the system andesite–$H_2O$–$CO_2$–$O_2$ at pressures to 10 kb: Geological Society of America, Bulletin, v. 84, pp. 2517–2532.

El–Hindi, M. A., 1977, Geology and mineralization of the southern part of the Pinos Altos area, Fort Bayard quadrangle, Grant County, New Mexico: Unpublished M.S. thesis, Colorado School of Mines, Golden

Ernst, W. G., 1976, Petrologic phase equilibria: Freeman Press, San Francisco.

Fridrich, C. J., and Mahood, G. A., 1987, Compositional layers in the zoned magma chamber of the Grizzly Peak Tuff. Geology, v. 15, pp. 299–303.

Gerlach, D. C., and Grove, T. L., 1982, Petrology of Medicine Lake highland volcanics: Characterization of endmembers of magma mixing: Contributions to Mineralogy and Petrology, v. 80, pp. 147–159.

Ghiroso, M. S., 1984, Activity/composition relations in the ternary feldspars: Contributions to Mineralogy and Petrology, v. 87, pp. 282–296.

Green, D. H., 1976, Experimental testing of equilibrium partial melting of peridotite under water saturated, high-pressure conditions: Canadian Mineralogist, v. 14, pp. 255–268.

Green, N. L., and Usdansky, S. I., 1986, Ternary-feldspar mixing relations and thermobarometry. American Mineralogist, v. 71, pp. 1100–1108.

Green, T. H., and Ringwood, A. E., 1968, Genesis of the calc-alkaline igneous rock suite: Contributions to Mineralogy and Petrology, v. 18, pp. 105–162.

Gregory, R. T., and Taylor, H. P., 1981, An oxygen isotope profile in a section of Cretaceous oceanic crust, Samail opiolite, Oman: Evidence for delta $^{18}O$ buffering of the oceans by deep (>5 km) seawater-hydrothermal circulation at mid-ocean ridges: Journal of Geophysical Research, v. 86, pp. 2737–2755.

Haggerty, S. E., 1976, Opaque mineral oxides in terrestrial igneous rocks: Mineralogical Society of America. Short Course Notes, v. 3, pp. 1–100.

Hammarstrom, J. M., and Zen, E An, 1986, Aluminum in hornblende: an empirical igneous geobarometer: American Mineralogist, v. 71, pp. 1297–1313.

Helz, R. T., 1979, Alkali exchange between hornblende and melt: A temperature-sensitive reaction: American Mineralogist, v. 64, pp. 953–965.

Henley, I. A., and McNabb, A., 1978, Magmatic vapor plumes and ground-water interactions in porphyry copper emplacement: Economic Geology, v. 73, pp. 1–20.

Hildreth, E. W., 1977, The magma chamber of the Bishop Tuff: Gradients in temperature, pressure, and composition: Unpublished Ph.D. dissertation, University of California, Berkeley, 328 pp.

Hildreth, E. W., 1979, The Bishop Tuff: Evidence for the origin of composition zonation in silicic magma chambers; *in* Chapin, C. E., and Elston, W. (eds), Ash-flow tuffs: Geological Society of America, Special Paper 180, pp. 43–75.

Holloway, J. R., 1976, Fluids in the evolution of granitic magmas: consequences of finite $CO_2$ solubility: Geological Society of America, Bulletin, v. 87, pp. 1512–1518.

Hyndman, D. W., 1981, Controls on source and depth of emplacement of granitic magma: Geology, v. 9, pp. 244–249.

Ishihara, S., 1977, The magnetite series and ilmenite series granitic rocks: Min. Geol. (Jpn.), v. 27, pp. 293–305.

James, R. S., and Hamilton, D. L., 1969, Phase relations in the system $NaAlSi_3O_8$–$KAlSi_3O_8$–$CaAl_2Si_2O_8$–$SiO_2$ at 1 kilobar water vapour pressure: Contributions to Mineralogy and Petrology, v. 21, pp. 111–141.

Jones, W. R., Hernon, R. M., and Moore, S. L., 1967, General geology of the Santa Rita quadrangle, Grant County, New Mexico: United States Geological Survey, Professional Paper 555, 144 pp.

Lindsley, D. H., 1976, The crystal chemistry and structure of oxide minerals as exemplified by the Fe–Ti oxides; *in* Oxide minerals: Mineralogical Society of America, Short Course Notes, v. 3, pp. L.1–L.88.

Lipman, P. W., 1971, Iron–titanium oxide phenocrysts in compositionally zoned ash-flow sheets from southern Nevada: Journal of Geology, v. 79, pp. 438–456.

Luth, W. C., 1967, Studies in the system $KAlSiO_4$–$Mg_2SiO_4$–$SiO_2$–$H_2O$: Inferred phase relations and petrologic applications: Journal of Petrology, v. 8, pp. 372–416.

Luth, W. C., Jahns, R. H., and Tuttle, O. F., 1964. The granite system at pressures of 4 to 10 kilobars: Journal of Geophysical Research, v. 69, pp. 759–773.

Maaloe, S., and Petersen, T. S., 1981, Petrogenesis of oceanic andesites: Journal of Geophysical Research, v. 86, pp. 10,273–10,286.

Marsh, B. D., 1978, On the cooling of ascending andesitic magma: Philosophical Transactions of the Royal Society of London, v. A288, pp. 611–625.

Marsh, B. D., 1981, On the crystallinity, probability of occurrence, and rheology of lava and magma: Contributions to Mineralogy and Petrology, v. 78, pp. 85–96.

Marsh, B. D., 1984, Mechanics and energetics of magma formation and ascension; *in* Boyd, F. R., Jr. (ed.), Explosive volcanism: Inception, evolution, and hazards: National Academy Press, Washington, DC, pp. 67–83.

Melson, W. G., and Hopson, C. A., 1981, Pre-eruption temperatures and oxygen fugacities in the 1980 eruption sequence; *in* Lipman, P. W., and Mullinax, D. R. (eds.), The 1980 eruption of Mount St. Helens, Washington: U.S. Geological Survey, Professional Paper 1250, pp. 641–648.

Mysen, B. O., and Boettcher, A. L., 1975, Melting of a hydrous mantle: II. Geochemistry of crystals and liquids formed by anatexis of mantle peridotite at high pressures and high temperatues as a function of controlled activities of water, hydrogen, and carbon dioxide: Journal of Petrology, v. 16, pp. 549–593.

Naney, M. T., 1977, Phase equilibria and crystallization in iron- and magnesium-bearing systems: Unpublished Ph.D. thesis, Stanford University, Stanford, California.

Naney, M. T., and Swanson, S. E., 1980, The effect of Fe and Mg on crystallization in granitic systems: American Mineralogist, v. 65, pp. 639–653.

Nekvasil, H., and Burnham, C. W., 1987, The calculated individual effects of pressure and water content on phase equilibria in the granite system; *in* Mysen, B. O. (ed.), Magmatic processes: Physicochemical principles: Special Publication 1, Geochemical Society, pp. 433–445.

Owen, J. A., 1983, A lithogeochemical survey of the Pinos Altos pluton, Silver City, New Mexico: Unpublished M.S. thesis, University of Georgia, Athens, 172 pp.

Potter, P. M., 1981, The geology of the Carlton quadrangle, Inner Piedmont, Georgia: Unpublished M.S. thesis, University of Georgia, Athens, 89 pp.

Presnall, D. C., and Bateman, P. C., 1973, Fusion relations in the system $NaAlSi_3O_8$–$CaAl_2Si_2O_8$–$KAlSi_3O_8$–$SiO_2$–$H_2O$ and generation of granitic magmas in the Sierra Nevada batholith: Geological Society of America, Bulletin, v. 84, pp. 3181–3202.

Price, J. G., 1985, Ideal site-mixing in solid solutions, with applications to two-feldspar geothermometry: American Mineralogist, v. 70, pp. 696–701.

Robertson, J. K., and Wyllie, P. J., 1971, Rock-water systems with special reference to the water-deficient region: American Journal of Science, v. 271, pp. 252–277.

Rutherford, M. J., Sigurdsson, H., Carey, S., and Davis, A., 1985, The May 18, 1980, eruption of Mount St. Helens: I. Melt composition and experimental phase equilibria: Journal of Geophysical Research, v. 90, pp. 2929–2947.

Smith, R. L., 1960, Zones and zonal variations in welded ash-flows: U.S. Geological Survey, Professional Paper 354–F, pp. 149–159.

Smith, R. L., 1979, Ash-flow magmatism; *in* Chapin, C. E., and Elston, W. (eds.), Ash-flow tuffs: Geological Society of America, Special Paper 180, pp. 5–27.

Smith, R. L., and Bailey, R. A., 1966, The Bandelier Tuff: A study of ash-flow eruption cycles from zoned magma chambers: Bulletin of Volcanology, v. 29, pp. 83–104.

Snoke, A. W., Quick, J. E., and Bowman, H. R., 1981, Bear Mountain igneous complex, Klamath Mountains, California: An ultrabasic to silicic calc-alkaline suite: Journal of Petrology, v. 22, pp. 501–552.

Sparks, R. S. J., and Marshall, L. A., 1986, Thermal and mechamical contrasts on mixing between mafic and silicic magmas: Journal of Volcanology and Geothermal Research, v. 29, pp. 99–124.

Spencer, K. J., and Lindsley, D. H., 1981, A solution model for coexisting iron–titanium oxides: American Mineralogist, v. 66, pp. 1189–1201.

Speer, J. A., 1977, Description of the Siloam Pluton in evaluation and targeting of geothermal energy resources in the southeastern United States: U.S. Department of Energy, Progress Report, Oct. 1977–Dec. 1977, Virginia Polytechnic Institute and State University, 4648–1, pp. A43–A63.

Speer, J. A., Becker, S. W., and Farrarr, S. S., 1980, Field relations and petrology of the post-metamorphic, coarse-grained granites and associated rocks in the Southern Appalachian Piedmont: Virginia Polytechnic Institute and State University, Memoir 2, IUGS Caledonide Volume, pp. 137–148.

Steiner, J. C., Jahns, R. H., and Luth, W. C., 1975, Crystallization of alkali feldspar and quartz in the haplogranite system $NaAlSi_3O_8$–$KAlSi_3O_8$–$SiO_2$–$H_2O$ at 4 kb: Geological Society of America, Bulletin, v. 86, pp. 83–93.

Stormer, J. C., Jr., 1975, A practical two-feldspar geothermometer: American Mineralogist, v. 60, pp. 667–674.

Stormer, J. C., Jr., 1983, The effects of recalculation on estimates of temperature and oxygen fugacity from analyses of multicomponent iron–titanium oxides: American Mineralogist, v. 68, pp. 586–594.

Stormer, J. C., Jr., and Whitney, J. A., (eds.), 1980, Geological, geochemical, and geophysical studies of the Elberton Batholith, eastern Georgia: Georgia Department of Natural Resources, Guidebook 19, 134 pp.

Stormer, J. C., Jr., and Whitney, J. A., 1985, Two feldspar and iron–titanium oxide equilibria in silicic magmas and the depth of origin of large volume ash-flow tuffs: American Mineralogist, v. 70, pp. 52–64.

Swanson, S. E., 1979, The effect of $CO_2$ on phase equilibria and crystal growth in the system $KAlSi_3O_8$–$NaAlSi_3O_8$–$CaAl_2Si_2O_8$–$SiO_2$–$CO_2$ to 8000 bars: American Journal of Science, v. 279, pp. 703–720.

Taylor, H. P., Jr., 1983, Oxygen and hydrogen isotope studies of hydrothermal interactions at submarine and subaerial spreading centers; *in* Rona, P. A., Bostrom, K., Lauber, L, and Smith, K. L., Jr. (eds.), Hydrothermal processes at seafloor spreading centers: NATO Conference Series 4, v. 12, pp. 83–139.

Thompson, A. B., 1982, Dehydration melting of pelitic rocks and the generation of $H_2O$-undersaturated granitic liquids: American Journal of Science, v. 282, pp. 1567–1595.

Thompson, A. B., and Algor, J. R., 1977, Model systems for anatexis of pelitic rocks: Theory of melting reactions in the system $KAlO_2$–$NaAlO_2$–$SiO_2$–$H_2O$: Contributions to Mineralogy and Petrology, v. 63, pp. 247–269.

Tuttle, O. F., and Bowen, N. L., 1958, Origin of granite in the light of experimental studies in the system $NaAlSi_3O_8$–$KAlSi_3O_8$–$SiO_2$–$H_2O$: Geological Society of America, Memoir 74, 153 pp.

Van Bergen, M. J., Ghezzo, C., and Ricci, C. A., 1983, Minette inclusions in the rhyodiacite lavas of Mt. Amiata (Central Italy): Mineralogical and chemical evidence of mixing between Tuscan and Roman type magmas: Journal of Volcanology and Geothermal Research, v. 19, pp. 1–35.

Vernon, R. H., 1983, Restite, xenoliths, and microgranitoid enclaves in granites: Journal of the Proceedings of the Royal Society N.S.W., v. 116, pp. 77–103.

Whitney, J. A., 1969, Partial melting relationships of three granitic rocks: Unpublished M.S. thesis, Massachusetts Institute of Technology, Cambridge, Massachusetts, 142 pp.

Whitney, J. A., 1972, History of granodioritic and related magma systems: An experimental study: Unpublished Ph.D. thesis, Stanford University, Stanford, Calif., 192 pp.

Whitney, J. A., 1975a, The effects of pressure, temperature, and $X_{H_2O}$ on phase assemblage in four synthetic rock compositions: Journal of Geology, v. 83, pp. 1–27.

Whitney, J. A., 1975b, Vapor generation in a quartz monzonite magma: A synthetic model with application to porphyry copper deposits: Economic Geology, v. 70, pp. 346–358.

Whitney, J. A., 1984a, Fugacities of sulfurous gases in pyrrhotite-bearing silicic magmas: American Mineralogist, v. 69, pp. 69–78.

Whitney, J. A., 1984b, Volatiles in magmatic systems: Reviews In Economic Geology, v. 1, pp. 155–175.

Whitney, J. A., 1988a, Composition and activity of sulfurous species in quenched magmatic gases associated with pyrrhotite-bearing silicic systems: Economic Geology, v. 83, pp. 86–92.

Whitney, J. A., 1988b, The origin of granite: The role and source of water in the evolution of granitic magmas: Geological Society of America, Bulletin, v. 100, pp. 1886–1897.

Whitney, J. A., Dorais, M. J., Stormer, J. C., Jr., Kline, S. W., Matty, D. J., 1988, Magmatic conditions and development of chemical zonation in the Carpenter Ridge Tuff, central San Juan volcanic field, Colorado: American Journal of Science, Wones Memorial Volume, v. 288-A, pp. 16–44.

Whitney, J. A., Jones, L. M., and Walker, R. L., 1976, Age and origin of the Stone Mt. Granite, Lithonia District, Georgia: Geological Society of America, Bulletin, v. 87, pp. 1067–1077.

Whitney, J. A., and Stormer, J. C., Jr., 1976, Geothermometry and geobarometry in epizonal granitic intrusions: a comparison of iron–titanium oxides and coexisting feldspars: American Mineralogist, v. 61, pp. 751–761.

Whitney, J. A., and Stormer, J. C., Jr., 1977a, Two-feldspar geothermometry, geobarometry in mesozonal granitic intrusions: Three examples from the Piedmont of Georgia: Contributions to Mineralogy and Petrology, v. 63, pp. 51–64.

Whitney, J. A., and Stormer, J. C., Jr., 1977b, The distribution of $NaAlSi_3O_8$ betweem coexisting microcline and plagioclase and its effect on geothermomtric calculations: American Mineralogist, v. 62, pp. 687–691.

Whitney, J. A., and Stormer, J. C., Jr., 1983, Igneous sulfides in the Fish Canyon Tuff and the role of sulfur in calc-alkaline magmas: Geology, v. 11, pp. 99–102.

Whitney, J. A., and Stormer, J. C., Jr., 1985, Mineralogy, petrology, and magmatic conditions from the Fish Canyon Tuff, central San Juan volanic field, Colorado: Journal of Petrology, v. 26, pp. 726–762.

Whitney, J. A., and Stormer, J. C., Jr., 1986, Model for the intrusion of batholiths associated with the eruption of large-volume ash-flow tuffs: Science, v. 231, pp. 483–485.

Wones, D. R., 1972, Stability of biotite: A reply: American Mineralogist, v. 57, pp. 316–317.

Wones, D. R., and Eugster, H. P., 1965, Stability of biotite: Experiment, theory and application: American Mineralogist, v. 50, pp. 1228–1272.

Wyllie, P. J., 1971, Experimental limits for melting in the earth's crust and upper mantle; *in* Heacock, J. G. (ed.), The structure and physical properties of the earth's crust: Geophysical Monograph 14, pp. 279–301.

Wyllie, P. J., 1979, Magmas and volatile components: American Mineralogist, v. 64, pp. 469–500.

Wyllie, P. J., Cox, K. G., and Biggar, G. M., 1962, The habit of apatite in synthetic systems and igneous rocks: Journal of Petrology, v. 3, pp. 238–243.

Yoder, H. S., Jr., Stewart, D. B., and Smith, J. R., 1957, Ternary feldspars: Carnegie Institute of Washington, Yearbook 56, pp. 207–214.

Zen, E–An, 1987, Wet and dry AFM assemblages of peraluminous granites and the usefulness of cordierite as the prime indicator of S-type granite (abs.): Geological Society of America, Abstracts with Program, v. 19, pp. 904

# Chapter 12

# Magmatic Ore-forming Fluids: Thermodynamic and Mass Transfer Calculations of Metal Concentrations

P. A. Candela

## INTRODUCTION

In this chapter I shall discuss some of the practical details involved in the implementation of thermodynamic and mass transfer calculations which model the chemical effects of magmatic vapor evolution (Candela, 1986a, 1986b, 1986c; Candela and Holland, 1984, 1986; Tacker and Candela, 1987; Candela, in review; Bouton et al., 1987; and Candela, Liu, and Piccoli, in prep.) particularly as they relate to the development of ore fluids in porphyry-type, and skarn-type ore environments. Further, the implications of some of the recently published melt/vapor data will be explored, and some of my previously published models will be altered to accommodate this information.

Candela (1986a) presents four basic algorithms which can be used to calculate the concentration of selected elements in an evolving magmatic aqueous fluid, and the efficiency with which these elements can be removed from a pluton by vapor/melt partitioning upon the completion of crystallization. The input data for these mass transfer calculations include the initial concentration of the elements of interest in the melt, the water concentration (initial, saturation, or final) in the melt, the chlorine/water ratio, pressure, temperature, and the vapor/liquid and solid/liquid partition coefficients. In future models, the peraluminosity of the melt and the proportion of hydrous phases crystallizing from the magma will be considered. The four cases covered by these algorithms include:

1) the partitioning of elements such as Mo, B, F, Cl, W(?) and Nb(?) which can be modeled as following a Nernstian law (constant partition coefficient under certain restricted conditions) into the vapor from a silicate melt during the polybaric rise of a magma (first boiling);
2) the partitioning of chloride-complexed elements into the vapor during the partitioning of chloride-complexed elements into the vapor during first boiling;
3) the partitioning of Nernstian elements into the vapor during isobaric crystallization of a water-saturated melt (second boiling); and
4) the partitioning of chloride-complexed elements into the vapor during second boiling.

In Candela (1986a), these four cases are referred to as Model I, {case I and case II}, and Model II, {case I and case II}, respectively. The vapor/melt partition coefficients are critical in these calculations. Therefore, before the models themselves are discussed, the thermodynamics of element partitioning must be dealt with.

## THE FORMULATION OF PARTITION COEFFICIENTS: STOICHIOMETRIC ANALYSIS

The formulation of partition coefficients is not a trivial matter, and some general aspects of the topic have recently been covered in great detail (Tacker and Candela, 1987; Candela, ms). However, it would be instructive to outline this procedure and expand on the failings of some formulations of vapor/melt partitioning equilibria.

In the case of the partitioning of sodium between a silicate melt and an aqueous fluid, the raw experimental data are usually presented in the form of a Nernst partition coefficient, $D^{v/l}_{Na} = C^{v}_{Na}/C^{l}_{Na}$. Holland (1972) and Urabe (1985) have determined $D^{v/l}_{Na}$ as a function of $C^{v}_{Cl}$, the concentration of chlorine in the aqueous phase, and have shown that the relation between these variables can be given by the expression

$$D^{v/l}_{\mathrm{Na}} = 0.46 C^{v}_{\mathrm{Cl}} \qquad [1]$$

at temperatures of 770–880°C. and pressures of 1.4 to 3.5 kb. The chloride dependence exhibited by $D^{v/l}_{Na}$ in equation [1] is interpreted to result from the chloride-complexing of sodium in the aqueous phase. Similarly, Candela and Holland (1984) determined the partition coefficient for copper and summarized the data by using the equation $D^{v/l}_{Cu} = 9.1 \cdot C^{v}_{Cl}$. Whereas it is convenient to present the data in this format, these expressions are not equilibrium constants. That is, no thermodynamically valid "balanced reaction," or mass action expression can be written in terms of independently variable components of the melt and aqueous phase which would have $D^{v/l}_{i}$ (where $i$ is any chloride-complexed cation) or some simple variant of equation [1] as the corresponding equilibrium constant. To illustrate this point, I will formulate the partitioning of sodium, copper and other substances between a silicate melt and an aqueous fluid by considering the model system $Na_2O$–$Al_2O_3$–$CuO_{0.5}$–$H_2O$–$Cl_2O_{-1}$. The component $Cl_2O_{-1}$ is an exchange component (Thompson, 1982); the corresponding chemical potential, $\mu_{Cl_2O_{-1}}$, can be expressed as

$$\mu_{Cl_2O_{-1}} = 2\mu_{NaCl} - \mu_{Na_2O} = 2\mu_{HCl} - \mu_{H_2O} = \ldots, \quad [2]$$

which is an efficient representation of the chlorine-bearing component of a melt–vapor system. The independently variable components of the melt and vapor phases can then be written as:

**MELT:** $Na_2O$–$Al_2O_3$–$CuO_{0.5}$–$H_2O$–$Cl_2O_{-1}$
**VAPOR:** NaCl–CuCl–$H_2O$–HCl.

The number of linearly independent equilibria in any system is given by $\Sigma_\phi C_\phi - C_s$ (the total number of phase components minus the number of system components) which, in this system, is equal to 10 − 6 = 4. That is, of the many possible "balanced chemical reactions" one can write within the constraints of the above system, only four are algebraically independent, and only four, therefore, need to be specified for mass transfer. Three of the four conditions of chemical equilibrium are easily found by inspection. These are:

$$\mu^l_{H_2O} = \mu^v_{H_2O}, \quad [3]$$

$$\mu^l_{CuO_{0.5}} + \mu^v_{NaCl} = \mu^v_{CuCl} + \mu^l_{NaO_{0.5}}, \quad [4]$$

and

$$\mu^l_{NaO_{0.5}} + \mu^v_{HCl} = \mu^v_{NaCl} + \frac{1}{2}\mu^l_{H_2O}. \quad [5]$$

Equation [3] represents the fundamental equilibrium of water saturation in a silicate melt, equation [4] is an example of the proper formulation of cation–sodium exchange equilibria, and equation [5] is the equilibrium which controls the HCl content of a magmatic vapor phase. Rather than leave the results of their experiments in the form of equation [1], Holland (1972), Candela and Holland (1984) and Urabe (1985) all reduce their data into the form given by equation [4]. There are a number of possible formulations of the fourth equilibrium. One that is consistent with the models presented in this paper is

$$\mu^l_{NaO_{0.5}} + \frac{1}{2}\mu^l_{Cl_2O_{-1}} = \mu^v_{NaCl}. \quad [6]$$

NaCl is a dependent component of the liquid and can be expressed as NaCl = $NaO_{0.5}$ + $1/2Cl_2O_{-1}$. In terms of this component, the equilibrium simplifies to

$$\mu^l_{NaCl} = \mu^v_{NaCl}. \quad [7]$$

The distribution coefficient (apparent equilibrium constant) which corresponds to equation [7] is

$$K' = C^v_{NaCl}/C^l_{NaCl}. \quad [7a]$$

Note that the partition coefficient in equation [1] is written in terms of *total* sodium concentration in the melt. However, the equilibrium constant, equation [7a], is written in terms of the concentration of sodium in the melt *which is bound to chlorine.*

A few examples will serve to illustrate how a non-thermodynamic expression such as equation [1] can fail. Consider the case of a melt–vapor system involving a hypothetical low sodium tonalitic melt. In such a system $CaCl_2^0$ would be the dominant aqueous species, and

$$C^v_{Cl} \approx 2C^v_{CaCl_2}. \quad [8]$$

The statement of equilibrium for Ca–Na exchange in a melt–vapor system is

$$CaO^l + 2NaCl^v = CaCl^v_2 + 2NaO^l_{0.5}. \quad [8a]$$

for which the corresponding apparent equilibrium constant is:

$$\left(\frac{C^v_{CaCl_2}}{C^l_{CaO}}\right) = K'^{v,l}_{Ca,Na}\left(\frac{C^v_{NaCl}}{C^l_{NaO_{0.5}}}\right)^2. \quad [9]$$

Rearranging this equation, and substituting equation [8] yields

$$\left(\frac{C^v_{NaCl}}{C^l_{NaO_{0.5}}}\right) = D^{v/l}_{Na} = \left(\frac{C^v_{Cl}}{2C^l_{CaO}K'^{v/l}_{Ca,Na}}\right)^{1/2}, \quad [10]$$

which is the form $D^{v,l}_{Na} = b \cdot (C^v_{Cl})^{1/2}$. The dependence of $D^{v/l}_{Na}$ on $C^v_{Cl}$ has changed from the first power dependence found by both Holland and Urabe in systems with Na ± K > Ca to a square root dependence in a hypothetical system with Na ± K << Ca. The same can be said to be true for copper or any other chlorine-complexed cation. On the other hand, the apparent equilibrium constants used by Holland (1972), Candela and Holland (1984) and Urabe (1985) are valid expressions of equilibrium which suffer only from variations in excess thermodynamic properties as a function of composition; that is, $K = K_{eq}(T,P \text{ only})/\pi\gamma_i^v i(T,P,X)$. Whereas $D_{Cu}^{v/l}$ varies as the first power of $C_{Cl}^v$ in normal felsic melts, it would suffer the same fate as $D_{Na}^{v/l}$ as the bulk composition of the melt and the associated vapor changed. It is important to point out that because of the non-thermodynamic basis of the Nernst partition coefficients, they change their intrinsic stoichiometry, something the exchange constants do not do. The empirical exchange constants may vary with composition due to non-ideality; however, the Nernst partition coefficients are susceptible to this problem in addition to the above-mentioned stoichiometric problems. Therefore, in the case of the chloride complexed elements (Table 12.1), the $K_{i,Na}^{v,l}$ formalism is strongly suggested as a mode of data presentation.

For some elements (i.e. molybdenum) the Nernst partition coefficient can be shown to be a constant under certain conditions. For example, an equation which may represent the vapor/melt partitioning of molybdenum near $f_{O_2}$ = NNO is

$$\frac{1}{2}O_2 + MoO_2^l + H_2O = H_2MoO_4^v. \qquad [11]$$

However, the oxidation state of molybdenum in these phases is probably a function of $f_{O_2}$ (Candela, this volume) and the oxidation state of molybdenum in the aqueous phase is somewhat uncertain and may be 5+ (Kudrin et al., 1980). The Nernst partition coefficient for molybdenum can be written as a function of the equilibrium constant for equation [11]

$$\left(\frac{C_{H_2MoO_4}^v}{C_{MoO_2}^l}\right) \equiv D_{Mo}^{v/l} = K_{11}' a_{H_2O} f_{O_2} \qquad [12]$$

and it is clear that if equation [11] is a reasonable representation of molybdenum melt–vapor equilibria, $D_{Mo}^{v,l}$ will be a constant (save for activity coefficient effects) at a constant temperature, pressure, water activity, and oxygen fugacity. The stoichiometric effects we have been discussing here are even more critical in the case of crystal/melt partitioning. Tacker and Candela (1987) and Candela (ms) have discussed

TABLE 12.1—Vapor-melt partition coefficients of selected elements.

| Element | n[a] | Melt[b] | $K_1$,Na ± 1σ | D ± 1σ | T,°C | P,kbar | Ref.[c] |
|---|---|---|---|---|---|---|---|
| K | 1 | high-silica rhyolite | 0.74 ± 0.06 | | 770–880 | 1.4–2.4 | 1,2 |
| Ca | 2 | high-silica rhyolite | 0.38[d] ± .09 | | 770–880 | 1.4–2.4 | 2 |
| Mg | 2 | high-silica rhyolite | 0.16[d] ± .04 | | 770–880 | 1.4–2.4 | 2 |
| Mn | 2 | high-silica rhyolite | 6.5[d] ± 1.6 | | 770–880 | 1.4–2.4 | 2 |
| Zn | 2 | high-silica rhyolite | 9.5[d] ± 2.4 | | 770–880 | 1.4–2.4 | 2 |
| | | aluminous high-silica rhyolite | 44 ± 13.1 | | 800 | 3.5 | 9 |
| | | peralkaline high-silica rhyolite | 0.39 ± .12 | | 800 | 3.5 | 9 |
| Pb | 2 | aluminous high-silica rhyolite | 10.6 ± 3.2 | | 800 | 3.5 | 9 |
| | | peralkaline high-silica rhyolite | 0.22 ± 0.07 | | 800 | 3.5 | 9 |
| Rb | 1[e] | $Q_{40}(Ab,Or)_{60}$ | 0.66 – 0.77 | | 800 | 2 | 3 |
| Cs | 1[e] | $Q_{40}(Ab,Or)_{60}$ | 0.67 – 1.0 | | 800 | 2 | 3 |
| Sr | 2[e] | $Q_{40}(Ab,Or)_{60}$ | 0.5 – 0.28 | | 800 | 2 | 3 |
| Ba | 2[e] | $Q_{40}(Ab,Or)_{60}$ | 0.43 – 0.29 | | 800 | 2 | 3 |
| Ce | 3 | spruce pine pegmatite | 2.56 ± 0.89 | | 800 | 1.25 | 4 |
| Yb | 3 | spruce pine pegmatite | 1.12 ± 0.39 | | 800 | 1.25 | 4 |
| Cu | 1 | high-silica rhyolite | 20 ± 5.6 | | 750 | 1.4 | 5 |
| Mo | 0 | high-silica rhyolite | | 2.5 ± 1.6 | 750 | 14 | 5 |
| Cl | 0 | granodiorite to high-silica rhyolite | | 40[d] ± 10 | 750–880 | 1–2.4 | 2,6 |
| F | 0 | haplogranite | | 0.2[d] ± 0.1 | 750–850 | 1 | 7 |
| B | 0[f] | haplogranite | | 3 ± 0.18 | 750–800 | 1 | 8 |

[a]n = ligation number in chloride media; experimentally determined unless noted.
[b]For more information on melt and aqueous phase composition, see reference listed as source of data.
[c](1) Gammon et al. (1969); Holland (1972); (3) Carron and Legache (1980); (4) Flynn and Burnham (1978); (5) Candela and Holland (1984); (6) Kilinc and Burnham (1972); (7) Dingwell and Scarfe (1983); (8) Pichavant (1981); (9) Urabe (1985).
[d]No error reported; general range of error estimated by inspection.
[e]n inferred from oxidation state.
[f]Not Determined, assumed zero for purpose of this study.

the failure of the Nerst partition coefficient in the case of the partitioning of molybdenum between magnetite and melt, and have shown that the equilibrium constants imply different $f_{O_2}$-dependencies than do the Nernst partition coefficients. In fact, we have shown in our papers that an analysis of the possible equilibria, together with experiments on the variation of the partition coefficients with respect to $f_{O_2}$, can be used to determine "model" oxidation states for molybdenum in silicate melts (Candela, this volume). The crystal/melt partition coefficients should be considered Nernstian (Tacker and Candela, 1987) only under certain specified conditions.

In summary, all partitioning equilibria are best formulated as equilibrium constants. In some cases, and under certain specified conditions, the apparent equilibrium constants may be reducible to Nernst partition coefficients. However, in many cases, they cannot be, as is true in the case of chloride-dependent, vapor/melt partitioning. Exchange constants are inherently more difficult to use in mass transfer models, but they most properly govern mass action. Unfortunately, the partitioning of chlorine has not yet been examined from a thermodynamic point of view, and we have no choice but to use the empirical Nernst partition coefficient. However, the equilibria controlling $D_{Cl}^{v,l}$ may, in fact, be simple (e.g. see equation [7]).

## SUMMARY OF THE MODELS OF CANDELA (1986A)

The first case to be considered involves the fractional evolution of an aqueous phase from a rising, decompressing water-saturated magma. Because the solubility of water in a silicate melt is a function of the square root of the pressure (to a first approximation), at pressures of a few kilobars or less, the saturation water concentration in a silicate melt decreases as the pressure decreases. Therefore, the rise of a water-undersaturated melt will tend to induce water saturation, and the rise of a water-saturated melt will lead to the progressive release of aqueous fluid. The polybaric release of an aqueous phase also leads to a rise in the solidus temperature of the melt, which may lead to crystallization (Burnham, 1967). For any given bulk composition and water concentration, there is therefore, a minimum temperature above which the magma temperature must lie to ensure the completion of polybaric decompression without complete crystallization. Further, if we wish to model the system as near- or superliquidus, then the necessary minimum temperature of the melt will be higher.

### The partitioning of elements with constant partitions during first boiling

The first set of equations is designed to accommodate Nernst partition coefficients only. These equations can be used to model the partitioning of elements such as chlorine, fluorine, molybdenum, boron, and possibly tungsten and niobium into an evolving magmatic aqueous phase from a rising and decompressing magma. The equation derived by Candela (1986a) which yields the instantaneous concentration of a Nernstian element in each successive aliquot of aqueous fluid expressed as a function of the vapor evolution progress variable $F = m_w^l/m_w^{l,o}$ (the ratio of the mass of water in the melt at any instant during the vapor evolution episode to the initial mass of water present), is

$$C_i^v(\boldsymbol{F}) = D_i^{v/l} C_i^{l,o}\{1 - [(1 - \boldsymbol{F})C_w^{l,o}]\}^{D_i^{v,l}-1}. \quad [13]$$

The concentration of the element $i$ in the associated melt, expressed as a function of $F$, is given by

$$C_i^l(\boldsymbol{F}) = C_i^v(\boldsymbol{F})/D_i^{v/l} \quad [14]$$

where $C_i^v(F)$ is from equation [13]. Given 1) the Nernst vapor/melt partition coefficient, 2) the initial concentration of $i$(i.e. the concentration of $i$ at the initiation of the vapor evolution episode), and 3) the water concentration in the melt at the initiation of vapor evolution, the concentration of $i$in the melt and the associated vapor can be calculated as a function of the progress of the vapor evolution event. The change of the mass of the melt which results from the loss of the aqueous phase is explicitly accounted for in the model. Equations [13] and [14] are derived in Candela (1986a). Assumptions utilized in the derivation include: 1) the vapor is fractionated from the melt as it is formed (i.e. the instantaneous concentration of $i$ in the vapor is given by ($dM_i^v/dM_w^v$), 2) no crystallization accompanies vapor evolution, 3) the vapor/melt partition coefficients are constant over at least some small increment of the vapor evolution process (so that they can be moved outside the integral when the differentials mentioned in assumption (1) above are cleared); and 4) equilibrium is attained between the melt and the infinitesimal amount of aqueous phase present at any time, and no chemical potential gradients are present within the melt phase.

The main limitation of this approach, aside from the assumption of equilibrium, involves the constancy of $D_i^{v,l}$. The first boiling process is necessarily polybaric and the partition coefficients are clearly pressure dependent (Shinohara et al., 1984; Webster et al., in press). An approximate solution to this problem is as follows: equations [13] and [14] can be solved for arbitrarily small increments of $F$, over which the partition coefficient is taken to be a constant. At each new increment of the vapor evolution progress variable, the final concentration of water in the melt and the final concentration of the element $i$in the melt from the previous increment are taken as initial values. Because the calculation is effectively starting anew, $F$ must be reset to unity. A new value for the partition coefficient which is commensurate with the pressure and associated water concentration in the melt, is used in this next increment.

The dependence of the chlorine vapor/melt partition coefficient upon pressure has been determined by Shinohara et al. (1984) at 800°C and for pressures from 0.6 kb to greater than 4 kb, and for haplogranitic melts with 1 molar HCl starting solutions. These authors present their data as Nernst melt/vapor partition coefficients plotted against pressure. On such a plot the data trace out a hyperbola, which the untrained eye might extrapolate to infinity as surface pressure is approached. Fig. 12.1 shows that the vapor/melt partition coefficient is a linear function of pressure, which is what one might expect if the melt/vapor partition coefficient is a hyperbolic function of pressure. A ten-point

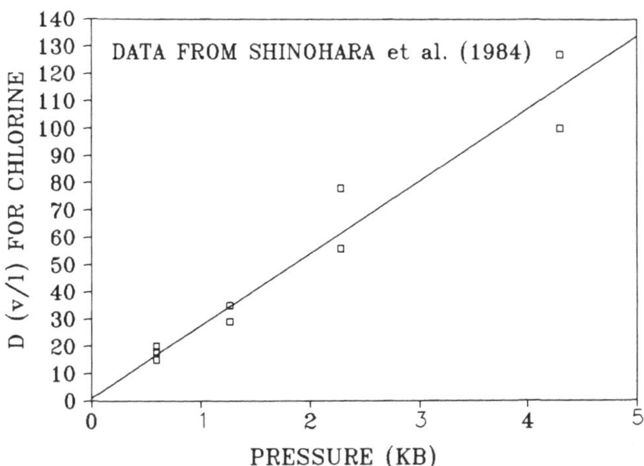


FIGURE 12.1—The experimentally determined Nernst **vapor/melt** partition coefficient for chlorine, as determined by Shinohara et al. (1984) at 800°C. for ternary minimum melts with HCl starting solutions, plotted against pressure. The line is a least squares line given by equation [17] in the text.

regression line ($r^2$ = 0.94) can be fit to the data plotted in Fig. 12.1 and yields the equation

$$E(i) = 1 - \left\{ \left( \frac{C_i^l(\boldsymbol{F}=0)}{C_i^{l,o}} \right) (1 - C_i^{l,o}) \right\}, \quad [15]$$

The vapor/melt partition coefficient given in equation [16] can be related to the water solubility in the melt by use of the approximation equation: $P$ (kb) = $568 \cdot (C_w^{l/s})^2$ where the solubility of water is given in weight fraction. This equation is generally consistent with the many studies which have reported on the solubility of water in granitic melts (cf. Burnham, 1979). For example, at 5.9 wt% water in the melt (weight fraction = 0.059), the equilibrium pressure of $H_2O$ is 2 kb. This equation can be combined with equation [16] to yield an expression for $D_{Cl}^{v/l}$ as a function of the water concentration in melt:

$$D_{\mathrm{Cl}}^{v/l} = 0.84 + 26.6P(\mathrm{kb}). \quad [16]$$

By using the polybaric equations [13], [14], and [17], the concentration of chlorine in the melt and in the evolving aqueous phase can be calculated. Fig. 12.2 illustrated the two extremes in the behavior of $C_i^{l \text{ or } v}$ as a function of the amount of water evolved from the melt during first boiling. Most elements will resemble the trend seen in Fig. 12.2 for the concentration of Cl in the melt during progressive polybaric vapor evolution without crystallization. Generally, the evolution of the magmatic aqueous phase will result in a depletion of the element in question in the melt, and a

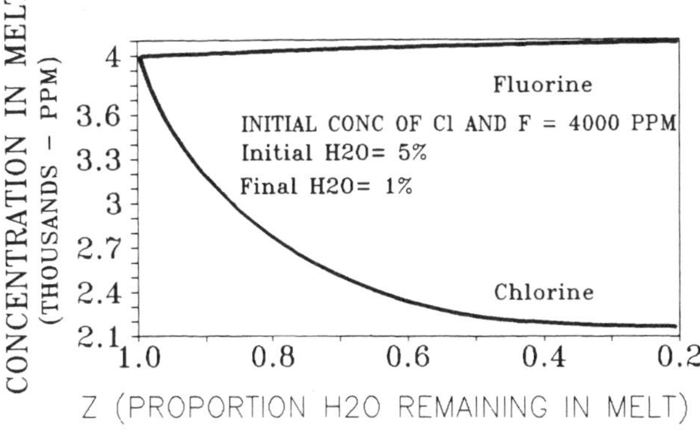


FIGURE 12.2a—The concentration of chlorine and fluorine in the melt phase of a magma undergoing decompressive vapor evolution (first boiling) from a depth of 5.4 km (5 wt% $H_2O$ at saturation) to the surface (with an arbitrarily chosen 1 wt% water remaining in the glass) plotted as a function of the proportion of water remaining in the melt. A concentration of 4000 ppm was chosen for the initial concentration of both Cl and F; however, the trends exhibited in this figure are independent of the initial concentration. $D_F^{v/l}$ = 0.2 and $D_{Cl}^{v/l}$ is given by equation [17] and decreases with pressure from a value of 39 at the start of the vapor evolution process to close to unity near the surface.

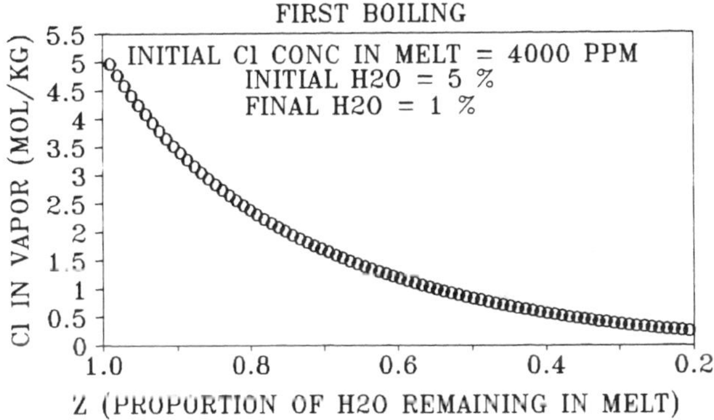


FIGURE 12.2b—Concentration of chlorine in the vapor evolved from the melt phase during first boiling as a function of the proportion of water remaining in the melt for the case given in Fig. 12.2a.

monotonic decrease in the concentration of the element in question in the aqueous phase as vapor evolution proceeds. By virtue of its low partition coefficient (see Table 12.1), very little fluorine is removed from the melt into the aqueous fluid during polybaric decompression without crystallization. In fact, the small reduction in the mass of melt due to the evolution of the aqueous phase is enough to produce a slight increase in the fluorine concentration in the melt as the vapor is evolved because of the ratio of fluorine lost to water lost is 1:5.

The efficiency of removal of element $i$ from the decompressing melt into the evolving aqueous phase, $E(i)$, is given by the ratio of the total amount of element $i$ removed into the aqueous phase divided by the amount of element $i$ originally present in the melt (at the commencement of vapor evolution in the case of first boiling). In this case, the most straightforward method for calculating $E(i)$ is to divide the mass of $(i)$ in the melt at $F = 0$ by the initial mass of $i$ in the melt, and to subtract this ratio from unity. Accounting for the reduction in the mass of the melt due solely to the loss of the aqueous phase, the ratio $m_i^{final}/m_i^{l,o}$ can be given by the quantity $(C_i^{final}/C_i^{l,o})\cdot(1 - C_w^{l,o})$. The efficiency of removal is, therefore, given by

$$D_{\text{Cl}}^{v/l} = 0.84 + 10^{4.18}\{C_w^{l/s}\}^2. \quad [17]$$

if we assume that all the water initially present in the melt is lost upon rise of magma. If all the water is not lost, i.e. if the magma is erupted and the glass still contains some water (because the rise of the magma was more rapid than the loss of vapor, or if the magma rises to a particular level in the subsurface and ceases to rise), then $C_w^{l,o}$ in equation [15] must be replaced by the difference $(C_w^{l,o} - C_w^{l,final})$, and the calculation would be terminated when $C_w^{l,o}$ reaches $C_w^{l,final}$.

The efficiency of removal of chlorine can be calculated as a function of the amount of vapor evolved from the system (see Fig. 12.3) Note that significant amounts of chlorine can be removed from the melt if a few weight percent of water is evolved. More accurate results are obtained if $D$ is modeled as a function of pressure. However, even with the assumption that $D_{Cl}^{v/l}$ is a constant, the same qualitative results are obtained. If the Nernst partition coefficients for the other elements (Mo, B, F, Nb) also decrease with decreasing pressure, then the calculations performed with high pressure partition coefficients yield an upper bound for the amount of these elements which can be removed by first boiling under upper-crustal conditions (See Fig. 12.4) Note that $E$(Mo) and $E$(F) are much lower than $E$(Cl) under similar conditions. $E$(B) is very close in magnitude to $E$(Mo). These data indicate that the elements Mo, B, and F cannot be strongly depleted in a melt by first boiling, at least with the high pressure partition coefficients listed in Table 12.1.

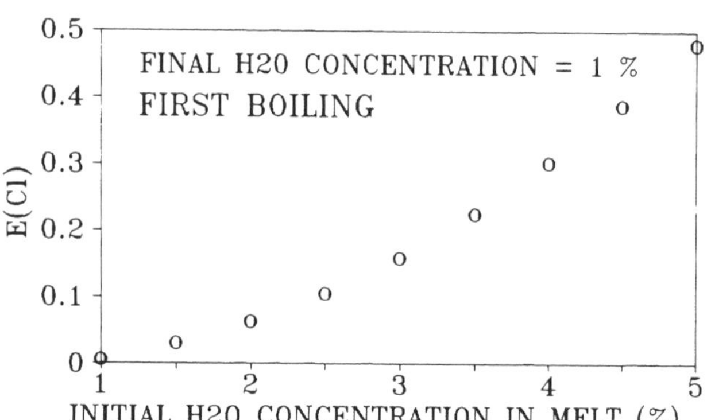


FIGURE 12.3—The efficiency of removal of chlorine from a melt into a fractionating vapor phase during first boiling with the partition coefficient for chlorine given by equation [17] plotted against the water concentration in the melt at the commencement of vapor evolution. The final water concentration in the melt is 1 wt%.

### Partitioning of chloride-complexed elements during first boiling

Instead of a Nernst partition coefficient, the vapor/melt partitioning of the chloride-complexed cations is governed by an exchange constant such as the following equilibrium constant written for equation [4]:

$$K_{\text{Cu,Na}}^{v,l} = \left(\frac{C_{\text{CuCl}}^v C_{\text{Na}}^l}{C_{\text{Cu}}^l C_{\text{NaCl}}^v}\right). \quad [18]$$

To perform this calculation, the concentration of chlorine in the aqueous phase is calculated from the previous model. The cations are then balanced against chlorine via a charge balance expression (written in terms of the cations sodium and potassium in this example):

$$C_{\text{Cl}}^v \cong C_{\text{Na}} + C_{\text{K}}. \quad [19]$$

Substituting an expression for $K_{K,Na}^{v/l}$ into equation [19] yields

$$C_{\text{Cl}}^v = C_{\text{Na}}^v \left(1 + K_{\text{K,Na}}^{v/l} \frac{C_{\text{K}}^l}{C_{\text{Na}}^l}\right), \quad [20]$$

where the term in brackets will henceforth be given by β when the concentration of alkali metals in the melt are held

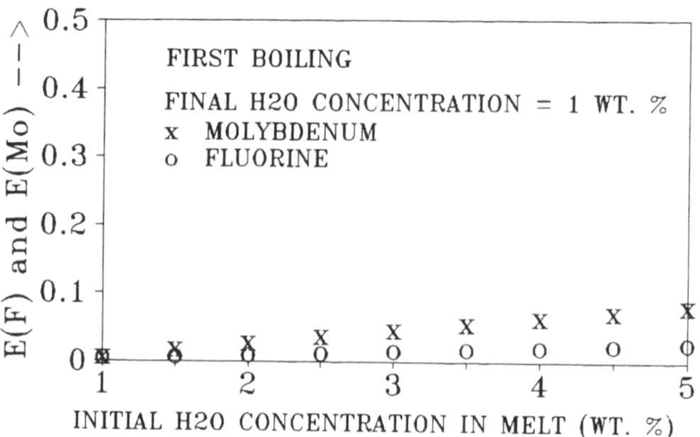


FIGURE 12.4—The efficiency of removal of fluorine and molybdenum from a melt into a fractionating vapor phase during first boiling, in accord with the partition coefficients given in Table 12.1, plotted against the water concentration in the melt at the commencement of vapor evolution. The final water concentration in the melt is 1 wt%. Note that these elements are removed from the melt less efficiently than is chlorine.

constant, and set equal to the concentrations prevailing at the initiation of vapor evolution. This is clearly a very simple form of the charge balance expression, but for the relatively dilute fluids associated with the polybaric rise of vapor-saturated felsic melts, equation [20] yields a good first approximation of the charge balance constraint. Using this relation, the concentration of sodium in the vapor phase is given by $C^v_{Na} = C^v_{Cl}/\beta$, with $C^v_{Cl}$ calculated at any $F$ from equation [13]. The absence of HCl from the charge balance equation is a limitation of the above discussion. This problem is rather trivial for first boiling which occurs within a pressure range of a few kilobars. However, the data of Shinohara et al. (1984) suggest that at 1 kb, upwards of 25% of the chlorine in the magmatic aqueous phase is complexed by hydrogen as opposed to other cations, and illustrate the functional dependence of the HCl/ΣCl ratio on pressure. At pressures of less than 2.3 kb, data from experiments wherein HCl solutions have been equilibrated with haplogranitic melts at a temperature of 800°C define a linear trend in terms of log(HCl/ΣCl) vs. pressure. If their datum at 4.3 kb is included, the trend becomes somewhat nonlinear. These data are preliminary; however, a line fit to the three data points at 0.6, 1.3, and 2.3 kb allows us to estimate the amount of HCl that should be included in the charge balance expression over this pressure range. The final melt compositions were not reported in the study of Shinohara et al., but they must have been decidedly peraluminous, particularly at elevated pressure, given that the starting solutions were 1 molar HCl. Therefore, the calculated HCl/ΣCl ratios are probably maxima at any given pressure (Candela, ms). By this approach an approximation of the proportion of chloride which is complexed by hydrogen in the magmatic aqueous phase can be obtained. The equation which results from an analysis of the data of Shinohara et al. is

$$\log_{10}(\mathrm{HCl}/\Sigma\mathrm{Cl}) = -0.11 - 0.47P(\mathrm{kb}). \qquad [21]$$

Solution of equation [21] at 2 kb indicates that less than 10% of the chlorine in the magmatic aqueous phase at this pressure is complexed by hydrogen. However, as a limiting case, equation [21] can be solved for surface conditions (P = 0.001 kb) which indicates that close to three quarters of the chlorine is complexed by hydrogen. A thermodynamic framework for the concentration of HCl in a magmatic aqueous phase at a given pressure, based on stoichiometric analysis, has been developed by Candela (ms), but will not be discussed in detail here. Based on the charge balance equation of the form

$$C^v_{\mathrm{Cl}} = (C^v_{\mathrm{Na}}\beta) + C^v_{\mathrm{H}}, \qquad [22]$$

and letting y = $\log_{10}$(HCl/ΣCl), the expression for the concentration of sodium in the magmatic aqueous phase can be given by

$$C^v_{\mathrm{Na}} = C^v_{\mathrm{Cl}}(1 - 10^y)/\beta. \qquad [23]$$

Modifying the equations of Candela (1986a) slightly, we can write for the concentration of selected chloride-complexed elements in the melt during first boiling

$$C^l_i(\boldsymbol{F}) = (C^{l,o}_i/f\{\boldsymbol{F}\})\exp\left[\left\{\frac{\alpha(i)}{\mathfrak{D}(i)}\right\}[f\{\boldsymbol{F}\}^{\mathfrak{D}(i)} - 1]\right] \qquad [24]$$

where

$$f\{\boldsymbol{F}\} = \{1 - [(1 - \boldsymbol{F})C^{l,o}_w]\},$$

$$\mathfrak{D}(i) = n(D^{v/l}_{\mathrm{Cl}} - 1), \quad \text{and}$$

$$\alpha(i) = K^{v/l}_{i,\mathrm{Na}}\left(\frac{D^{v/l}_{\mathrm{Cl}}C^{l,o}_{\mathrm{Cl}}(1 - 10^y)}{C^{l,o}_{\mathrm{Na}}\beta}\right).$$

To obtain the concentration of the element $i$ in the evolving vapor at any value of $F$, the concentration of $i$ in the melt at $F$, ($C^l_i(F)$ from equation [24]), and the concentration of chlorine in the vapor at the given value of $F$, ($C^v_{Cl}(F)$ from equation [13]), must be substituted into the equation

$$C^v_i(\boldsymbol{F}) = C^l_i(\boldsymbol{F})K^{v/l}_{i,\mathrm{Na}}\left(\frac{C^v_{\mathrm{Cl}}(\boldsymbol{F})(1 - 10^y)}{\beta C^{l,o}_{\mathrm{Na}}}\right)^n. \qquad [25]$$

In order to calculate the efficiency with which any given element $i$ can be removed from a rising, decompressing melt that is evolving an aqueous phase, equations [13], [17], [24], and [25] are solved for the specified range of $C^{l,o}_w$ [illegible] $C^{l,final}_w$, and the calculated final melt concentration of the element $i$ is then substituted into equation [15].

As an example of the solution of these equations, the efficiency of removal of the elements Cl, Cu, Mn, Zn, Ce, Rb, Sr are calculated given $C^{l,o}_w$ = 5 wt%, $C^{l,final}_w$ = 1 wt%, Cl/$H_2O$ (by wt.) = 0.08, β = 1.5, $C^{l,o}_{Na}$ = 1 mole/kg of melt and the equilibrium constants listed in Table 12.1. The calculations indicate that if a magma rises from 5000 meters to a depth of 200 meters, 37% of the copper, and 49% of the chlorine will be removed from the melt into the aqueous phase. Other calculated efficiencies include $E$(Zn) = 27%, $E$(Mn) = 20%, $E$(Ce) = 15%, $E$(Rb) = 4%, $E$(Sr) = 5%, and $E$(Mo) = 8%. Under these hypothetical conditions, the elements with high vapor/melt partition coefficients partition rather strongly in favor of the vapor phase during first boiling given a reasonably high Cl/$H_2O$ ratio. This class of elements includes Cu, Zn and Mn. These elements are classified at "Cl-vapor compatible" elements because of their potential to be removed into a Cl-bearing aqueous fluid to a significant degree in the absence of crystallization. Although the numerical value of the exchange constants for the rare earth elements appear much lower than those of the Cl-vapor compatible elements, significant amounts of the rare earths can be mobilized at high $Cl/H_2O$ because of their relatively high ligation numbers. On the other hand, the relatively low exchange constants and ligation numbers of the alkali and alkali earth cations lead to rather low efficiencies of removal for these elements at any reasonable $Cl/H_2O$ ratio. Mo, B, and F (and to some extent, Rb, Cs, etc.) are examples of "melt-compatible" elements. These elements can be expelled from the melt only by the power of crystallization or devitrification. The conclusion to be reached from this exercise is that first boiling can, under certain conditions (a few weight percent water evolved and $Cl/H_2O$ wt ratios on the order of 0.1), remove tens of percent of the Cl-vapor compatible elements into a magmatic aqueous phase, whereas the melt-compatible elements possess efficiencies of removal of generally less than 10%.

The physical details of how this chemical mass transfer is effected remain to be determined. If the fluid is evolved from the system as it is formed, then it is unlikely that the metal-bearing fluids would be concentrated spatially. Protracted zones of anomalous mineralization would result that are spread over depths of many kilometers. This has not, to my knowledge, been observed. On the other hand, the physical release of the aqueous phase from the magma is probably a rather sluggish process, and the vapor which is evolved from the melt over a range of depths may be released from the magma at the cessation of rise of the body at a relatively shallow level. At this point vapor would be evolved, significant fracturing would result from the vapor evolution episode (Burnham and Ohmoto, 1980) and the remaining melt would crystallize in a second boiling (q.v.) mode to yield a fine-grained texture, and possibly a porphyritic one if some crystallization occurred upon rise. The irreversibly released fluid would be relatively rich in base metals (as discussed above) with only modest amounts of the so-called incompatible elements (including the melt-compatible class discussed here) contained in the fluid. Given a melt Cu/Mo ratio in the melt of 50, the overall ratio of Cu/Mo in the released fluid would generally be greater than 100, as long as a significant amount of crystallization does not occur simultaneously. When crystallization occurs upon cessation of the intrusive event, the second boiling systematics, (to be discussed presently), most properly describe the partitioning systematics. Generally, under these latter conditions, higher Mo/Cu ratio result, especially for significant degrees of crystallization before vapor evolution. The froth of melt, crystals and ore metal- and chlorine-charged vapor may be propelled upward by its low bulk density. This process may explain why some rather small, satellitic intrusions have cupolas containing sizable amount of mineralization.

These equations can be used to back-calculate the composition of erupted rocks. For example, in an interesting paper on the Spor Mountain Topaz rhyolites, Webster et al. (1987) convincingly argue that a vitrophyre from that locality had a water content at the time of eruption of approximately 5 wt%, based on the observed crystallization sequence in the rock and the phase equilibria determined on the same material. However, the rock presently contains on the order of 0–1 wt% water. Therefore, this system evolved on the order of 4 wt% water. This corresponds approximately to the calculation above. The percentage of trace elements removed from the melt according to the above calculation could then be added back into the melt to calculate its pre-eruption composition. Similar calculations could be performed on other glassy rocks which have been shown to have evolved a given amount of water based on phase equilibrium and petrographic data, as above, or based on isotopic data (Taylor et al., 1983).

## SECOND BOILING

Upon crystallization of a water-undersaturated magma, the water concentration rises due to the lower water content of the hypersolidus-prevaporus crystalline assemblage relative to the bulk magma. Before vapor saturation, ore metals are partitioned among the melt and hypersolidus minerals. Elements which partition strongly into crystalline phases are sequestered during this prevaporus stage and will be dispersed throughout the crystallized pluton upon cessation of the plutonic stage of igneous activity. When the concentration of water in the melt rises to the concentration of water at saturation $C_w^{l,s}$ (which is a function of *P*, *T*, and bulk composition, see MacMillan and Holloway, 1987) a separate aqueous phase will form (barring kinetic barriers to the nucleation and growth of vapor bubbles). Further crystallization leads to the progressive evolution of vapor as quartz and feldspar crystallize. At this stage, ore metals are partitioned among the melt, crystalline, and vapor phases. Upon final plutonic consolidation, the sum total of any given ore metal (or other element in question) removed from the plutonic system relative to the initial magmatic concentration of the element is called the efficiency of removal. A similar function appeared in our analysis of the first boiling process. The efficiency function can be calculated for the removal of an element into any phase. However, in the case of ore systems associated with felsic igneous rocks, the magmatic aqueous phase is most likely the active "mineralizing phase" by virtue of its fluidity and buoyancy in a magmatic setting, and therefore only the efficiency of removal of elements into a (potentially) ore-forming aqueous phase from the melt will be calculated.

In Candela (1986a) and Candela and Holland (1986), equations are derived which can be used to model the evolution of a magmatic aqueous phase during second boiling. A revised derivation of these equations is presented in the appendix to this paper. As opposed to first boiling, which is fundamentally a polybaric process, second boiling can be isobaric, and that is how the process will be treated here. In this model, the vapor phase and the various crystalline phases will be fractionated as they are formed. That is, as in the case of the vapor formed during first boiling, only an infinitesimal amount of vapor or any crystalline phase will be in equilibrium with the melt at any time, and the concentration of element *i* in the phase will be given by the ratio of the infinitesimal quantities, $dM_i^{\phi}/dM_{\phi}$. In this model, chlorine will partition according to a Nernstian law, and will be treated as a perfectly incompatible element with respect to crystalline phases. For simplicity, we will discuss the systematics of the partitioning process at 800°C and 2 kb because of the problems associated with the narrow range of temperature and pressure covered by the melt–vapor data base. In some cases, calculations will be extended beyond the data base to illustrate the utility of the calculations. When this is done, the assumptions involved will be noted.

The master variable in these calculations is the vapor evolution progress variable, Z, which is the ratio of the mass of water dissolved in the melt at any time to the mass of water dissolved in the melt at the initiation of vapor saturation. Z therefore varies from unity at the initiation of vapor evolution to 0 at the cessation of crystallization and evolution of the vapor.

### Partitioning of elements with constant partition coefficients during second boiling

The calculations are performed in the following manner: for each increment of vapor evolution, the concentration of chlorine in the associated vapor is calculated from the equation:

$$C_i^v(Z) = \frac{D_i^{v/l} C_i^{l,s} Z^{\overline{\mathbb{D}}-1}}{\{1 - \sum_j X_j(Z)\}^{1/2(\overline{\mathbb{D}}-1)}} \exp\left\{ D_i^{v/l} C_w^{l,s} \int_1^Z \sqrt{1 - \sum_j X_j(Z)} \frac{dZ}{Z} \right\}. \quad [26]$$

(All of the second boiling equations have been derived in Appendix 1; these derivations and equations are meant to supersede those in Candela, 1986a.) The square root term accounts for the reduction of the activity of water (from unity) in the melt–vapor system due to the presence of solutes in the aqueous phase. Many workers have shown that the solubility of water in the melt phase varies (to the first approximation) as the square root of the fugacity or activity of water in the vapor (Burnham and Davis, 1974; Stolper, 1982) at low pressures. Therefore, the reduction in the solubility of water in the melt is proportional to the square root of the mole fraction of water in the aqueous phase. The mole fraction of water in the aqueous phase is expressed as one minus the mole fraction of the solutes in the vapor, $\{1 - \Sigma_j X_j(Z)\}$, where $\Sigma_j X_j$ is the sum of the total chloride, fluoride, borate and molybdate concentrations. The total $SiO_2$ and dissolved feldspar components in the aqueous phase, independent of the above components, are probably limited to below 2 wt% (approximately one-third of a mole per kg of solution, for $SiO_2$) at the pressures considered in this paper (~2 kb), (Burnham and Nekvasil, 1986). This reduction in the activity of water is significant in the end-stages of crystallization where the concentration of the "melt-compatible" elements reaches high levels. The melt-compatible elements (e.g. Mo) partition preferentially into neither the crystalline phases nor the magmatic aqueous phase. It is important to note that, because the equilibrium concentration of water in the melt at saturation is reduced at this stage, smaller increments of water are evolved as the reaction progresses. In nature, these effects will be tempered to some extent by the precipitation of minerals containing these substances.

To solve equation [26] it should be noted that the integral must be solved numerically, Further, equation [26] must be solved simultaneously for Cl, B, F, Mo and any other independent elements which may achieve high concentrations at late stages in the vapor evolution process, because the numerical solution of the integral for each element requires the total molal solute concentration. That is, their concentrations are not truly independent, and the equations are recursive in nature.

The calculation is begun by assuming that the square root term is unity, and the process is iterated until the desired agreement is reached between iterations. It is best to use at least 20 subdivisions (nodes) of Z from 1 toward zero. Note that the integral is solved at every node, from one to the value of Z at the current node, to obtain the concentration of the elements Cl, B, F, and Mo at Z. the integrals are solved using a modified form of Simpson's rule including adaptive quadrature which splits the last node ($Z = 0.05 >>>> 0$) twice into small intervals so that the rapid changes in $C_{Cu}^{l \text{ or } v} = f(Z)$ can be accurately accounted for in the integration. Further, the integral in equation [26] cannot be evaluated at zero, so the calculation is taken to some arbitrarily small value of Z. The geological significance of this lower limit is that it marks the latest stage at which vapor is actually removed from the pluton. At present this is an unknown quantity, so it would be instructive to run these calculations at various lower limits of Z to ascertain the effect of terminating the release of vapor from the system at different stages.

When the term in $\Sigma_j X_j$ is small, the square root terms may be neglected, and the integral in [26] may be solved in closed form. Then, equation [26] reduces to

$$C^v(Z) = D_i^{v/l} C_i^{l,s} Z^{\{D_i^{v/l} C_w^{l,s} + \overline{\mathbb{D}} - 1\}}. \quad [27]$$

For convenience, (and for the sake of calculating efficiencies of removal later) the concentration of $i$ in the melt at the initiation of vapor saturation, $C_i^{l,s}$, can be replaced by the initial concentration of $i$ in the melt, $C_i^{l,o}$, by employing a variant of the simple Rayleigh fractionation equation, (Candela and Holland, 1986),

$$C_i^{l,s} = C_i^{l,o} \left( \frac{C_w^{l,o}}{C_w^{l,s}} \right)^{\overline{\mathbb{D}}-1}. \quad [28]$$

Equation [28] assumes no water is taken up by crystallizing phases, so that the increase in the concentration of water before vapor saturation is a quantitatively accurate measure of the progress of crystal fractionation before vapor saturation.

### Partitioning of chloride-complexed elements during second boiling

To calculate the concentration of the chloride-complexed metals such as Na, K, Cu, Sr, Yb, etc., we must explicitly account for chloride-charge balance in the aqueous phase (Holland, 1972) and incorporate the vapor/melt exchange constants for the chloride-complexed cations into the mass transfer calculation. These two related factors complicate the form of the equations, and makes it necessary to resort to numerical integration. The concentration of any given chloride-complexed cation in the magmatic aqueous phase as a function of Z is given by the equation

$$C_i^v(Z) = \frac{K_{i,\mathrm{Na}}^{v/l}[C_{\mathrm{Na}}^v(Z)]^n C_i^{l,s} Z^{\overline{\mathbb{D}}-1}}{[1-\sum_j X_j(Z)]^{1/2(\overline{\mathbb{D}}-1)}(C_{\mathrm{Na}}^{l,s})^n} \exp\left\{ \frac{K_{i,\mathrm{Na}}^{v/l}}{(C_{\mathrm{Na}}^{l,s})^n} \int_1^Z \frac{[C_{\mathrm{Na}}^v(Z)]^n C_w^l(Z)}{Z} \, dZ \right\}. \quad [29]$$

Further, in combination with a generalized expression for the vapor/melt exchange constant,

$$C_i^l = C_i^v (C_{\mathrm{Na}}^{l,s})^n / \{(C_{\mathrm{Na}}^v)^n K_{i,\mathrm{Na}}^{v/l}\}, \quad [30]$$

equation [29] can be solved for the concentration of the element *i*in the melt. Candela and Holland (1986) discuss the possibility that the bulk solid/melt partition coefficient for copper, $\mathbb{D}_{\mathrm{Cu}}$, could change at the time of vapor saturation due to the destabilization of magmatic sulfide. This topic is discussed in depth in my other chapter in this volume. A change in $\mathbb{D}_i$ at the time of vapor evolution can be effected in these calculations by using one value for the bulk partition coefficient (before vapor evolution) in equation [28], and another value for the bulk partition coefficient (for crystallization during vapor evolution) in either equation [26], [27], or [29] as appropriate. In all the example calculations given in this paper the bulk partition coefficient for copper is changed from 2.0, before vapor saturation, to 0.1 after saturation.

Equations [29] and [30] must be solved simultaneously with equation [26], the equation for the partitioning of chlorine (and the other Nernstian elements). The terms $C_w^l$ and $C_{Na}^v$ are the links between the above equations. $C_w^l$ is a function of the total solute of the aqueous phase (calculated by equation [26] solved individually but iteratively for the elements Cl, F, Mo, and B) and $C_{Na}^v$ is related to the concentration of chlorine through the charge balance constraint. Therefore, the calculation is started, as stated above, with the iterative solution of the Nernstian equation. This yields the concentration of Cl, F, Mo, and B in the aqueous fluid at selected values of Z (nodes) between 1 and 0. The concentration of sodium in the vapor for this increment is obtained by solution of the charge balance equation of which equation [20] is a simplified form expressed in terms of the aqueous Na and K concentrations only. Note that in the charge balance equation, the ration $C_{Cl}^v/C_{Na}^v$ is expressed as a function of the exchange constants (for those elements which are likely to attain high concentrations in the aqueous fluid), the sodium concentration in the melt (which is assumed constant over the course of these calculations), and the calculated concentrations of the cations in the melt from the previous increment (node). This aqueous sodium concentration is then input into equation [29] to calculate the new melt and vapor concentrations of the elements in question.

The procedure is iterated until the sodium concentration in the aqueous fluid agrees to within a few percent of the value from the previous interation. As with the Nernstian equation, equation [29] includes a running integration of every Z for each element. For any value of Z, there are a number of iterations and numerical integrations being performed during the calculation, and intermediate results of the integrations are used to calculate the concentration of each element in the aqueous phase at any specified value of Z.

### The efficiency integral

One of the integrals being calculated in the above procedure is the efficiency integral (Candela, 1982; Candela and Holland, 1986). The efficiency function is defined by the ratio of the total quantity of *i* removed into the vapor, $M_i^{v,T}$, divided by the initial quantity of *i* in the melt, $M_i^{l,o}$:

$$E(i) = M_i^{v,T} / M_i^{l,o}. \quad [30a]$$

The numerator of equation [30a] is proportional to the integral of the instantaneous concentration of *i* over Z. The full expression for the efficiency integral is then,

$$E(i) = (C_w^{l,o}/C_i^{l,o}) \int_0^1 C_i^v(Z) \, dZ. \quad [31]$$

As stated previously, the integral is not actually carried out to zero but to some lower limit of Z, between 0.1 and $10^{-5}$. The optimum value of Z to which calculations are carried in our program, and the value most commonly employed in this paper, is Z (lower limit) = 0.002.

In one case, the integral in [31] can be solved in closed form. When the simplified approximation, equation [27], is used to express the concentration of a Nernstian element in the evolving aqueous phase as a function of Z, and is substituted into [31], the efficiency of removal of Nernstian element *i* can be expressed as

$$E(i) = \left(\frac{C_w^{l,o}}{C_w^{l,s}}\right)^{\overline{\mathbb{D}}_i} \left(1 + \frac{\overline{\mathbb{D}}_i}{C_w^{l,s} D_i^{v/l}}\right)^{-1}. \quad [31a]$$

A number of constants must be specified in order to solve the above set of equations including: the concentration of sodium in the melt, the initial concentrations of chlorine, water and the elements of interest in the melt, the partition coefficients for both crystal–melt and vapor–melt equilibria, and the ligation number for the chloride-complexed cations. Further, the concentration of water in the melt at the time of vapor saturation must be specified. This variable is related to the pressure, and for a given geobaric gradient, the depth of vapor evolution. The full implications involved in varying the ratio of the initial and saturation water concentrations in the melt will be explored in another chapter.

### Status of iron in the magmatic aqueous phase

Published partitioning data concerning the distribution of iron between melt and vapor is surprisingly poor given the importance of this element in magmatic-hydrothermal systems. Rather than try to estimate the vapor/liquid partition coefficient for this element, I have chosen to model the concentration of iron from a crystal-vapor rather than a liquid-vapor model. I propose, after the NaCl and HCl concentrations (Candela, 1986a, and above) are calculated and the oxygen fugacity of the system is defined that the iron concentration in each successive aliquot of magmatic vapor be calculated from the reaction of Frantz et al. (1981):

$$H_2 + 6HCl + Fe_3O_4 = 3FeCl_2 + 4H_2O. \quad [32]$$

A problem arises given that the original data was collected up to only 600°C. However, if the data is extrapolated to 800°C, then the estimated error is probably much less than that which would result from ignoring the concentration of $FeCl_2$ in the magmatic aqueous phase or estimating it by some other means. In progressing from peralkaline through metaluminous to peraluminous systems, the HCl/ΣCl ratio increases (Candela, ms). At a given $f_{O_2}$ and fugacity of water, equation [22] indicates that the $FeCl_2$/ΣCl ratio would also increase. For a magmatically derived ore fluid with a given, integrated concentration of copper, (and a given $f_{O_2}$) this increase in aqueous iron with increasing melt peraluminosity would, for example, probably yield a lower Cu/Fe ratio in the resulting mineralized rock.

## MODEL CONCENTRATIONS OF METALS AND CHLORINE IN MAGMATIC-HYDROTHERMAL FLUIDS

In this section, I shall present example solutions of the above equations. Both the concentration of the elements of interest as a function of Z, and the efficiencies of removal of these elements will be calculated. Table 12.2 lists the many constants which are used in these calculations.

Fig. 12.5 shows a plot of the concentration of chlorine in the melt as a function of the vapor evolution progress variable, Z. In this figure, $C_w^{l,s}$ = 0.06. Under these conditions, six grams of water are evolved for each 100 grams of melt crystallized, and this quantity of water released is more than sufficient to cause a progressive decrease in the concentration of chlorine in successive aliquots of the magmatic aqueous phase in spite of e fact that chlorine is behaving as a perfectly incompatible element with respect to crystallizing phases.

The previously accepted value for the partition coefficient of chlorine is 40 (Holland, 1972; Kilinc and Burnham, 1972), and the error estimated for this parameter is ±25% or 40±10. For a further discussion of this and other partition coefficients see Candela (1986a) and Candela (1986c). Recent studies by Webster et al., (ms) and Delboff et al., (1987) show a concentration dependence for $D_{Cl}^{v,l}$ which is probably a result of non-Henrian behavior of chlorine in the melt, but a discussion of this effect is beyond the scope of this paper. According to the study of Shinohara et al. (1984), the chlorine partition coefficient is on the order of 55 at 2 kb.

Candela and Holland (1986) point out that the critical parameter which indicates whether an element following a Nernstian vapor/melt partitioning law will increase or decrease progressively in successive aliquots of fluid (and the associated melt) is the product of the partition coefficient and the water concentration at saturation (in weight fraction), symbolized by (see Candela, 1986c, and Candela and Holland, 1986). When $\Gamma_i$ = 1, the element $i$ will be depleted in successive aliquots of the aqueous fluid and the associated melt, even if the element $i$ is behaving as a perfectly incompatible element with respect to crystallizing phases. For the case given in Fig. 12.1, Γ = 2.4; if the revised partition coefficient from Shinohara is used, then Γ is even larger. In any case, $C_{Cl}^{l}$ strongly decreases as crystallization proceeds at 2 kb and $C_w^{l,s}$ = 0.06, and the difference produced by substituting an even larger value for the chlorine partition coefficient is not very significant. At lower $C_w^{l,s}$, a smaller proportion of water is released from the melt for a given amount of crystallization, and, according to the results of Shinohare et al., the partition coefficient is smaller than at higher pressures. For example, at 500 bars, the equilibrium water solubility in the melt is on the order of 0.03 (3 wt% $H_2O$), and according to equation [17], the chlorine partition coefficient is on the order of 15. In this case, Γ = 0.45, and chlorine progressively builds in the melt during vapor-saturated crystallization (See Fig. 12.6). By the same arguments, the critical value for $C_w^{l,s}$ at which Γ = 1 is approximately 0.04. Therefore, at shallow levels in the earth's crust, (pressures < 1000 bars) chlorine may progressively increase as vapor evolution proceeds.

Returning to the case where $C_w^{l,s}$ = 0.06 and Γ > 2.4, it is instructive to examine the behavior of chloride-complexed ore metals such as copper during vapor-saturated crystallization of a felsic melt. The concentration of copper in the vapor and the melt as a function of Z is shown in Figs. 12.7a

TABLE 12.2—Bulk partition coefficients and initial melt concentrations used in calculations.

| | Cu | Mn | Zn | Ca | Mg | Ce | K | Na | B | Mo | F | Cl |
|---|---|---|---|---|---|---|---|---|---|---|---|---|
| $\mathbb{D}_i^b$, $\mathbb{D}_i^a$ | 2, 0.1 | 0.5, 0.5 | 1, 1 | 1, 1 | 1, 1 | 0.19, 0.19 | 1, 1 | 1, 1 | 0.1, 0.1 | 0.1, 0.1 | 0.1, 0.1 | 0, 0 |
| $C_i^{l,o}$ | 50^P | 500^P | 50^P | 1.5^w | 0.3^w | 50^P | 1.0^w | 1.0^w | 30^P | 5^P | 500^P | * |

$\mathbb{D}_i$ are bulk Nernst solid/melt partition coefficients: [b]indicates before vapor evolution and [a]indicates after vapor evolution.
Initial concentrations in the melt $C_i^{l,o}$ are given in ppm (P) and in weight percent (w) *of the element.*

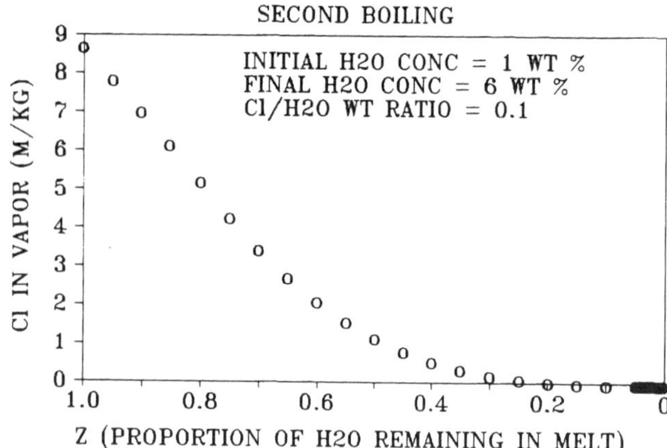


FIGURE 12.5—The concentration of chlorine in successive aliquots of vapor evolved from the melt phase during second boiling as a function of the proportion of water remaining in the melt, for $C_w^{l,o}$ = 1 wt%, $C_w^{l,s}$ = 6 wt% and a Cl/$H_2O$ wt. ratio = 0.1. $D_{Cl}^{v/l}$ is given by equation [17]. The trend exhibited for the concentration of chlorine in successive aliquots of magmatic vapor in this figure holds qualitatively for vapor evolution at depths greater than 3–4 km.

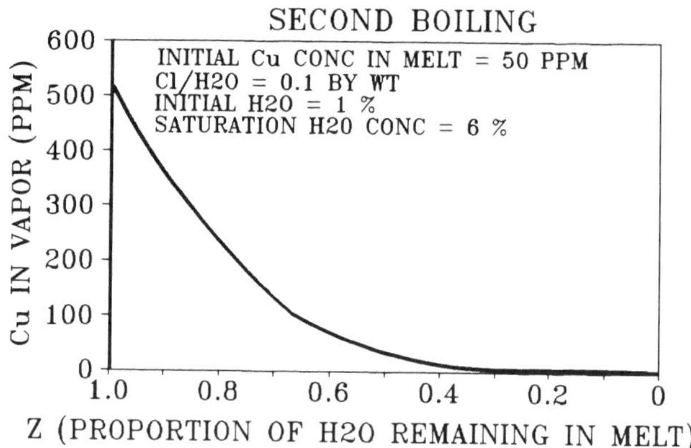


FIGURE 12.7a—The concentration of copper in successive aliquots of vapor evolved from a melt during second boiling as a function of the proportion of water remaining in the melt for $C_w^{l,o}$ = 1 wt%, $C_w^{l,s}$ = 6 wt% and a Cl/$H_2O$ wt. ratio = 0.1. $D_{Cl}^{v/l}$ is given by equation [17]. $C_{Cu}^{v}$ decreases monotonically as a function of $-Z$ because of the monotonically decreasing concentration of chlorine under these conditions, and because the evolution of the aqueous fluid depletes the melt in copper up to a point, as second boiling proceeds. The initial cooper concentration in the melt = 50ppm; the maximum concentration of copper in the aqueous fluid under these conditions of deep (depths > 3–4km) vapor evolution is between 500 and 600 ppm, and occurs early in the second boiling event.

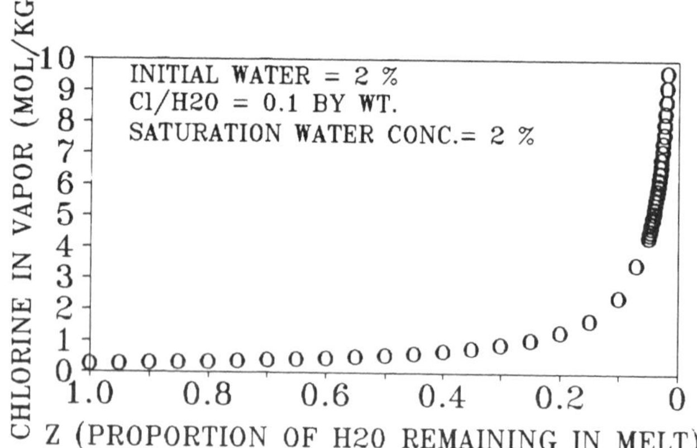


FIGURE 12.6—The concentration of chlorine in successive aliquots of the vapor evolved from the melt phase during second boiling as a function of the proportion of water remaining in the melt, for $C_w^{l,o}$ = 2 wt% and $C_w^{l,s}$ = 2 wt% and a Cl/$H_2O$ wt. ratio = 0.1. $D_{Cl}^{v/l}$ is given by equation [17]. Note that the trend of $C_{Cl}^{v}$ as a function of $-Z$ is opposite of that found in Fig. 12.5. This is the characteristic trend for vapor evolution at a shallow level (depths less than 3–4 km).

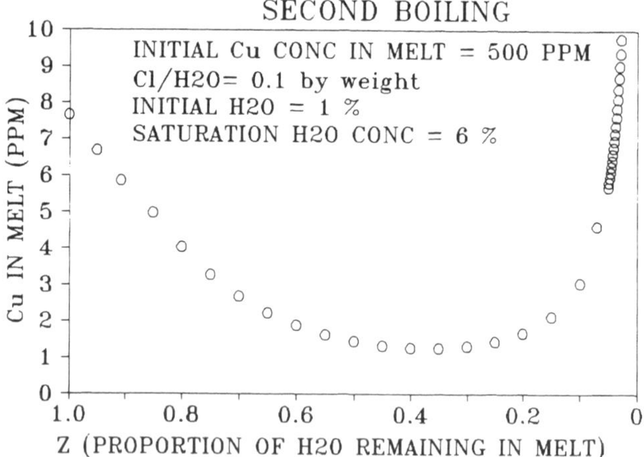


FIGURE 12.7b—The concentration of copper in a melt undergoing second boiling, as a function of the proportion of water remaining in the melt. The conditions are the same as those listed in Fig. 12.7a. Note that the copper concentration in the melt first decreases as vapor evolution proceeds, when the chlorine concentration is relatively high, but begins to increase again as crystallization of non-sulfide minerals proceeds in the presence of a monotonically decreasing chlorine concentration in the aqueous phase.

and 12.7b, respectively, for the conditions specified in Table 12.2. The most important characteristics of this figure include both the actual range of concentrations of Cu in the vapor and the details of the variation in $C_{Cu}^{l\text{ or }v}$ with Z. At the onset of vapor evolution (Z = 1) the concentration of chlorine in the vapor is approximately sixty times the concentration of copper in the melt. That is, if the concentration of copper in the melt at the time of vapor saturation is 8 ppm, then the concentration of copper in the vapor is on the order of 500 ppm. However, it is important to point out that this concentration is the maximum attained by the fluid (because $\Gamma > 1$). As the melt crystallizes, the concentration of chlorine in the melt and associated vapor decreases as described above, and this results in a monotonic decrease in the concentration of the chloride-complexed cations including copper, manganese, and zinc. Because of the monotonic decrease in the instantaneous copper concentration, the ability of the aqueous phase to drive a progressive depletion of copper in the melt is steadily reduced, and at a value of Z ~ 0.1, the concentration of copper begins to rise in the melt. The only possible terminous to this rise is the precipitation of a primary copper phase such as is (Candela, 1982; Whitney and Stormer, 1983) in the melt. However, because the Cl concentration has steadily decreased, the rise in $C^{l}_{Cu}$ is not reflected in $C^{v}_{Cu}$. The end result of this scenario may include the precipitation of Cu–Fe–S minerals from the intercrystalline residua.

Under these same conditions, the concentration of Mo, F, and B monotonically increases in the melt due to their low vapor/melt partition coefficients, which result in $\Gamma < 1$ for these elements under all reasonable values of $C^{l,s}_{w}$. Fig. 12.8 illustrates this point in the case of molybdenum. In this model calculation, the initial melt concentration for Mo is 5 ppm (an "elevated" concentration), and the final 2.5% of the vapor evolved has a model concentration of Mo > $10^{-2}$ molal, or approximately $10^3$ ppm. For melts with more "normal" (lower) initial Mo concentrations, these late fluids will most likely contain a few hundred ppm molybdenum. The model concentrations of fluorine and boron in these aqueous fluids is even greater. These results hold qualitatively for vapor evolution at pressures on the order of 1000 bars or greater. However, it must be stated at this point that these are model concentrations. These fluids, at some point, become metastable. If these solutions saturate with respect to crystalline phases at concentrations below those calculated her for fluids in equilibrium with melts, saturation of the melt with respect to the phase in question is indicated.

In summary, for this general case of $\Gamma_{Cl} > 1$, chlorine and the chloride-complexed cations reach their maximum "ore fluid" concentration in the early stage of the vapor evolution process, whereas elements such as Mo, F and B reach their maxima in the late stages. W may behave in a manner similar to that of molybdenum.

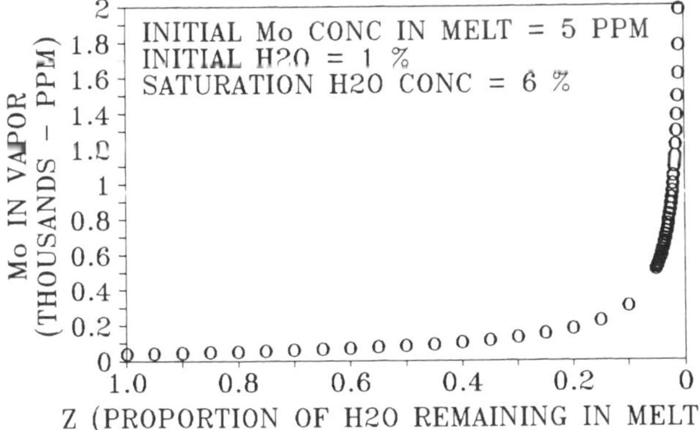


FIGURE 12.8—The concentration of molybdenum in successive aliquots of vapor evolved from a melt phase during second boiling as a function of the proportion of water remaining in the melt, for $C^{l,o}_{w}$ = 1 wt% and $C^{l,s}_{w}$ = 6 wt%. $D^{v/l}_{Mo}$ is given in Table 12.1. The initial molybdenum concentration in the melt is equal to 5 ppm. Note that, under these conditions, concentrations over 1000 ppm in the vapor are achieved in the last few percent of vapor evolved, assuming that the melt–vapor–crystal system does not saturate with respect to a primary Mo phase.

TABLE 12.3—Efficiencies of removal, E(i), for selected elements: Cl/$H_2O$ = 0.1, lower limit of Z = 0.002, partition coefficients are given in Table 12.1, initial water conc. = 1wt%, saturation water conc. = 6wt%; other parameters are given in Table 12.2. Fe-free system is on the left, Mt-saturated system is on the right. Max E(i) = 1.

| (A) | | (B) | |
|---|---|---|---|
| E(Cu) | = 0.025 | E(Cu) | = 0.019 |
| E(Mn) | = 0.30 | E(Mn) | = 0.078 |
| E(Zn) | = 0.13 | E(Zn) | = 0.03 |
| E(Ca) | = 0.012 | E(Ca) | = 0.002 |
| E(Mg) | = 0.0053 | E(Mg) | = 0.0007 |
| E(Ce) | = 0.55 | E(Ce) | = 0.05 |
| E(K) | = 0.011 | E(K) | = 0.005 |
| E(Na) | = 0.016 | E(Na) | = 0.007 |
| E(B) | = 0.59 | E(B) | = 0.59 |
| E(Mo) | = 0.53 | E(Mo) | = 0.53 |
| E(F) | = 0.069 | E(F) | = 0.068 |

Table 12.3a lists the efficiencies of removal for selected elements under the conditions of $C^{l,o}_{w}$ = 0.01, $C^{l,s}_{w}$ = 0.06 and a chlorine water ratio of 0.1. Note that $E$(Cu) is low compared to the other ore metals. Inspection of Table 12.3a shows that the crystal/melt partition coefficient for copper is larger than those of Mn, Zn, or Mo (Candela and Holland, 1986). Therefore, most of the copper which was initially present in the melt was sequestered into the crystalline phases before vapor saturation. The critical factors which lead to the strong sequestering of copper and a low efficiency of removal are the presence of a phase into which copper partitions heavily (such as a sulfide phase) and a low ratio of $C^{l,o}_{w}/C^{l,s}_{w}$. This ratio will be dealt with in greater detail presently. The efficiencies of removal which are listed in Table 12.3a are rather sensitive to the total "cation" equivalents in the melt, that is, the sum total of the number of moles of cations in the melt, which can be chloride-complexed in the vapor, per kilogram, each divided by its respective ligation number, $n$. Obviously, as this sum increases, the average efficiency of removal must decrease for a given initial chlorine concentration in the melt. Further, if $i$ is present in high concentration in the vapor, due to a high $E(i)$ or $C^{l,o}_{i}$, then the efficiencies of other cations will be suppressed. The efficiencies listed in Table 12.3b are for a melt which contains magnetite as a saturating phase, and therefore possesses iron as a major component of the vapor. If the calculations are carried out for an iron free melt, then the chlorine present in the system can be charge-balanced by other cations, and the efficiencies are much higher, as can be seen in Table 12.3b. This effect is dealt with in greater depth in Candela (1986a).

When the conditions are changed so as to model a very shallow level of intrusion ($C^{l,s}_{w}$ = 0.02 or a pressure of 250 bars), rather different trends and efficiencies are observed for some elements. Under these conditions, the equations are being used outside of the data base, and any conclusions drawn are highly speculative. Due to the difficulty involved in the calculation of magnetite solubility at these variable pressures, iron has been left out of these calculations and

TABLE 12.4—Efficiencies of removal, E(i), for selected elements: $Cl/H_2O$ = 0.1, lower limit of Z = 0.002, partition coefficients are given in Table 12.1, initial water conc. = 2wt%, saturation water conc. = 2wt%; other parameters are given in Table 12.2. System is Fe-free. Max E(i) = 1.

| | |
|---|---|
| E(Cu) | = 0.95 |
| E(Mn) | = 0.27 |
| E(Zn) | = 0.12 |
| E(Ca) | = 0.0075 |
| E(Mg) | = 0.0032 |
| E(Ce) | = 0.58 |
| E(K) | = 0.015 |
| E(Na) | = 0.021 |
| E(B) | = 0.33 |
| E(Mo) | = 0.28 |
| E(F) | = 0.026 |

the efficiencies calculated under the conditions (Table 12.4) should be compared with those of Table 12.3a. Further, the high-salinity fluids (e.g. Fig. 12.6) may in fact be metastable with respect to unmixing into steam and a coexisting hydrosaline melt. However, the qualitative results probably shed light on the gross processes which occur in the shallow magmatic-hydrothermal environment. According to the study of Shinohara et al. (1984), $D_{Cl}^{v,l}$ increases significantly and HCl becomes more significant in charge-balancing chlorine in the aqueous phase as pressure decreases. Incorporation of this data improves the accuracy of the calculations, but it is important to point out that they do not affect the substance of the conclusions expressed in this paper or the conclusions reached in my previous papers in which I have assumed a constant chlorine partition coefficient and neglected HCl in the charge balance equations.

The low pressure results of the efficiency calculations are shown in Table 12.4. Under these conditions, $C_w^{l,o} = C_w^{l,s}$; this is a limiting case wherein no crystallization occurs prior to vapor evolution. Therefore, the efficiency of removal of copper is very high (~95%). This is the case even though the chlorine partition coefficient is on the order of 7, and 3/5 of the chlorine is balanced by HCl. Iron was ignored in this low pressure calculation and its incorporation would lower the *E*(*i*) of the other chloride-complexed cations. The efficiency of removal of molybdenum has been halved (*E*(Mo) = 0.27) relative to E (Mo) = 0.54 at $C_w^{l,o}$ = 0.01 and $C_w^{l,s}$ = 0.06.

The "button-hook" shape of the $C_{Cu}^v = f(Z)$ function (Fig. 12.9) is rather distinctive and is typical of the trend found for a number of chloride-complexed elements under low pressure conditions. This is the converse of the one we examined previously, where $C_w^{l,s}$ = 0.06. The low chlorine partition coefficient and the small proportion of water released from the melt for a given proportion of melt crystallized allow copper to build in concentration in the melt during vapor-saturated crystallization of the melt. However, because $\Gamma_{Cl} < 1$, the concentration of chlorine increases monotonically in the melt and the associated vapor phase. This induces a steadily increasing "effective partition coefficient" for copper, which eventually leads to a late-stage depletion of copper in the melt and the associated vapor phase at low Z. The concentrations of copper reached by these fluids is rather

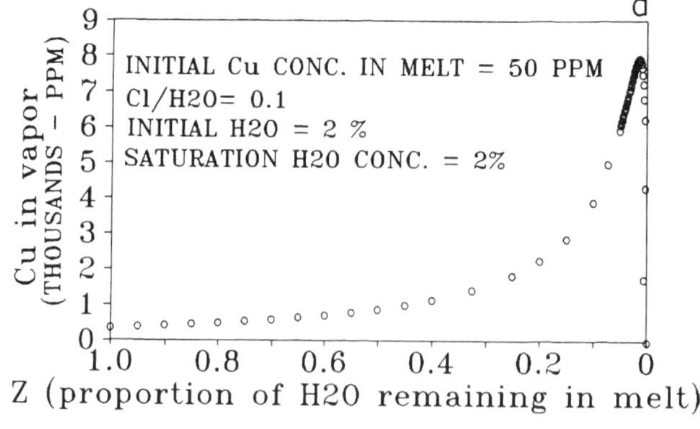


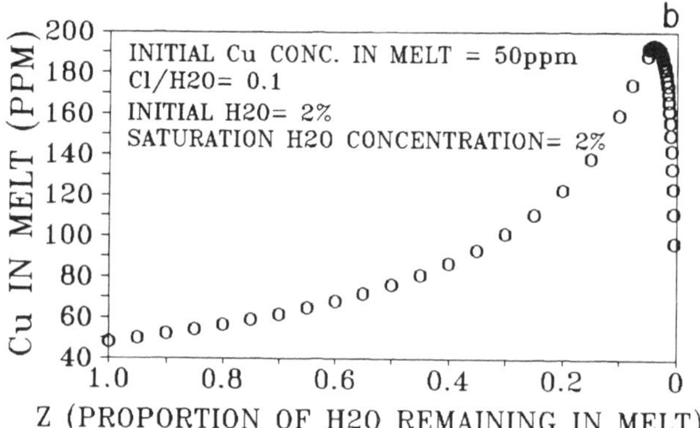


FIGURE 12.9a,b—The concentration of copper in successive aliquots of vapor evolved from a melt during second boiling as a function of the proportion of water remaining in the melt for $C_w^{l,o}$ = 2 wt%, $C_w^{l,s}$ = 2 wt% and a $Cl/H_2O$ wt. ratio = 0.1. $D_{Cl}^{v/l}$ is given by equation [17]. $C_{Cu}^v$ increases progressively in both the vapor (a) and the melt (b) as a function of −Z because of the low proportion of vapor evolved from the melt per given amount of crystallization, until the chlorine concentration in the vapor is high enough to deplete successive portions of the melt in copper. The concentration of copper in the vapor continues to increase to lower Z values, because of the rate of rise of the chlorine concentration, until the precipitous drop in the concentration of copper in the melt induces a decrease in the concentration of chlorine in successive aliquots of the aqueous phase. Under these conditions, and because the evolution of the aqueous fluid depletes the melt in copper, up to a point, as second boiling proceeds. The initial copper concentration in the melt = 50ppm; the maximum concentration of copper in the aqueous fluid under these conditions of deep (depths > 3–4km) vapor evolution is between 500 and 600 ppm, and occurs early in the second boiling event.

high, and these are close to the optimum conditions for achieving high concentrations of copper in an unmixed orthomagmatic fluid. Inspection of Fig. 12.9 shows that a concentration of over 8000 ppm copper is reached in the aqueous phase at low Z at the top of the "button-hook" given a rather average initial melt concentration of 50 ppm Cu. However, it is clear from Fig. 12.9 that such a high concentration is transient, and that the average concentration of copper in the aqueous phase is closed to (50 ppm Cu in the melt)/(10,000 ppm $H_2O$ in the melt), or 2500 ppm in the resulting aqueous phase. Qualitatively, the trend in the concentration of Mo, and the other "melt-compatible" elements in the same at low and high pressure. That is, the latest aqueous fluids will be enriched in Mo, B, and F, and probably also in W and Nb. This may also be true for the crystal incompatible elements Rb, Cs, Li, etc. for conditions of $\Gamma_{Cl} > 1$, but not at shallow levels wherein the high aqueous chloride concentrations may cause late stage depletion in the elements resulting in an alteration halo enriched in these elements and an intercrystalline residua in the pluton depleted in these elements.

The most important factors calculated in this paper, from the point of view of the economic geologist, are the efficiencies of removal of Cu and Mo. Examination of Tables 12.3 and 12.4 shows that the ratio of these efficiencies, $E$(Mo)/$E$(Cu), increases by a factor of 70 from ~0.3 at $C_w^{l,o} = 0.06$. Both calculations were performed at a Cl/$H_2O$ weight ratio of 0.1. The exact figures change as the Cl/$H_2O$ ratio is varied, but the general results remain unchanged. In general, as the $C_w^{l,o}/C_w^{l,s}$ ratio of the melt decreases, more of the crystal-compatible element, copper, is taken up by crystallizing phases before vapor evolution, and $E$(Cu) decreases. On the other hand, molybdenum behaves as a crystal-incompatible element, and its efficiency removal increases with a decreasing $C_w^{l,o}/C_w^{l,s}$ ratio. A low $C_w^{l,o}/C_w^{l,s}$ ratio results when, for example, a melt with a few weight percent water crystallizes at a deep level or by when a relatively dry melt crystallizes at a shallow level. Therefore, the removal of molybdenum is most efficient, relative to that of copper, when an initially dry melt evolves at shallow depths, or when a melt evolves water at a great depth. In either case, the evolution of vapor during crystallization is more effective in removing Mo from the melt than any reasonable first boiling scenario. This effect will be dealt with in greater detail in another chapter.

ACKNOWLEDGMENTS—Many people have contributed to this study and to the preparation of this paper. I would especially like to thank my doctoral students Philip Piccoli and Huifang Liu for their many and varied contributions including their work on derivations, experimentation, calculations, computer science, editing, drafting and typing. Without them, this paper would not have been possible. Further, I would like to thank my wife, Mary Ostrowski, and our children Emily, Alison, and Nicholas for carrying on the family life while I labored over this paper and the other paper is this volume. Proofreading by Mary Ostrowski has improved the manuscript. I would also like to thank the Magma Energy Project of the Department of Energy, and the National Science Foundation (Grants EAR–8319101 and EAR–8706198) for funding this research.

## REFERENCES

Bouton, S. L., Candela, P. A., Weidner, J. R., and Joseph, J., 1987, Experimental determination of the partitioning of tungsten and molybdenum between a high silica melt and Ti-bearing minerals: Geological Society of America, Abstracts with Programs, v. 19, p. 596.

Burnham, C. W., 1981, Physiochemical constraints on porphyry mineralization; *in* Dickenson, W. R., and Payne, W. D. (eds.), Relations of tectonics to ore deposits in the southern Cordillera: Arizona Geological Society, Digest, v. 14, pp. 71–77.

Burnham, C. W., 1967, Hydrothermal fluids at the magmatic stage; *in* Barnes, H. L. (ed.), Geochemistry of hydrothermal ore deposits: New York, Holt Rinehart and Winston, pp. 34–76.

Burnham, C. W., and Davis, N. F., 1974, The role of $H_2O$ in silicate melts: II. Thermodynamic and phase relations in the system $NaAlSi_3O_8$–$H_2O$ to 10 kilobars, 700° to 1100°C: Am. J. Sci., v. 274, pp. 902–940.

Burnham, C. W., and Nekvasil, H., 1986, Equilibrium properties of granite pegmatite magmas: Am. Min., v. 71, pp. 239–263.

Burnham, C. W. and Ohmoto, H., 1980, Late-stage processes of felsic magmatism: Min. Geol. Spec. Issue, v. 8, pp. 1–11.

Candela, P. A., 1982, Copper and molybdenum in silicate melt–aqueous fluid systems: Ph.D. thesis, Harvard University, Cambridge, Massachusetts, 138 pp.

Candela, P. A., 1986a, Generalized mathematical models for the fractional evolution of vapor from magmas in terrestrial planetary crusts; *in* Saxena, S. K. (ed.), Advances in Physical Geochemistry, v. 6: New York, Springer–Verlag, pp. 362–396.

Candela, P. A., 1986b, The evolution of aqueous vapor from silicate melts: effects on oxygen fugacity: Geochim. Cosmo. Acta., v. 50, pp. 1205–1211.

Candela, P. A., 1986c, Towards a thermodynamic model for the halogens in magmatic systems: an application to melt–vapor–apatite equilibria: Chem. Geol., v. 57, pp. 289–301.

Candela, P. A., and Holland, H. D., 1984, The partitioning of copper and molybdenum between silicate melts and aqueous fluids: Geochim. Cosmo. Acta., v. 48, pp. 373–380.

Candela, P. A., and Holland, H. D., 1986, A mass transfer model for copper and molybdenum in magmatic hydrothermal systems: the origin of porphyry-type ore deposits: Economic Geology, v. 81, pp. 1–19.

Carron, J. P., and LaGache, M., 1980, Etude experimentale du fractionnement des elements Rb, Cs, Sr, et Ba entre feldspaths alcalins, solutions hydrothermals et liquides silicates dans le systeme Q.Ab.Or.$H_2O$ a 2K bar entre 700 et 800°C,: Bull. Mineral., v. 703, pp. 571–578.

Cygan, G. L., and Chou, I-ming, 1987, Calibration of the $WO_2$–$WO_3$ buffer: Am. Geophys. Union (EOS), v. 68, p. 451.

Delboff, F., Lebedev, Ye. B., and Malinin, S. D., 1986, Chloride-ion behavior and cation exchange in a magma-fluid system: Geochem. Int., pp. 28–36.

Dingwell, D. B., and Scarfe, C. N., 1983, Major element partitioning in the system haplogranite–HF–$H_2O$: implications for leucogranites and high-silica rhyolites: Am. Geophys. Union (EOS), v. 64, p. 342.

Flynn, R. T., and Burnham, C. W., 1978, An experimental determination of rare earth partition coefficients between a chloride-containing vapor phase and silicate melts: Geochim. Cosmochim. Acta., v. 42, pp. 685–701.

Frantz, J. D., Popp, R. K. and Boctor, N. Z., 1981, Mineral-solution equilibria—V. Solubilities of rock-forming minerals in supercritical fluids: Geochim. Cosmochim. Acta., v. 45, pp. 69–77.

Gammon, J. B., Borcsic, M., and Holland, H. D., 1969, Potassium–sodium ratio in aqueous solutions and co-existing silicate melts: Science, v. 163, pp. 179–181.

Holland, H. D., 1972, Granites, solutions, and base metal deposits: Econ. Geol., v. 67, pp. 281–301.

Kilinc, I. A., and Burnham, C. W., 1972, Partitioning of chloride between a silicic melt and coexisting aqueous phase from 2 to 8 kilobars: Econ. Geol., v. 67, pp. 231–235.

Kurdin, A. V., Rekharsky, V. I., and Khodakovsky, I. L., 1980, Experimental study of solubility of Tugarinovite $MoO_2$ in water solutions at 250 and 450°C: Geokhimia, v. 5, pp. 1825–1833.

Manning, D. A. C., and Henderson, P., 1984, The behaviour of tungsten in granitic melt–vapour systems: Contrib. Mineral. Petrol., v. 86, pp. 286–293.

McMillan, P. F., and Holloway, J. R., 1987, Water solubility in aluminosilicate melts: Contrib. Mineral. Petrol., v. 97, pp. 320–332.

Pichavant, M., 1981, An experimental study of the effect of boron on a water satuated haplogranite at 1 kbar vapor pressure: Contrib. Mineral. Petrol., v. 76, pp. 430–439.

Shinohara, H., Iiyama, J. T., and Matsuo, S., 1984, Geochemistry—Behaviour of chlorine in the system granitic magma-hydrothermal solution (in French): C. R. Acad. Sc. Paris, t. 298, Serie II, n° 17, pp. 741–743.

Stolper, E., 1982, The speciation of water in silicic melts: Geochim. Cosmochim. Acta., v. 46, pp. 2609–2620.

Tacker, R. C., and Candela, P. A., 1987, Partitioning of molybdenum between magnitite and melt: A preliminary experimental study of partitioning of ore metals between silicic magmas and crystalline phases: Econ. Geol., v. 82, pp. 1827–1838.

Taylor, B. E., Eichelberger, J. C., and Westrich, H. R., 1983, Hydrogen isotopic evidence of rhyolite magma degassing during shallow intrusion and eruption: Nature, v. 306, pp. 541–545.

Thompson, J. B., Jr., 1982, Composition space: An algebraic and geometric approach; *in* Ferry, J. M. (ed.), Characterization of metamorphism through mineral equilibria: Mineral. Soc. Amer., pp. 1–31.

Urabe, T., 1985, Aluminous granite as a source of hydrothermal ore-deposits: an experimental study: Econ. Geol., v. 80, pp. 148–157.

Webster, J. D., and Holloway, J. R., 1987, Partitioning of halogens between topaz rhyolite melt and aqueous and aqueous-Carbonic fluids: Geol. Soc. Amer., Abstracts with Programs, v. 19, p. 884.

Webster, J. D., Holloway, J. R., and Hervig, R. L., 1987, Phase equilibria of a Be, U and F-enriched vitrophyre from Spor Mountain, Utah: Geochim. Cosmochim. Acta, v. 51, pp. 389–402.

Whitney, J. A., and Stormer, J. C., Jr., 1983, Igneous sulfides in the Fish Canyon Tuff and the role of sulfur in calc-alkaline magmas: Geology, v. 11, pp. 99–102.

## APPENDIX I

In this appendix the equations that govern the chemistry of magmatic vapor during second boiling are developed. The statement of mass conservation for the element in question is,

$$0 = dM_i^l + dM_i^v + \sum_{\phi}^{p} dM_i^{\phi}. \qquad [A-1]$$

The sum $\Sigma_{\phi}^{p} M_i^{\phi}$ is taken over all crystalline phases which accept the element $i$ ($p$ in number). Assuming no hydrous phases are crystallizing,

$$0 = dM_w^l + dM_w^v. \qquad [A-2]$$

Dividing Eq. (A–1) by Eq. (A–2) yields

$$dM_i^v/dM_w^v \equiv C_i^v = dM_i^l/dM_w^l + \sum_{\phi}^{p} dM_i^{\phi}/dM_w^l. \qquad [A-3]$$

The concentration of $i$ in the vapor, $C_i^v$, is given by

$$C_i^v = D_i^{v/l} C_i^l = D_i^{v/l} \left( \frac{M_i^l}{M_l} \right), \qquad [A-4]$$

or incorporating $C_w^l = (M_w^l/M_l)$,

$$C_i^v = D_i^{v/l} \frac{C_w^l}{M_w^l} M_i^l. \qquad [A-5]$$

Substituting Eq. (A–5) into (A–3) and rearranging yields

$$C_w^l dM_w^l \frac{D_i^{v/l}}{M_w^l} = \frac{dM_i^l}{M_i^l} + \sum_{\phi}^{p} \frac{dM_i^{\phi}}{M_i^l}. \qquad [A-6]$$

If the crystalline phases are fractionated, then it can be shown (Candela and Holland, 1986) that:

$$\sum_{\phi}^{p} \frac{dM_i^{\phi}}{M_i^{\phi}} = -\overline{\mathbb{D}}_i \frac{dM_l}{M_l} \qquad [A-7]$$

Where $\overline{\mathbb{D}}_i^{s/l}$ is the Nernst bulk solid-liquid partition coefficient. Rewriting Eq. (A–5) in light of Eq. (A–6) yields

$$\left( C_w^l D_i^{v/l} \frac{dM_w^l}{M_w^l} \right) + \left( \overline{\mathbb{D}}_i^{s/l} \frac{dM_l}{M_l} \right) = \frac{dM_i}{M_i^l}. \qquad [A-8]$$

Because $C_w^l$ deviates from $C_w^{l,s}$ during vapor evolution (see text), the reduction in the solubility of water in the melt is proportional to the square root of the mole fraction of water in the aqeous phase. Therefore,

$$C_w^l(M_w^l) = C_w^{l,s} \sqrt{1 - \sum_j X_j^v(M_w^l)}, \qquad [A-9]$$

where $\Sigma_j X_j^v$ is the mole fraction of solutes in the vapor, which may also be written as:

$$\sqrt{1 - \sum_j X_j^v(M_w^l)} = \frac{M_w^l M_l^s}{M_l M_w^{l,s}} = \frac{(M_w^l / M_w^{l,s})}{(M_l / M_l^s)}. \qquad [A-10]$$

The numerator of the extreme right-hand side of Eq. (A–10) is the vapor evolution progress variable, $Z = (M_w^l/M_w^{l,s})$ (see text). The denominator, $(M_l/M_l^s)$ may be termed the crystallization progress variable, $Z_{xl}$. Therefore,

$$Z_{xl} = Z \Big/ \sqrt{1 - \sum_j X_j(M_w^l)}. \qquad [A-11]$$

Noting that $dZ/Z = dM_w^l/M_w^l$, and that $dZ_{xl}/Z_{xl} = dM_l/M_l$, equation (A–8) may be set up for integration:

$$\int_{M^{l,s}}^{M_i^l} \frac{dM_i^l}{M_i^l} = \int_1^{Z_{xl}} \overline{\mathbb{D}}_i \frac{dZ_{xl}}{Z_{xl}} + D_i^{v/l} C_w^{l,s} \int_1^Z \left[1 - \sum_j X_j(Z)\right]^{1/2} \frac{dZ}{Z}. \qquad [A-12]$$

Partial integration, and substitution of Eq. (A–11) for $Z_{xl}$ yields

$$M_i^l = \frac{M_i^{l,s} Z^{\overline{\mathbb{D}}}}{[1 - \sum_j X_j(Z)]^{\overline{\mathbb{D}}/2}} \exp\left\{ D_i^{v/l} C_w^{l,s} \int_1^Z \left[1 - \sum_j X_j^v(Z)\right]^{1/2} \frac{dZ}{Z} \right\}.$$

which, when combined sith Eq. (A–4) and (A–10), yields

$$C_i^v(Z) = \frac{D_i^{v/l} C_i^{l,s}}{\left[1 - \sum_j X_j^v(Z)\right]^{1/2(\overline{\mathbb{D}}_i - 1)}} Z^{\overline{\mathbb{D}}-1} \exp\left\{ D_i^{v/l} C_w^{l,s} \int_1^Z \left[1 - \sum_j X_j^v(Z)\right]^{1/2} \frac{dZ}{Z} \right\}. \qquad [A-13]$$

To derive the fundamental equation for the mass transfer of chloride-complexed cations, Eq. (A–3) is combined with the definition of $K_{i,Na}^{v/l}$ (see text) to yield

$$\frac{dM_i^l}{M_i^l} + \sum_\phi^p \frac{dM_i^\phi}{M_i^l} = \frac{dM_w^l}{M_l} K_{i,\mathrm{Na}} \left(\frac{C_{\mathrm{Na}}^v}{C_{\mathrm{Na}}^l}\right)^n, \qquad [A-14]$$

where $C_i^l$ was expressed as $(M_i^l/M_l)$. Utilizing Eq. (A–7) to introduce $\overline{\mathbb{D}}_i$ and letting $C_{Na}^l = C_{Na}^{l,s}$, we can set up an integral in a manner similar to the procedure used to set up the integral (A–12):

$$\int_{M^{l,o}}^{M_i^l} \frac{dM_i^l}{M_i^l} = \int_1^{Z_{xl}} \overline{\mathbb{D}}_i \frac{dZ_{xl}}{Z_{xl}} + \frac{K_{i,\mathrm{Na}}^{v,l}}{(C_{\mathrm{Na}}^{l,s})^n} \int_1^Z \{[C_{\mathrm{Na}}^v(Z)]^n C_w^l(Z)\} \frac{dZ}{Z}. \qquad [A-15]$$

Partial integration, and substitution of Eq. (a–11) of $Z_{xl}$ yields

$$M_i^l = \frac{M_i^{l,s} Z^{\overline{\mathbb{D}}_i}}{\left[1 - \sum_j X_j(Z)\right]^{\overline{\mathbb{D}}_i/2}} \exp\left\{\frac{K_{i,\mathrm{Na}}}{(C_{\mathrm{Na}}^{l,s})^n} \int_1^Z \frac{[C_{\mathrm{Na}}^v(Z)]^n C_w^l(Z)}{Z} dZ\right\}. \qquad [A-16]$$

This equation yields the number of moles of *i* present in the magma at a given stage of isobaric crystallization-controlled vapor evolution (second boiling). When Eq. (A–16) is combined with Eq. (A–10) and the definition of $K_{i,Na}^{v,l}$ (see text), the final equation for the concentration of chloride-complexed element in the evolving magmatic vapor as a function of $Z$ is obtained:

$$C_i^v(Z) = \frac{K_{i,\mathrm{Na}} [C_{\mathrm{Na}}^v(Z)]^n C_i^{l,s}}{1 - \sum_j X_j^v(Z)} \frac{Z^{\overline{\mathbb{D}}_i^{s/l} - 1}}{(C_{\mathrm{Na}}^{l,s})^n} \exp\left\{\frac{K_{i,\mathrm{Na}}}{(C_{\mathrm{Na}}^{l,s})^n} \int_1^Z \frac{[C_{\mathrm{Na}}^v(Z)]^n C_w^l(Z)}{Z} dZ\right\}. \qquad [A-17]$$

## Appendix II

### Symbols used in this paper

$C_i^\phi$ = concentration $i$ in phase $\phi$ in moles/kg of the phase

$D_i^{v/l}$ = Nernst partition coefficient for $i$ expressed as the concentration of $i$ in the vapor over the concentration of $i$ in the melt (liquid)

$\overline{\mathbb{D}}_i$ = bulk crystal/melt partition coefficient for $i$ (Candela and Holland, 1986)

$E(i)$ = efficiency with which $i$ is removed from the silicate melt into an aqueous fluid

$\boldsymbol{F}$ = vapor evolution progress variable during first boiling

$K_{i,\mathrm{Na}}^{v/l}$ = empirical $i$ – Na vapor/liquid exchange constant

$K_{\mathrm{eq}}$ = equilibrium constant (true)

$l$ = superscript indicating the liquid phase

$M_i^\phi$ = moles of $i$ in phase $\phi$

$n$ = aqueous phase chloride-ligation number

$o$ = superscript indicating initial conditions

$P$ = pressure

$v$ = superscript indicating the vapor phase

$s$ = superscript indicating conditions at water saturation

$T$ = temperature, or as a superscript, indicates total

$w$ = subscript-$H_2O$

$Z$ = vapor evolution progress variable during second boiling

$\beta$ = (see equation 20)

$\gamma_i$ = activity coefficient for $i$

$\Gamma_i = C_w^{l,s} \cdot D_i^{v/l}$

$\mu_i^\phi$ = chemical potential of $i$ in phase $\phi$

$\nu_i$ = stoichiometric coefficient for $i$ in a balanced chemical reaction

$\Pi$ = multiplication operator

## Chapter 13

# Felsic Magmas, Volatiles, and Metallogenesis

P. A. Candela

## INTRODUCTION

Many hypotheses for the formation of porphyry and skarn deposits and other deposits which are spatially and temporally associated with felsic igneous rocks suggest that the magmatic "volatile constituents" exert strong controls on the quantity, chemistry and style of the mineralization found in these systems. Many workers have summarized the data on the relationships between metallization and parameters such as rock type, ore metal content, and halogen content of the associated pluton(s). More recently, thermodynamic parameters such as $f_{O_2}$ and the fugacities of water, HF, and HCl have been related to the nature of the associated deposits.

In our laboratory, my coworkers and I have examined the partitioning of molybdenum and tungsten between water-saturated, high silica melts and minerals such as magnetite, ilmenite and rutile as a function of $f_{O_2}$ at 800°C and 1 kb (Tacker and Candela, 1987; Bouton et al., 1987; Bouton and Candela, in prep.). In this paper, our experimental results will be combined with previous work on felsic igneous rocks and associated ores in an effort to shed light on the role of $f_{O_2}$ in magmatic hydrothermal ore genesis. Further, some of the models discussed in Candela (this volume) will be used to suggest links between the volatile constituents of magmas, their source regions, and the ore systems they spawn.

This chapter is not meant to be a review of the work done to date on the igneous petrology of the volatile constituents or their relationship to ore genesis. Rather, I will explore the implications of some selected recent ideas, particularly with respect to the role played by water, fluorine, chlorine and $f_{O_2}$ in ore genesis.

## THE EFFECTS OF INITIAL MAGMATIC WATER AND FLUORINE CONCENTRATIONS AND THE DEPTH OF VAPOR EVOLUTION ON THE FORMATION OF MAGMATIC-HYDROTHERMAL ORE DEPOSITS

The formation of a porphyry or skarn-type mineral deposit by magmatic-hydrothermal processes involves the sum total of the mechanisms operative from the initial genesis of the parental magma, through a complex and protracted contamination and crystallization history, to vapor evolution and ultimate plutonic consolidation. Of the volatiles present in these systems, water is, by far, the most important in upper crustal environments.

### The working model

Following the final emplacement of a hydrous, felsic magma, characterized by an initial water concentration, $C_w^{l,o}$, into the epizone of Earth's crust, crystallization occurs due to the loss of heat through the walls of the chamber, and water increases in concentration in the bulk melt as quartz and feldspar crystallize. As crystallization proceeds, minor and trace elements partition among the melt and the crystallizing solids. At some point(s) in space and time within the magma, bubbles of a magmatic aqueous phase will nucleate and grow when the water concentration in the melt rises to the water concentration at saturation, $C_w^{l,s}$, by virtue of the crystallization of anhydrous minerals. As crystallization of the water-saturated melt proceeds, water is progressively released from the magma so as to maintain the water concentration at $C_w^{l,s}$, and the components of the magmatic system are partitioned among the melt, aqueous and crystalline phases.

Obviously there are physical problems associated with the separation of the aqueous phase from the pluton. As the crystal/melt ratio increases in any arbitrary volume of the magma, the interconnectedness of the melt and/or vapor phase will decrease to a point where the removal of magmatic vapor from the system ceases. At this stage in the evolution of the magma, the vapor will be trapped along cooling-induced microfractures and along grain boundaries, and the magmatic vapor evolution process will cease. Before this stage is reached, however, significant quantities of some elements, including ore elements, may be partitioned into a mobile magmatic aqueous phase and removed from the pluton.

### Ore-metal sequestering in crystallizing phases: the role of volatiles in ore-metal dispersal

As I have shown previously, (Candela, this volume) the quotient $\{C_w^{l/o}/C_w^{l/s}\}$, which is the ratio of the initial water concentration in the melt, $C_w^{l/o}$ to the solubility of water in the melt (i.e. the saturation water concentration), $C_w^{l/s}$, is a master variable in the calculation of the proportion of any given ore metal removed from the magma into an ore fluid. In this model, hydrous phases account for an insignificant proportion of magmatic water, and the ratio $\{C_w^{l/o}/C_w^{l/s}\}$ is characterized by the domain $\mathfrak{D} = [0 < \{C_w^{l/o}/C_w^{l/s}\} \leq 1]$. Generally, a low $\{C_w^{l/o}/C_w^{l/s}\}$ ratio indicates a large amount of crystallization before water saturation, i.e. a significant amount of crystallization occurs before the water concentration is increased to the saturation value. Therefore, a low $\{C_w^{l/o}/C_w^{l/s}\}$ ratio results in the sequestering of crystal-compatible elements in the minerals of the hypersolidus assemblage and the concentration of incompatible elements in the residua. A high $\{C_w^{l/o}/C_w^{l/s}\}$ ratio allows compatible elements to gain access to the magmatic aqueous phase by inducing the early evolution of vapor in the magmatic system.

The behavior of an element during magmatic fractionation can greatly affect its availability for ore formation. *However, it is not the case that if any element is to be found in a magmatic-hydrothermal ore deposit that it must behave as an incompatible element before the magmatic system in question saturates with respect to an aqueous fluid.* Copper is a case in point. The available evidence suggests that copper behaves as a compatible element during fractional crystallization in a wide range of magmas (Andiambololona and Dupuy 1978; Candela and Holland, 1986). Further, sulfide is the most likely candidate to serve as a host phase for copper. It is not uncommon for the small amounts of magmatic pyrrhotite which occur in many volcanic rocks to contain large concentrations of copper (Anderson, 1974; Luhr et al., 1984). Au may be sequestered in a manner similar to that of Cu (Hattori, 1987). The study of Carroll and Rutherford (1987) suggests that pyrrhotite exhibits a minimum solubility in a felsic melt in the vicinity of $f_{O_2}$ = NNO, with pyrrhotite breaking down in Ca-bearing magmas to yield anhydrite-bearing assemblages at oxygen fugacities 1 to 2 log units above $f_{O_2}$ = NNO. At higher oxygen fugacities, copper should behave as an incompatible element. Further, pyrrhotite appears to be out of equilibrium with many erupted glasses (Whitney and Stormer, 1983). Candela (1982) and Candela and Holland (1986) discussed the possibility of a reaction relationship during magmatic vapor evolution, which removes pyrrhotite from the hypersolidus crystalline assemblage, of the form

$$FeS^{po} + {}^{3}/_{2}O_2 = FeO^{l} + SO_2^{v}. \quad [1]$$

This reaction shifts to the right as the $SO_2$-bearing aqueous phase is fractionally removed, but may not be very effective in removing po as a saturating phase at oxygen fugacities much below the $H_2O = SO_2$ isofugacity line in $\log f_{O_2}$–$T$ space. The efficiencies of removal (see Candela, this volume) of copper from a melt into an evolving vapor discussed in this paper are calculated assuming that copper behaves as a compatible element before water saturation, and becomes an "incompatible" element after vapor saturation, consistent with the observations that:

1) copper decreases in concentration with increasing differentiation index in most observed systems (Andiambololona and Dupuy, 1978);
2) copper does not partition strongly in favor of any crystalline silicate or oxide phase relative to the melt;
3) high concentrations of copper are found in some pyrrhotite in felsic lavas; and
4) pyrrhotite is rarely found in contact with the glass phase in lavas suggesting that the melts became undersaturated with respect to pyrrhotite at some stage.

The data for elements such as molybdenum and tungsten are much poorer. In general, these elements seem to be enriched in rocks which are high in the classical "incompatible elements" such as Ta, Nb, Rb, etc. This is true for systems ranging from felsic extrusive rocks to the Thingmuli suite, Iceland (Carmichael, et al., 1974).

When copper is modeled as a compatible element and molybdenum and tungsten are modeled as incompatible elements, then for a given Cl/$H_2O$ ratio, $\{E(\text{Mo or W})/E(\text{Cu})\}$ will vary inversely with $\{C_w^{l/o}/C_w^{l/s}\}$. $E(i)$ is the overall efficiency with which element $i$ is removed from the silicate melt in the aqueous fluid. Fig. 13.1 is a plot of $\{E(\text{Mo})/E(\text{Cu})\}$ as a function of $\{C_w^{l/o}/C_w^{l/s}\}$. Note that the efficiency ratio (Mo/Cu) decreases as the $\{C_w^{l/o}/C_w^{l/s}\}$ increases. A high $\{C_w^{l/o}/C_w^{l/s}\}$ ratio may be produced by either vapor evolution from a magma at a shallow level (< 3–4 km) or by vapor evolution at depth (> 3–4 km, but probably < 8 km) from a wet magma. This is consistent with the general observation that porphyry copper deposits are relatively shallow deposits.

Conversely, deposits more enriched in molybdenum or tungsten relative to copper can result from lower values of $\{C_w^{l/o}/C_w^{l/s}\}$, which translate into either a deeper level of emplacement of the associated stocks and/or drier magmas relative to the porphyry copper systems. This conclusion is supported by a number of lines of evidence. Woodcock and Hollister (1978), in a summary of the characteristics of porphyry molybdenum deposits of the North American Cordillera, state (p. 10): "Most porphyry molybdenum deposits . . . are deep-seated compared to porphyry copper deposits; they do not commonly occur within the extrusive volcanic phases of their related magma systems." The theoretical analysis here is consistent with the observations and generalizations of Newberry and Swanson, (1986) who state, in their summary paper on scheelite skarn granitoids (from the abstract): "Scheelite and Cu skarn-associated granitoids are generally similar. Textural and bulk compositional data, however, suggest that scheelite skarn granitoids are different from Cu skarn granitoids by virtue of greater degree of differentiation and by crystallization in a comparatively deep plutonic environment. Consideration of relevant phase equilibria indicates that magmatic water does not exsolve until very late in the crystallization of a scheelite skarn granitoid." Their summary of the pressures of skarn formation, which includes data based on stratigraphic reconstructions, sphalerite geobarometry and calc-silicate phase equilibria, indicates that the scheelite skarns formed at pressures generally greater than 1.3 kb. On the other hand, Einaudi et al. (1981) report that copper skarns form generally at pressures less than 500 bars.

The situation is somewhat more complex in the case of Climax-type Mo deposits. These deposits seem to have formed over a range of depths, or pressures. For example, Keith and Shanks (1988) suggest that the Pine Grove system formed at pressures of approximately 3–4 kb. On the other hand, the Climax and Urad–Henderson deposits formed under relatively shallow conditions (White et al., 1981). The depth estimates for Climax and Henderson deposits rest upon reconstructions of lithologic sections and the projection of

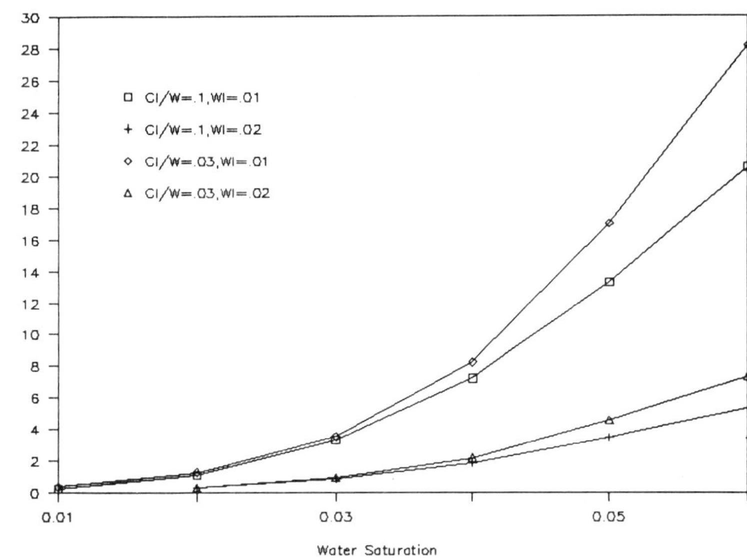


FIGURE 13.1—$E$(Mo)/$E$(Cu) as a function of the concentration of water in the melt at saturation, $C_w^{l,s}$, for given initial water concentrations, $C_w^{l,o}$, and Cl/$H_2O$ ratios for granitic melts. The effect of oxygen fugacity on this system is neglected. Molybdenum is modeled as an incompatible melt ($\bar{\mathbf{D}}$ = 0.1) and copper is modeled as compatible before vapor saturation ($\bar{\mathbf{D}}$ = 2) but incompatible after ($\bar{\mathbf{D}}$ = 0.1). For further details on carrying out these calculations, see Candela, this volume).

Tertiary erosion surfaces. In the case of the Climax Mo Deposit, the lithologic reconstruction is performed by summing the the thicknesses of overlying Precambrian rocks, Paleozoic strata, and early Tertiary sills. This analysis places the top of the central mass of the Climax stock at a depth of between 2.3 and 3.2 km below the Oligocene surface. Assuming a lithostatic gradient, these depths imply pressures of crystallization of 0.6 to 0.84 kb.

Turning to the Henderson deposit, White et al. (1981) citing the arguments of Ogden Tweto, estimate that the top of Red Mt. was approximately 0.6 km beneath the surface at the time of ore formation based on the interpretation that the mountain tops in the vicinity of Red Mountain represent a pre-Oligocene erosion surface. The Henderson ore body is 1.1 to 1.7 km below the top of Red Mountain; therefore, the Henderson system formed at depths of 1.7 to 2.3 km, or lithostatic pressures of 0.4 to 0.6 kb. These pressures are not very high, and do not suggest any particular range of values for the master variable $\{C_w^{l/o}/C_w^{l/s}\}$. It is interesting to note, however, that these systems are among the most fluorine-rich known (Munoz, 1984; Ague and Brimhall, 1988a).

According to the model presented by Burnham (1979, 1981), the $H_2O$ + F in hydrous minerals fluxes the melting of rock in magma source regions. The vapor-absent melting of, for example, biotite source rock yields melts with approximately 3.3 wt% combined $H_2O$ + F. Fluorine is combined with $H_2O$ in this model because HF has essentially the same effect per unit mass as $H_2O$ on melting relations (Burnham, 1981). If a source region has been previously melted, (Burt et al., 1982) or has undergone significant dehydrative metamorphism (Holloway, 1977) then the F/$H_2O$ ratio may be high in the source region and in the melt derivative derived therefrom. Relative to melts derived from sources with lower F/$H_2O$ ratios but with similar mineralogies, these melts will be relatively dry, that is, they will possess lower $C_w^{l,o}$. Therefore, these magmas will exhibit depressed $\{C_w^{l,o}/C_w^{l,s}\}$ ratios, even at relatively shallow levels, and will evolve a vapor phase at a later stage of crystallization relative to magmas derived rom source regions with lower F/$H_2O$ ratios. For whereas F fluxes melting, it does not promote vapor saturation. This suggests that magmas produced by the partial melting of a source region with a high F/$H_2O$ ratio will be impoverished in water, and will therefore experience a protracted crystallization history before vapor evolution, allowing Mo (and other incompatible elements) to concentrate in the melt relative to compatible elements such as copper. A high $E$(Mo)/$E$(Cu) ratio would result from such a second boiling episode. This effect may explain why fluorine-rich systems yield ore fluids with high ratios of incompatible elements to compatible elements, such as Climax-type Mo deposits, some skarn deposits and the peraluminous granitoid-associated, incompatible element (Be, F, W, Sn) deposits of the Western U.S. recently described by Barton (1987). Boron may have an effect not unlike that of fluorine because of its ability to delay vapor saturation by increasing the solubility of water, $C_w^{l,s}$, in silicate melts (London, 1986) which in turn reduces the $\{C_w^{l/o}/C_w^{l/s}\}$ ratio. High boron concentrations would be precluded by high Fe + Mg concentrations in the melt, as tourmaline precipitation would result (London, 1986).

The restriction of most magmatic-hydrothermal ores (and hydrothermal ore deposits in general) to the epizonal environment is rather striking, and a number of arguments have been put forward in the literature for why this should be so. Burnham and Ohmoto (1980) suggest that at shallow levels (2–8 km) sufficient $P\Delta V$ energy is released from the reaction

melt —> crystals + vapor

to cause the extensive fracturing which serves as host for the mineralization in many of the deposits in this class. As pressure increases, the molar volume of the aqueous phase, (the most compressible of the phases in the above reaction), decreases more rapidly than the molar volumes of the other phases and the overall $\Delta V$ of the reaction, (and hence the $P\Delta V$ energy), is reduced. Thus, the work available for fracturing is diminished, and ground preparation for mineral deposition within a restricted volume, necessary for the production of bulk-tonnage ore grades, is not achieved. Further, recent studies of the effect of pressure upon the partitioning of ore metals between melt and vapor (Urabe, 1987; Webster et al., ms) suggest that as pressures are increased beyond 3–4 kb, the melt/vapor partition coefficients for many elements decrease rather substantially.

### Caveat emptor

The precipitation of large quantities of ore substance in restricted volumes is dependent not only upon proper ground preparation and on efficient separation of ore constituents from source rocks or magmas, but also upon the existence of a suitable driving force for precipitation. Many times a reduction in temperature and/or the mixing of magmatic aqueous fluids with fluids of meteoric origin are important in promoting irreversible reactions including the precipitation of ore metals in a restricted volume. At greater depths, higher ambient temperatures and reduced permeability lessen the temperature gradients surrounding plutons and restrict the access of large quantities meteoric fluids to the potential sites of ore deposition, respectively. Large changes in free energy are still possible if the magmatic fluid encounters rocks which are greatly different in composition from the crystallizing magma, such as when a magmatic fluid interacts with carbonate rocks in the production of a skarn deposit. This mode of deposition may still be limited by the lower efficiencies of removal found at greater depths (depths > 10–15 km).

At this point, a cautionary word is necessary to prevent misinterpretations of the "efficiency of removal," $E(i)$. A high $E(i)$ does not necessarily translate into a large absolute amount of $i$ being removed into the aqueous phase. If the element $i$ is present in *low* concentrations in the initial melt then, even with very efficient removal of $i$ from the melt into the aqueous phase, the integrated concentration of $i$ in the aqueous fluid may be rather modest. In specific instances the calculated instantaneous aqueous concentration, or calculated total tonnage (calculated from a given $E(i)$, a model $C_w^{i,o}$ and mass of magma) must be elevated with respect to specific precipitation reactions or observed total ore tonnages, respectively (see Candela and Holland, 1986). Conversely, a rather low $E(i)$ does not necessarily translate into a "low" concentration or activity. For example, upon rise of a magma, vapor may be evolved without accompanying crystallization, as supported by studies including the analysis of hydrogen isotopes in obsidian glass by Taylor et al. (1983), and the phase equilibrium analysis of a vitrophyre from Spor Mountain by Webster et al. (1987). According to the model I presented in Candela (this volume), the fluorine concentration in such a system would remain approximately constant. With a vapor/melt partition coefficient for fluorine equal to approximately 0.2, a melt with 400 ppm fluorine (similar to that found, for example, in the trachyandesite of El Chichon (Varekamp et al., 1984)) would be in equilibrium with an aqueous vapor containing approximately 80 ppm fluorine. This translates to 0.0042 molal F, or log ($HF/H_2O$) $\approx -4$. This is the fluorine/water fugacity ratio suggested for the moderately contaminated magmas (I–MC) of arc environments (Ague and Brimhall, 1988a), the class of felsic plutonic rocks which is most commonly associated with porphyry copper deposits. If a very high fluorine concentration in the melt is present (e.g. on the order of 1 wt%, or 10,000 ppm), as in the systems associated with topaz rhyolites (Christiansen et al., 1983; Webster et al., 1987) or Climax-type porphyry systems, the polybaric rise of the magma may result in the exsolution of a vapor with fluorine concentrations on the order of 2000 ppm F, or log ($HF/H_2O$) > $-3$, consistent with the log ($HF/H_2O$) calculated by Ague and Brimhall (1988) for the most highly evolved systems they studied, which included the igneous rocks of the Climax and Henderson porphyry Mo deposits. In both cases, the efficiency of removal of F from the melt by first boiling is on the order of 3% or less under upper crustal ($P \leq 2$ kb) conditions (Candela, this volume); obviously, fluorine is evolved from some melts in an amount sufficient to produce significant fluorine alteration and the deposition of fluorine minerals. However, the proportion of the fluorine originally present in the melt which is released to the evolving aqueous fluid is rather low. When forced out of the melt by crystallization (second boiling), less than 10% F is removed from the system after 99.9% crystallization assuming a rather modest bulk solid/melt partition coefficient for fluorine equal to 0.1. Therefore arguments based on "efficiency of removal" must always be conditioned on the initial concentration of the element in question in the melt phase, and the aqueous activities of the element which are necessary to produce a given mineral assemblage.

### Summary

Late stage vapor evolution and accompanying crystallization favors the partitioning of "melt-compatible elements" such as Mo, B, F, W, etc. into the buoyant and low viscosity ore-forming aqueous fluid; crystal compatible elements, including copper and other first row transition elements, are sequestered into crystallizing phases and are dispersed throughout the pluton. Late evolution of vapor is promoted by vapor evolution in the deeper levels of the epizone, or, as in the case of the Climax-type deposits, by a low activity of water in the magma source region (which reflects a high mole fraction of F/OH in the hydrous phases). Conversely, the partitioning of compatible elements, such as Cu, into the magmatic aqueous phase is enhanced by high initial water concentrations in the melt or shallow (< 3–4 km) levels of intrusion.

## THE EFFECT OF $f_{O_2}$ ON THE EFFICIENCY OF REMOVAL OF METALS FROM MAGMAS

In recent years a number of authors have noted a relationship between ore deposit type and the inferred or deduced oxidation state of the associated magmatic system (e.g. Ishihara, 1977; Burnham and Ohmot, 1980; Keith and Shanks, 1988, and Ague and Brimhall, 1988a). In a seminal paper published in 1974, Chappell and White established a new perspective from which granitic rocks could be viewed. Later workers have refined, restricted and elaborated upon these original definitions, so that today we understand some granitoids (I-type) to be derived from a source region composed of rocks which are dominantly igneous or meta-igneous in nature, and other granitoids (S-type) to be derived from the melting of sedimentary rocks or their metamorphic equivalents. In general, I-type granitoids appear to be more oxidized than S-type granitoids. Numerous other properties characterize these classes of rocks, and it is beyond the scope of this paper to treat this topic in great detail. Rather, I would like to discuss the methods and conclusions of some recent papers by Ague and Brimhall (1987,1988a,1988b) which are predicated on the I/S model of Chappell and White (1974).

### The Ague–Brimhall model, and the $f_{O_2}$ of intrusive systems

In their papers, Ague and Brimhall studied biotite Mg/Fe and F/OH ratios. They relate the Mg/Fe ratio to the fugacity of oxygen, and the F/OH ratio to the relative fugacities of HF and $H_2O$. Quoting from Ague and Brimhall, (1988b):

> Rocks containing biotite with log $(X_{Mg}/X_{Fe}) > -0.21$ are divided into three subgroups based upon increasing F/OH:
> 1) I–WC type (weakly contaminated I-type),
> 2) I–MC type (moderately contaminated I-type), and
> 3) I–SC type (strongly contaminated I-type).
> Reduced rocks containing biotite with log $(X_{Mg}/X_{Fe}) < -0.21$ are classified as I–SCR type (strongly contaminated and reduced I-type). The term contamination is used here in a broad sense to refer to interactions of mafic "I-type" magmas derived from the upper mantle or subducted slabs with continental crustal source components . . . by such processes as partial melting, magma mixing, and assimilation.

In fact, the I–SCR rocks possess many of the characteristics of the S-type granitoids of Chappell and White, including, as we shall see, a low $f_{O_2}$.

The value of log $(X_{Mg}/X_{Fe}) = -0.21$ was chosen by Ague and Brimhall as the dividing line between the reduced and non-reduced rocks because of the virtual absence of muscovite, garnet, and tourmaline and the presence of titanite at higher values of log $(X_{Mg}/X_{Fe})$. (See Fig. 13.2). Further, they quantified their results by calculating log $(f_{HF}/f_{H_2O})$ using the model of Gunow et al. (1980) which involves the use of the equation

$$\log(f_{HF}/f_{H_2O}) = -2100/T - 1.523X_{Phl} - 0.416X_{Ann} - 0.2X_{Sid} + \log(X_F/X_{OH})^{Bi} \quad [2]$$

where $X_{Phl}$, $X_{Ann}$, and $X_{Sid}$ are the mole fractions of the biotite end-members phlogopite, annite and siderophyllite calculated following Gunow et al. (1980).

The calculation of $T$–$f_{O_2}$ relations in plutonic systems has been problematic. The oxygen fugacities which prevail in plutonic systems have been inferred, in a general manner, from the oxygen fugacities of extrusive rocks based on the compositions of coexisting magnetite–ilmenite pairs. Most attempts to apply the magnetite–ilmenite geothermometer to plutons have met with failure because of the re-equili-

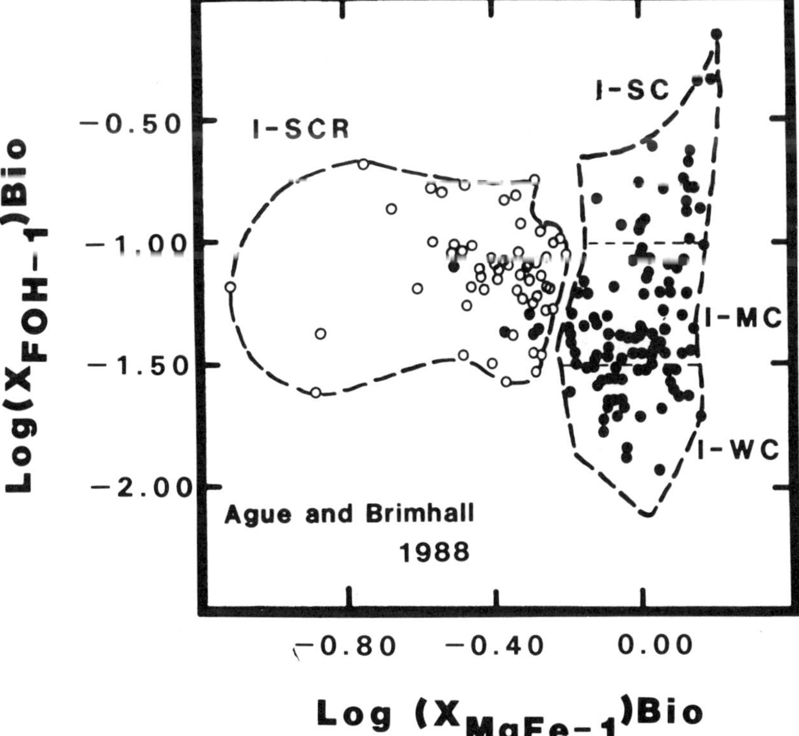


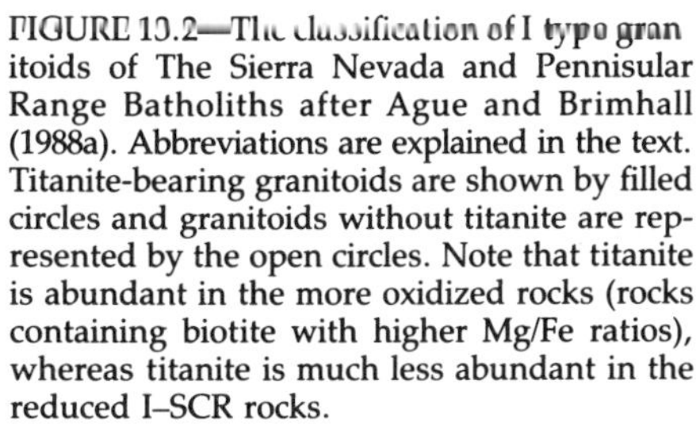
FIGURE 13.2—The classification of I type granitoids of The Sierra Nevada and Pennisular Range Batholiths after Ague and Brimhall (1988a). Abbreviations are explained in the text. Titanite-bearing granitoids are shown by filled circles and granitoids without titanite are represented by the open circles. Note that titanite is abundant in the more oxidized rocks (rocks containing biotite with higher Mg/Fe ratios), whereas titanite is much less abundant in the reduced I–SCR rocks.

bration of the magnetite–ilmenite pair. An interesting attempt to solve this problem has been presented by Ague and Brimhall. Based on the assumption that the ilmenite in a magnetite–ilmenite pair does not change its composition significantly upon cooling (while the magnetite certainly does), they reason that the $T$–$f_{O_2}$ of a modified "granite buffer" indicator curve, represented by the equilibrium

$$KFe_3AlSi_3O_{10}(OH)_2^{bio} + 0.75O_2 = KAlSi_3O_8^{K\text{-}Fsp} + 1.5Fe_2O_3^{ilmenite} + H_2O \quad [3]$$

(where magnetite has been replaced by the hematite component of ilmenite), will cross the ilmenite isopleth in $\log_{10}f_{O_2}$–$T$ space near the original intersection of the magnetite–ilmenite isopleths in the model of Buddington and Lindsley (1964). Limitations of this technique include the estimation of $a_{H_2O}$, and the possible presence of $Fe^{3+}$ in biotite. Ague and Brimhall consider that the error of $a_{H_2O} < 1$ will approximately offset the lowered activity of annite which results from the presence of some of the iron as Fe (III). Further, the pressure of crystallization must be stated before this calculation proceeds. Based on the pressure estimates made by Ague and Brimhall (using the technique of Hammarstrom and Zen (1986)) on the rocks which were the subjects of their study, a pressure of 3 kb was found to be an average pressure of equilibration. A sample calculation, kindly provided by Jay Ague, is given in Appendix I. When $f_{O_2}$–$T$ are calculated for rocks of the Sierra Nevada and Peninsular Ranges Batholiths by Ague and Brimhall (using a uniform $a_{H_2O} = 1$, $a_{K\text{-spar}} = 0.9$ and a pressure of 3 kb) the data generally plot at temperatures between 700–900°C and at oxygen fugacities which are similar to those commonly found in extrusive rocks. I have replotted their data on an $f_{O_2}$–$T$ plot (Fig. 13.3) along with the buffer curves for HM, NNO, QFM, $WO_2$–$WO_3$, and MW, and an $Sn^{2+} = Sn^{4+}$ isoconcentration curve for alkali silicate melts (Johnston, 1965; Johnston and Chleko, 1966). The implications of the range of oxygen fugacities suggested by Ague and Brimhall will be discussed in the following sections of this paper.

### Experimental and field data bearing on the role of $f_{O_2}$ in ore-metal sequestration

Candela (1982) pointed out that $Sn^{4+}$, $Mo^{4+}$, and $W^{4+}$ have ionic radii which are very similar to the ionic radius of $Ti^{4+}$, and indicated that this fact may be important in the sequestering of these metals during magmatic crystallization. $Sn^{4+}$, $Mo^{4+}$, and $W^{4+}$ may be incorporated into titanium-bearing minerals or into minerals which accept significant quantities of titanium. This idea has been expanded upon by Tacker and Candela (1987) and Bouton et al., (1987). In these studies, the mineral/melt partition coefficient for molybdenum was found to decrease with increasing $f_{O_2}$. Tacker and Candela (1987) suggest that the dominant oxidation state for molybdenum varies from 4+ to 3+ between NNO + 1/2 and the graphite–methane buffer, and indicate that, at least for sequestering reactions involving magnetite, the partitioning of molybdenum into the solid phase is enhanced at low $f_{O_2}$. The observed rate of decrease of $D_{Mo}^{mt/l}$ with increasing $f_{O_2}$ was such that it was difficult to extrapolate the observed decrease in the *mt/l* partition coefficient with increasing $f_{O_2}$ to other mineral/melt systems, and Tacker and Candela state that further experimentation on other mineral/melt systems would be needed if the problem of the dependence of a general, bulk solid/melt partition coefficient on oxygen

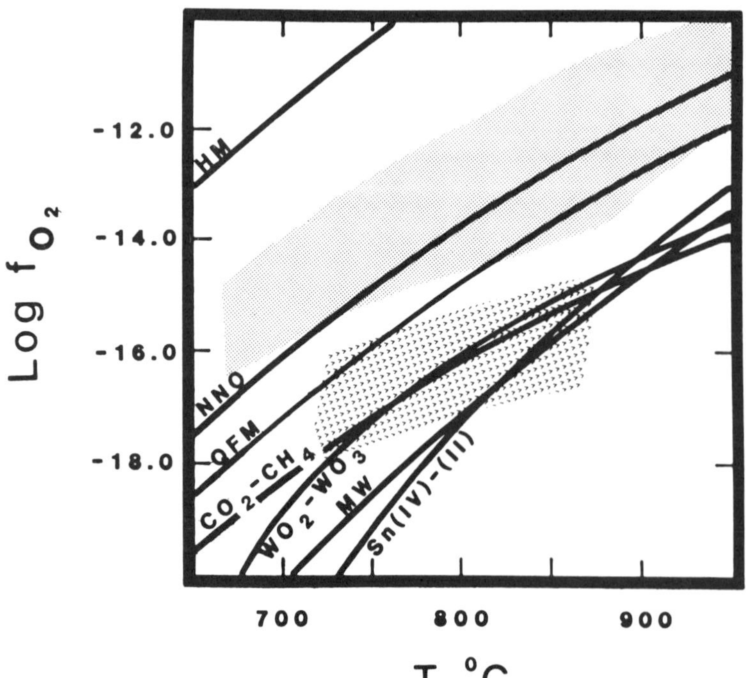


FIGURE 13.3—Log $f_{O_2}$–$T$ plot illustrating the relationship between the magnetite-bearing, I-WC, I-MC and I-SC rocks (gray pattern) and the I-SCR rocks (dotted pattern) from the study of Ague and Brimhall (1988a). For reference, the $WO_2$–$WO_3$ buffer from Cygan and Chou (1987) is shown, in addition to the standard reference buffer curves. Additionally, an isoconcentration line has been calculated (Candela, 1982) for the condition $Sn^{2+} = Sn^{4+}$ in sodium silicate melts from the data of Johnston (1965) and Johnston and Chelko (1966).

fugacity is to be solved. The partitioning reactions involved indicate that the *mt/l* partition coefficient for molybdenum decreases with increasing $f_{O_2}$ for Mo oxidation states in the melt equal to 6+, 5+, and 4+ and 3+. For example, if the oxidation state in the melt is 6+ or 3+, the pertinent model reactions are, respectively,

$$2Fe_3O_4^{mt} + 3MoO_3^{melt} = 3Fe_2MoO_4^{mt} + 2.5O_2, \quad [4]$$

and

$$2Fe_3O_4^{mt} + 1.5Mo_2O_3^{melt} = 3Fe_2MoO_4^{mt} + 0.25O_2. \quad [5]$$

$Fe_2MoO_4$ was chosen by Tacker and Candela (1987) as the most probable Mo-bearing phase component in Magnetite based on phase equilibria in the system Fe–Mo–O (Drabek, 1982). Note that, in both cases, an increase in the oxygen fugacity results in a decrease in the mole fraction of the molybdenum-bearing phase component in magnetite at a given concentration of molybdenum in the melt. The oxygen fugacity dependence of reaction [5] is controlled by the fact that iron, not only molybdenum, changes its oxidation state during the partitioning reaction so that molybdenum, changes its oxidation state during the partitioning reaction so that molybdenum may be present as a stoichiometric entity in magnetite. Therefore, the study of Tacker and Candela was immediately succeeded by a study of molybdenum partitioning between ilmenite and melt. In that study, which involves changes in the oxidation state of molybdenum (or tungsten) only, the above-mentioned ambiguity is absent. The results of Bouton et al. (1987) demonstrate that the trend found by Tacker and Candela can indeed be extended to systems beyond magnetite, and also show that the partition coefficients for molybdenum are on the same order for both ilmenite and magnetite, with $D_{Mo}^{ilm/l} = 0.42$ at the same oxygen fugacity of the graphite–methane buffer, and $D_{Mo}^{ilm/l} = 0.28$ at the oxygen fugacity of the nickel–nickel oxide buffer. The oxygen fugacity dependence of these partition coefficients can probably be extended to more complex crystal–melt equilibria, (Tacker and Candela, 1987; Bouton and Candela, in prep.) and is probably indicative of the oxygen fugacity dependence of the bulk partition coefficient for Mo in more complex natural systems. The results in these model systems therefore indicate that the bulk partition coefficient for molybdenum varies inversely with oxygen fugacity between graphite–methane and nickel–nickel oxide. This, in turn, leads to an increase in the calculated efficiency of removal of Mo with increasing oxygen fugacity. Further, this effect may be compounded by an increase in the *vapor/melt* partition coefficient for Mo if the oxidation state of Mo is higher in the aqueous phase than it is in the melt (Candela, ms); this hypothesis awaits future experimentation.

In the case of tungsten, the results of Bouton et al. (1987) indicate that the *ilm/l* partition coefficient for W does not decrease with increasing oxygen fugacity. There is a significant amount of scatter in the data, but the partition coefficient exhibits a slight, but statistically insignificant, increase with increasing oxygen fugacity. In any event, it is clear that the ratio $\overline{\mathbb{D}}_{Mo}/\overline{\mathbb{D}}_W$ varies inversely with $f_{O_2}$ indicating that, other factors being equal, high magmatic oxygen fugacities should lead to a higher Mo/W ratio in the associated mineralized zones if the results for tungsten can also be applied to more complex mineral/melt systems. The phases most likely to accept Mo and W during crystallization would be Ti-bearing accessories such as sphene, ilmenite, and especially rutile, and phases which contain significant Ti, such as biotite or magnetite.

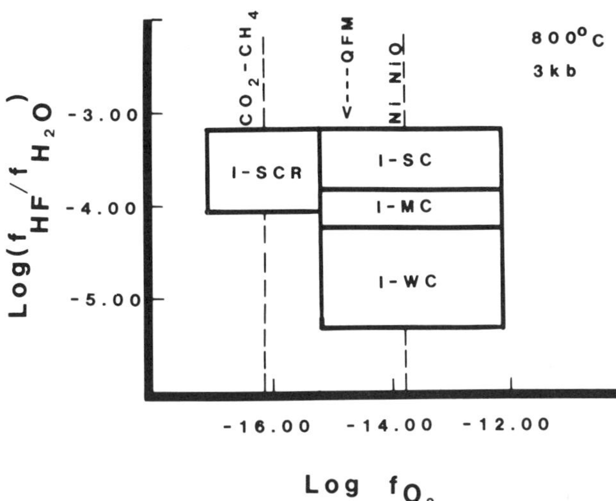


FIGURE 13.4—(After Ague and Brimhall, 1988a). The classification system of Ague and Brimhall plotted on a log fugacity ratio of HF to $H_2O$ vs. log oxygen fugacity. These authors have shown that: 1) some tungsten deposits of the Sierra Nevada are related to I–SCR granitoids; 2) the Climax porphyry Mo deposits are associated with I–SC granitoids; and 3) porphyry copper deposits are generally associated with I–MC granitoids.

The foregoing experimental data can be evaluated in light of data on natural systems, and some of what is presented in the next few section results from the study of Bouton and Candela (in prep). The reduced granitoids of Ague and Brimhall (the I–SCR class) include the intrusions which are spatially associated with the Pine Creek and Strawberry tungsten skarn deposits of the Eastern Sierra Nevada. These deposits are rather reduced, and contamination of the magmas by assimilation has been proposed to account for their low $f_{O_2}$ (see Figs. 13.3 and 13.4) based on their spatial association with graphitic roof pendants. Further, Keith and Van Middlelaar (1987) indicate that a rather high W/Mo ratio occurs in the relatively reduced MacTung and CanTung deposits. However, some tungsten deposits appear to be related to somewhat more oxidized systems although these are though to be smaller than the large reduced deposits mentioned above (Keith and Van Middlelaar (1987). On the other hand, the Climax Molybdenum deposits plot in the more oxidized region of Fig. 13.4 (Ague and Brimhall, 1988a). Additionally, Keith and Shanks (1988) suggest that the ash flow tuff associated with the Pine Grove system, which the authors relate genetically to the mineralizing pluton, yields rather high oxygen fugacities based on coexisting ilmenite–magnetite populations ($f_{O_2} \approx$ QFM + 2). Keith and Shanks (1988) and Keith and Van Middlelaar (1987) also point out that the oxygen fugacities of the topaz rhyolites, which are thought to be genetically related to Climax-type deposits (Burt et al.,

1982), are significantly lower ($f_{O_2} \approx$ QFM; Christiansen et al., 1983) than the oxygen fugacities inferred from data on the Climax systems. A cautionary note is required here. It is possible that oxygen fugacity may also effect the reactions involved in the deposition of W and Mo. The arguments presented in this study suggest a control of $f_{O_2}$ on the efficiency of removal of W and Mo, but they do not necessarily negate other hypotheses for the control of $f_{O_2}$ on metal ratios. This study should be seen as a test of the hypothesis concerning the effect of magmatic $f_{O_2}$ on the sequestering of W relative to Mo.

Porphyry copper deposits are generally associated with plutons which appear to fit the I-type classification (Burnham and Ohmoto, 1980), and the magnetite series classification of Ishihara (1977). This is suggestive of a high oxygen fugacity ($f_{O_2} >$ QFM) in porphyry copper deposits. According to Ague and Brimhall, porphyry copper deposits belong to their moderately contaminated I-type classification, plotting at values of $\log(X_{Mg}/X_{Fe})$ which are generally indicative of oxygen fugacities equal to or greater than NNO.

Experiments have yet to be performed on the partitioning of copper between sulfide minerals and silicate melts, but as I stated in a previous section of this chapter, all of the available evidence suggests that copper behaves as a compatible element during vapor-undersaturated fractional crystallization under a wide range of magmatic conditions, and it is clear that sulfide is the most likely candidate for sequestering copper. Further, it is suggested that if the oxygen fugacity is elevated much above NNO, then pyrrhotite may be destabilized. Therefore, high oxygen fugacities probably lead to elevated efficiencies of removal of Cu (and Au). This line of reasoning is consistent with the conclusions reached by Hattori (1987) concerning the relationship between Archean gold deposits and magnetite-series granites in Canada.

### Summary

According to our experimental results, the crystal incompatibility of molybdenum is enhanced with increasing $f_{O_2}$ (Tacker and Candela, 1987; Bouton et al., 1987). The data also suggest that $\mathbb{D}_{Mo}/\mathbb{D}_{W}$ decreases as $f_{O_2}$ increases (Bouton and Candela, in prep). Field data on the $f_{O_2}$ of magmatic-hydrothermal Mo–W mineralization indicate that W deposits are more reduced than Mo deposits, suggesting that differential sequestration of W and Mo may be an important control on the Mo/W ratio in magmatic-hydrothermal deposits. This is a very general conclusion which must be modified in light of the variable hypersolidus mineral assemblages found in granitoids when considering specific deposits. In as much as sulfides are destabilized at high $f_{O_2}$, the efficiency of removal of copper and possibly Au may also be enhanced under conditions of elevated (>NNO) oxygen fugacity.

## DISCUSSION: SOURCE ROCKS, VOLATILES, AND METALLOGENIC PROVNCES

The formation of a magmatic-hydrothermal mineral deposit involves the effects of processes which include magma generation and contamination crystallization, vapor evolution, mineral deposition and plutonic solidification. These processes appear, on a large scale, to be determinate in nature. The existence of metallogenic provinces is evidence of this. The mineralized zones present in any given geographic area can be related to the tectonic environments to which the given arbitrary volume of crust in question has been subject, and to the level of erosion exposed as the present day surface. As the tectonic processes which have affected the volume element in question has been subject, and to the level of erosion exposed as the present day surface. As the tectonic processes which have affected the volume element in question leave different marks at different levels in the crust, the nature of mineralization which we see is the result of the interplay between tectonics *and* the present stage of erosion in any given area.

The following hypothetical examples summarize how volatile chemistry and source region characteristics combine to yield particular deposit types which serve as motifs in a given metallogenic province. Consider a case wherein mantle-derived magma is periodically input into a large tensional region within old crust over geologic time, and underplates deep crustal granitoids. Partial melting of these rocks may be induced by the presence of the hot, mantle-derived mafic magma, and the deep crustal rocks can therefore be considered as potential source rocks for upper crustal magmas. Further, assume that the source rocks are characterized by:

1) an $f_{O_2} \approx$ NNO;
2) an elevated $F/H_2O$ ratio;
3) a low $Cl/H_2O$ ratio relative to subducted sea-floor basalt; and,
4) a low compatible/incompatible element ratio relative to mantle- or "subducted slab"-derived melts.

These source rock characteristics can be accounted for in the following manner. The oxidized nature of these source rocks relative to the mantle may be attributed to the crustal residence of their protoliths in the crust and their possible derivation from rocks, such as pre-existing oceanic crust, which have been weathered on the surface of the earth (Wones, 1980). This "distillation" from more mafic material would result in a higher ratio of incompatible elements (including Mo, W, and F) to compatible elements (such as copper) in the potential granitic source rocks relative to mafic rocks. Further, if these granitoids have been subjected to either high-grade metamorphism or a previous melting event, then the remaining hydrous minerals would be rather rich in F (Burt, 1981; Holloway, 1977). High $Cl/H_2O$ ratios, such as those reported by Gill (1981) for arc magmas (0.035–0.1 by weight) are probably restricted to rocks derived from the melting of rocks which have been hydrothermally altered by sea water. The deep-crustal granitoids possess lower $Cl/H_2O$ ratios.

Due to the relatively high $F/H_2O$ ratio of the source rocks considered in this example, the magma will experience a rather protracted crystallization history before water saturation. Little copper, relative to Mo and W, will be available for partitioning into the vapor because of the initially low Cu/Mo ratio in the magma relative to magmas derived from the melting of more mafic rocks, *and* the relatively large degree of crystallization before vapor evolution. Additionally, the relatively high oxygen fugacity allows a progressive increase in the Mo/W ratio in the melt during crystallization. During second boiling, some of the copper which remains in the rest-melt is mobilized into the ore fluid, but only to a modest degree as the $Cl/H_2O$ ratio in the source region and hence in the melt is rather low compared to arc-derived magmas. The relatively low Fe, Mg, and Ca concentration in the magma results in little sequestration of elements such as F or B in hydrous phases (Barton, 1987) or tourmaline (London, 1986), respectively. Further, by virtue of its low iron concentration (more precisely, a low $Fe/H_2O$, particularly during later stages when vapor saturation occurs) the melt has a relatively low $f_{O_2}$-buffer capacity. Therefore, the fractionation of $H_2$ into the vapor, which is necessary if the vapor is to maintain oxidative equilibrium with the melt, causes a progressive oxidation of the melt and the associated vapor phase (Candela, 1986b). [Note that this loss of $H_2$ is a necessary condition required by chemical equilibrium; the oft quoted "$H_2$ diffusion" idea, wherein the melt and the associated vapor are oxidized by $H_2$ diffusion out of either phase, is a rather vacuous hypothesis which cannot be demonstrated with any generality.] The resulting oxidation further enhances the partitioning into the aqueous phase of molybdenum relative to tungsten. The result of the foregoing process is a relatively oxidized magmatic-hydrothermal deposit which is rich in Mo and F and other incompatible elements, including subordinate W, but which is poor in Cu. This scenario illustrates how magmatic parameters may affect the formation of porphyry Mo deposits and related systems.

In other regions, the primary melts may be produced by partial melting of subducted, altered oceanic crust or by the fluxing of melting in the mantle wedge above a dehydrating subducted plate. A magma produced in this environment would have a higher ratio of compatible to incompatible elements and a depressed $F/H_2O$ ratio. Further, melts derived from this source possess a higher $Cl/H_2O$ ratio relative to melts produced from the partial melting of deep crustal granitoids. By virtue of the circulation of seawater through hot oceanic crust at the mid-ocean ridges, chlorine is added to altered basalt, where locally it may form high-Cl amphiboles (Vanko, 1986), or may precipitate as chlorides along grain boundaries during low water/rock hydration reactions (Sanford, 1981). Chlorine added to the oceanic crust in this way may be transported into the mantle during subduction, and would become a fugative constituent of the magma. Elevated oxygen fugacities will also be found in this environment, due to the oxidative alteration of the source rock protolith on the seafloor (Wones, 1980). These magmas rise close to the surface in arc setting. The very shallow intrusions of this class experience low degrees of crystallization prior to vapor evolution. Therefore, little copper could be sequestered into the solids before vapor evolution, and second boiling under these conditions leads to a high $E(Cu)/E(Mo)$ ratio. Deposits formed under these conditions would be rich in copper relative to Mo or W, would be relatively oxidized (due to the oxidized nature of the source region), and would not be very fluorine rich.

Variations on this theme are numerous. The assimilation of sedimentary material by the above mentioned magmas would produce melts which are more aluminous, higher in their initial $H_2O$ concentrations (Burnham, 1981), and possibly lower in $f_{O_2}$ (if graphitic sediments are involved) relative to the magmas derived from the melting of igneous precursors. The lower oxygen fugacities would lead to an increase in $E(W)/E(Mo)$. Further, the relationship between the $Sn^{4+}/Sn^{2+}$ equilibria and the reduced (I–SCR) type granitoids of Ague and Brimhall (1988a), (See Fig. 13.3) suggests that Sn may be present in the divalent state in low $f_{O_2}$ melts. Under these conditions, Sn may behave as an incompatible element which in turn, may yield a higher $E(Sn)$ than at elevated oxygen fugacities. However, much more work on Sn is necessary before we can speak with any certainty of its behavior in magmatic-hydrothermal systems. Assimilation of metasedimentary material by primitive I-type magmas probably results in a monotonically decreasing $Cl/H_2O$ ratio (Ague and Brimhall, 1988b) as a high $Cl/H_2O$ arc-derived melt is progressively contaminated with lower $Cl/H_2O$ metasedimentary materials. This would tend to reduce the efficiency of removal of chloride-complexed cations.

When viewed on a small scale in either space or time, igneous or hydrothermal activity appears almost indeterminate. It is when one broadens one's scope of view to the province scale that these processes appear determinate. Many factors act in concert to produce the variations in metal ratios observed in deposits of different metallogenic provinces: $\{C_w^{l,o}/C_w^{l,s}\}$, $Cl/H_2O$, $F/H_2O$ and $f_{O_2}$ and the initial melt ratios of the magma and the source region. However, within any given volume of crust, and at any given erosion level within that crust, the confluence of the geological forces acting over time produce a common theme: Mo deposits in the Climax–Henderson area of the Colorado Mineral Belt, porphyry copper deposits in the southwestern U.S., etc. In this simplified discussion I have neglected the subtleties of partial melting, considerations of batch versus fractional melting, the stability of ore-metal host phases in the source region, etc. Further, this chapter has been skewed toward magmatic effects on the chemistry of ore systems. This should not be interpreted to mean that the purely hydrothermal aspects of ore formation are superfluous. In many cases, the absence of a given element in a deposit may result from non-deposition in the ore zone rather than its absence from the ore fluid. Rather, I have tried to develop a generalized picture of the magmatic processes involved, and I have touched on some of the hypotheses which have been suggested to explain the variations in metal chemistry found within the class of deposits we call magmatic-hydrothermal. Probably no single factor is important in producing the characteristics of the deposits discussed in this paper, or the metallogenic provinces characterized by these deposits. For example, variations in initial metal ratios of source regions probably cannot account for the variations found in the metal ratios in ore deposits. Halogen and water concentrations in the melt, depth, initial metal concentrations in the melt, and oxidation state all work in concert to produce the bulk chemistries we find in the large, world class magmatic-hydrothermal deposits.

ACKNOWLEDGEMENTS—I would like to thank the present and past graduate students of the Laboratory for Mineral Deposits Research at the University of Maryland, including R. Chris Tacker, Steve Bouton, Phil Piccoli, and Hulfang Liu, who have contributed to our understanding of the chemistry of the magmatic aqueous phase and the origin of magmatic-hydrothermal ore deposits, and who have contributed in many ways to the ideas expressed herein. I would like to thank my family for their support while writing this paper, especially my wife Mary, who also provided editorial assistance. Phil Piccoli and Hulfang Liu are also gratefully acknowledged for their assistance in putting this paper together. I am also greatly indebted to Jay Ague and Jim Webster who very kindly provided me with preprints of their recent papers. I am grateful to Jeff Keith for an insightful review of this paper. The support of the Magma Energy Project of the Department of Energy, and the National Science Foundation (Grants EAR–8319101 and EAR–8706198) is gratefully acknowledged.

## REFERENCES

Ague, J. J., and Brimhall, G. H, 1987, Granites of the batholiths of California: products of local assimilation and regional-scale crustal contamination: Geology, v. 15, pp. 63–66.

Ague, J. J., and Brimhall, G. H., 1988a, Diverse regional controls on magmatic arc asymmetry and distribution of anomalous plutonic belts in the batholiths of California: Effects of assimilation, cratonal rifting, and depth of crystallization: Geological Society of America, Bull. (in press).

Ague, J. J., and Brimhall, G. H., 1988b, Regional variations in bulk chemistry, mineralogy, and the compositions of mafic and accessory minerals in the batholiths of California: Geological Society of America, Bull. (in press).

Anderson, A. T., 1974, Chlorine, sulfur, and water in magmas and oceans: Geological Society of America, Bull., v. 85, pp. 1485–1492.

Andiambololona, R., and Dupuy, C., 1978, Repartion et comportement des elements de transition dans les roches volcaniques I. Cuivre et zinc: Bur. Recherche Geol. Min. [Paris] Bull., v. 2, pp. 121–138.

Barton, M. D., 1987, Lithophile-element mineralization associated with Late Cretaceous two-mica granites in the Great Basin: Geology, v. 15, pp. 337–340.

Bouton, S. L., Candela, P. A., Weidner, J. R., and Joseph, J., 1987, Experimental determination of the partitioning of tungsten and molybdenum between a high silica melt and Ti-bearing minerals: Geological Society of America, Abstracts with Programs, v. 19, p. 596.

Buddington, A. F., and Lindsley, D. H., 1964, Iron–titanium minerals and synthetic equivalents: Journal of Petrology, v. 5, pp. 310–357.

Burnham, C. W., 1981, Physiochemical constraints on porphyry mineralization; *in* Dickenson, W. R., and Payne, W. D. (eds.), Relations of tectonics to ore deposits in the southern Cordillera: Arizona Geological Society Digest, v. 14, pp. 71–77.

Burnham, C. W., 1979, Magmas and hydrothermal fluids; *in* Barnes, H. L. (ed.), Geochemistry of hydrothermal ore deposits: New York, John Wiley and Sons, 2nd ed., pp. 71–136.

Burnham, C. W., and Ohmoto, H., 1980, Late-stage processes of felsic magmatism: Min. Geol., Spec. Issue, v. 8, pp. 1–11.

Burt, D. M., Bikun, J. V., and Christiansen, E. H., 1982, Topaz rhyolites—distribution, origin, and significance for exploration: Econ. Geol., v. 77, pp. 1818–1836.

Candela, P. A., 1989, Theoretical constraints on the chemistry of the magmatic aqueous phase; *in* Stein, H. J., and Hannah, J. (eds.), Granites and related ore deposits: Geological Society of America, Spec. Paper, accepted.

Candela, P. A., 1986, The evolution of aqueous vapor from silicate melts: Effect on oxygen fugacity: Geochim. Cosmochim. Acta, v. 50, pp. 1205–1211.

Candela, P. A., 1982, Copper and molybdenum in silicate melt–aqueous fluid systems: Unpublished Ph.D. thesis, Harvard University, Cambridge, Massachusetts, 138 pp.

Candela, P. A., and Holland, H. D., 1986, A mass transfer model for copper and molybdenum in magmatic hydrothermal systems: the origin of porphyry-type ore deposits: Geochim. Cosmo. Acta., v. 81, pp. 1–19.

Carmichael, I. S. E., Turner, F. J., and Verhoogen, J., 1974, Igneous petrology: New York, McGraw–Hill, 739 pp.

Carroll, M. R., and Rutherford, M. J., 1987, The stability of igneous anhydrite: experimental results and implications for sulfur behavior in the 1982 El Chichon trachyandesite and other evolved magmas: Journal of Petrology, v. 28, pp. 781–801.

Chappel, B. W., and White, A. J. R., 1974, Two contrasting granite types: Pacific Geology, v. 8, pp. 173–174.

Christiansen, D. H., Burt, D. M., Sheridan, M. F. and Wilson, R. T., 1983, Petrogenesis of topaz rhyolites: Contrib. Mineral. Petrol., v. 83, pp. 16–30.

Drabek, M., 1982, The system Fe–Mo–S–O and its geologic application: Econ. Geol., v. 77, pp. 1053–1056.

Geraghty, E. P., Carten, R. B., and Walker, B. M., 1987, Northern Rio Grande rift tectonics: tilting of Urad–Henderson and Climax porphyry molybdenum deposits: Geological Society of America, Abstracts with Programs, v. 19, p. 277.

Gill, J. B., 1981, Orogenic andesites and plate tectonics: New York, Springer, 390 pp.

Gunow, A. J., Ludington, S., and Munoz, J. L., 1980, Fluorine in micas from the Henderson molybdenite deposit, Colorado: Econ. Geol., v. 75, pp. 1127–1137.

Hammarstrom, J. M., and Zen, E–an, 1986, Aluminum in hornblende: an empirical igneous geobarometer: American Mineralogist, v. 71, pp. 1297–1313.

Hannah, J. L., and Stein, H. J., 1986, Oxygen isotope compositions of selected Laramide–Tertiary granitoid stocks in the Colorado Mineral Belt and their bearing on the origin of Climax-type granite–molybdenum systems: Contrib. Mineral. Petrol., v. 93, pp. 347–358.

Hattori, K., 1987, Magnetic felsic intrusions associated with Canadian Archean gold deposits: Geology, v. 15, pp. 1107–1111.

Holloway, J. R., 1977, The effect of fluorine on dehydration equilibria: Geological Society of America, Abstracts with Programs, v. 9, p. 1021.

Ishihara, S., 1977, The magnitite-series and ilmenite-series granitic rocks: Mining Geology, v. 27, pp. 293–305.

Johnson, W. D., 1965, Oxidation–reduction equilibria in molten $Na_2O$–$SiO_2$ glass: J. Amer. Ceram. Soc., v. 48, pp. 184–190.

Keith, J. D., and Shanks, W. C., III, 1988, Chemical evolution and volatile fugacities of the Pine Grove porphyry molybdenum and ash flow tuff system, southwestern Utah: *in* Taylor, R. P., and Strong, D. F., eds., Recent advances in the geology of granite-related mineral deposits: Canadian Institute of Mining and Metallurgy, Special Volume 39, pp. 402–423.

Keith, J. D., and Van Middelaar, W. T., 1987, Inferred oxygen and halogen fugacities of granitoids related to W and Mo deposits in the North American Cordillera: Geological Society of America, Abstracts with Programs, v. 19, p. 723.

London, D., 1986, Magmatic-hydrothermal transition in the Tanco rare-element pegmatite: Evidence from fluid inclusions and phase-equilibrium experiments: Am. Min., v. 71, pp. 376–395.

Luhr, J. F., Carmichael, I. S. E., and Varekamp, J. C., 1984, The 1982 eruptions of El Chichon volcano, Chiapas, Mexico: Mineralogy and petrology of the anhydrite-bearing pumices: J. Volcan. Geoth. Research, v. 23, pp. 69–108.

Munoz, J. L., 1984, F–OH and Cl–OH exchange in micas with applications to hydrothermal ore deposits; *in* Bailey, S. W. (ed.), Micas: Min. Soc. Am., pp. 469–493.

Newberry, R. J., and Swanson, S. E., 1986, Scheelite skarn granitoids: An evaluation of the roles of magmatic source and process: Ore Geology Reviews, v. 1, pp. 57–81.

Sanford, R. F., 1981, Mineralogical and chemical effects of hydration reactions and applications to serpentinization: Am. Min., v. 66, pp. 290–297.

Tacker, R. C., and Candela, P. A., 1987, Partitioning of molybdenum between magnitite and melt: A preliminary experimental study of partitioning of ore metals between silicic magmas and crystalline phases: Econ. Geol., v. 82, pp. 1827–1838.

Taylor, B. E., Eichelberger, J. C., and Westrich, H. R., 1983, Hydrogen isotopic evidence of rhyolite magma degassing during shallow intrusion and eruption: Nature, v. 306, pp. 541–545.

Tsusue, A., Nedachi, M., and Hashimoto, K., 1981, Geochemistry of apatites in the granitic rocks of the molybdenum, tungsten and barren provinces of southwest Japan: J. Geochem. Explor., v. 15, pp. 285–294.

Urabe, T., 1987, The effect of pressure on the partitioning ratios of lead and zinc between vapor and rhyolite melts: Econ. Geol., v. 82, pp. 1049–1052.

Vanko, D. A., 1986, High-chlorine amphiboles from oceanic rocks: product of highly saline hydrothermal fluids?: Am. Min., v. 71, pp. 51–59.

Varekamp, J. C., Luhr, J. F., and Prestegaard, K. L., 1984, The 1982 eruptions of El Chichon Volcano (Chiapas, Mexico): Character of the eruptions, ash-fall deposits, and gas phase: J. Volcanol. Geotherm. Res., v. 23, pp. 39–68.

Webster, J. D., Holloway, J. R., and Hervig, R. L., 1987, Phase equilibria of a Be, U and F-enriched vitrophyre from Spor Mountain, Utah: Geochim. Cosmochim. Acta., v. 51, pp. 389–402.

White, W. H., Bookstrom, A. A., Kamilli, R. J., Ganster, M. W., Smith, R. P., Ranta, D. E., and Steininger, R. C., 1981, Character and origin of Climax-type molybdenum deposits: Econ. Geol., 75th Anniv. Vol., pp. 270–316.

Whitney, J. A., and Stormer, J. C., Jr., 1983, Igneous sulfides in the Fish Canyon Tuff and the role of sulfur in calc-alkaline magmas: Geology, v. 11, pp. 99–102.

Wones, D. R., 1981, Mafic silicates as indicators of intensive variables in granitic magmas: Min. Geol., v. 34, pp. 191–212.

Woodcock, J. R., and Hollister, V. F., 1978, Porphyry molybdenite deposits of the North American cordillera: Minerals Sci. Eng., v. 10, pp. 3–18.

## Appendix I

### Calculation of $f_{O_2}$ in plutonic rocks by the method of Ague and Brimhall, 1988

The value of $\log f_{O_2}$ for reaction [3] is given by

$$\log f_{O_2} = 2\log a_{\mathrm{hm}}^{\mathrm{ilmenite}} + 4/3\log a_{\mathrm{kspar}}^{\mathrm{feld}} - 4/3\log a_{\mathrm{ann}}^{\mathrm{biot}} - \log K + 4/3\log a_{\mathrm{w}},$$

where

1) $a_{\mathrm{hm}}^{\mathrm{ilmenite}} = hX_{\mathrm{hm}}^{\mathrm{ilmenite}}$ with $\log h = \big(1332/T(\mathrm{K})\big) - 1.124$ (Nakada, 1983);
2) $a_{\mathrm{kspar}}^{\mathrm{feld}} = 0.9$;
3) $a_{\mathrm{w}} = 1$;
4) an ideal model is used for calculating the activity of annite; and
5) $\log K = \big(20977/T(\mathrm{K})\big) - 5.73$ (at 3kb).

When the above substitutions are made, and data from biotite and ilmenite compositions are included, an equation results in which $\log f_{O_2}$ is expressed as a function of $T$. The resulting curve can be plotted on a Buddington-Lindsley diagram, and the $\log f_{O_2} - T$ where the curve intersects the appropriate ilmenite isopleth gives the calculated conditions of formation.

## Chapter 14

# Granitoid Textures, Compositions, and Volatile Fugacities Associated with the Formation of Tungsten-dominated Skarn Deposits

J. D. Keith, W. van Middelaar, A. H. Clark, and C. J. Hodgson

### INTRODUCTION

Perhaps a dominant theme that has been expressed in the last few years concerning the petrogenesis and metallogeny of granitoids in western North America concerns the *relative proportions* and *composition* of various types of crustal material that have been incorporated in batches of magma (Keith et al., 1985a; Barton, 1987; Farmer and DePaolo, 1983; Ague and Brimhall, 1985, 1986, 1987; Christiansen et al., 1986; Stein and Hannah, 1985; Newberry and Swanson, 1986). Many workers suggest that these processes of crustal melting or contamination by crustal components exert one of the strongest controls on the character or grade of mineralization that ultimately may be produced by that magma. However, broad distinctions between classes of mineralization (i.e. porphyry Cu versus W skarn) may largely be a function of depth of crystallization and correlative water content of the magma (Einaudi et al., 1981; Burt et al., 1982; Barton et al., 1988; Newberry and Swanson, 1986). The controversy over how metal ratios in a deposit are related to granitoid composition is particularly applicable to W, Mo, and Sn deposits because of their close spatial association to granitoids and strong evidence that the dominant portion of the metals and the hydrothermal fluid are derived from the magma. This being the case, the question that must be asked is which magmatic characteristics are set (or modified) during assimilation of crustal material, that have some control on the metal ratios or grade of mineralization, versus those which are "inert" in terms of effecting mineralization. Perhaps the leading answers to this question that have been proposed are: 1) high concentrations of metals (W, Mo, or Sn); 2) high or low magmatic water content; 3) an appropriate depth of crystallization; 4) a geochemically "specialized" magma (usually implying a granitic major element composition and various trace element enrichments); 5) appropriate fugacities of F, Cl, or O; or 6) combinations of the above.

In order to address this question we (Keith et al., 1985b; van Middelaar, 1988) have examined the field relations of plutons and petrography of granitoids associated with most of the major W-skarn deposits (~30)in western North America, including the granitoids adjacent to the 5 large deposits that are capable of production (given slightly improved markets for W): CanTung, MacTung, Pine Creek, Springer (Mill City), and Emerson (Tem Piute) (Fig. 14.1). The objective of this short course and paper is to help define alteration and magmatic processes and granitoid compositions which may be fundamental to the development of *unusually large and/or high-grade* W-skarn deposits. As part of the contrast needed in our initial study, some "barren" and Mo-related granitoids were examined, as well as some very small, low-grade tungsten skarn systems. Approximately 130 samples of granitoids were selected from the spectrum of deposits with particular regard for defining fresh, representative suites. During the last year, we have also completed a more detailed study of the granitoids (120 samples; van Middelaar, 1988; van Middelaar and Keith, 1988) related to the CanTung deposit, Northwest Territories (the highest grade deposit in North America) which complements the previous work done on this deposit (Zaw and Clark, 1978; Mathieson and Clark, 1984; Bowman et al., 1985).

### FIELD RELATIONS OF PLUTONS

Several lines of evidence and reasoning suggest that the ore fluid is dominantly magmatic (or "intrusive" as discussed later) for most tungsten skarns. If this is true, then some evidence of fluid transport through the pluton carapace (defined as the early crystallized rind of the pluton), by means of either veins and fractures or intracrystalline transport, should be present (Mathieson and Clark, 1984). Abundant evidence from field relationships noted in our studies (and from others) indicate that the magmatic fluid may transect a large volume of granitoid carapace prior to skarn formation. Such evidence of the pathways of magmatic fluids, previously, has not been examined and summarized for the spectrum of W-skarn-related granitoids of western North America (although some data are presented by Newberry, 1982 and Kerr, 1946 and others).

It was noted during our studies that fluid transport by means of veins and fractures or along grain boundaries can only be correlated roughly with independent estimates of depth of emplacement (Barton et al., 1988) and other features summarized below. Synthesis of the data concerning

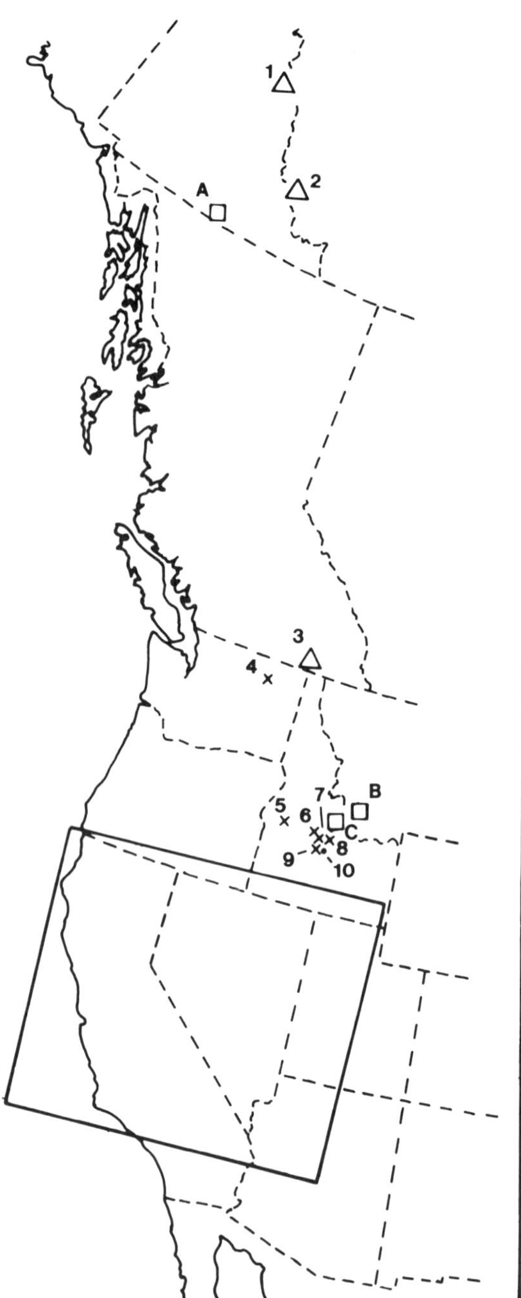


△ Large Deposit
○ Moderate Deposit
• Small Deposit
× Trivial Deposit or Barren
□ Large Deposit not examined

1. Mactung
2. Cantung
3. Emerald
4. Mt. Tolman
5. Idaho Batholith-1
6. Idaho Batholith-2
7. White Cloud
8. Trail Creek
9. Little Falls
10. Wildhorse
11. Bloody Run
12. Osgood Mtns.
13. Rose Creek
14. Mill City
15. Oreana
16. Nightingale
17. Ragged Top
18. Toy
19. Quik-Tung
20. N. Sierras-1
21. N. Sierras-2
22. Deep Creek
23. Tungstonia
24. Cherry Creek
25. Monte Cristo
26. Nevada Scheelite
27. Osceola
28. Notch Peak
29. Pilot Peak
30. Black Rock
31. Round Valley
32. Pine Creek
33. Little Sister
34. Tem Piute
35. Garnet Dike
36. Consolidated
37. Tungstore
38. Atolia
39. Shadow Mtn.
40. Mirage Lake

A. Logtung
B. Brown's Lake/Lost Creek
C. Tungsten Jim
D. Strawberry

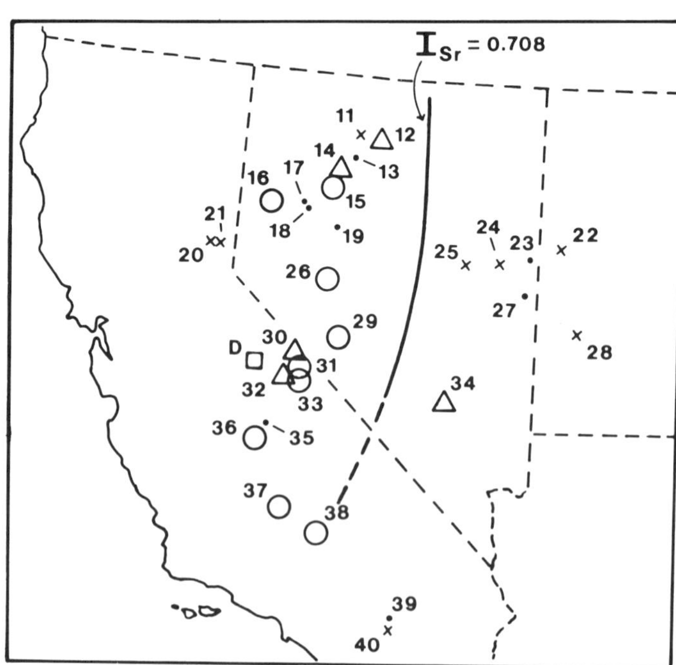

FIGURE 14.1—Location of plutons in western North America examined and sampled for this study. Deposits of tungsten associated with these plutons are ranked according to tonnage (production + reserves); large deposits consist of >500,000 tons, moderate deposits contain <500,000 and >15,000 tons; and small deposits are <15,000 tons ore. The $Sr_i$ = 0.708 line (Farmer and De Paolo, 1983) separates plutons to the east with higher initial strontium isotope values (underlain by Precambrian crust) from plutons to the west with lower values (little or no Precambrian crust). The following is a list of sample prefixes (for this report) and correlative location names and numbers used on this index map as well as references which pertain to each deposit.

| Sample Prefix | Map Name, State, and Number | | Reference |
|---|---|---|---|
| Atol | Atolia, Calif. | 38 | Kerr, 1946 |
| Black | Black Rock, Calif. | 30 | Lemmon, 1941 |
| Bldy | Bloody Run, Nev. | 11 | Kerr, 1946 |
| Can | CanTung, N.W.T. | 2 | Mathieson and Clark, 1982 |
| Cher | Cherry Creek, Nev. | 24 | Kerr, 1946 |
| Cons | Consolidated, Calif. | 36 | Newberry, 1980 |
| Deep | Deep Creek, Ut. | 22 | Kerr, 1946 |
| Emer | Tem Piute, Nev. | 34 | Buseck, 1967 |
| Garn | Garnet Dike, Calif. | 35 | Newberry, 1980 |
| Gold | Gold Hill, Ut. | | Kerr, 1946 |
| Idaho | Idaho Batholith | 5–6 | |
| Kern | Tungstonia, Ut. | 23 | Best et al., 1974 |
| Litt | Little Falls, Idaho | 9 | Cook, 1956 |
| Mac | MacTung, Yukon, N.W.T. | 1 | Anderson, 1983 |
| Mill | Mill City, Nev. | 14 | Kerr, 1934 |
| Mont | Monte Cristo, Nev. | 25 | Sonnevil, 1979 |
| Nite | Nightingale, Nev. | 16 | Smith and Guild, 1942 |
| Notc | Notch Peak, Ut. | 28 | Kerr, 1946 |
| Orea | Oreana, Nev. | 15 | Kerr, 1938 |
| Osgo | Osgood Mtns., Nev. | 12 | Silberman et al., 1974 |
| Pilot | Pilot Peak, Nev. | 29 | Kerr, 1946 |
| Pine | Pine Creek, Calif. | 32 | Newberry, 1982 |
| Quik | Quik-Tung, Nev. | 19 | Sigurdson, 1974 |
| Ragt | Ragged Top, Nev. | 17 | Kerr, 1946 |
| Rose | Rose Creek, Nev. | 13 | Roberts, 1943 |
| Roun | Round Valley, Calif. | 31 | Bateman, 1965 |
| Sasp | Nevada Scheelite | 26 | |
| Shad | Shadow Mtn., Calif. | 39 | Kerr, 1946 |
| Shaw | Shawave Mtns., Nev. | | |
| Sier | N. Sierras -1, -2 | 20–21 | |
| Stan | Toy, Nev. | 18 | Kerr, 1946 |
| Tolm | Mt. Tolman, Wash. | 4 | |
| Tral | Trail Creek, Idaho | 8 | Cook, 1956 |
| Tung | Little Sister, Calif. | 33 | Bateman, 1965 |
| Tust | Tungstore, Calif. | 37 | Kerr, 1946 |
| White | White Cloud, Idaho | 7 | Schmidt et al., 1979 |
| Wild | Wildhorse, Idaho | 10 | Cook, 1956 |

field relations, pluton size, granitoid texture and alteration suggests that classification of W-related plutons into three categories may be appropriate. These categories *may* be fundamentally related to distinct variations in depth of emplacement or variations in magmatic water content. Variations in these two factors may be significant enough to have influenced the timing of fluid release and how intimately the magmatic fluid transects the carapace and deposits or leaches W and other metals in the process. However, absolute values for depth of emplacement are often the least well known data for any pluton; consequently, depth of emplacement is not a fundamental criteria used in our classification scheme (although some speculations are included). A summary of our classification scheme (Keith et al., 1985b) for plutons associated with W skarn deposits of the western Cordillera is presented below.

Petrographic and field observations suggest that there are three general types of plutons associated with tungsten skarn deposits in the western Cordillera. Those plutons most similar to the Cirque Lake stock of MacTung are classified as one type (Fig. 14.2). The most important characteristic of this type is that a small (~1.0 km in diameter), steep-walled, highly fractured granodiorite pluton shows the closest proximity to the skarn. Megascopic veins and fractures may be as closely spaced as 10–20 cm apart with near random orientation. Fractures and veins may be lined with pyrite, tourmaline, and muscovite as well as sparse molybdenite and scheelite (common components with no divisions of assemblages inferred). These relationships may cause Mo and W whole-rock analyses of the stock to be generally elevated and have prompted some mine geologists to refer to the scheelite and molybdenite as being "disseminated". How-

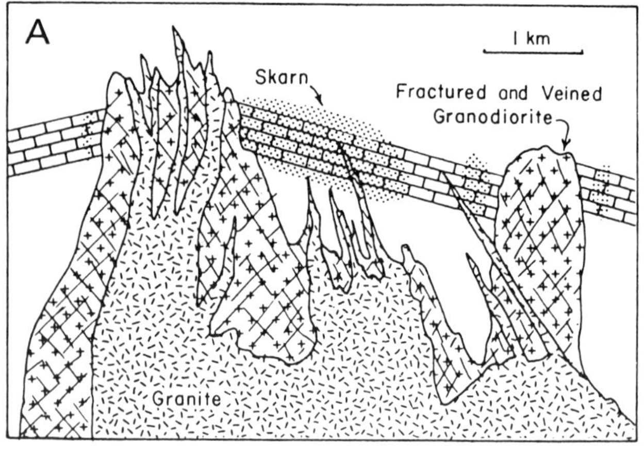


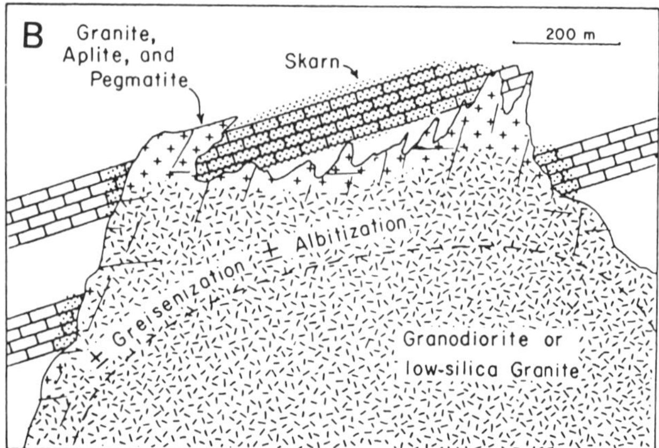


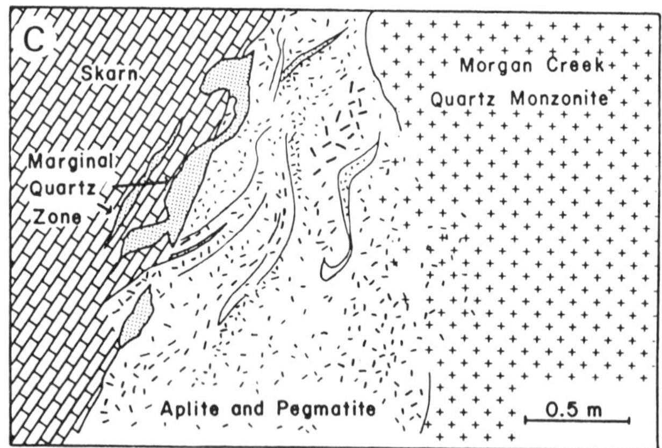


FIGURE 14.2—Characteristic field relations of plutons associated with W skarn deposits. **A.** Cartoon section of a granodiorite-type pluton associated with tungsten skarn. Important characteristics of this pluton type are that: 1) highly fractured and veined granodiorite dikes and stocks show the closest proximity to skarn; 2) the skarn may be located a substantial distance from the granitoid that provided the magmatic fluid; 3) a younger granitic pluton is emplaced within the granodiorite stock or adjacent to it; and 4) dikes of each phase may parallel each other, but at the level of skarn formation the less differentiated granitoid is volumetrically more important. Fractures and veins exert the greatest control on location of skarn, but the source of the fluids may have been the granite intrusion. **B.** Cartoon section of a batholith-type pluton associated with tungsten skarn. The pluton is not strongly veined, but the veins that are present may have transported some of the ore fluid. The marginal lecocratic zone is best developed in apophyses or irregularities along the contact and may show close proximity to skarn. Possibly skarn is not as well developed along extended portions of the contact that are devoid of the leucocratic zone. A carapace of greisenization and albitization may often be present along the upper portion of the pluton. **C.** Schematic section of the marginal quartz zone and marginal leucocratic zone in the Pine Creek mine. The marginal leucocratic zone consists of mixed aplites, pegmatites, and leuco-granodiorite. Texture, composition, and width of this zone is extremely variable. It varies from 0–30 m thick and may or may not be gradational with typical Morgan Creek granodiorite. Pegmatitic zones occasionally contain disseminated molybdenite, pyrite, and chalcopyrite. Flow foliation is common, but not persistent. Internal contacts and significant variations in biotite content are also present in the leucocratic zone. The marginal quartz zone has an irregular distribution, but is generally found near the contact.

ever, the distribution of mineralization is almost certainly controlled by small fractures and veinlets. Commonly, small apophyses or dikes of a less-veined, more-differentiated granitoid phase are also present crosscutting, or adjacent to, the strongly veined stock. Ore fluids were probably derived from a larger underlying volume of the more-differentiated granitoid phase and subsequently channeled along fractures and veins in the carapace to reach the zone of skarning. Geologists working in southeast China (Wu Yongle and Mei Yongwen, 1982; Liu Jiayuan and Shen Jili, 1982) recently have reported a very similar type of pluton associated with scheelite mineralization which they have aptly named a "granodiorite", "transitional" or "porphyry" type. Plutons that correspond to this model are those associated with the MacTung, Tem Piute, Pilot Peak, Mill City, and Rose Creek deposits among others (Fig. 14.1). According to the estimated depths of emplacement of Barton et al. (1988) for Mesozoic plutons in the western United States, most of these W skarn systems would be shallow (4–8 km) relative to other Mesozoic plutonism related to W skarns. The applicability of the terms "porphyry" or "granodiorite" seem appropriate (and are used for this class of pluton in the remainder of this report), because the stock tends to be more mafic and shallowly emplaced than the average W-skarn related granitoid (common for shallow Mesozoic intrusions; Barton et al., 1988). In addition, the size and shape of the stock, the amount of veining, and to some extent, the metal ratios are analogous to porphyry Cu–Mo systems.

Plutons associated with CanTung and Pine Creek define another major type of tungsten skarn-related intrusion referred to as the "batholith" type in this report (Fig. 14.2). Characteristically these plutons comprise a portion of a more extensive batholith and are therefore less stock like and less veined and altered. They are granodiorite or low-silica granite (68–72% SiO2; Keith et al., 1985b) plutons that often exhibit marginal leucocratic granitoids, aplites, or pegmatites and/or quartz zones along the walls or roof of the intrusion (Fig. 14.2). Quartz veins and aplite dikes crosscutting the pluton are generally not abundant (CanTung having more than usual), but greisening, albitization, or myrmekitic-granoblastic quartz fabrics are often present. Some of these features may indicate that magmatic fluids related to mineralization moved predominantly through a network of high-temperature, sub-solidus crystal interstices and a smaller proportion of the fluid migrated through distinctly later veins and fractures. Deposits at Black Rock, Tungsten Hills, Round Valley, Ragged Top, Tungstore, Consolidated, and possibly Emerald also have plutons of this type (Fig. 14.1). A wide range of depths of emplacement are probably represented in this category (although some are distinctly deeper, ~8–12 km); However, the association of plutons in this group with batholiths may provide a slower cooling history (due to heat added to the crust from neighboring plutons) and less rapid volatile release which might inhibit development of abundant fractures and veins (related to volatile collection and dissipation). Consequently, "absolute" depth of emplacement may be less critical than the relationships between cooling history and fluid- and rock-pressures in creating these veining and textural relationships.

Finally, two-mica granitoids that are related to a distinctive suite of lithophile-element (Be, F, W, Mo, Sn, Zn) mineralization in the Great Basin (not illustrated in Fig. 14.2) are generally emplaced at a deeper level than the previous types (Barton et al., 1988; Barton, 1987). This deep level of emplacement may be related to the parental magma being cooler, wetter, and crustally derived. This type may be equated with the muscovite-bearing "crustal-type" granite pluton identified as the main alternative to the "granodiorite" type in southeast China (Wu Yongle and Mei Yongwen, 1982; Liu Jiayuan and Shen Jili, 1982). This type has not proven to be as economically important in the United States; however, the Cirque Lake stock associated with the MacTung deposit exhibits some similarities to this type as well.

## GRANITOID COMPOSITIONS

Isotopic evidence and field relations indicate that the dominant portion of the metals and the hydrothermal fluid related to W-skarn deposits are derived from a nearby granitoid pluton. However, the wide range of granitoid compositions which occur in plutons adjacent to W deposits (data from Keith et al., 1985b plotted and discussed in Kwak, 1987) makes the assertion seem plausible that granitoid composition is *not* an important control on mineralization (Newberry and Swanson, 1986). Inspection of the differentiation indices for productive plutons from our sample suite reveals that a wide range is present (roughly from gabbro to granite, Fig. 14.3). Not surprisingly, the gross compositional range of granitoids from our sample suite which are barren of W mineralization, or related to only small deposits, mimics the range of the more productive granitoid suite very well. Consequently, a most critical question that is being examined in our work is what *more subtle* relationships, if any, exist between magma composition and high-grade W mineralization.

The field relationships of the "porphyry-type" W-related plutons often lend themselves to a generalization that the younger, more differentiated member of a granitoid suite is the most probable source of the magmatic component of the ore fluid (although important exceptions exist). Precisely what magmatic characteristics are attained by the differentiated end products that may be critical to ore formation are poorly known (but widely speculated about). In addition, whether the more differentiated granitoids achieve increased ore-forming capabilities by means of processes such as crystal fractionation or progressive crustal contamination must be investigated. For example, Newberry and Swanson (1986) discount the relative importance of assimilation of crustal material in effecting the mineralizing ability of Cordilleran granitoids. They correctly point out that no sedimentary country rocks (yet found in North America) are abnormally enriched in W enough to cause a significant change in magmas which assimilate them (controversy exists over potential sedimentary tungsten enrichment elsewhere in the world such as the Rwanda–SW Uganda belt and the Sea of Okhotsk). In addition, they note that no clear pattern of granitoid $Sr_i$ and skarn size or abundance exists. However, the recent work of Ague and Brimhall (1987; see review by Candela, this volume) documents changes in magmatic accessory minerals and volatile fugacities which accompany progressive crustal contamination in the Sierra Nevada batholith. Such mineralogical and volatile fugacity variations may have subtle influences on the grade of mineralization (as discussed later) rather than skarn location or size (Keith et al., 1985a).

## LITHOPHILE METAL CONTENT OF GRANITOIDS

The possible importance of magmatic W enrichment by means of crystal fractionation and crustal contamination should be considered. Data concerning the W content of granitoids and sedimentary rocks from south China indicate that both are substantially enriched in W compared to worldwide averages (Liu Yimao et al., 1982; Yu Shoujun, 1982). However, in cases such as this, it is difficult to decipher whether the high values are related to "disseminated" mineralization or initially high magmatic values resulting from long-term W anomalies in the crust or mantle. Although the later interpretation is favored by many authors, if primary metal-rich magmas exist, then examples of fresh metal-rich

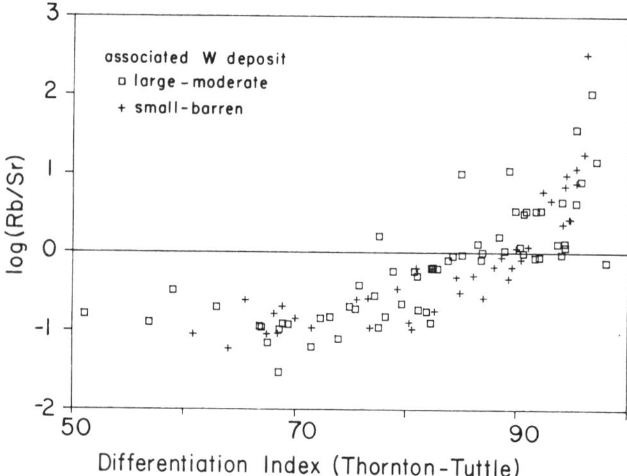


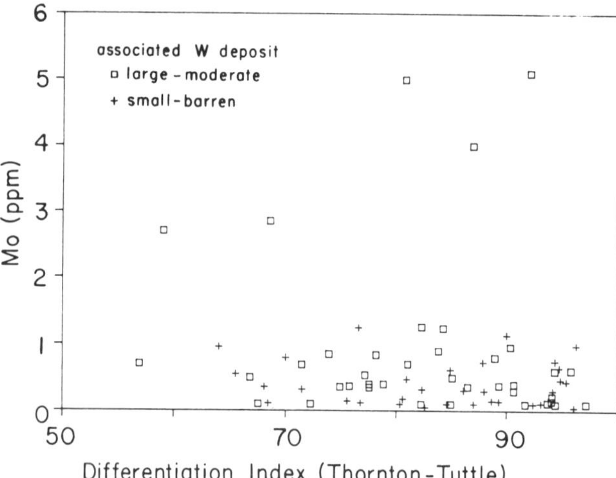


FIGURE 14.3—Whole-rock Rb/Sr ratios and Mo content versus the Thornton–Tuttle differentiation index of North American granitoid suite. Granitoids associated with moderate to large deposits (squares) are contrasted with granitoids associated small deposits or barren of mineralization (crosses). Values above 2 ppm Mo are from visibly mineralized samples. See FIG. 14.1 for summary of tonnage of large, moderate, and small deposits. Petrographic descriptions, sample locations, and major- and trace-element composition of each sample are available upon request from the authors.

volcanic rocks should also occasionally be found. However, such metal-rich volcanic rocks with major-element compositions appropriate for correlation with W-related granitoids are rare or nonexistent. The unusually lithophile-rich ash-flow tuffs and volcanic glass from the Macusani volcanic field in Peru and topaz rhyolites from the western United States have been variously proposed as being the extrusive equivalents of Sn, W, or Mo granitoids (Clark et al., 1987; Noble et al., 1984; Burt et al., 1982; Christiansen et al., 1986). Clark et al. (1987) have documented that the Palca XI W-polymetallic vein system (one of the world's richest W deposits) is genetically related to the Macusani Tuff. Pichavant et al. (1987) have found elevated concentrations of W (60–90 ppm) and Sn (150–200 ppm; also Smith, 1985) in nonhydrated glass from Macusani. Although these strongly peraluminous, crustally derived volcanic rocks (and associated sillimanite-rich granites) do not have appropriate bulk compositions for correlation with typical W-related granitoids, they document that significant magmatic W enrichment can occur. They may represent end-member extremes of crustal contamination and crystal fractionation processes that may lead to magmatic W enrichment.

Magmatic enrichment of W in metaluminous systems may be more modest than in Macusanite and the enrichment processes may operate predominantly after 50 percent crystallization of the magma (beyond which point of crystallization, volcanic eruptions are almost precluded). High-quality W analyses of granitoids from the Sierra Nevada batholith demonstrate that a slight increase in W content occurs with increasing differentiation index (Simon, 1972). The best analytical work on the tungsten content of unaltered granitoid rocks from North America suggest average values of ~0.2–0.5 ppm for granodiorites and ~0.5–2.0 ppm for granites (Simon, 1972; Terashima, 1980; Campbell and Aruscavage, 1982; Aruscavage and Campbell, 1978). At present, there is insufficient evidence to document that parental magmas for unaltered granitoids associated with W ore deposits in the North American cordillera are generally enriched relative to these values (Newberry and Swanson, 1986). However, a large part of the problem with detecting the smaller enrichment trends that do exist is that most analytical methods (and published W analyses) have inappropriate detection limits and precision (1 ± 1 ppm). Nevertheless, evidence of relatively small (but significant) amounts of magmatic enrichment of W in some granitoid (or volcanic) suites may exist and needs to be investigated and documented.

Indirect evidence of magmatic (or granitoid) enrichment of W comes from the granitoids of southeast China (noteworthy for the amount of related W deposits; Liu Yimao et al, 1982). Although no mention is made of excluding obviously mineralized samples from their data base, Liu Yimao et al. (1982) show a clear increase in average W content with differentiation index from granodiorite (1.3 ppm) to biotite granite (8.2 ppm). They also suggest that two-mica and albite granites have average values of 14 ppm and 107 ppm respectively (although hydrothermal enrichment may be important in these cases as will be discussed later). They note that these values may be strongly dependent upon the age and "geochemical province" of the granitoids as well.

The only potential evidence of similar W-enrichment trends being present in North American granitoids comes from the plutons of the Selwyn Basin. Van Middelaar (1988) notes that the W content of the fresh, least differentiated granitoids from the Mine stock at CanTung (Table 14.1) is in the 2–4 ppm range (Fig. 14.4). Anderson (1983) found somewhat higher values (8–12 ppm W) for the two-mica granitoid of the Cirque Lake Stock adjacent to the MacTung deposit. However, the occurrence of rare "disseminated" molybdenite and other alteration products in this plutonic phase make it difficult to be certain if these values reflect magmatic or hydrothermal enrichment. Anderson (1983) found only "normal" (1–2 ppm) W contents in the other plutonic phases at MacTung and in other samples of the Selwyn Basin plu-

TABLE 14.1—Granitoid alteration and reequilibration textures and processes (and approximated temperatures) summarized from examination of granitoids (van Middelaar, 1988) and the North American granitoid suite (Keith et al., 1985b). These lists are not intended to be comprehensive or universally applicable.

| Subsolidus reequilibration 700–600 °C | Disseminated alteration 600–500 °C | Vein-controlled alteration 500–400 °C | Greisen 400–300 °C |
|---|---|---|---|
| Myrmekite | Hydrothermal oligoclase | Quartz | Quartz |
| Granoblastic quartz | Hydrothermal biotite | Sericite/muscovite | Muscovite/sericite |
| K-spar megacrysts | Biotite to muscovite, pyrrhotite, rutile | Rutile/titanite | Titanite |
| Graphic intergrowths | Ilmenite to pyrrhotite, rutile | Tourmaline | Tourmaline |
| Titanomagnetite to magnetite | | Pyrrhotite/pyrite | Pyrite/pyrrhotite |
| Microclinitization | | Calcite/scheelite | Calcite/scheelite |
| Perthite | | | Albite |

tonic suite. Additional work would be needed to determine if any consistent magmatic enrichment of W is present in differentiated granitoids of the Selwyn Basin.

Considering the similarities in geochemical behavior between W and Mo, one might anticipate similar magmatic enrichments for these two elements, if such enrichment occurs. Our (incomplete) data concerning the Mo content of granitoids related to moderate and large W deposits versus barren granitoids or granitoids related to small W deposits indicate that there is no significant difference between the two groups (Fig. 14.3). Even relatively fresh granitoids associated with large Mo or W deposits in this sample suite show no evidence of general magmatic enrichment of Mo (despite excellent accuracy and precision of the analytical method; Keith and Shanks, 1987).

## SUBSOLIDUS RE-EQUILIBRATION AND HYDROTHERMAL ALTERATION

Thermodynamic calculations of the solubilty of tungsten in typical skarn-forming magmatic fluids suggest that the total W content in the fluid *potentially* may increase 3 to 4 log units as the temperature decreases from that of the solidus to about 400°C (Fig. 14.5; Newberry, 1980; Newberry and Einaudi, 1981). Whether or not the fluid attains any W increase through this temperature interval may depend on: 1) how intimately it interacts with a "leachable" source of W; and 2) the nature of subsolidus reactions or hydrothermal alteration which occurs. Consequently, late-stage magmatic fluids which interact with the granitoid through this temperature range may be important agents in scavenging W from the crystallized carapace of the pluton.

Prior to examining evidence of subsolidus equilibration and alteration of W-related granitoids, the rationale for considering that this may be a significant process in altering the composition of the magmatic fluid should be examined. First, the bulk of fluid inclusion homogenization temperatures from well-studied W skarns is often at or below about 500°C (Kwak, 1987; Einaudi et al., 1981, Bowman et al., 1985) which is well below solidus values. Consequently, the magmatic fluid, originally exsolved at a substantially higher temperature (~700°C), reacts or "equilibrates" down-temperature with the granitoid to become an "intrusive" fluid which ultimately forms the skarn. Intrusive fluid, as used in this paper, is a water-rich phase which has equilibrated (isotopically and chemically) with the granitoid at subsolidus temperatures to attain a composition which is distinct from the magmatic fluid (which equilibrated with the magma at solidus temperatures).

Intrusive fluid, passively released from a slowly cooling pluton, may stand in significant chemical contrast to the magmatic fluid involved in the mineralization of many porphyry deposits (where some metals are deposited near the magma-carapace interface). Although some W-skarn granitoids are transitional to porphyry type deposits, some important general distinctions can be made. For instance, most W-skarn granitoids are medium-grained rocks and lack a fine-grained matrix indicative of a wide-spread rapid volatile release; the last dregs of melt were allowed to crystallize more slowly and the rock then "stewed in its own juices". At what sub-solidus temperature is the fluid permanently drained off and when does exchange between the granitoid and fluid cease? Undoubtedly a wide variety of "stewing" times are possible. Oxygen isotope data from the CanTung granitoid suggest that quartz, biotite, and feldspar equilibrated down temperature to about 420–520°C (Bowman et al., 1985). Wesolowski et al. (1988) noted that biotite and feldspar from the granitoid adjacent to the King Island W skarn in Tasmania re-equilibrated to subsolidus temperatures of about 600°C. Examination of the structural state and exsolution of K-feldspars, Ti-content of titanomagnetite, and numerous other evidences of mineral re-equilibration suggests that it is far more common (in all granitoids) to re-equilibrate to approximately these temperatures than to preserve compositions indicative of magmatic conditions.

However, an equally important question is whether or not the W content of the granitoid minerals can be accessed by "intrusive" fluids through this subsolidus temperature range. The following discussion examines some of the subsolidus or hydrothermal processes which may be important in this regard.

Potentially, a variety of subsolidus re-equilibration processes might allow a magmatic fluid to scavenge W (Table 14.1). We systematically examined our plutonic suite for evidence of myrmekite, graphic intergrowths, K-feldspar megacrysts, and granoblastic quartz. Although many of these

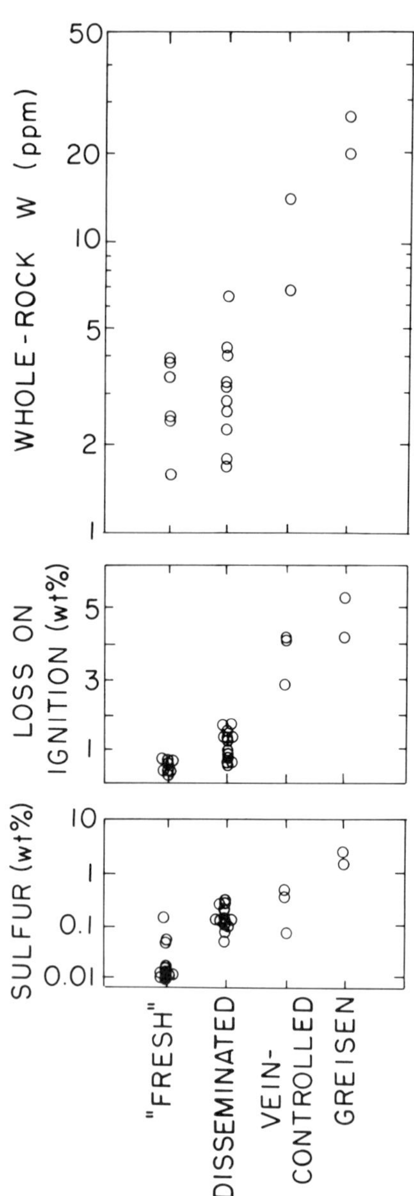


FIGURE 14.4—Whole-rock W content, loss on ignition, and sulfur content of CanTung granitoids versus the type of granitoid alteration (data from van Middelaar, 1988). Loss on ignition and sulfur contents of granitoids exhibit the best analytical correlations with the type of alteration noted for each sample (as listed in Table 14.1). Granitoids with disseminated alteration contain notable amounts of pyrrhotite with no apparent increase or decrease in W content. Granitoid samples which exhbit vein-controlled or greisen alteration have slightly to significantly elevated W concentrations.

features are fairly common in the plutons and samples we examined, none showed a consistent increase in abundance in association with large or high grade deposits. The processes that create these textures likely occur very near the solidus temperatures and may contribute some W to the fluid; however, this amount may be minor because the solubility limit of W in the hydrothermal fluid may have changed very little from the slightly lower (?) magmatic value (Fig. 14.5).

The only two common magmatic minerals that show a strong tendency to concentrate W are titanite and ilmenite (Simon, 1972; Candela, this volume). These data and other considerations suggest that W often substitutes for Ti in many minerals (or occurs in concentrations directly related to the Ti content). Consequently, how closely might the geochemical behavior of W be coupled to Ti during hydrothermal alteration or liberation of the magmatic component of the ore fluid? Several Ti-liberating processes deserve attention. For example, Van Middelaar (1988) has documented that most of the ilmenite in the Mine stock at CanTung (in a widespread assemblage) has been converted to pyrrhotite (or more rarely cubanite resulting from a Cu–Fe–S intermediate solid solution) at a relatively high temperature (Table 14.1). Whole-rock analyses suggest that the net Ti content of the altered rocks is not significantly reduced from the value of fresher rocks; however, even weakly altered granitoids contain disseminated clots of rutile as well as rare vein-controlled titanite. Enough titanite is present in the CanTung skarn (and occasionally intergrown with scheelite) that it creates a problem during benefication of the ore (van Middelaar, 1988). Mass balance calculations which derive skarn-forming constituents from the granitoid suggest that possibly less than 1% of the Ti in the granitoid would need to be mobilized and transported into the skarn to account for its approximate Ti content. Einaudi et al. (1981) have noted that the presence of titanite is closely associated with higher grade W skarns elsewhere as well.

However, most of the Ti in the CanTung granitoid is located in biotite. Conversion of biotite to muscovite (or K-feldspar) and pyrrhotite is one Ti-liberating reaction that has been noted in the CanTung granitoids (Table 14.1; van Middelaar, 1988). Perhaps more commonly, conversion of Ti-bearing biotite to a Ti-free biotite or mica is noted (Table 14.1). Several methods of analysis are being employed to examine this process. XRF analysis of many biotite separates from our North American granitoid suite demonstrate that on the *average* about a 12% reduction in Ti content of biotites often occurs as Ti-poor biotite (or chlorite in some cases) is produced around the margins of the "magmatic" crystal. High-resolution SEM elemental mapping of the altered margins of biotite crystals demonstrates that when the Ti does not completely leave the biotite structure, it is stabilized (and concentrated?) by the introduction of Ca into the same domain of the crystal. Under some conditions, tungsten could be retained in the crystal in a similar manner, because very small inclusions of unknown W–Sn-bearing phases or Sn–W-rich titanite have been found in some micas from elsewhere in the world (Taylor, 1979; Il'in and Ivanova, 1972). After liberation from a mineral lattice, W may be more effectively scavenged by deuteric fluids than Ti (although some Ti is certainly introduced into the skarn during metasomatism).

Another Ti-(and W?)-liberating process that may be important in magnetite-bearing granitoids is the almost universal conversion of titanomagnetite to magnetite during cooling of a granitoid from relatively high temperatures. Simon (1972) also noted that there was strong evidence for the inclusion of submicroscopic W-rich phases, such as wol-

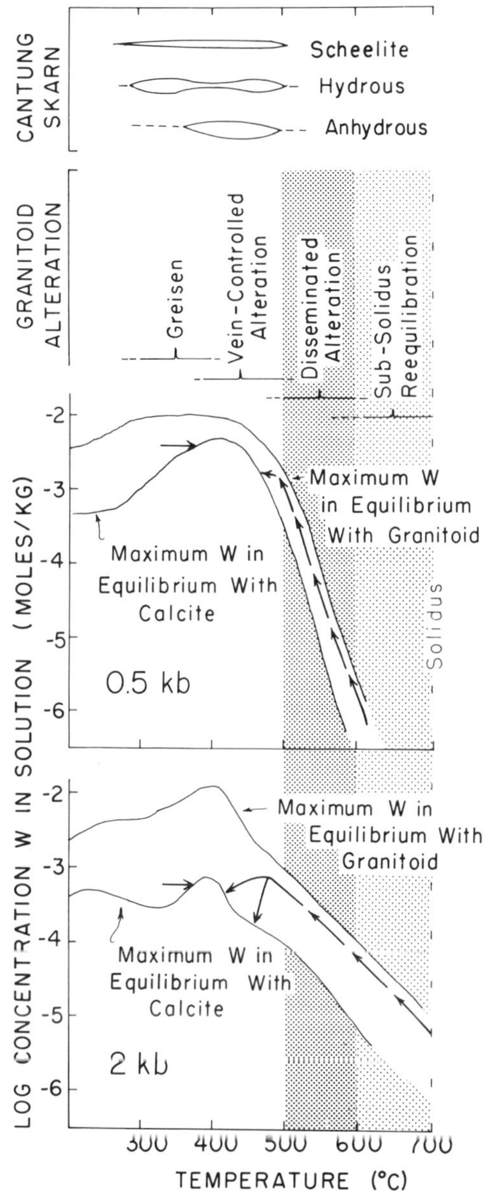


FIGURE 14.5—Log concentration W in solution versus temperature, types of granitoid alteration, and skarn development at CanTung. Calculations showing maximum W in equilibrium with granitoid and calcite as a function of temperature and pressure are from Newberry and Einaudi (1981; see Newberry, 1980 for a complete description of the methods of calculation). The patterned areas highlight the approximate termperature intervals of disseminated alteration and sub-solidus equilibration during which the solubility of W is strongly retrograde. Consequently, magmatic fluid evolved near solidus temperatures is theoretically capable of following a pathway indicated by arrows whereby W content of the fluid dramatically increases (assuming a leachable source of W is available in the granitoid) prior to escape of the fluid from the carapace (see FIG. 14.6 for a schematic representation of this process). Interaction of that fluid with calcite would reduce the solubility of W and cause scheelite deposition. Such a model compares favorably with the dominant development of scheelite-rich skarn at CanTung (and many other W skarns) beginning at about 500°C rather than at solidus temperatures. Scheelite deposition within the granitoid is also restricted to lower temperature (often calcite-bearing) alteration assemblages. Not considered in these calculations is the possibility of wolframite deposition which often may occur in porphyry Mo environments at higher temperatures (E. Seedorff, personal communication). Such deposition may be related to the higher salinities of their hydrothermal fluids and resulting higher Fe concentrations (wolframite deposition may be especially favored as the relative strength of the Fe chloride complex drops at temperatures below about 600°C; Whitney et al., 1985).

framite, in some of the mineral separates he analyzed (i.e. magnetite, Table 14.2). If W behaves predominantly as an incompatible element during crystallization (Candela, this volume), then "magmatic" submicroscopic crystals of wolframite might often occur along grain boundaries. As the solubility of W increases with declining temperature, such crystals would be easily accessible and almost certainly removed by a deuteric fluid. Other alteration processes which do not noticeably affect Ti, such as microclinization or "oligoclasization" (Table 14.1), may be important in releasing W to a deuteric fluid. Inasmuch as feldspars (or submicroscopic phases within feldspar) often contain the dominant portion of the W in the granitoid (Table 14.2), their tendency to release W (or W-rich phases), although difficult to measure and evaluate, may be very significant. Individually, any one of these processes may not have a substantial effect on helping to create an ore fluid, but an alteration assemblage that combines several of these potential W-liberating processes may often be part of the skarn-forming process (Figs. 14.5 and 14.6).

If the solubility of W is retrograde (i.e. it increases with decreasing temperature from magmatic temperatures to 400°C, Fig. 14.5) and some of these processes of W leaching do occur, then the occurrence of leached zones beneath W-skarn deposits may be anticipated. However, by analogy, no such leaching can be documented (analytically) for other major skarn-forming elements with retrograde solubility. For example, Whitney et al. (1985) have demonstrated the retro-

TABLE 14.2—Tungsten content of minerals from a Sierra Nevada batholith granite. Data from Simon (1972). W content of magnetite in this sample is anomalously high compared to other Sierra Nevada granitoids.

| Mineral | % Mineral in rock | W content (ppm) |
|---|---|---|
| Plagioclase | 34 | 0.17 |
| Quartz | 33 | 0.36 |
| K-feldspar | 32 | 0.44 |
| Biotite | 2 | 2.1 |
| Hornblende | 1 | 3.1 |
| Magnetite | 0.5 | 19 |
| Hypersthene | 0.5 | 3.2 |
| Sum of minerals | 103 | 0.50 |
| Whole-rock value | | 0.27 |

grade solubility of Fe chloride from 700°C to about 600°C. Undoubtedly significant amounts of Fe can be leached from the granitoid by intrusive fluids and transported into the skarn environment. Despite the fact that textural evidence of such leaching is present in certain granitoid lithologies (van Middelaar, 1988), the net removal of Fe is not analytically significant (as was also noted for Ti).

Absence of such leached zones (for Fe, Ti, and W) also may be related to the water/rock ratio and overprinting of later alteration assemblages. If only a small amount of magmatic fluid is exsolved by the granitoid, its subsequent path through the granitoid may be tortuous and slow (low water/rock ratio) prior to collection along widely spaced fractures (Fig. 14.6). If the solubility of W also gradually increases during this cooling interval, then the amount of W leached from any unit volume of granitoid may be small, despite a substantial increase in the W content of the fluid. Therefore, it may be difficult to detect "leached" granitoids by a procedure of analyzing various altered or "fresh" lithologies. For example, although apparent removal of 1 ppm of W from a granitoid may seem trivial in terms of supplying metal for mineralization, mass balance calculations suggest that much less than 1 $km^3$ of granitoid would need to be leached of 1 ppm W to account for all of the W in any of the moderate or small deposits studied for this project. (This also emphasizes the point that W-rich magmas are not needed for the mineralization process.) The scatter in the W content of CanTung granitoids which are effected mainly by higher temperature, disseminated alteration assemblages (Fig. 14.4) excedes the analytical uncertainty and may attest to some W leaching and redistribution. Ti analyses exhibit the same scatter in even a more pronounced way (van Middelaar, 1988). Analysis of granitoids beneath or adjacent to W skarn deposits has probably rarely been done with sufficient care and precision to document subtle variations in the W content of the granitoid and individual mineral phases.

Another potential problem with the identification of zones of "leached" granitoid concerns the weak mineralization of granitoids associated with lower temperature alteration assemblages (Keith et al., 1985b; van Middelaar, 1988). Several different lines of evidence suggest that granitoids with obvious vein-controlled alteration, albitization, or lower-temperature greisen often exhibit high W contents due to deposition of traces of scheelite, wolframite, or other W-rich phases. In particular, the anomalously high W content of granitoids from CanTung appears to be directly correlated with the amount of low-temperature (~400°C) alteration (quartz–muscovite–albite–calcite–pyrite assemblage) present (Fig. 14.4). Calcite-bearing greisen samples contain abundant visible scheelite (van Middelaar, 1988). In addition, this may be the explanation for the anomalously high W contents (average of 107 ppm W; Liu Yimao et al., 1982) of muscovite albite granitoids of China. Weak sericitic alteration may be roughly contemporaneous with greisen development and create analogous enrichments (Table 14.1); consequently, if traces of calcite- and scheelite-bearing sericitic alteration are overprinted on the granitoid (Fig. 14.6), it may be difficult to find samples and document W contents which are truly "fresh" or affected only by higher temperature disseminated alteration.

## RELATIVE VOLATILE FUGACITIES

Regardless of whether the W content of the magma is 0.1 ppm or 100 ppm, the W content of the magma would have no influence on initiating or affecting magmatic processes of fluid collection related to mineralization (but would have an influence on the *grade* of the skarn). However, the relative fugacities of volatile components (including oxygen) would critically influence the timing of fluid collection, the efficiency of leaching W from the magma or granitoid, the order of mineral crystallization, the partitioning behavior of W, the rheolgy of the magma, diffusion rates, and possibly the oxidation state of W (Candela, this volume). Magmatic volatile fugacities obviously may be strongly influenced by differentiation or assimilation of crustal material.

The two volatile components whose role in the formation of W, Sn, or Mo deposits is most often debated are F and Cl. Munoz (1984; Munoz and Swenson, 1981) has used experimental data to calibrate micas as tools for investigating the relative (and absolute) fugacities of F and Cl in fluids or magmas which equilibrated with those micas. Perhaps the best way to compare the relative fugacities of these two volatiles from one deposit or granitoid to another is by means of the F/Cl intercept value (Munoz, 1984). This value is directly related to the $f_{HCl}/f_{HF}$ of the fluid with which the micas equilibrated and is independent of the temperature of equilibration and the OH content of the mica (both of which are often difficult to determine). Inspection of the F/Cl intercept values (IV(F/Cl)) for the granitoid suite from the western cordillera reveals that no significant difference is present between barren and productive granitoids (Table 14.3, Fig. 14.7). Consequently, with this preliminary data, neither high F nor high Cl fugacities seem positively correlated with W skarns. Such a result is not unexpected. Zaw and Clark (1978) suggested that there is no marked enrichment of F associated with the CanTung skarns. Munoz (1984) noted that the F/Cl intercept values for a variety of Sn–W–Be deposits were intermediate in value between the F-rich biotites associated with the Henderson Mo deposit and the more Cl-rich biotites associated with porphyry Cu deposits. This correlation might also be related to the fact that, with increasing differentiation, the relative fugacity of F in granitoid magmas generally increases (lower IV(F/Cl) value; Fig. 14.7). In

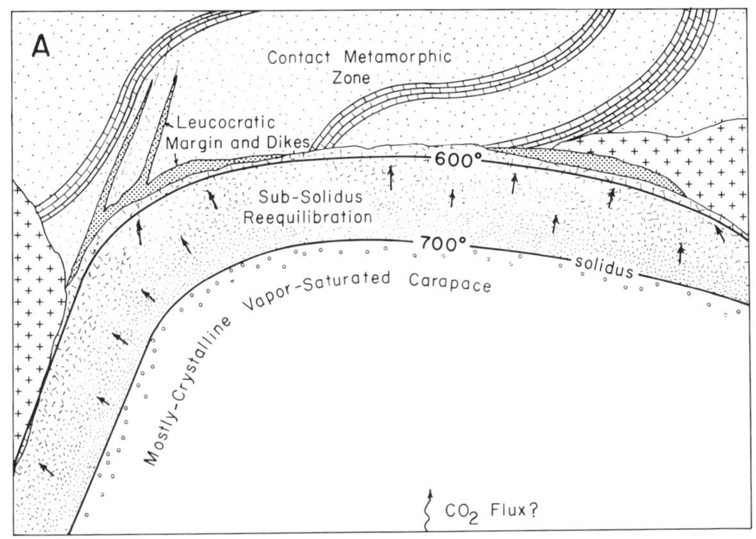


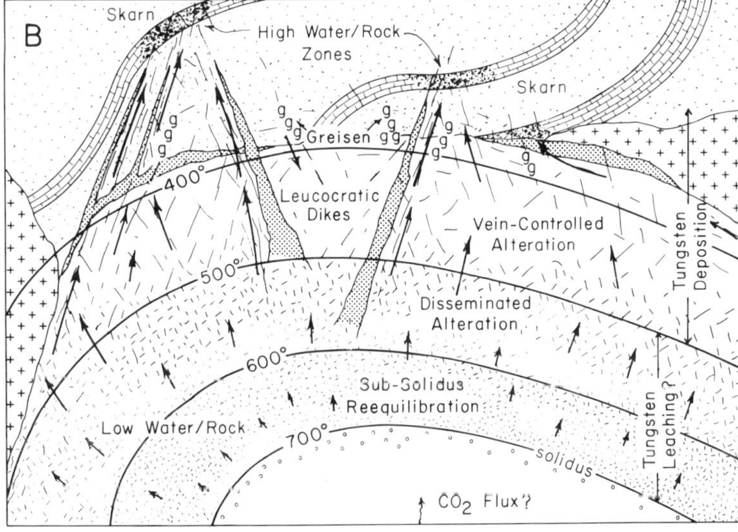


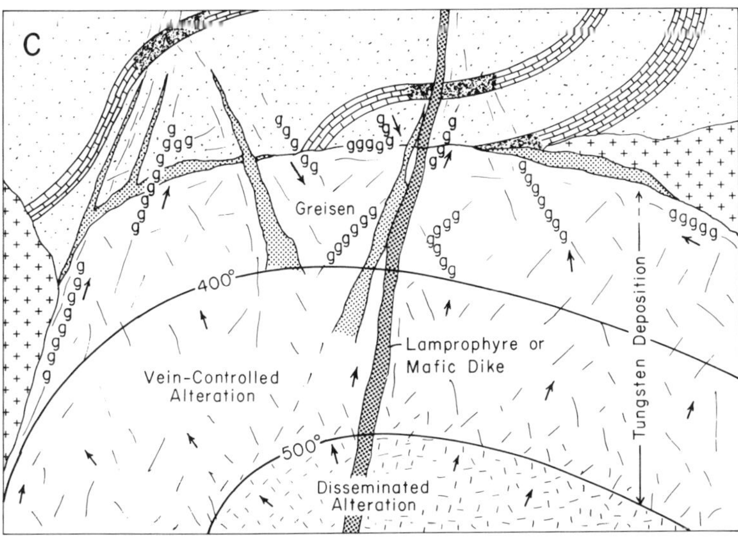


FIGURE 14.6—Schematic representation of skarn development from intrusive fluid of a cooling pluton versus alteration development. The depth and scale of these cartoons are omitted to eliminate the need for rigorous placement of isotherms. Consequently, some time may be compressed into these otherwise sequential views, from A to C, of a cooling pluton emplaced in a shallow batholithic terrane below a shale/limestone sequence. **A.** As the crystalline margin of the pluton develops, magmatic fluid (arrows) is evolved, but not efficiently nor rapidly collected to a potential skarn-forming unit. Initially, fluid flow is dominantly intracrystalline on a pervasive scale rather than by persistent fractures. Sub-solidus re-equilibration (Table 14.2, Fig. 14.5) may begin to change the composition of the granitoid minerals and the magmatic (now "intrusive") fluid. Contact metamorphism and bimetasomatism are the dominant processes in the country rock. **B.** Continued crystallization of the carapace is accompanied by intermittent intrusion of aplite or leucocratic porphyry dikes and development of fractures in the cooler more brittle portion of the carapace. Down-temperature from the solidus, fluid flow becomes more channeled by widely spaced fractures or veins. Vein-controlled alteration and greisen (linear "g" pattern) begin to be overprinted on disseminated alteration assemblages as isotherms move downward. The increasing length of the arrows upwards indicates higher water/rock ratios along major fractures and dike margins. The location of major skarn development is in part controlled by the location of abundant dikes, fractures, and veins (as is very apparent at MacTung; Dorthy Atkinson, personal communication). Tungsten leaching may accompany higher temperature alteration; whereas scheelite deposition may occur with lower temperature assemblages. **C.** The downward incursion of lower temperature alteration assemblages is arrested by complete crystallization of the pluton and diminished flow of intrusive fluid. Lamprophyre dikes may be dominantly post-mineralization, but still exhibit evidence of some alteration. Sericitic alteration may be the less intensely developed, more widely dispersed equivalent of fracture-controlled greisen.

TABLE 14.3—Chemistry of biotites from Cordilleran granitoids.

| Sample | Can 10 | Can 14 | Emer 5 | Mac 4 | Mac 8 | Mill 2 | Mill 3 | Pine 4 |
|---|---|---|---|---|---|---|---|---|
| $SiO_2$ | 33.94 | 34.70 | 35.31 | 35.76 | 34.51 | 36.38 | 35.91 | 36.09 |
| $Al_2O_3$ | 20.17 | 19.56 | 14.92 | 16.43 | 21.10 | 15.34 | 15.43 | 14.61 |
| $TiO_2$ | 3.20 | 3.05 | 3.65 | 3.48 | 1.46 | 3.97 | 3.33 | 2.49 |
| $Fe_2O_3$ | 1.02 | 0.63 | 0.00 | 1.37 | 1.15 | 2.84 | 2.69 | 2.57 |
| FeO | 21.93 | 21.59 | 18.51 | 21.68 | 24.24 | 17.22 | 16.75 | 21.52 |
| MnO | 0.75 | 0.48 | 0.81 | 0.56 | 1.25 | 0.53 | 0.47 | 0.50 |
| MgO | 5.26 | 5.83 | 11.23 | 7.24 | 2.09 | 10.48 | 10.55 | 8.12 |
| CaO | 0.01 | 0.16 | 0.12 | 0.11 | 0.21 | 0.01 | 0.17 | 0.08 |
| $Na_2O$ | 0.10 | 0.10 | 0.10 | 0.10 | 0.10 | 0.10 | 0.10 | 0.10 |
| $K_2O$ | 9.41 | 9.27 | 9.33 | 9.20 | 9.39 | 9.45 | 9.44 | 9.42 |
| BaO | 0.00 | 0.00 | 0.00 | 0.00 | 0.00 | 0.70 | 0.77 | 0.00 |
| F | 0.15 | 0.24 | 1.04 | 0.55 | 0.92 | 0.05 | 0.04 | 0.89 |
| Cl | 0.13 | 0.15 | 0.05 | 0.09 | 0.09 | 0.04 | 0.05 | 0.06 |
| $H_2O$(calc) | 2.46 | 2.23 | 2.60 | 2.28 | 1.55 | 2.58 | 2.53 | 2.28 |
| SUM | 98.42 | 97.99 | 97.67 | 98.85 | 98.06 | 99.69 | 98.23 | 98.73 |
| O=F,Cl | 0.09 | 0.13 | 0.45 | 0.25 | 0.41 | 0.03 | 0.03 | 0.39 |
| Total | 98.33 | 97.85 | 97.22 | 98.60 | 97.66 | 99.66 | 98.20 | 98.34 |
| Recalculation based on 6 cations in octahedral site | | | | | | | | |
| Si | 5.42 | 5.54 | 5.58 | 5.68 | 5.63 | 5.65 | 5.66 | 5.76 |
| Al(IV) | 2.58 | 2.46 | 2.42 | 2.32 | 2.37 | 2.35 | 2.34 | 2.24 |
| Al(VI) | 1.21 | 1.22 | 0.36 | 0.75 | 1.69 | 0.47 | 0.53 | 0.52 |
| Ti | 0.38 | 0.37 | 0.43 | 0.42 | 0.18 | 0.46 | 0.39 | 0.30 |
| Fe(3+) | 0.12 | 0.08 | 0.00 | 0.16 | 0.14 | 0.33 | 0.32 | 0.31 |
| Fe(2+) | 2.93 | 2.88 | 2.45 | 2.88 | 3.31 | 2.24 | 2.21 | 2.87 |
| Mn | 0.10 | 0.06 | 0.11 | 0.08 | 0.17 | 0.07 | 0.06 | 0.07 |
| Mg | 1.25 | 1.39 | 2.65 | 1.71 | 0.51 | 2.43 | 2.48 | 1.93 |
| Ca | 0.00 | 0.03 | 0.02 | 0.02 | 0.04 | 0.00 | 0.03 | 0.01 |
| Na | 0.03 | 0.03 | 0.03 | 0.03 | 0.03 | 0.03 | 0.03 | 0.03 |
| K | 1.92 | 1.89 | 1.88 | 1.86 | 1.95 | 1.87 | 1.90 | 1.92 |
| Ba | 0.00 | 0.00 | 0.00 | 0.00 | 0.00 | 0.04 | 0.05 | 0.00 |
| F | 0.08 | 0.12 | 0.52 | 0.28 | 0.47 | 0.02 | 0.02 | 0.45 |
| Cl | 0.04 | 0.04 | 0.01 | 0.02 | 0.02 | 0.01 | 0.01 | 0.02 |
| OH(calc) | 2.42 | 2.29 | 2.70 | 2.34 | 1.62 | 2.59 | 2.58 | 2.37 |
| Oct. Cat. | 6.00 | 6.00 | 6.00 | 6.00 | 6.00 | 6.00 | 6.00 | 6.00 |
| Int. Cat. | 1.95 | 1.95 | 1.93 | 1.91 | 2.02 | 1.95 | 2.01 | 1.96 |
| CATSUM | 15.95 | 15.95 | 15.93 | 15.91 | 16.02 | 15.95 | 16.01 | 15.96 |
| Mg/(Mg+Fe*) | 0.29 | 0.32 | 0.52 | 0.36 | 0.13 | 0.49 | 0.50 | 0.38 |
| X(Mg) | 0.21 | 0.23 | 0.44 | 0.29 | 0.08 | 0.40 | 0.41 | 0.32 |
| X(Sid) | 0.69 | 0.63 | 0.31 | 0.45 | 0.81 | 0.32 | 0.32 | 0.33 |
| X(An) | 0.10 | 0.13 | 0.25 | 0.27 | 0.10 | 0.28 | 0.26 | 0.35 |
| IV(F) | 2.00 | 1.81 | 1.55 | 1.56 | 0.87 | 2.82 | 2.91 | 1.42 |
| IV(Cl) | −3.57 | −3.70 | −3.56 | −3.58 | −3.36 | −3.40 | −3.52 | −3.47 |
| IV(F/Cl) | 5.58 | 5.52 | 5.11 | 5.14 | 4.23 | 6.22 | 6.44 | 4.89 |
| T, °C | 350 | 350 | 350 | 350 | 350 | 350 | 350 | 350 |
| log(HF/HCl) | 0.70 | 0.63 | 0.23 | 0.26 | −0.65 | 1.34 | 1.56 | 0.01 |

Note: Mole fractions of the phlogopite, siderophyllite and annite endmembers in biotite are defined as:

$X(Mg) = Mg/(\text{total octahedral cations})$

$X(Sid) = [(3 - Si/Al)/1.75]\,[1 - X(Mg)]$

$X(Ann) = 1 - X(Mg) - X(Sid)$

Intercept value of fluorine and chlorine in biotite is defined as:

$IV(F) = 1.52X(Mg) + 0.42X(Ann) + 0.20X(Sid) - \log[X(F)/X(OH)]$

$IV(Cl) = -5.01 - 1.93X(Mg) - \log[X(Cl)/X(OH)]$

The fluorine-chlorine intercept value is defined as:

$IV(F/Cl) = IV(F) - IV(Cl)$

The fluorine-chlorine fugacity ratio in the fluid equilibrating with the biotite is defined as:

$\log(HF/HCl) = -3051/T(°K) + IV(F/Cl)$

(After Munoz, 1984)

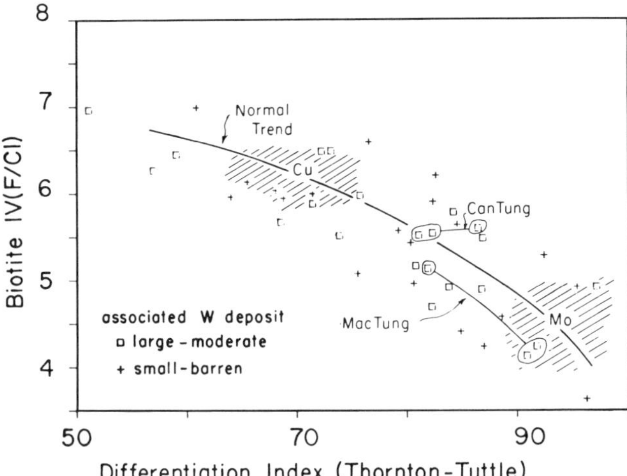


FIGURE 14.7—F/Cl intercept value (IV(F/Cl)) of granitoid biotites versus the Thornton–Tuttle differentiation index of granitoid suite (van Middelaar and Keith, in prep.). The decrease of the intercept value (higher relative fluorine) with increasing differentiation reflects the apparent normal igneous trend. The intercept values of biotites from various phases of MacTung and CanTung granitoids are indicated, as well as the average intercept values for porphyry copper and molybdenum deposits (Munoz, 1984; common values for the differentiation indices of Climax-type porphyry Mo deposits and porphyry Cu deposits are estimated and calculated from a variety of sources to show approximate values). Note that these values follow the expected igneous trend. Granitoids associated with moderate to large deposits (squares) are contrasted with granitoids associated small deposits or barren of mineralization (crosses). See Fig. 14.1 for summary of tonnage of large, moderate, and small deposits. Petrographic descriptions, sample locations, and major- and trace-element composition of each sample are available upon request from the authors.

agreement with this observation, the differentiation indices of our granitoid population also are generally intermediate to those that would be expected for porphyry Cu deposits and the Henderson Mo deposit (Gunow et al., 1980).

This does not mean that F and Cl play no part in the magmatic hydrothermal fluid associated with W skarn deposits. However, their relative importance needs to be carefully evaluated by other means. Experimental work which measured the partitioning of W between an aqueous phase and silicate melt suggests that W is more strongly partitioned into the aqueous phase as the chloride content is increased (Manning and Pichavant, 1988). Conversely, increasing the fluoride content of the system decreases the partitioning of W into the fluid phase. Examination of the biotites from CanTung granitoid and skarn reveal that a relatively constant range of IV(F/Cl) values are present (van Middelaar and Keith, in prep.). The implication is that the same intrusive fluid that equilibrated with the main phase of the pluton also equilibrated with (and likely formed) the skarn.

## INFERRED OXYGEN FUGACITY

Important changes in the magmatic and hydrothermal geochemical behavior of W and Mo might occur as a function of the relative oxidation state of the magma. Candela (this volume) reviews experimental evidence of changes in magmatic partitioning of these elements (and subsequent metal ratios in the deposit) due to changes in oxygen fugacity. Keith (1982) and Keith and Shanks (1988) suggest that high oxygen fugacities are important for development of Climax-type porphyry Mo deposits because under these conditions only a small proportion of the Mo is partitioned into magmatic phenocrysts.

Perhaps a significant facet of our preliminary data concerns a weaker correlation of high grade tungsten deposits with relatively low hydrothermal (and magmatic) oxygen fugacities. Granitoids from our suite which are most nearly devoid of Ti-bearing phases such as titanite, ilmenite, and titanomagnetite and exhibit a low $Fe^{3+}/Fe^{2+}$ ratio in biotite appear to be correlated with higher grade W deposits (Keith et al., 1985a). This relationship may be caused by the partitioning of W into Ti sites in early crystallizing mafic and accessory phases which are more abundant in the oxidized granitoids related to the lower grade deposits. In addition, the general mobility of W may be influenced in other ways by changes in the composition of the hydrothermal fluid which may be influenced by the oxidation state of the granitoid. For example, magmas and hydrothermal fluids with very low oxygen fugacities might allow sulfidation of Fe–Ti oxides and biotite (such as the disseminated pyrrhotite at CanTung) at higher temperatures than oxidized systems (such as porphyry Cu or porphyry Mo systems). This might allow the hydrothermal fluid to more effectively scavenge any W released in the process during the temperature interval of strong retrograde solubility of W (Table 14.1, Fig. 14.5). Dick and Hodgson (1982) noted that the skarn mineral assemblages from CanTung and MacTung indicated a very reduced hydrothermal fluid and proposed that this may be linked to the high grades of these deposits and the assumed W-content of the metasomatic fluid. Gerstner et al. (in prep.) recently found that fluid inclusions from all facies and stages of the MacTung skarns are methane-bearing and contain no detectable $CO_2$, which again documents a very low oxygen fugacity for the skarn-forming fluids (dominantly magmatic or intrusive; Bowman, 1986). Keith et al. (1985a) have extended this relationship by noting that the CanTung and MacTung granitoids are the most reduced of our granitoid suite and were responsible for buffering the metasomatic solutions to a low $f_{O_2}$ value (and high W content?).

These data raise several questions about the relative importance of the granitoid (and magmatic fluid) versus the country rocks and depths of emplacement in setting the relative oxidation state and W content of the skarn and the metasomatic fluid. Einaudi et al. (1981) proposed a classification scheme for W skarns that recognizes oxidized and reduced varieties (based on the mineralogy and ferric/ferrous ratio of the skarn). They propose that the oxidation state of the skarn is set by oxidation state of the country rocks or the depth of fluid evolution, but recognize that the granitoid may impose an oxidation state on the skarn. It may be difficult to decipher whether the level of granitoid emplacement or the oxidation state of the granitoid is responsible for buffering the oxidation state of the skarn-forming magmatic fluids, because oxidized granitoids are often shallowly emplaced and reduced granitoids are generally deeper. This may be related to the fact reduced magmas are often largely derived from crust and are consequently cooler and wetter with less latent heat to rise to shallow levels in the crust.

Consequently, in order to investigate the relative importance of these factors in determining the oxidation state of the skarn, reduced skarns associated with shallowly emplaced plutons (or vice versa) should be examined. Our data concerning the ferrous/total iron ratio of granitoid and skarn-forming biotites at CanTung show that both are similar and indicate very reducing conditions despite the shallow level of pluton emplacement. Consequently, the level of emplacement does not appear to be the controlling factor of oxidation state in this case. However, the country rocks are in part graphitic and may have contributed in buffering the skarn-forming fluids to a relatively low oxygen fugacity.

How effectively a pluton might impose the granitoid oxidation state on the skarn may be related to volume of magmatic water that interacts with the country rock as well as the buffering capacity of the fluid and the rock. Bowman et al. (1985) used stable isotope data to demonstrate that the high-grade CanTung E-zone skarn experienced a high ratio of magmatic water to unit volume of country rock during mineralization; whereas, a few lower grade deposits investigated had progressively lower values. Consequently, the grade (and size) and oxidized or reduced character of the skarn often may be related to the oxidation state of the granitoid as well as to the volume of intrusive water channeled to the location of skarn formation (Fig. 14.6).

## CONCLUSIONS

Our data suggest that neither the bulk composition of the granitoid nor the F/Cl ratio of the intrusive fluid are of fundamental importance in determining whether or not a pluton may develop a W skarn deposit. The apparent lack of any special granitoid or fluid compositions for W skarn formation suggests that the *processes* of (unusually efficient) fluid collection and concentration at the site of skarn formation are perhaps the most critical factors in the mineralizing process (excluding the obvious need for a favorable carbonate host rock, Fig. 14.6). Our classification of W skarn related plutons is based, in part, upon the apparent dominant mechanisms of fluid transport through the margin of the pluton (veins and fractures versus intracrystalline transport).

The grade of the W skarn deposit (and/or the efficiency of removal and transport of W) can likely be affected by the relative oxygen fugacities of the magma and subsequent granitoid and intrusive fluid. Our data suggest that the highest grade W deposits are often associated with the most reduced granitoids (Keith et al., 1985a). The oxidation states for granitoids from our suite (inferred from ferrous/total iron ratios of biotites and types of Fe–Ti oxides; Keith, unpublished data) are generally correlative with the inferred oxidation state of the skarn (according to the classification of Einaudi et al., 1981). This implies that the oxidation state of the granitoid may often be imposed on the skarn; the alternatives are that the oxidation state of the skarn is set by the depth of intrusion or the reducing/oxidizing capability of the wallrocks (Einaudi et al., 1981).

Several lines of reasoning suggest that intrusive fluid (fluid which equilibrates chemically and isotopically with some granitoid minerals at subsolidus temperatures) may play a significant role in the formation of W skarn deposits. Some of these reasons include:

1. The calculated geologic solubility of W has been shown to strongly increase in intrusive fluid as it cools from solidus values near 700°C to about 400°C (Fig. 14.5; Newberry and Einaudi, 1981).
2. Oxygen isotope, petrographic, and microprobe studies of W skarn granitoids suggest some minerals pervasively re-equilibrate with intrusive fluid during this temperature interval.
3. Ti is often released from some mafic minerals and Fe–Ti oxides during subsolidus re-equilibration and alteration processes and deposited in the skarn. The somewhat analogous geochemical behavior of W and Ti suggests that W may be mobilized by hydrothermal fluids in the same manner.
4. Temperatures of fluid inclusion homogenization and oxygen isotope equilibration for W skarn-forming minerals from most well-studied systems suggest major contributions from an intrusive fluid (400°C–600°C range) rather than a magmatic fluid (~700°C). Skarn formation from an early magmatic fluid may occur, but often be thoroughly overprinted by a later intrusive fluid. However, an intrusive fluid is theoretically capable of carrying and depositing 1 to 3 orders of magnitude more W than a magmatic fluid (Fig. 14.5). Perhaps not coincidentally, the MacTung deposit (Fig. 14.1) contains 1 to 3 orders of magnitude more W than the other deposits in North America (excepting CanTung) and is dominated by skarn formed at unusually low temperatures near 430°C (average fluid inclusion homogenization temperature; Gerstner et al., in prep.) The W solubility of an intrusive fluid would be maximized near this temperature (Fig. 14.5) prior to reaction with a Ca-rich rock during skarn formation.

ACKNOWLEDGEMENTS—Reviews of this paper by P. Candela, E. Christiansen, and R. Newberry are gratefully acknowledged as well as the editorial efforts and comments of J. Whitney.

## REFERENCES

Ague, J. J., and Brimhall, G. H., 1985, Compositions of biotite from granitoids of the Sierra Nevada batholith: Constraints on magmatic source rocks: Geological Society of America, Abstracts with Programs, v. 17, p. 510.

Ague, J. J., and Brimhall, G. H., 1986, Two types of granite in the batholiths of California: Products of local and regional scale contamination: Geological Society of America, Abstracts with Programs, v. 18, p. 523.

Ague, J. J., and Brimhall, G. H., 1987, Granites of the batholiths of California: Products of local assimilation and regional-scale crustal contamination: Geology, v. 15, pp. 63–66.

Anderson, R. G., Armstrong, R. L., Parrish, R., and Bowman, J. R., 1983, Potential of the SE Selwyn plutonic suite for W-skarn deposits: Geological Association of Canada–Mineralogical Association of Canada–Canadian Geophysical Union, Program with Abstracts, v. 8, p. 2.

Aruscavage, P. J., and Campbell, E. Y., 1978, Spectrophotometric determination of tungsten in rocks by using zinc-dithiol: U.S. Geol. Survey, Journal of Research, v. 6, pp. 697–699.

Barton, M. D., 1987, Lithophile-element mineralization associated with Late Cretaceous two-mica granites in the Great Basin: Geology, v. 15, pp. 337–340.

Barton, M. D., Battles, D., Bebout, G. E., Capo, R., Christensen, J. N., Davis, S. R., Hanson, R. B., Michelson, C. J., and Trim, H. E., 1988, A review of Mesozoic contact metamorphism in the western U.S.; *in* Ernst, W. G. (ed.), Metamorphism and crustal evolution, western coterminous United States: Rubey Volume VII, Prentice–Hall, pp. 110–178.

Bateman, P. C., 1956, Economic geology of the Bishop district, California: California Div. Mines, Spec. Rept. 47, 87 pp.

Bateman, P. C., 1965, Geology and tungsten mineralization of the Bishop district, California: U. S. Geol. Survey, Prof. Paper 470, 208 pp.

Best, M. G., Armstrong, R. L., Granstein, W. C., Embree, G. F., and Ahlborn, R. C., 1974, Mica granites of the Kern Mountains pluton, eastern White Pine County, Nevada: Remobilized basement of the Cordilleran miogeosyncline?: Geological Society America, Bulletin, v. 85, pp. 1277–1286.

Bowman, J. R., 1986, Groundwater influx coupled with transition from lithostatic to hydrostatic fluid pressure during contact skarn development: EOS Transactions, American Geophysical Union, v. 67, no. 16, p. 272.

Bowman, J. R., Covert, J. J., Clark, A. H., and Mathieson, G. A., 1985, The CanTung E zone scheelite orebody, Tungsten, Northwest Territories: oxygen, hydrogen, and carbon isotope studies: Economic Geology, v. 80, pp. 1872–1895.

Burt, D. M., Sheridan, M. F., Bikun, J. V., and Christiansen, E. H., 1982, Topaz rhyolites—distribution, origin, and significance for exploration: Econ. Geol., v. 77, pp. 1818–1836.

Buseck, P. R., 1967, Contact metasomatism and ore deposition: Tem Piute, Nevada: Econ. Geol., v. 62, pp. 331–353.

Campbell, E., and Aruscavage, P. J., 1982, Molybdenum and tungsten contents of five CRPG and eight ANRT geochemical reference materials: Geostandards Newsletter, v. 6, pp. 229–231.

Christiansen, E. H., Sheridan, M. F., and Burt, D. M., 1986, The geology and geochemistry of Cenozoic topaz rhyolites from the western United States: Geological Society of America, Special Paper 206, 82 pp.

Clark, A. H., Yamamura, B. K., and Taipe Alejandro, J., 1987, Tungsten mineralization associated with subvolcanic pegmatite: the Palca XI deposit, Puno, southeastern Peru: Geological Association of Canada/Mineralogical Association of Canada, Program with Abstracts, v. 12, p. 32.

Clark, A. H., Kontak, D. J., and Farrar, E., 1982, Metallogenetic and geochronological relationships in the northern part of the central Andean tin belt, SE Peru and NW Bolivia: Collected Abstracts of 6th Symposium of the International Association of Genesis of Ore Deposits, Tbilisi, Soviet Georgia, September, 1982, pp. 64–65.

Cook, E. F., 1956, Tungsten deposits of south-central Idaho: Idaho Bur. Mines Geology, Pamphet 108, 40 pp.

Dick, L. A., and Hodgson, C. J., 1982, The MacTung W–Cu(Zn) contact metasomatic and related deposits of the northeastern Canadian cordillera: Economic Geology, v. 77, pp. 845–867.

Einaudi, M. T., Meinert, L. D., and Newberry, R. J., 1981, Skarn deposits: Economic Geology, 75th Anniversary Volume, pp. 317–391.

Farmer, G. L., and DePaolo, D. J., 1983, Origin of Mesozoic and Tertiary granite in the western U. S. and implications for pre-Mesozoic crustal structure: I. Nd and Sr isotopic studies in the geocline of the northern Great Basin: Journal of Geophysical Research, v. 88, pp. 3379–3401.

Gerstner, M. R., Bowman, J. R., and Pasteris, J. D., (in prep.), Skarn formation at the Macmillan Pass tungsten deposit (MacTung), Yukon and District of MacKenzie, Canada: I. P–T–X–V characterization of methane-bearing, skarn-forming fluids: submitted to Canadian Mineralogist.

Govett, G. J. S., 1983, Rock geochemistry in mineral exploration: Elsevier, New York, 461 pp.

Gower, S. J., Clark, A. H., and Hodgson, C. J., 1985, Tungsten–molybdenum skarn and stockwork mineralization, Mount Reed–Mount Haskin district, northern British Columbia, Canada: Canadian Journal of Earth Science, v. 22, pp. 728–747.

Gunow, A. J., Ludington, S., and Munoz, J. L., 1980, Fluorine in micas from the Henderson molybdenite deposit, Colorado: Economic Geology, v. 75, pp. 1127–1137.

Huanzhang, L., Jixi, S., Yu, C., and Shengjiao, X., 1982, Geologic and fluid inclusion studies of ten types of tungsten ore deposits in South China: Geochemistry, v. 1, no. 2, pp. 200–219.

Il'in, N. P., and Ivanova, G. F., 1972, X-ray microanalysis of muscovite from zones with tungsten mineralization: Geochemistry International, v. 9, pp. 186–193.

Keith, J. D., Clark, A. H., and Hodgson, C. J., 1985a, Characterization of granitoid rocks associated with tungsten skarn deposits of the North American Cordillera: Geological Society of America, Abstracts with Programs, v. 17, p. 625.

Keith, J. D., Clark, A. H., and Hodgson, C. J., 1985b, Characterization of granitoid rocks associated with tungsten skarn deposits of the North American Cordillera: unpublished report submitted to the Geological Survey of Canada, 129 p.

Keith, J. D., and Shanks, W. C., III, 1988, Chemical evolution and volatile fugacities of the Pine Grove porphyry molybdenum and ash flow tuff system, southwestern Utah: *in* Taylor, R. P., and Strong, D. F., eds., Recent advances in the geology of granite-related mineral deposits: Canadian Institute of Mining and Metallurgy, Special Volume 39, pp. 402–423.

Kerr, P. F., 1946, Tungsten mineralization in the United States: Geological Society of America, Memoir 15, 241 pp.

Kerr, P. F., 1934, Geology of the tungsten deposits near Mill City, Nevada: University of Nevada, Bulletin, v. 28, pp. 1–26.

Kerr, P. F., 1938, Tungsten mineralization at Oreana, Nevada: Economic Geology, v. 33, pp. 390–427.

Kwak, T. A. P., 1987, W–Sn skarn deposits and related metamorphic skarns and granitoids: Elsevier, New York, 451 pp.

Lemmon, D. M., 1941, Tungsten deposits of the Benton Range, Mono County, California: U. S. Geol. Survey, Bulletin 922–5, pp. 581–583.

Liu Jiayuan and Shen Jili, 1982, Metallogenetic-magmatic systems of tungsten in Jiangxi Province; *in* Hepworth, J. V., and Yu Hong Zhang, (eds.), Tungsten geology, Jiangxi, China: ESCAP/RMRDC (United Nations), Bandung, Indonesia, pp. 513–526.

Liu Yiamao, Yang Qishun, Zhu Yunzie, and Jiang Qingsong, 1982, Tungsten abundance and its evolution in granitoid rocks of south China; *in* Hepworth, J. V., and Yu Hong Zhang (eds.), Tungsten geology, Jiangxi, China: ESCAP/RMRDC (United Nations), Bandung, Indonesia, pp. 349–358.

Manning, D. A. C., and Henderson, P., 1984, The behavior of tungsten in granitic melt-vapor systems: Contributions to Mineralogy and Petrology, v. 86, pp. 286–293.

Manning, D. A. C., and Pichavant, M., 1988, Volatiles and their bearing on the behavior of metals in granitic systems; *in* Taylor, R. P., and Strong, D. F. (eds.), Recent advances in the geology of granite-related mineral deposits: Canadian Institute of Mining and Metallurgy, Special Volume 39, pp. 13–24.

Mathieson, G. A., and Clark, A. H., 1984, The CanTung E-zone scheelite skarn orebody, Tungsten, Northwest Territories: a revised genetic model: Economic Geology, v. 79, pp. 883–901.

Meinert, L. D., 1983, Variabilty of skarn deposits: Guides to exploration; *in* Boardman, S. J., (ed.), Revolution in the earth sciences—advances in the past half-century: Kendall–Hunt, Iowa, pp. 301–316.

Moreton, P., Clark, A. H., and Peatfield, G. R., 1983, The Boya claim group: A tectonically transported W–Mo stockwork-skarn occurrence, NE Bitish Columbia: Geological Association of Canada/Mieralogical Association of Canada/Canadian Geophysical Union Joint Annual Meeting, Program with Abstracts, v. 8, p. A48.

Munoz, J. L., 1984, F–OH and Cl–OH exchange in micas with applications to hydrothermal ore deposits; *in* Bailey, S. W. (ed.), Micas, reviews in mineralogy, v. 13: Mineralogical Society of America, pp. 469–493.

Munoz, J. L., and Swenson, A., 1981, Chloride–hydroxyl exchange in biotite and estimation of relative HCl/HF activities in hydrothermal fluids: Economic Geology, v. 76, pp. 2212–2221.

Newberry, R. J., 1982, Tungsten-bearing skarns of the Sierra Nevada: I. The Pine Creek mine, California: Economic Geology, v. 77, pp. 823–844.

Newberry, R. J., 1980, The geology and chemistry of skarn formation and tungsten deposition in the central Sierra Nevada, California: Unpub. Ph.D. thesis, Stanford Univ., 325 pp.

Newberry, R. J., and Einaudi, M. T., 1981, Tectonic and geochemical setting of tungsten skarn deposits in the Cordillera: Arizona Geological Society, Digest, v. 14, pp. 99–112.

Newberry, R. J., and Swanson, S. E., 1986, Scheelite skarn granitoids: An evaluation of the roles of magmatic source and process: Journal of Ore Geology Reviews, v. 1, pp. 57–81.

Noble, D. C., Vogel, T. A., Peterson, P. S., Landis G. P., Grant N. K., Jezek, P. A., McKee, E. H., 1984, Rare-element-enriched, S-type ash-flow tuffs containing phenocrysts of muscovite, andalusite, and sillimanite, southeastern Peru: Geology, v. 12, pp. 35–39.

Pichavant, M., Herrera, J. V., Boulmier, S., Briquieu, L., Joron, J., Juteau, M., Marin, L., Michard, A., Sheppard, S. M. F., Treuil, M., and Vernet, M., 1987, The Macusani glasses, SE Peru: evidence of chemical fractionation in peraluminous magmas; *in* Mysen, B. O., (ed.), Magmatic processes: Physicochemical principles: The Geochemical Society, Special Publication No. 1, pp. 359–373.

Roberts, R. J., 1943, The Rose Creek tungsten mine Pershing County, Nevada: U.S. Geol. Survey, Bulletin 940–A, 14 pp.

Schmidt, E. A., Worthington, J. E., and Thomssen, R. W., 1979, K–Ar dates for mineralization in the White Cloud–Cannivan porphyry molybdenum belt of Idaho and Montana—a discussion: Economic Geology, v. 74, pp. 698–699.

Seedorff, E., 1988, Cyclic development of hydrothermal mineral assemblages related to multiple intrusions at the Henderson porphyry molybdenum deposit, Colorado; *in* Taylor, R. P., and Strong, D. F. (eds.), Recent advances in the geology of granite-related mineral deposits: Canadian Institute of Mining and Metallurgy, Special Volume 39, pp. 367–393.

Sigurdson, D. R., 1974, Mineral parogenesis and fluid inclusion thermometry of four western U. S. tungsten deposits: Unpub. Ph.D. thesis, Univ. California, Riverside, 214 pp.

Silberman, M. L., Berger, B. R., Koski, R., 1974, K–Ar age relations of granodiorite emplacement and tungsten and gold mineralization near the Getchell mine, Humboldt County, Nevada: Economic Geology, v. 69, pp. 646–656.

Simon, F. O., 1972, The distribution of chromium and tungsten in the rocks and minerals from the southern California batholith: Unpub. Ph.D. thesis, Univ. Maryland, 130 pp.

Smith, R. L., 1985, Tin, molybdenum, and other metal enrichments in silicic magmas and their relation to fluorine and chlorine distribution; *in* Krafft, K. (ed.), USGS research on mineral resources—1985 program and abstracts: U.S. Geol. Survey, Circular 949, pp. 52–53.

Smith, W. C., and Guild, P. W., 1942, Tungsten deposits of the Nightingale district, Pershing Co., Nev.: U.S. Geol. Survey, Bulletin 936–B, pp. 39–58.

Sonnevil, R. A., 1979, Evolution of skarn at Monte Cristo, Nevada: Unpub. M.S. thesis, Stanford Univ., 83 pp.

Stein, H. J., and Hannah, J. L., 1985, Movement and origin of ore fluids in Climax-type systems: Geology, v. 13, pp. 469–474.

Tacker, R. C. and Candela, P. A., 1987, Partitioning of molybdenum between magnetite and melt: A preliminary experimental study of partitioning of ore metals between silicic magmas and crystalline phases: Economic Geology, v. 82, pp. 1827–1838.

Taylor, R. G., 1979, Geology of Tin Deposits: Elsevier, Amsterdam, 543 pp.

Terashima, S., 1980, Spectrophotometric determination of molybdenum and tungsten in forty geological reference samples: Geostandards Newsletter, v. 4, pp. 9–12.

van Middelaar, W. T., 1988, Alteration and mica chemistry of the granitoid associated with the CanTung scheelite skarn, Tungsten, Northwest Territories, Canada: Unpublished Ph.D. thesis, University of Georgia, Athens, 215 pp.

van Middelaar, W. T. and Keith, J. D., 1987, Albitization and greisening: Possible pathways of the hydrothermal fluid in the granitoid associated with scheelite skarn, Tungsten, NWT, Canada Cordilleran: Geological Society of America, Abstracts with Programs, v. 19, no. 6, p. 459.

van Middelaar, W. T., and Keith, J. D., 1990, Mica chemistry as an indicator of oxygen and halogen fugacities in the CanTung and other W-related granitoids in the North American Cordillera: *in* Stein, H. J., and Hannah, J. L. (eds.), Ore-bearing granite systems: Petrogenesis and mineralizing processes: Geological Society of America, Special Paper 246.

Wesolowski, D., Cramer, Jan J., and Ohmoto, Hiroshi, 1988, Scheelite mineralization in skarns adjacent to Devonian granitoids at King Island, Tasmania: *in* Taylor, R. P., and Strong, D. F. (eds.), Recent advances in the geology of granite-related mineral deposits: Canadian Institute of Mining and Metallurgy, Special Volume 39, pp. 234–251.

Wu Yongle and Mei Yongwen, 1982, Multi-phase intrusion and multi-phase mineralization and their evolution in Xihuashan tungsten ore field: *in* Hepworth, J. V., and Yu Hong Zhang (eds.), Tungsten geology, Jiangxi, China: ESCAP/RMRDC (United Nations), Bandung, Indonesia, pp. 533–542.

Xu Kequin, Hu Shouci, Sun Minzhi, Zhang Jingrong, Ye Zun, and Li Huiping, 1982, Regional factors controlling the formation of tungsten deposits in south China: *in* Hepworth, J. V., and Yu Hong Zhang (eds.), Tungsten geology, Jiangxi, China: ESCAP/RMRDC (United Nations), Bandung, Indonesia, pp. 473–488.

Yu Shoujun, 1982, A preliminary study of metallogenetic relations between the tungsten ore deposits and the early Yanshanian multistage granitic bodies in the Nanling region: *in* Hepworth, J. V., and Yu Hong Zhang (eds.), Tungsten geology, Jiangxi, China: ESCAP/RMRDC (United Nations), Bandung, Indonesia, pp. 533–542.

Zaw, U. K., 1976, The CanTung E-zone orebody, Tungsten, Northwest Territories: A major sceelite skarn deposit: Unpub. M. S. thesis, Queen's University, Kingston, Ontario, 327 pp.

Zaw, U. K., and Clark, A. H., 1978, Fluoride–hydroxyl ratios of skarn silicates, CanTung E-zone sceelite orebody, Northwest Territories: Canadian Mineralogist, v. 16, pp. 207–221.